Lecture Notes in Chemistry

Volume 93

The Lecture Notes in Chemistry

The series Lecture Notes in Chemistry (LNC) reports new developments in chemistry and molecular science-quickly and informally, but with a high quality and the explicit aim to summarize and communicate current knowledge for teaching and training purposes. Books published in this series are conceived as bridging material between advanced graduate textbooks and the forefront of research. They will serve the following purposes:

- provide an accessible introduction to the field to postgraduate students and nonspecialist researchers from related areas,
- provide a source of advanced teaching material for specialized seminars, courses and schools, and
- be readily accessible in print and online.

The series covers all established fields of chemistry such as analytical chemistry, organic chemistry, inorganic chemistry, physical chemistry including electrochemistry, theoretical and computational chemistry, industrial chemistry, and catalysis. It is also a particularly suitable forum for volumes addressing the interfaces of chemistry with other disciplines, such as biology, medicine, physics, engineering, materials science including polymer and nanoscience, or earth and environmental science.

Both authored and edited volumes will be considered for publication. Edited volumes should however consist of a very limited number of contributions only. Proceedings will not be considered for LNC.

The year 2010 marks the relaunch of LNC.

More information about this series at http://www.springer.com/series/632

Biswanath Dinda

Essentials of Pericyclic and Photochemical Reactions

Springer

Biswanath Dinda
Department of Chemistry
Tripura University
Agartala, Tripura
India

and

Department of Chemistry
NIT Agartala
Jirania
India

ISSN 0342-4901 ISSN 2192-6603 (electronic)
Lecture Notes in Chemistry
ISBN 978-3-319-45933-2 ISBN 978-3-319-45934-9 (eBook)
DOI 10.1007/978-3-319-45934-9

Library of Congress Control Number: 2016951666

Printed on acid-free paper

This Springer imprint is published by Springer Nature
The registered company is Springer International Publishing AG
The registered company address is: Gewerbestrasse 11, 6330 Cham, Switzerland

Dedicated
to
my
parents and teachers

Preface

The part of pericyclic and photochemical reactions is the cornerstone of organic chemistry of the 20th century. Critical understanding of the principles of these reactions will be useful to design the synthesis of enormous organic compounds with high yields maintaining regio- and stereoselectivity. In this book, utilizing my long teaching experience, I have aimed to present the basic principles of pericyclic and photochemical reactions in the student's comprehension by citing numerous examples with references to develop a thorough and sound sense of actuality on the subject. Literature citations throughout the text will be helpful to the students and teachers, who want to get the access to the original work of the factual material. This book is not designed to be comprehensive with respect to the experimental details and evidences on which the reaction mechanisms are based. The main objectives of this book are to develop a broad understanding and scientific thinking of the students on the subject. The book will help teachers to motivate students in their scientific imagination on the subject for new application in industrial fields avoiding hazardous chemicals. A large number of excellent and representative problems at the end of each chapter and their answers in Appendix-1 of the book will help the students for their self-evaluation on the lessons of the chapter.

This book is basically designed for the students of postgraduate and M. Phil levels. However, the students of upper undergraduate levels in chemistry may use it for advancement of their knowledge on the subject. The book will also be useful for students to compete for different qualifying examinations after postgraduation.

I have consulted three excellent books, Advanced Organic Chemistry by F. A. Carey and R. T. Sundberg, Pericyclic Reactions by I. Fleming and Principles and Applications of Photochemistry by B. Wardle at several points in writing this book.

I wish to acknowledge the technical assistance of my students, Dr. Saikat Das Sarma, Dr. Rajarsi Banik, Dr. Indrajit Sil Sarma, Dr. Prasenjit Rudrapaul, Smt. Ankita Chakraborty, Sri Sukhen Bhowmik, Sk. Nayim Sepay, Sri Subhadip Roy, Sri Arnab Bhattacharya and my son, Dr. Subhajit Dinda for typing of the major part of the manuscript.

I would appreciate to receive the letters from teachers and students on errors, questions, criticisms and suggestions on this book so that I may improve this book in the forthcoming edition.

Finally, I like to acknowledge to my wife, Chitralekha, and our children, Subhajit and Manikarna, and son-in-law Shekhar for their constant encouragement and patient endurance. I am grateful to my publishers for their support and interest in this endeavour.

Agartala, Tripura, India
January 2016

Biswanath Dinda

Contents

Abbreviations

acac	Acetylacetonate
BINAP	Bis-(2,2'-diphenylphosphinyl)-1,1'-binaphthalene
BINOL	Binaphthol
Boc	*Tertiary*-butoxycarbonyl [Me_3COCO]
BOX	Bisoxazoline
Bz	Benzyl [$PhCH_2$]
DBMP	6-di-*tert*-butyl-4-methyl phenol
DBP	Dibutyl phthalate
DBU	Diazabicycloundecane
DMA	Dimethylallene
DPM	Di-π-methane
ee	Enantiomeric excess
Et	Ethyl [C_2H_5]
FVP	Flash vacuum pyrolysis
HMPA	Hexamethylphosphoramide
HOMO	Highest occupied molecular orbital
hv	Ultraviolet or visible irradiation
IL	Ionic liquid
i-Pr	Iso-propyl[Me_2CH]
KHMDS	Potassium hexamethyldisilazane or potassium bis(trimethylsilyl) amide [$(Me_3Si)_2NK$]
LDA	Lithium diisopropylamide [$LiNi\text{-}Pr_2$]
LUMO	Lowest unoccupied molecular orbital
Me	Methyl [CH_3]
MTAD	N-methylthiazolinedione
N,N-DEA	N, N-diethanolamine [$NH(CH_2CH_2OH)_2$]
n-Pr	Normal-propyl [$MeCH_2CH_2$]
ODPM	Oxa-di-π-methane
PET	Photo-induced electron transfer
Ph	Phenyl [C_6H_5]

PhH	Benzene
P_i	Phosphate, inorganic
Py	Pyridine
rt	Room temperature
sens	Sensitizer
SOMO	Singly occupied molecular orbital
TADDOL	α,α,α,α-tetraaryl-1,3-dioxolane-4,5-dimethanol
TBDPS	*Tert*-butyldiphenylsilyl
TBS	*Tert*-butylmethyl silyl
t-Bu	Tertiary-butyl [Me_3C]
TCB	Tetracyanobenzene
THF	Tetrahydrofuran
TMS	Trimethylsilyl[Me_3Si]
Ts	Tosyl [4-MeC_6H_4]
TS	Transition structure

List of Figures

List of Tables

List of Schemes

Part I
Pericyclic Reactions

Chapter 1
General Aspects of Pericyclic Reactions

1.1 Introduction

Reactions in Organic Chemistry are broadly classified into three major categories—ionic, radical, and pericyclic. Ionic reactions involve the formation of ionic intermediates by movement of pair of electrons in one direction of a covalent bond. In a unimolecular reaction, it occurs by ionization process and in a bimolecular reaction, it occurs when one component acts as a nucleophile (or electron pair donor) and another component as electrophile (or electron pair acceptor). For example,

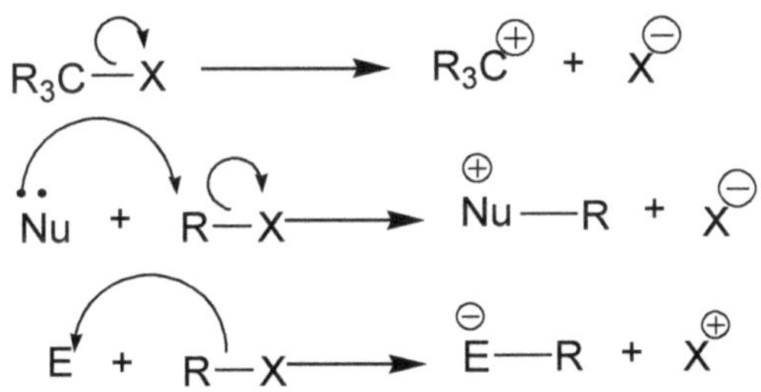

Radical reaction involves the homolytic cleavage of a covalent bond by movement of single electrons in opposite directions. The movement of a single electron is represented by fish hook arrow. For example,

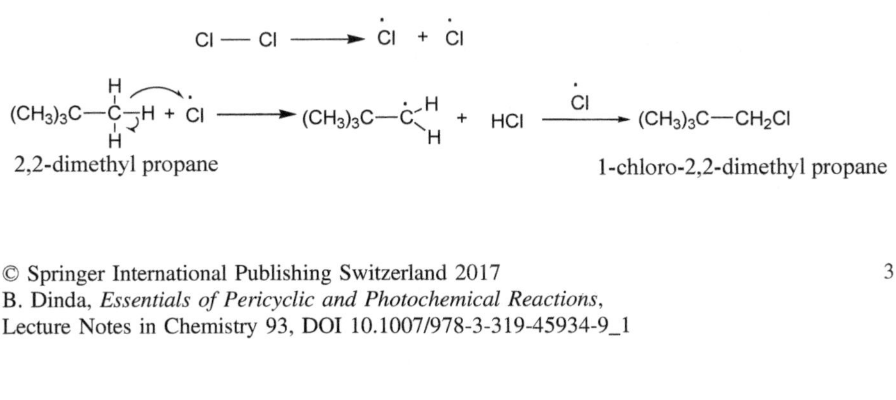

B. Dinda, *Essentials of Pericyclic and Photochemical Reactions*,
Lecture Notes in Chemistry 93, DOI 10.1007/978-3-319-45934-9_1

Pericyclic reactions involve the continuous flow of electrons in cyclic transition states (TS) by breaking and making of bonds in a concerted process, without formation of an intermediate. Hence, these reactions are known as concerted reactions [1]. These reactions are insensitive to solvent polarity and free radical initiators or inhibitors. These reactions are activated by heat (thermal) or light (photochemical). Detailed study of the mechanisms of these reactions by Woodward and Hoffmann [2] predicted that these reactions occur by the maintenance of symmetry properties of the orbitals of reactant(s) and product(s). The Diels–Alder reaction is a typical example.

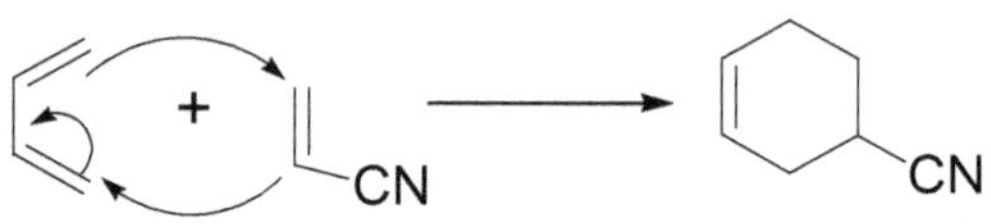

A Diels-Alder reaction

1.2 Molecular Orbitals and Their Symmetry Properties

In pericyclic reactions, the reactivity of the reactions can be explained on the basis of Perturbational Molecular Orbital (PMO) theory [3]. The basic postulate of PMO theory is that a chemical reaction takes place by the perturbation of molecular orbitals (MOs) of reactants on heating and on irradiation with light. The degree of perturbation is a function of degree of overlapping interactions of the atomic orbitals in an MO. These interactions are strongest among the orbitals close in energies. These orbital overlapping interactions produce degenerate MOs of different energies. The suitable degenerate MOs take part in the reactions to give products through cyclic TSs in a concerted process. The interactions of two atomic orbitals will produce two MOs, one of them will be stabilized and other will move to higher energy. The linear combination of atomic orbitals is known as LCAO theory or PMO theory. The shapes of the MOs that are formed by the linear combination of atomic orbitals (LCAO) are related to the shapes of atomic orbitals. The MOs are denoted by ψ (psi) and atomic orbitals by Φ (phi). Dewar and Hoffmann first developed a general PMO method to explain the reactivity in organic chemistry. To illustrate the idea, let us consider a diatomic molecule where an MO formed by the combination of atomic orbitals of A and B is represented as

$$\psi = \Phi_A + \Phi_B$$

Quantum mechanics shows that the linear combination of two wave functions gives two combinations and hence two MOs are generated from two-component atomic orbitals. One MO is bonding orbital, more stable than the component atomic orbitals and other one is an antibonding orbital, less stable than the component orbitals.

$$\psi_+ = \Phi_A + \Phi_B \text{ (bonding MO)}$$
$$\psi_- = \Phi_A - \Phi_B \text{ (antibonding MO)}$$

The MOs that have resulted from overlap of various kinds of atomic orbitals are shown in Fig. 1.1.

Similarly, the linear combination of three p-orbitals in allyl system will give rise to three new MOs ψ_1, ψ_2 and ψ_3. The antibonding interactions increase the energy of the MO. Thus, the energy of ψ_2 is higher than that of ψ_1 and ψ_3 is of higher energy than ψ_2. The wave functions and their symmetry in relation to the mirror plane are shown in Fig. 1.2.

Similarly, the linear combination of four atomic π orbitals in 1, 3-butadiene will generate four MOs ψ_1, ψ_2, ψ_3, and ψ_4 (Fig. 1.3).

Next, the linear combination of six atomic p-orbitals of 1,3,5-hexatriene will give rise to six MOs. The wave functions and symmetry properties of these MOs are shown in Fig. 1.4.

On the basis of the above examples of polyene systems, the symmetry properties of MO, ψ_n of a linear conjugated polyene are summarized in Table 1.1.

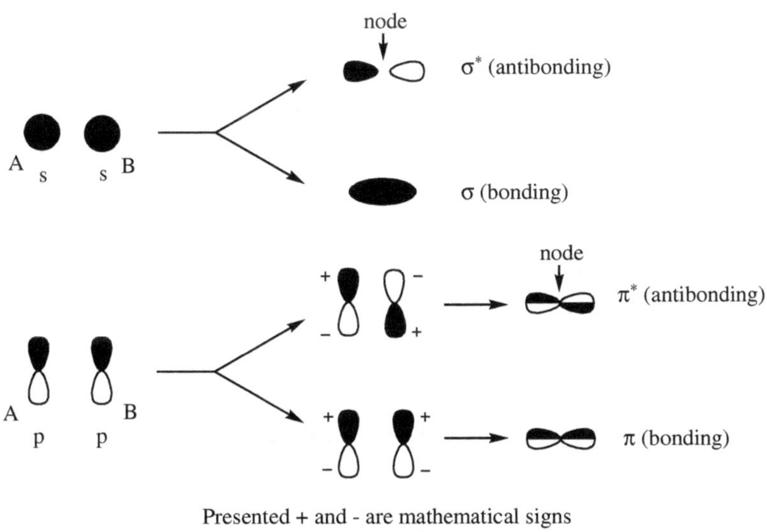

Fig. 1.1 Formation of bonding and antibonding orbitals

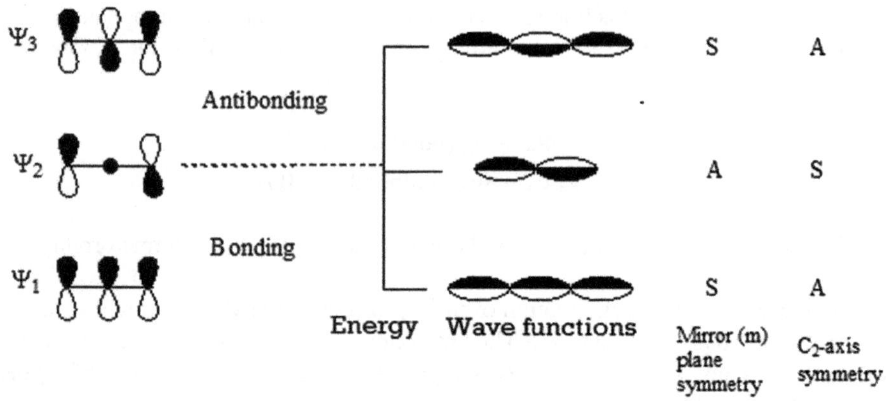

Fig. 1.2 Molecular orbitals formation in allyl systems

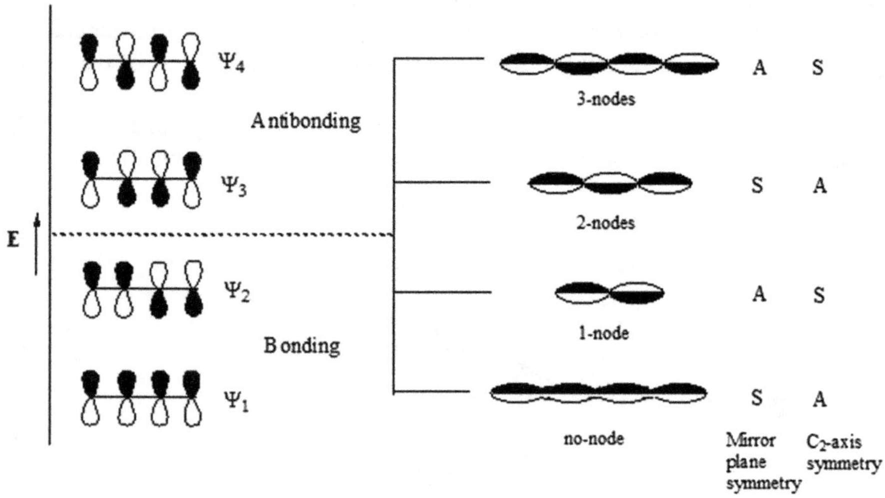

Fig. 1.3 Molecular orbitals of 1,3-butadiene and their symmetry properties. (S means symmetric and A means antisymmetric)

1.3 Classification of Pericyclic Reactions

Pericyclic reactions are classified into four classes. These are electrocyclic reactions, cycloadditions, sigmatropic rearrangements, and group transfer reactions.

Electrocyclic reactions are characterized by the creation of a ring from an open-chain conjugated system with the formation of a new σ bond at the ends of the conjugated system or its reverse process.

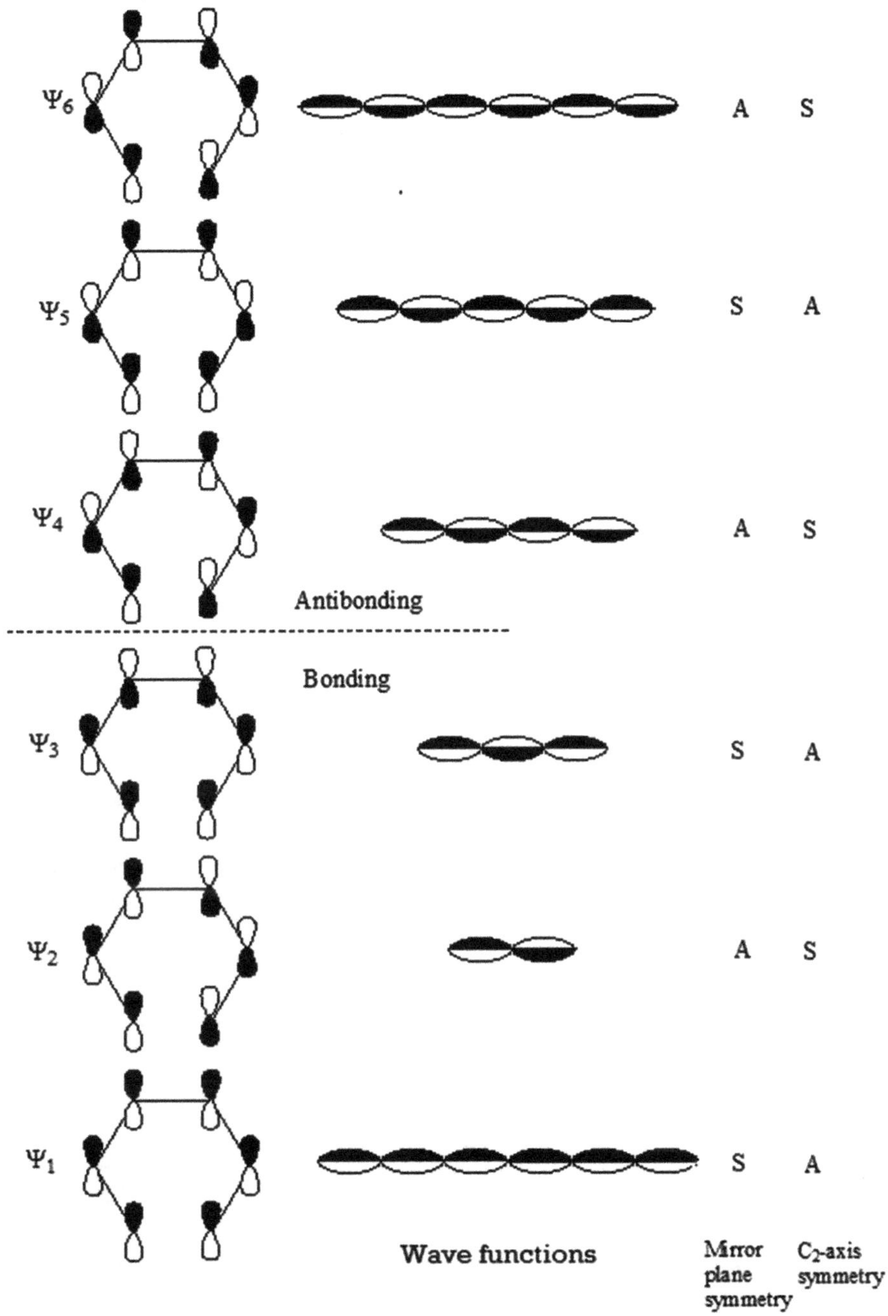

Fig. 1.4 Molecular orbitals of 1,3,5-hexatriene and their wave functions and symmetry properties

Table 1.1 Symmetry properties of the orbital ψ_n of a linear conjugated polyene

Wave function	Nodes	m-symmetry	C2-symmetry
ψ_{odd}: ψ_1, ψ_3, ψ_5	0 or even	S	A
ψ_{even}: ψ_2, ψ_4, ψ_6	odd	A	S

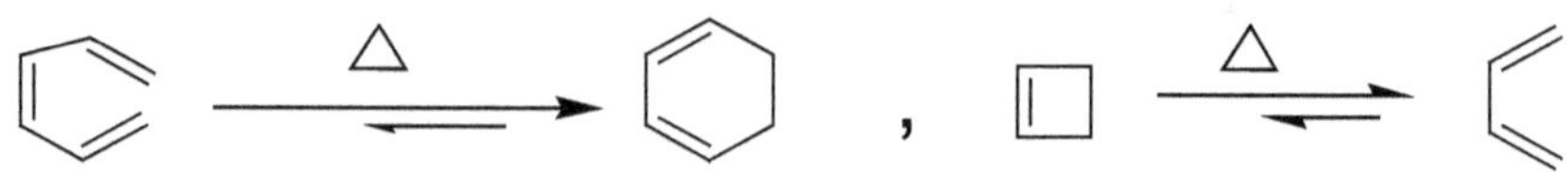

Cycloaddition reactions are characterized by the addition of two π-systems by the formation of two new σ bonds, at the ends of both components, with the reduction of one π-bond from each component, e.g., Diels–Alder reactions.

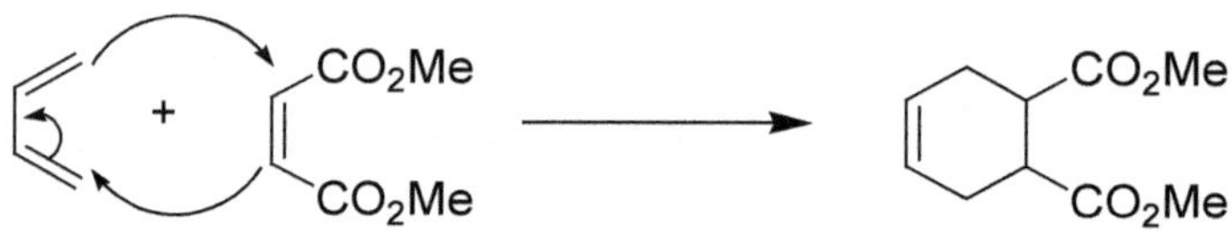

1, 3-Dipolar cycloadditions are another family member of cycloaddition reactions, e.g.,

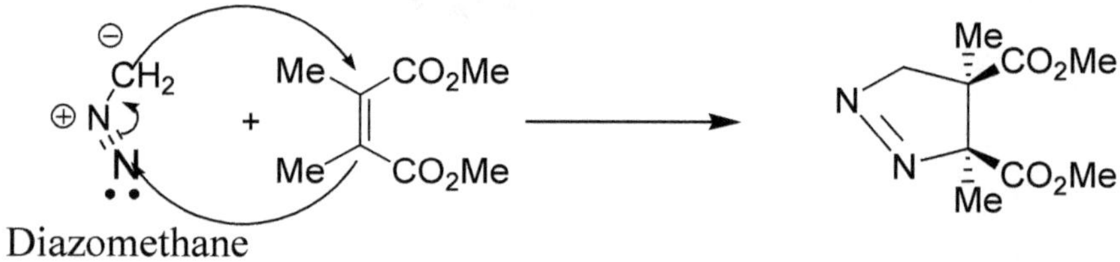

Cheletropic reactions are a special group of cycloadditions or cycloreversions in which two σ bonds are made or broken from the same atom.

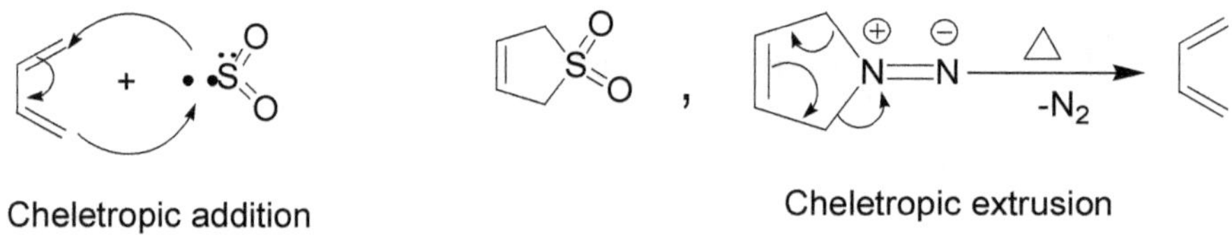

Sigmatropic rearrangements are characterized by the movement of a σ bond to a more distant terminus of an adjacent π system followed by movement of the π system to accommodate this new σ bond, e.g.,

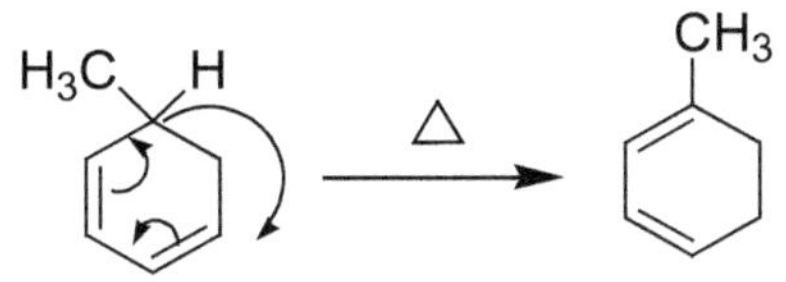

Only a few reactions belong to the class of group transfer reactions. Most common is the ene reaction, where allylic hydrogens are transferred to an electrophilic alkene, referred as an enophile. For example,

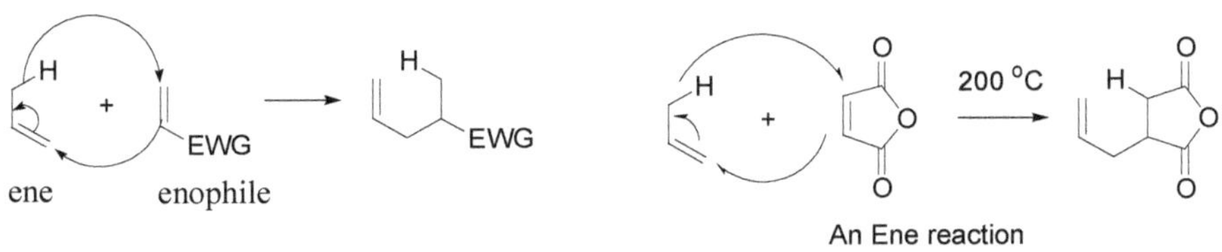

Another type of group transfer reaction is the transfer of hydrogens from diimide to an alkene or alkyne.

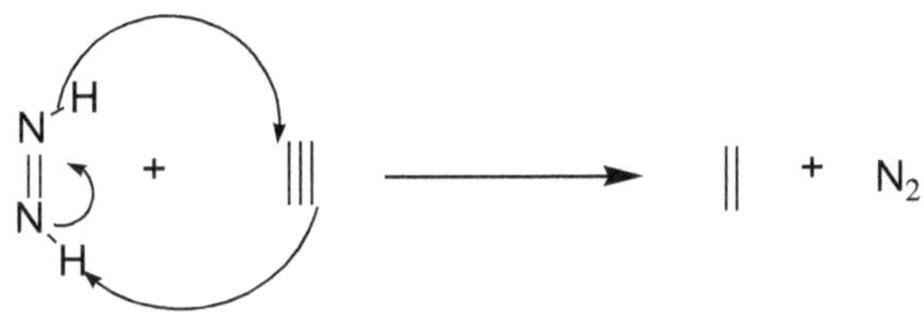

1.4 Concertedness of Pericyclic Reactions

High-level MO calculations suggest that these reactions proceed in concerted pathways through the lowest energy TSs. Isotope effect studies also support the concertedness of pericyclic reactions.

1.5 Orbital Symmetry Property of Pericyclic Reactions

All pericyclic reactions conserve a definite orbital symmetry property throughout the course of their reactions, i.e., the reactants, TS, and products of a reaction have the same orbital symmetry.

The allowedness and forbiddenness of a pericyclic reaction toward heat or light depends on this symmetry property. Usually two types of element of symmetry, namely mirror (m) plane symmetry and C_2-axis of symmetry are considered to correlate the orbital symmetry properties of reactants, TSs, and products of a pericyclic reaction. There are three ways of applying the orbital symmetry principle in these reactions—the frontier molecular orbital (FMO) method, the Mobius–Huckel aromatic TS method, and the orbital correlation diagram method.

The Frontier Molecular Orbital Method

The FMO method was introduced by Fukui, Woodward and Hoffmann [1, 2, 4]. According to this method, an electrocyclic reaction will be allowed when the termini orbitals of the highest occupied MO (HOMO) of a conjugated π-system (reactant) overlaps between them in such a way that a positive (shaded) lobe overlaps with another positive lobe or a negative lobe with another negative lobe. Similarly, a cycloaddition reaction will be allowed when the HOMO of one reactant and the lowest unoccupied MO (LUMO) of the other overlap among themselves in such a way that a positive lobe overlaps with another positive lobe and a negative lobe with another negative lobe. A sigmatropic reaction will be allowed when a σ-bonded atom or group moves from a positive lobe to another positive lobe or from a negative lobe to another negative lobe. Woodward and Hoffmann predicted some orbital symmetry rules for these pericyclic reactions for concertedness of these reactions.

The Mobius–Huckel aromatic transition state method

The Mobius–Huckel method is based on the PMO method and is applicable for prediction of an allowed or forbidden pericyclic reaction. It was developed by Dewar and Zimmerman on the basis of Huckel aromaticity rule for a cyclic conjugated system of π-electrons [5]. In this method, the allowedness of a pericyclic reaction is considered on the basis of arrangement of p-orbitals in the TS. When the p-orbitals are arranged in the TS with zero or an even number of sign inversions (node) of positive or negative lobes, the system is called the Huckel system. A system of arrangement of p-orbitals in the TS with an odd number of sign inversions is known as Mobius system. A thermal pericyclic reaction involving Huckel system is allowed for a total number of 4n+2 π electrons, whereas a thermal pericyclic reaction involving a Mobius system is allowed for a total number of 4n π electrons. For photochemical pericyclic reactions, these rules are reversed for their allowedness. For example, a thermal [2+2]-cycloaddition is forbidden and a thermal [4+2]-cycloaddition is allowed as per Huckel system of TS (Fig. 1.5).

The Correlation Diagram Method

The orbital correlation diagram was introduced by Longuet-Higgins and Abrahamson to predict the allowedness of a pericyclic reaction [6]. In this method, the orbital symmetry properties of both reactants and products are considered. The symmetry elements of the MOs are evaluated and the MOs of reactants and products are arranged in a diagram in two columns. In an allowed pericyclic reaction, the ground-state MO of the reactants and the products has the same element of symmetry.

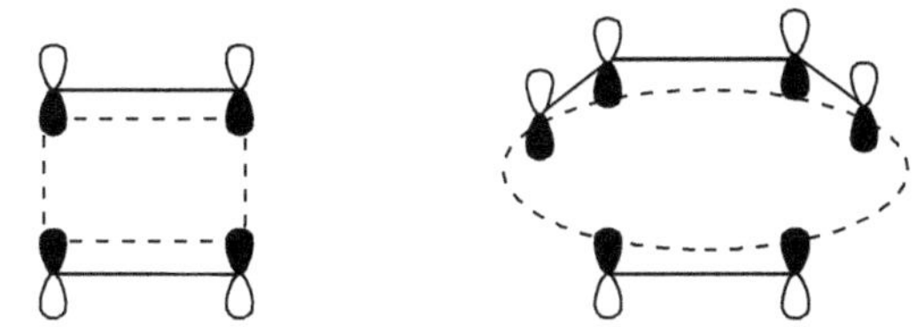

Fig. 1.5 Huckel TS for thermal cycloaddition reactions

1.6 Further Reading

1. Gilchrist L, Storr RC (1979) Organic reactions and orbital symmetry, 2nd edn. Cambridge University Press, Cambridge
2. Marchand AP, Lehr RE (1972) Orbital symmetry. Academic Press, New York
3. Woodward RB, Hoffmann R (1970) The conservation of orbital symmetry. Academic Press, New York

References

1. Woodward RB, Hoffmann R (1970) Conservation of orbital symmetry. Academic Press, New York
2. Woodward RB, Hoffmann R (1965) J Am Chem Soc 87: 395
3. Coulson CA, Longuet-Higgins HC (1947) Proc R Soc London Ser A 192: 16; Hoffmann R (1963) J Chem Phys 39:1397; Dewar MJS, Dougherty RC (1975) The PMO theory of organic chemistry. Plenum, New York
4. Fukui K, Yonezama T, Shingu H (1952) J Chem Phys 20:722; Fukui K, Fujimoto (1967) Bull Chem Soc Jpn 40: 2018, (1969) 42: 3399; Fukui K (1971) Acc Chem Res 4: 57; Fukui K (1982) Angew Chem Int Ed Engl 21:801; Fleming I (1978) Frontier orbitals and organic chemical reactions. Wiley, London, pp 29–109
5. Zimmerman HE (1966) J Am Chem Soc 88:1566; idem (1971) Acc Chem Res 4:272; Dewar MJS(1971) Angew Chem Int Ed Engl 10: 761
6. Longuet- Higgins HC, Abrahamson EW (1965) J Am Chem Soc 87:2045

Chapter 2
Electrocyclic Reactions

2.1 Introduction

An electrocyclic reaction is defined as the thermal or photochemical conversion of an acyclic conjugated system into a ring system by formation of a σ bond between the ends of the conjugated system in a concerted process, or the reverse of this reaction. These reactions are reversible in nature.

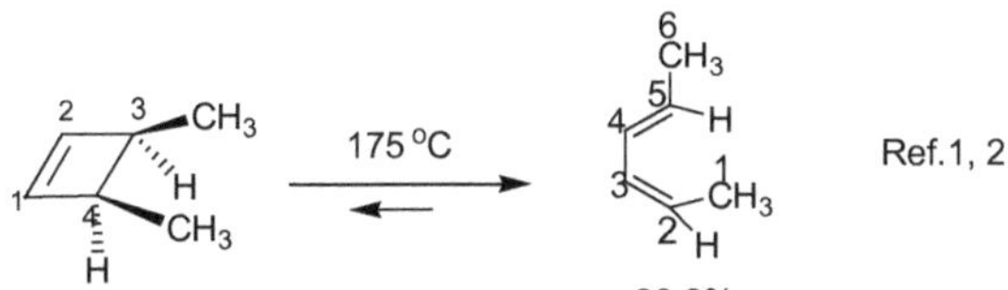

Ref.1, 2

99.9%

cis-3,4-dimethyl cyclobutene → *cis, trans*-2,4-hexadiene

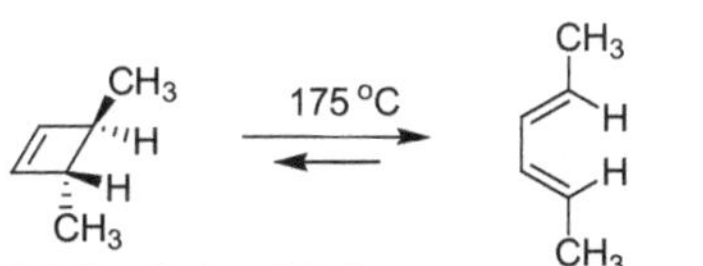

Ref.1, 2

trans-3,4-dimethyl cyclobutene → *trans, trans*-2,4-hexadiene

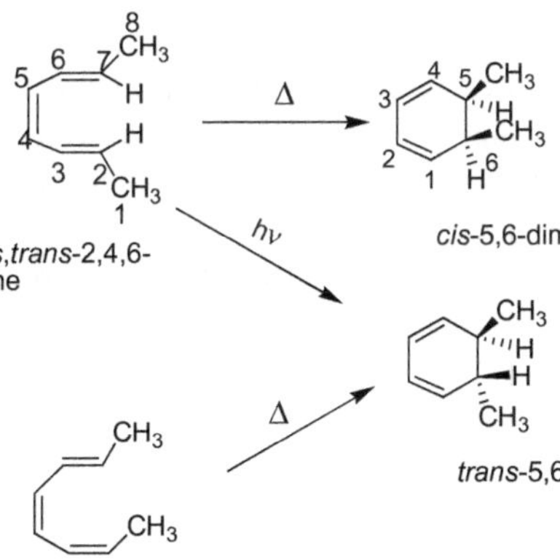

trans,cis,trans-2,4,6-octatriene

cis-5,6-dimethyl-1,3-cyclohexadiene — Ref.3

trans-5,6-dimethyl-1,3-cyclohexadiene — Ref.4

trans,cis,cis-2,4,6-octatriene

B. Dinda, *Essentials of Pericyclic and Photochemical Reactions*,
Lecture Notes in Chemistry 93, DOI 10.1007/978-3-319-45934-9_2

2.2 Orbital Symmetry Basis for Stereospecificity

Woodward and Hoffmann [5] suggested a complete description of the mechanisms of these reactions by introducing the terms 'conrotatory' and 'disrotatory' motions of the groups at the termini of acyclic conjugated system or of the groups at the sp^3 carbons of a ring system. The motion of the substituents in the same direction clockwise or anticlockwise is known as conrotatory mode of motion, while the motion of the substituents in opposite direction is known as disrotatory mode of motion. They suggested that the stereochemistry of these reactions is controlled by the symmetry properties of HOMO (highest occupied molecular orbital) of the open-chain conjugated system. It was also supported by the frontier molecular orbital (FMO) theory [6].

For thermal reactions of 4nπe conjugated systems, ψ_2 would be HOMO because it contains lowest number of nodes (one node) and provides a transition state of lowest energy similar to Mobius topology as per perturbation molecular orbital (PMO) theory. While for photochemical reactions of 4nπe systems, ψ_3 would be HOMO because it is the first excited state of ground-state ψ_2. Therefore, in thermal reactions of 4nπ systems, conrotatory motion of the groups in the terminal carbons of the open-chain π system brings the lobes of the same phase for bonding with a Mobius type TS and is orbital symmetry allowed process, while disrotatory motion brings the lobes of the opposite phase for antibonding formation and is said to be orbital symmetry forbidden process. While for photochemical reactions of 4nπe systems, disrotatory motion brings the lobes of same phase, and hence, the reaction will proceed with low activation energy and is said to be orbital symmetry allowed process. On the other hand, conrotatory motion brings the lobes of opposite phase, and hence, reaction is unfavourable for its high activation energy and is referred as symmetry forbidden reaction path (Fig. 2.1).

In case of thermal reactions of 4n+2 πe conjugated system, ψ_3 would be HOMO as it has minimum even nodes (two nodes) and is related to Huckel topology (as per PMO theory). Hence, the disrotatory motion of the groups of terminal carbons brings the lobes of same phase for bonding, involving a Huckel-type transition state and is said to be symmetry allowed path, while conrotatory motion would be symmetry forbidden path as it leads to a TS of high activation energy. For their photochemical reactions, ψ_4 would be HOMO and hence conrotatory motion would be symmetry allowed path (Fig. 2.2).

Woodward–Hoffmann rules for electrocyclic reactions are summarized in Table 2.1.

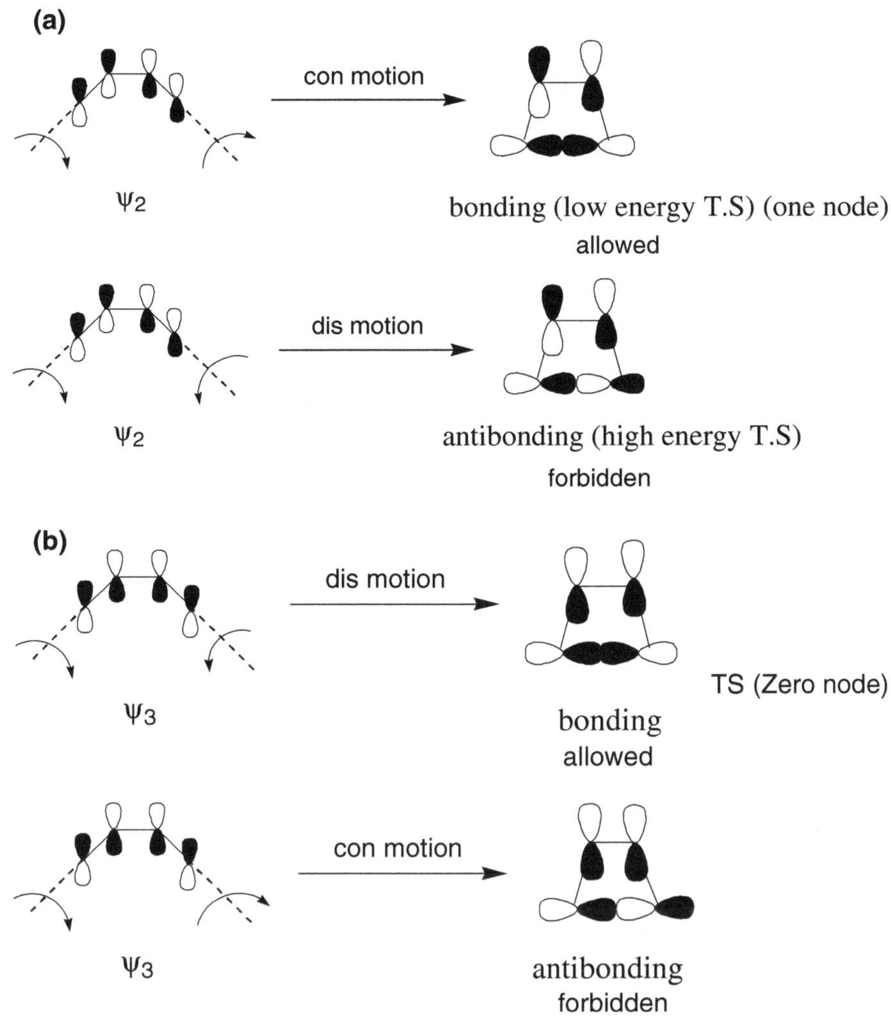

Fig. 2.1 **a** Thermal electrocyclization of 4nπe conjugated system; **b** photochemical electrocyclization of 4nπe conjugated system

2.3 The Orbital Correlation Diagrams of Reactants and Products

Longuet-Higgins and Abrahamson [6] suggested that in any concerted process, the orbitals of the starting material and product have the same symmetry. This is also supported by Woodward and Hoffmann [5]. The cyclobutene–butadiene interconversion may be considered as an example to verify the fact by construction of a correlation diagram. For cyclobutene, the bonding orbitals are σ and π, while the

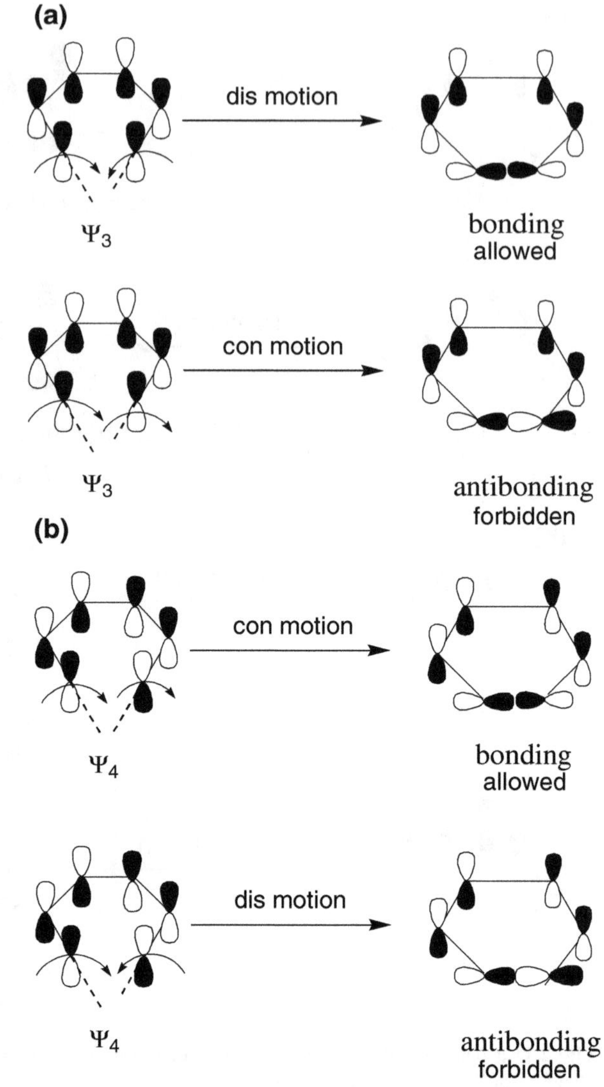

Fig. 2.2 **a** Thermal electrocyclization of 4n+2 πe conjugated system; **b** photochemical electrocyclization of 4n+2 πe conjugated system

Table 2.1 Woodward–Hoffmann rules for electrocyclic reactions

Acylic conjugated system	Reaction allowed	Motion
4n πe	Thermal	Conrotatory
	Photochemical	Disrotatory
(4n+2) πe	Thermal	Disrotatory
	Photochemical	Conrotatory

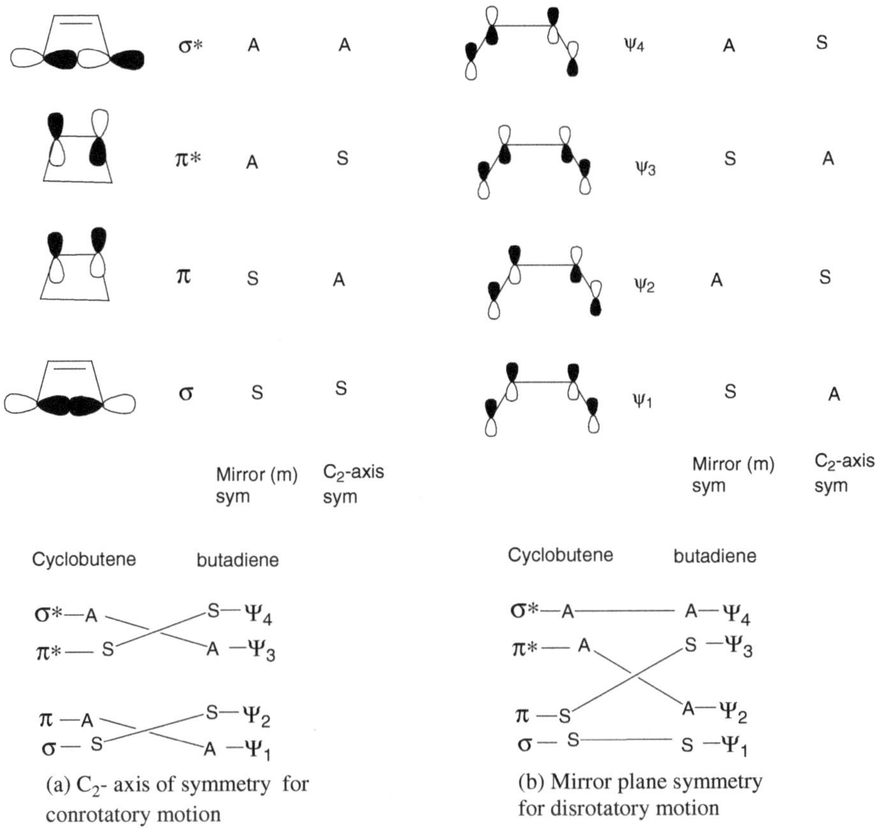

Fig. 2.3 **a** C_2-axis of symmetry is maintained in thermal conversion of cyclobutene to butadiene; **b** mirror plane symmetry is maintained in photochemical conversion of cyclobutene to butadiene

antibonding orbitals are σ* and π* (Fig. 2.3). For butadiene, the bonding orbitals are ψ_1 and ψ_2, and antibonding orbitals are ψ_3 and ψ_4. In thermal reaction, conrotatory ring opening of cyclobutene to butadiene, C_2 (twofold) axis of symmetry is maintained throughout the reaction, while for photochemical reaction, disrotatory ring opening, a mirror plane (m) symmetry is maintained throughout the reaction (Fig. 2.3).

Next, consider the thermal conversion of a 1,3,5-hexatriene to a 1,3-cyclohexadiene by the disrotatory motion where mirror (m)-symmetry is maintained in the orbitals of the reactant and product (Fig. 2.4). In photochemical conversion of 1,3-cyclohexadiene into 1,3,5-hexatriene or vice versa, the C_2-axis of symmetry is maintained in conrotatory motion of the termini groups (Fig. 2.4).

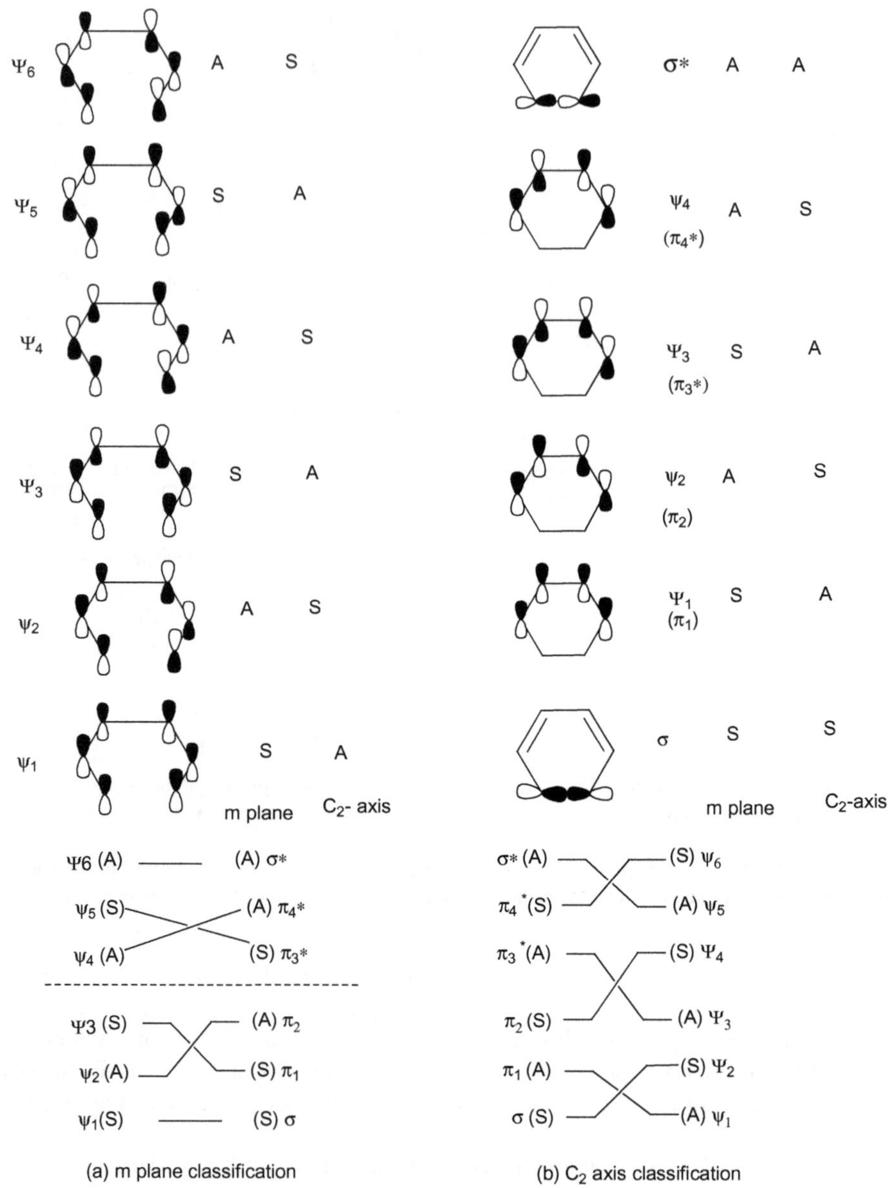

Fig. 2.4 **a** Mirror plane (m) symmetry is maintained in thermal conversion of 1,3,5-hexatriene into 1,3-cyclohexadiene; **b** C_2-axis of symmetry is maintained in photochemical conversion of 1,3-cyclohexadiene into 1,3,5-hexatriene or vice versa

2.4 Applications of Neutral Conjugated Systems in Electrocyclic Reactions

Electrocyclic reaction of *E*, Z-1, 3-cyclooctadiene leads to *cis*-bicyclo[4.2.0]-oct-7-ene because of strain associated with *trans* double bond.

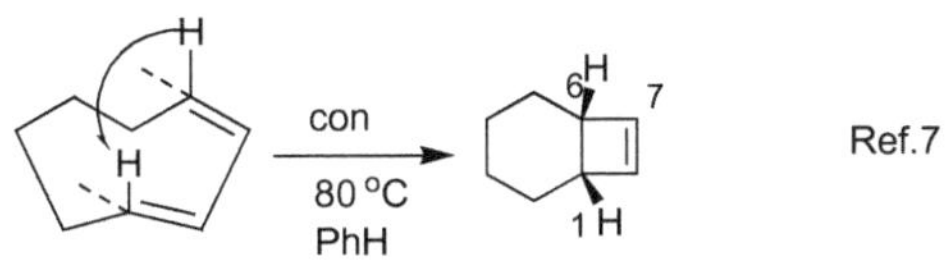

Although cyclobutenes are converted into butadienes on heating to get relief of ring strain, *cis*-bicyclo[3.2.0]-hept-6-ene on heating gave *Z,Z*-1,3-cycloheptadiene by forbidden disrotatory motion. This anomaly of the Woodward–Hoffmann rules can be accounted for by the stability of the product formed. In this case, allowed conrotatory motion gives the strained *E,Z*-1,3-cycloheptadiene, which is less stable due to ring strain and hence rapidly isomerizes to *Z,Z*-isomer at the reaction temperature in low yield.

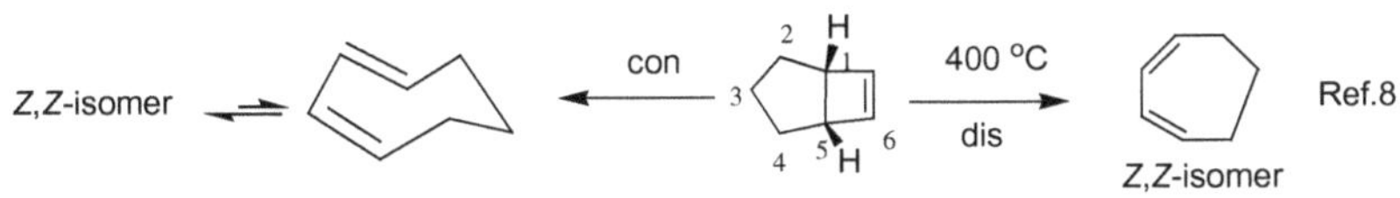

It was supported by the thermal conversion of *cis*-1,6-dideuterio-bicyclo[4.2.0]-oct-2,7-diene to *cis*-3,8-dideuterio-isomer via *trans*-isomer.

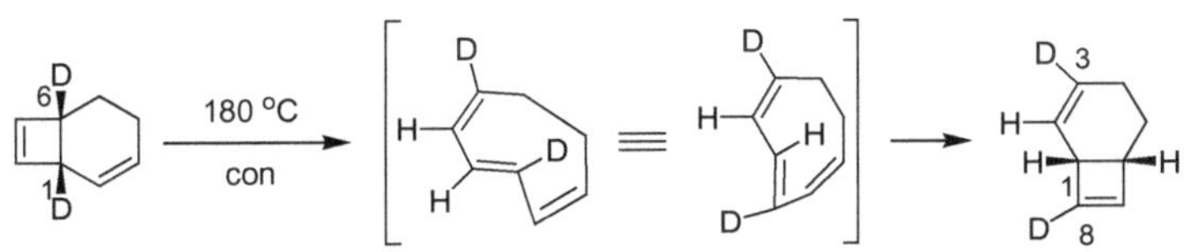

Cyclobutene-3-carboxylic acid on heating in acidic and basic solutions gives isomeric *cis*- and *trans*-pentadienoic acids [9]. In thermal ring opening of cyclobutenes, electron-donating substituents tend to move outward of the butadiene chains to minimize the repulsive interaction with π system of butadiene in the TS, while the π electron-accepting substituents tend to move inward of butadiene chains to stabilize the HOMO TS by interaction with the donor lobes of p-orbitals of breaking C(3)–C(4) σ bond. This preferential direction of movement of C(3) substituent in cyclobutene ring opening is called torquoselectivity and it works in ranges far beyond the 4πe-electrocyclic system [9].

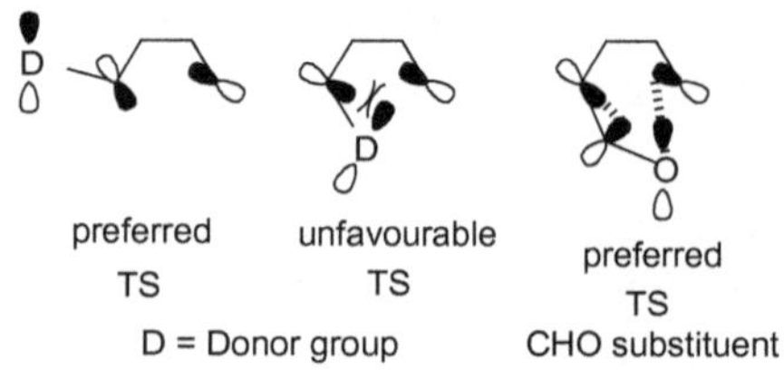

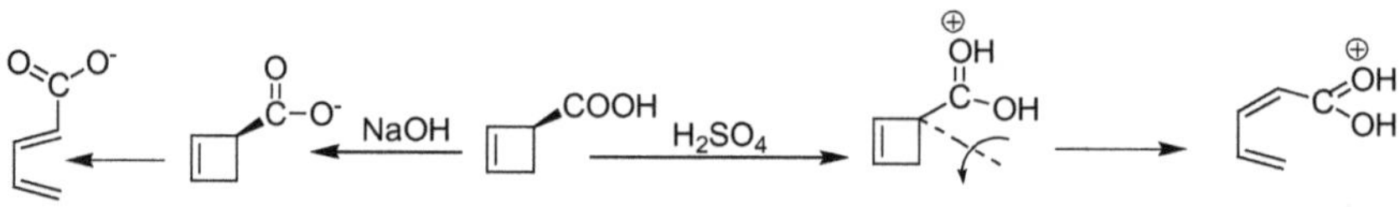

Study of the electrocyclic ring opening of *cis*- and *trans*-3,4-dichlorocyclobutenes indicated that *trans*-isomer reacts at lower temperature. This is due to ring opening by outward conrotatory motion of donor chlorine substituents while in case of *cis*-isomer, activation energy is higher as one of the chlorines rotates inward.

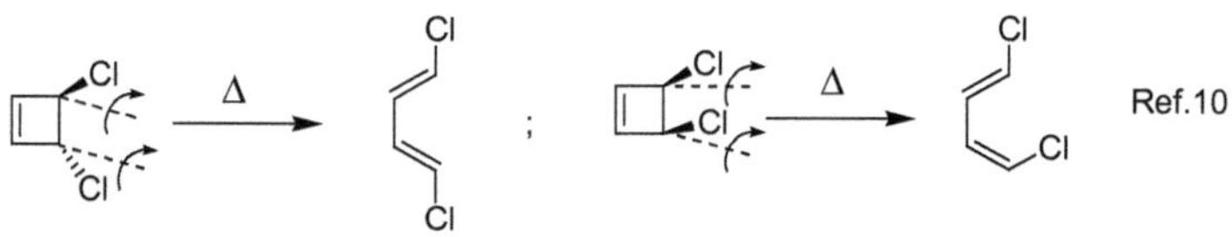

When a cyclobutene ring contains both electron-donating and electron-accepting substituents, conrotatory outward or inward motion of the substituents depends on the size of the substituents. Thus, **1** gives **2**.

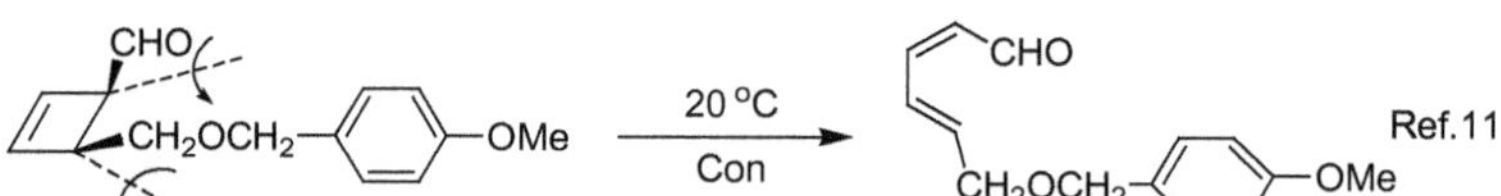

1 86% **2**

When a cyclobutene ring contains two electron-donating substituents at C-3 position, the substituent with greater donor ability will move outward [12]. For example, **3** gives **4**.

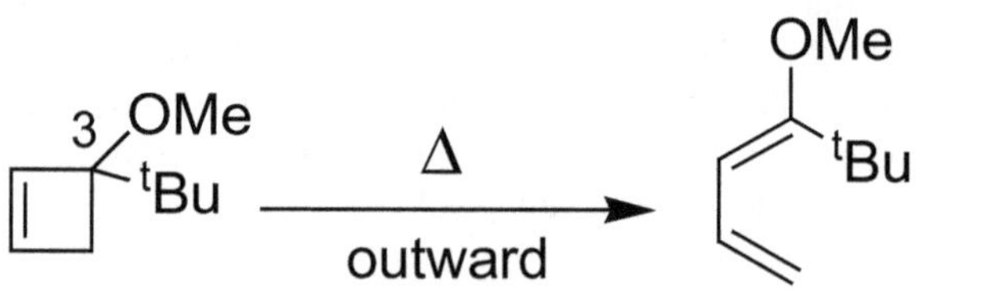

3 **4** > 99%

3-Hydroxy-2-methylcyclobutene **5** on electrocyclic ring opening undergoes keto-enol tautomerization to afford the product **6**.

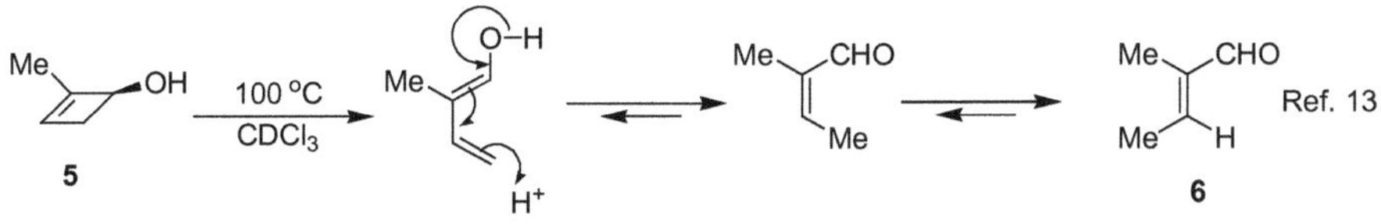

When a cyclobutene ring contains both donor hydroxyl and olefinic substituents at C-3, the inward con-motion of the olefinic substituent occurs preferentially to increase the stability of the TS. For example, **7** gives **8**.

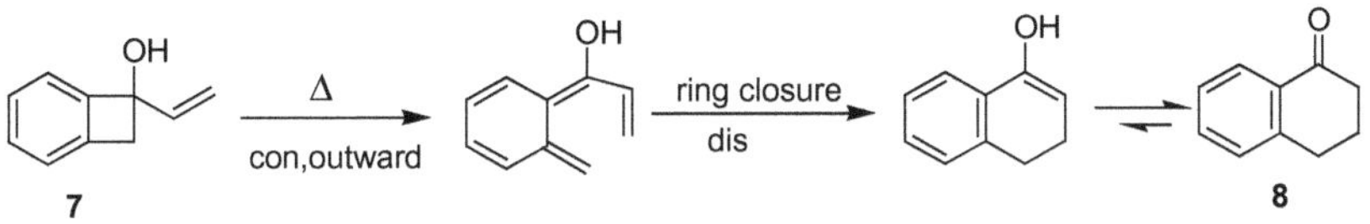

Cyclobutene **9** having olefinic function at C-3 or C-4 position undergoes inward ring opening from the olefinic substituent site followed by ring closure to give the product. This inward motion of the olefinic substituent stabilizes the HOMO of the TS by π-orbital interaction of the substituent with the donor lobes of p-orbitals of the breaking σ bond of the ring carbons [14].

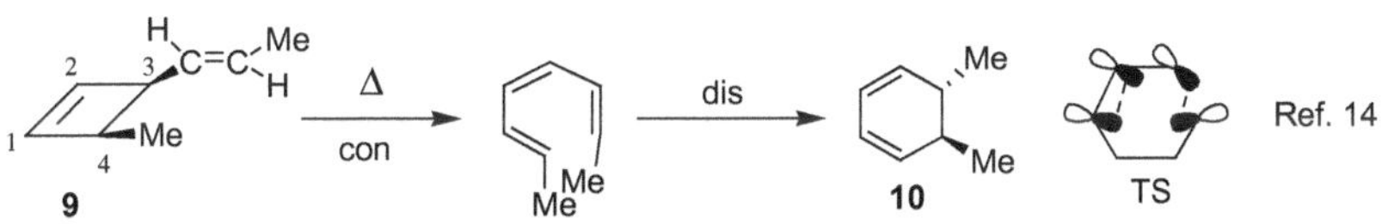

Similarly, when both electron-withdrawing substituents are present at C-3, inward conrotatory motion of the less bulky substituent takes place. When alkyl and olefinic substituents are present at C-3 or C-4 position, the olefinic substituent will move inward to provide more stable TS by closer interaction with the π-orbitals of C-1 and C-2. The following examples are illustrative [12].

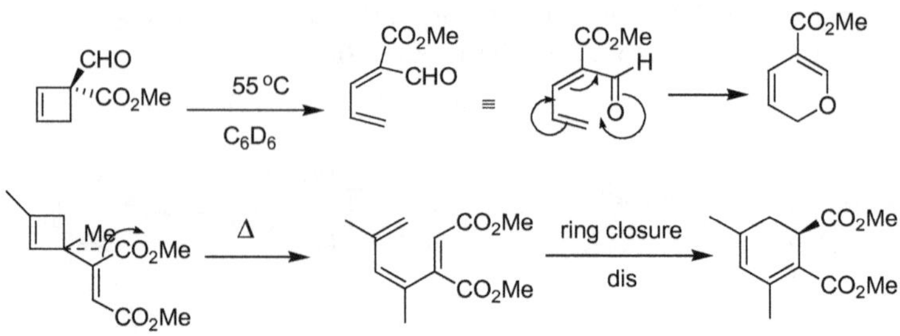

Cyclobutene **11** containing two *ortho*-alkoxy groups at C-3 and C-4 positions gives major product **12** by inward movement of smaller alkoxy group [15].

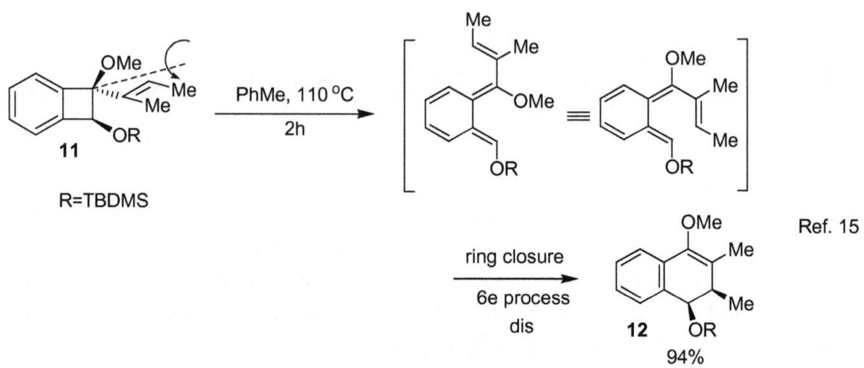

Cyclobutene fused with a carbocyclic ring gives isomeric product by more than one electrocyclic processes. For example, cyclobutene **13** gives **14** and cyclobutene **15** gives **16**.

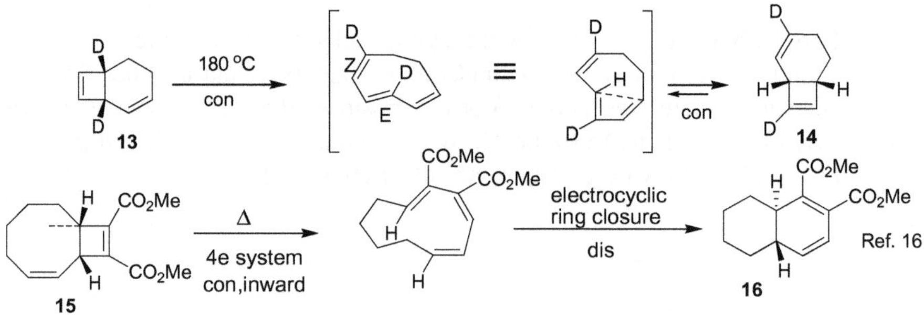

Dewar benzene having two cyclobutene rings on heating gives benzene rather than the expected product **17** from an allowed conrotatory opening. This is due to the presence of strained *E*-double bond in the expected product, which rapidly isomerizes to benzene.

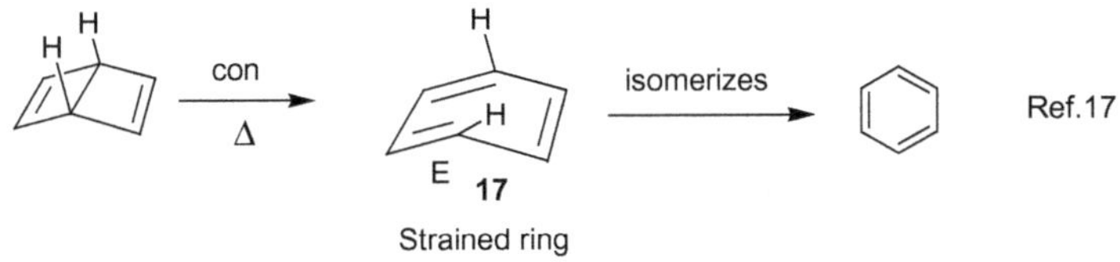

Heptatriene **18** having electron-withdrawing substituents, undergoes rapid conversion into bicyclo[4.1.0]-hepta-2,4-diene **19** and both the isomers remain in equilibrium because of low activation energy Ea < 10 kcal/mol. This transformation is known as valence tautomerism [18].

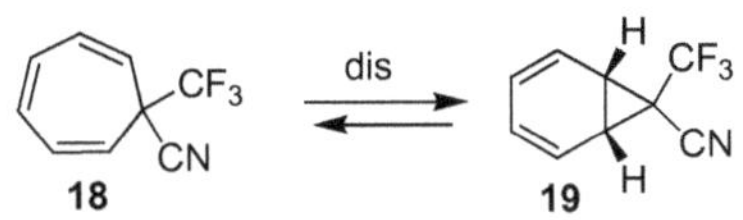

Triphenylhexatrienes **20** and **21** undergo ring closures by dis-motion to give **22** and **23,** respectively. Similarly, cyclic trienes **24** and **25** undergo allowed disrotatory reactions to give products.

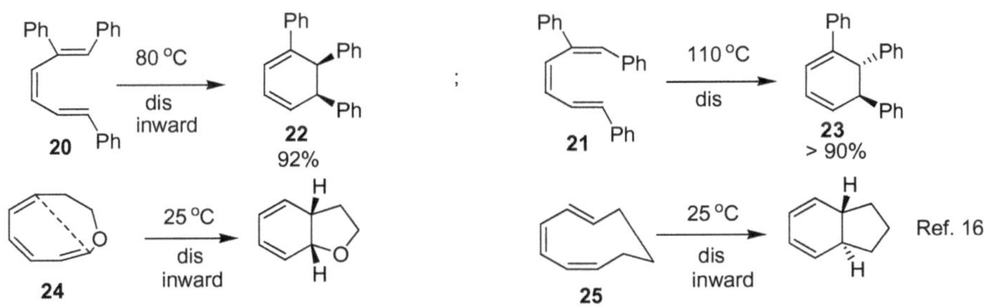

Substituted 1,3,5-hexatriene **26** on electrocyclization, followed by isomerization gives **27**.

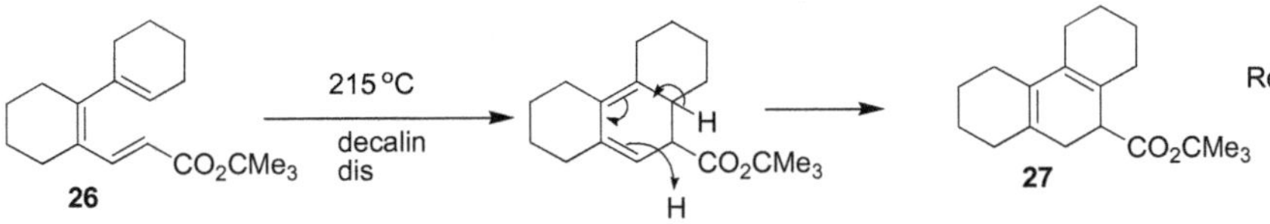

Bicyclic conjugated dienes **28** and **29** undergo ring opening by electrocyclic process.

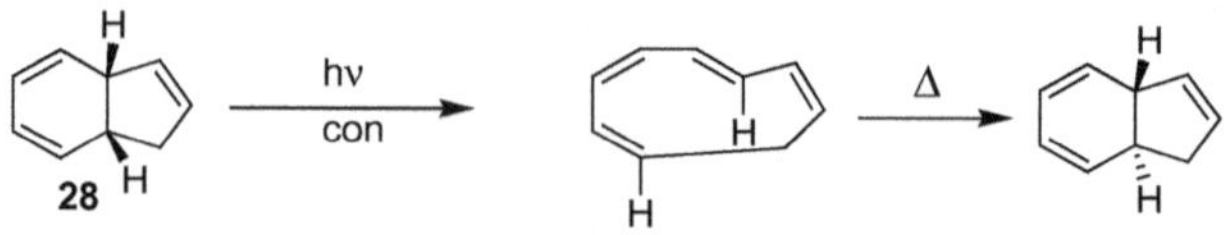

28

Ref.20

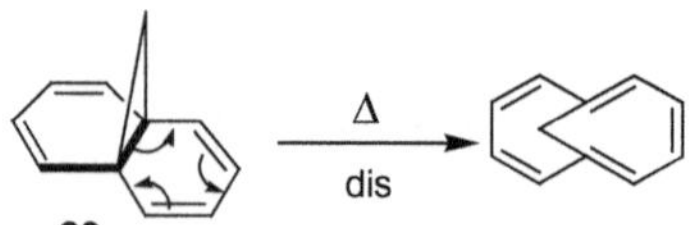

29

Some photochemical electrocyclic reactions take place to yield products of much higher energy than the starting materials [21]. For example, pyrone **30**, oxocin **31**, cyclic ketone **32** and *cis*-stilbene **33** undergo photo-induced allowed electrocyclization to give products of higher energy.

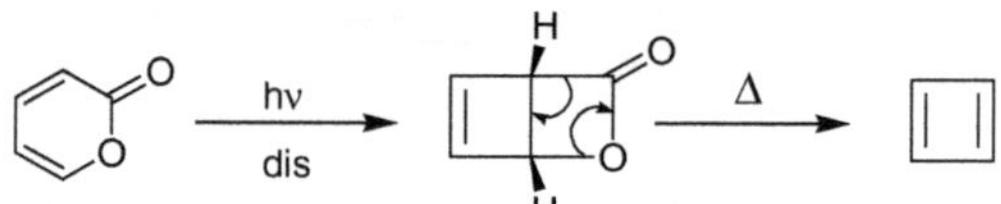

30

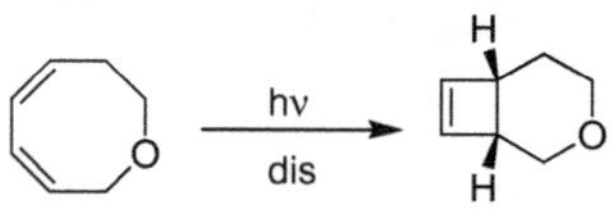

31

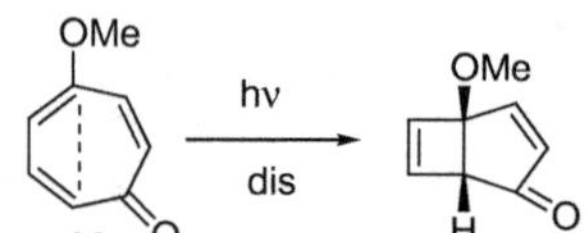

32

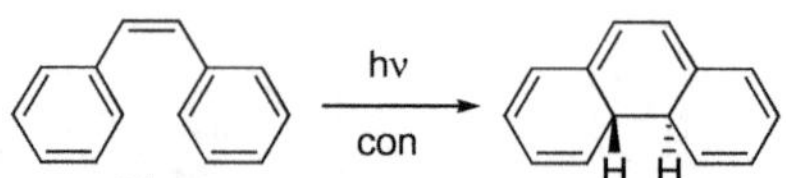

33

Light-induced electrocyclization of bicyclo-[4.2.1]-nona-2,4-dienes, **34** and **35** gives both *endo*- and *exo*-isomers. Direct irradiation of the unsubstituted diene gives *endo*-isomer as major product through a singlet excited state. The presence of heavy atom, such as chlorine in the diene system facilitates the ISC by spin-orbit coupling and increases the percentage of *exo*-isomer through triplet excited state. Use of photosensitizer gives *exo*-isomer as the major product [22].

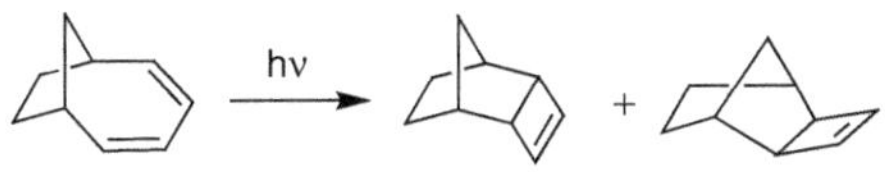

34 70 : 30

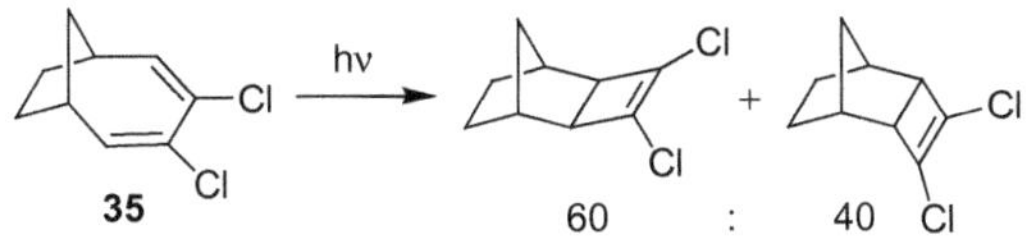

35 60 : 40

1, 3, 5-Trienes, **36**, **37** and **37a** in a ring system undergo double electrocyclic processes to yield stable products.

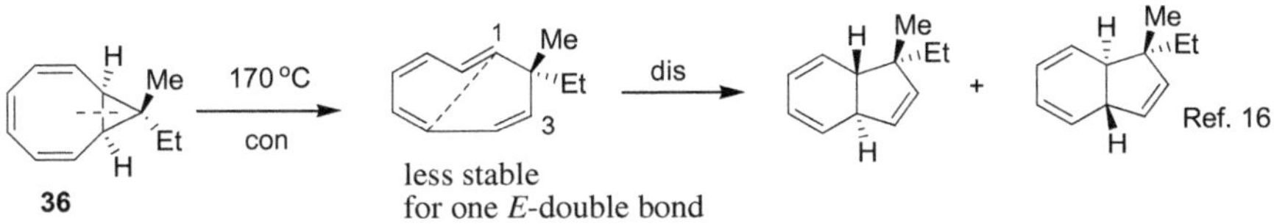

36

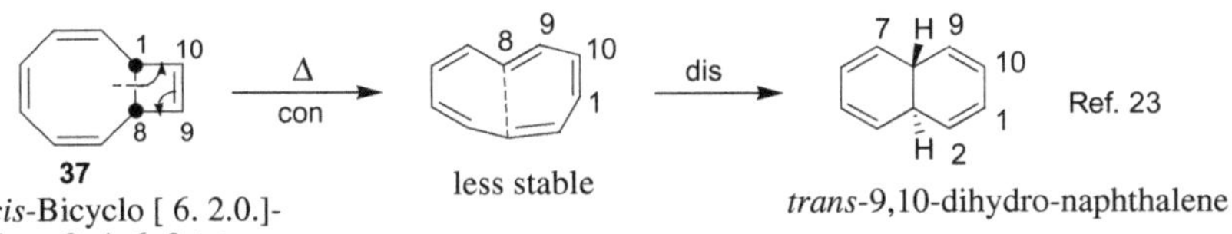

37
cis-Bicyclo [6. 2.0.]-
deca-2, 4, 6, 9-tetraene

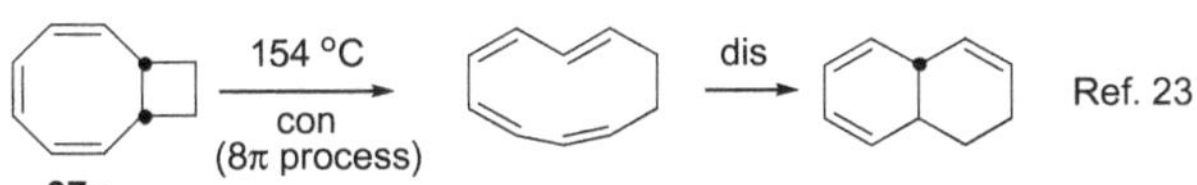

37a

The presence of methyl substituents in cyclic 1, 3, 5-triene **38** causes cyclization involving 4π electrons. Due to steric interaction, one of the olefinic double bonds remains out of plane of other olefinic double bonds.

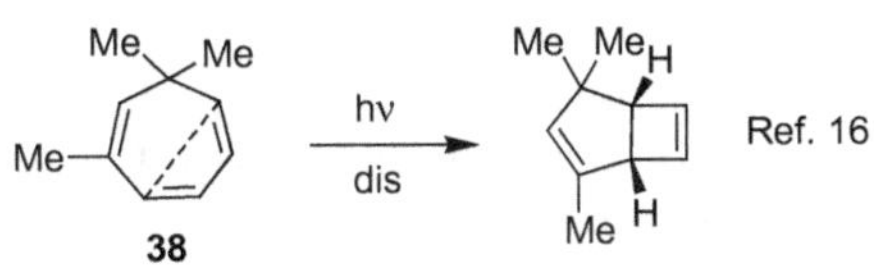

38

2, 4, 6, 8-Decatetraenes **39** and **40** undergo electrocyclic reactions near room temperature and maintain an equilibrium favouring the cyclooctatriene products. At slightly more elevated temperatures, the cyclooctatriene system undergoes another cyclization to produce bicyclo[4.2.0]-octa-2,4-diene.

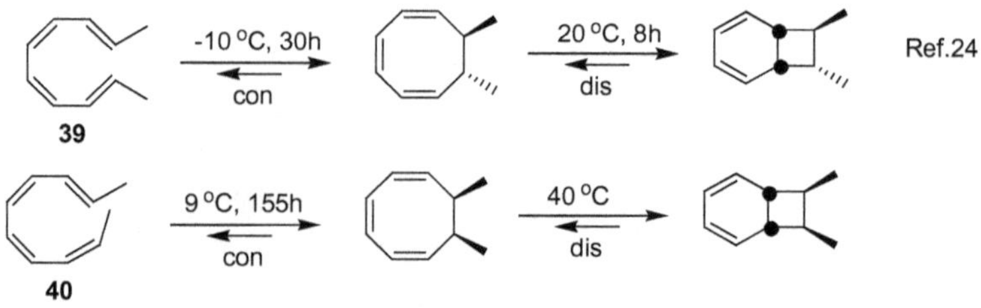

Conjugated decapentaene **41** undergoes electrocyclic ring closure using 6πe because of nonplanarity of 10πe.

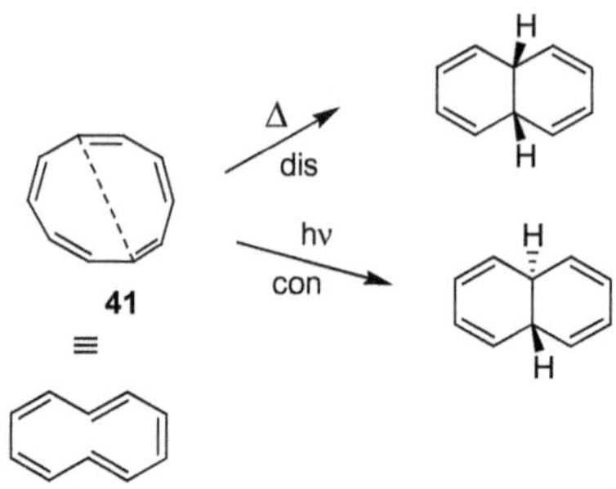

Oxonin **42** and azonine **43** having a tetraene system undergo cyclization using their triene system to bicyclo[4.3.0] systems due to interaction of the lone pair of oxygen and nitrogen, respectively, with one double bond [25].

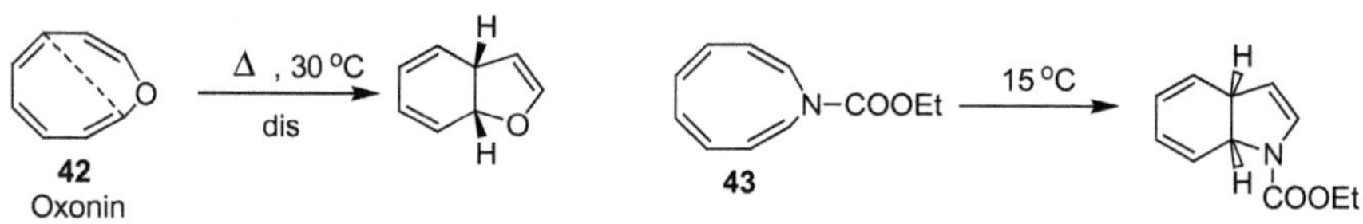

Acyclic conjugated enyne **44** in singlet excited state undergoes skeletal rearrangement via photoelectrocyclization to a highly strained 1,2-cyclobutadiene, followed by ground-state ring opening [26]. Calculation of energies of 1,2-cyclobutadiene suggested its planar geometry of C_2-symmetry with the vinylic hydrogens twisted 6° out of plane.

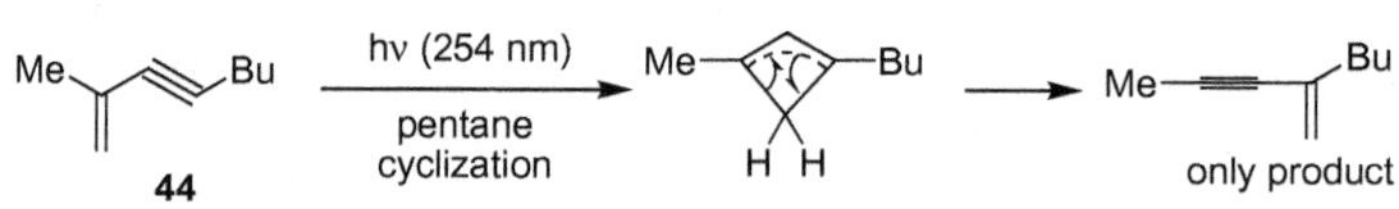

2.5 Applications of Ionic Conjugated Systems in Electrocyclic Reactions

The acetolysis of isomeric cyclopropyl tosylates (**45–47**) at 100 °C takes place by concerted electrocyclic ring opening and ionization. The loss of tosyloxy group in ionization step is assisted by the electron density of the developing p-orbitals that are *trans* to the leaving group. Hence, the isomer **47** undergoes much faster acetolysis than the other isomers [27].

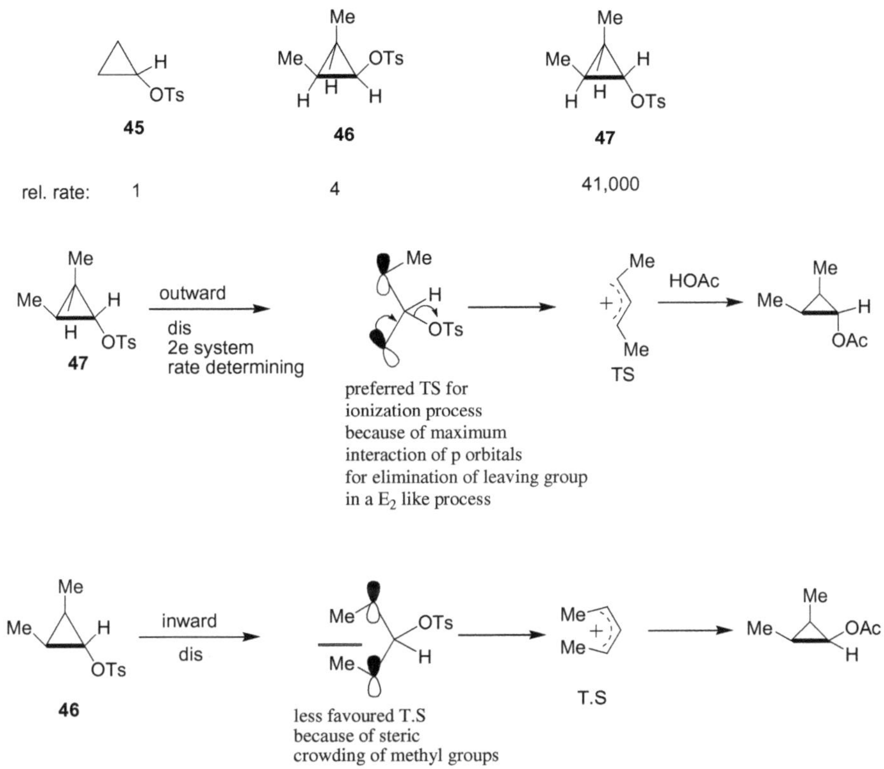

The rate of acetolysis of bicyclic tosylates **48** and **49** with acetic acid at 150 °C depends on the geometry of the generated allyl cation. Isomer **49** reacts about 2×10^6 times faster than **48** because in the former the reaction proceeds via the formation of stable cis-cyclohexenyl cation [28].

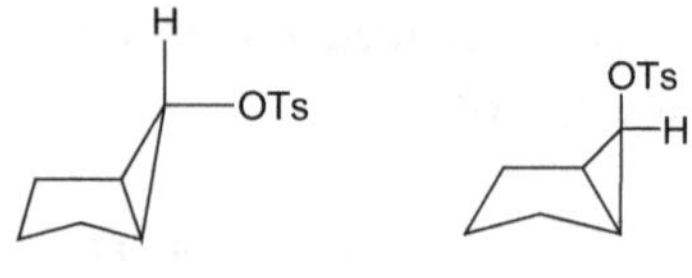

48 49

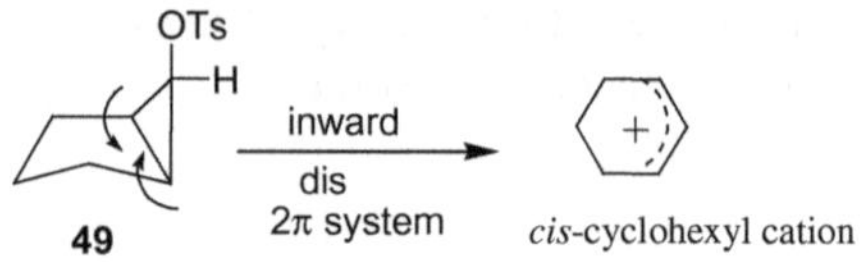

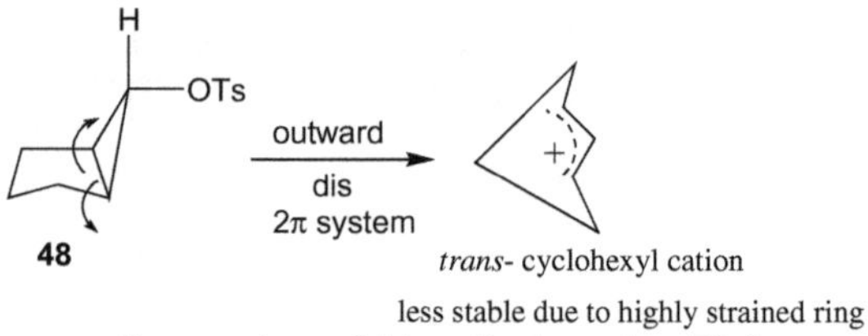

The similar electrocyclic reaction of bicyclic bromide **50** in aqueous dioxane at 100 °C gives *trans*-cyclooctene-3-ol **51** [29]. The generated *p*-orbitals from the breaking of a σ bond in cyclopropanone ring by inward dis-motion participate in the removal of bromine atom in an E_2-like process. The outward ring opening will provide less stable *trans*-cyclooctenyl cation.

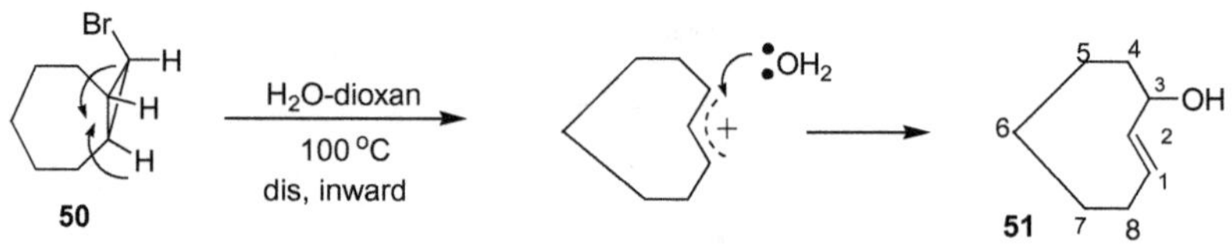

Methanolysis of bicyclic dibromide **52** with MeOH in the presence of $AgClO_4$ gives *trans* product **53** [30].

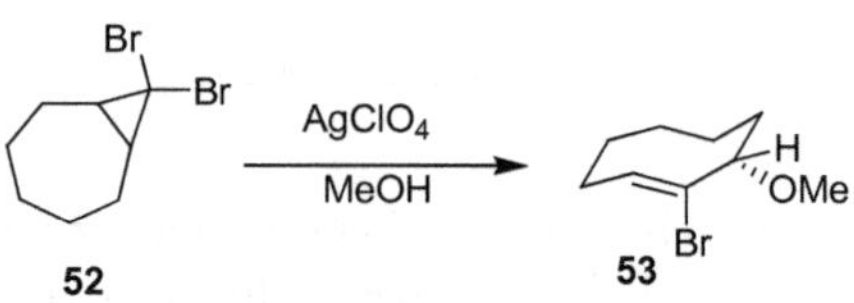

Hydrolysis of *endo*-2-chloro-*exo*-2-bromo-bicyclo[3.1.0]-hexane **54** gives 2-bromo-3-cyclohexenol **55** [16]. The inward disrotatory motion of ring opening is preferred because it provides the movement of *endo*-chlorine atom inside the cyclopropane ring to acquire an antigeometry with respect to electron-rich lobes of the p-orbitals generated by cleavage of a sigma bond of cyclopropane ring and helps to participate in an E_2-like elimination process for faster elimination. Moreover, it provides a more stable *cis*-cyclohexenyl cation [16].

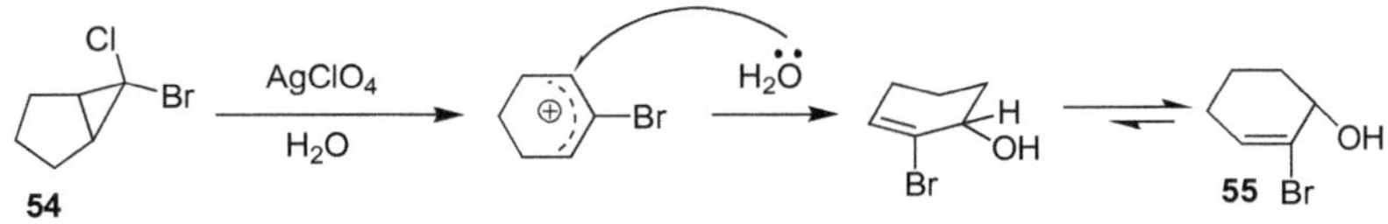

54 55

Acid-catalyzed cyclization of divinyl ketone **56–57** occurs by conrotatory cyclization of 3-hydroxy pentadienyl cation [31]. This type of cyclization reaction is known as Nazarov cyclization reaction [31].

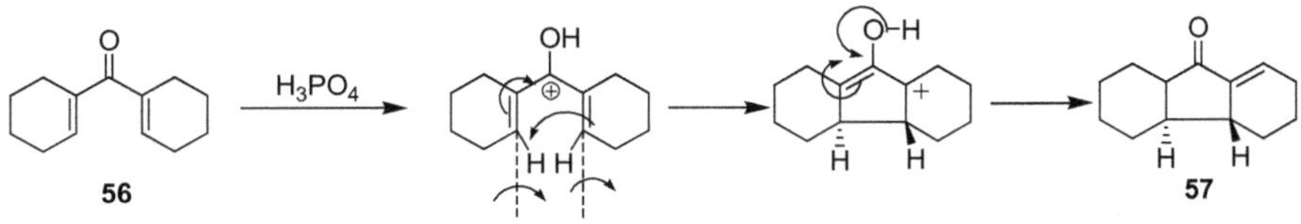

56 57

Divinyl ketone **58** undergoes Lewis acid-catalyzed electrocyclization to give **59** [21].

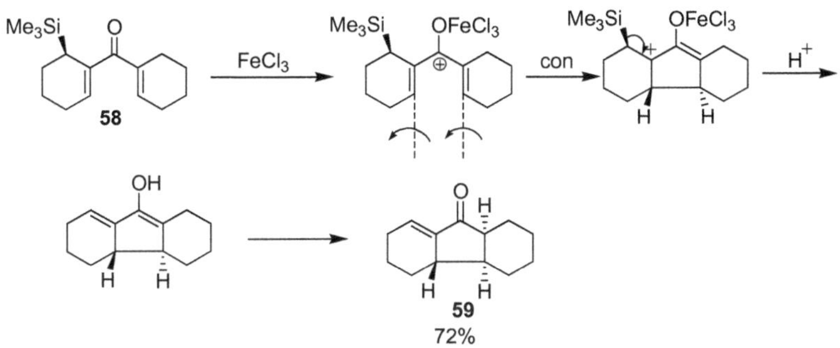

58 59

Similar electrocyclic ring closure occurs in aryl vinyl ketone **60** by strong acid.

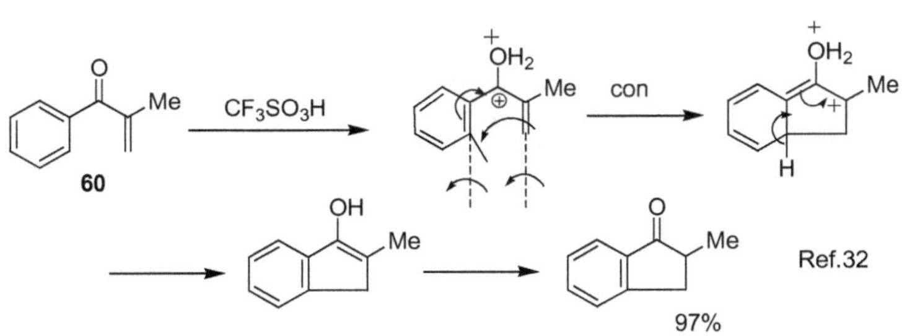

60

Cyclooctadienyl lithium **61** having pentadienyl anion undergoes cyclization by disrotatory motion at 35 °C [33].

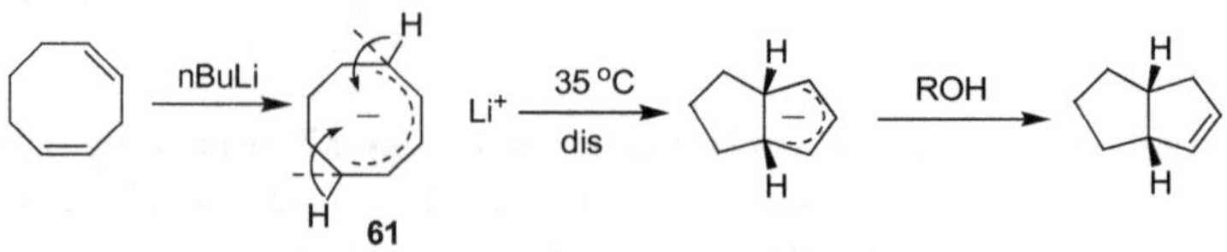

Cyclooctadienyl anion generated from cyclopentene derivative **62** undergoes ring opening in the presence of a strong base to give **63** and **64**.

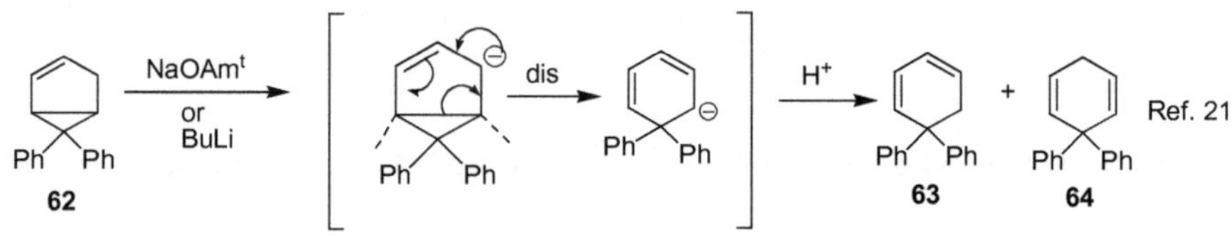

Cyclooctatetraene dianion **65** on treatment with acetyl chloride gives 3, 5, 7, 9-dodecatetraene-2,11-dione **66** by electrocyclic ring opening along with other products [34].

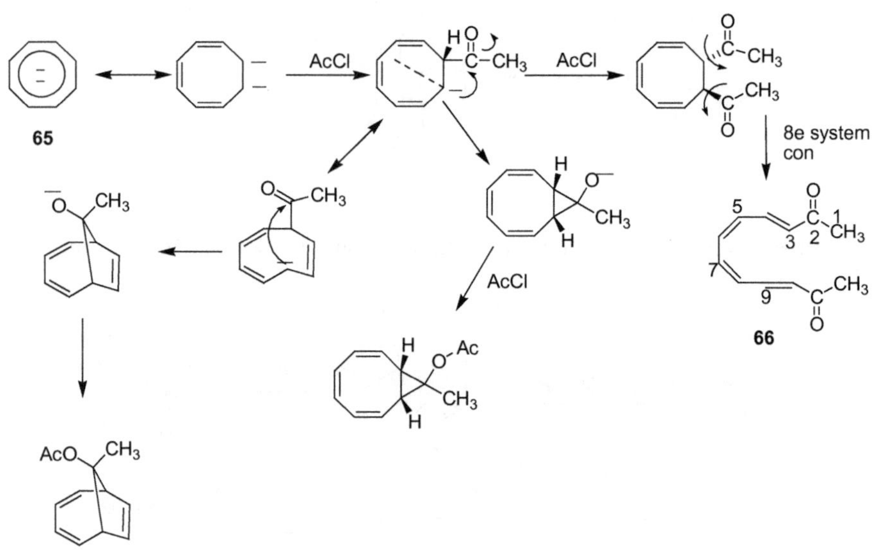

Cyclooctadienyl anion obtained from carbolithiation of 3-methylene-1,4-cyclooctadiene **67**, on electrocyclization gives a *cis*-bicyclo-product **68**, which on trapping with an electrophile gives *exo*-isomer **69** as major product [33].

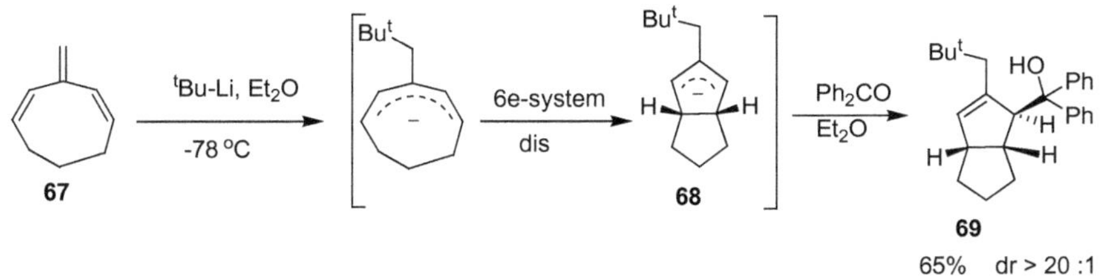

Butadienyl pyridinium ylide **70** on electrocyclization gives stereoselective 1, 2-annulated-2,3-dihydroazepine **71** as major product [35].

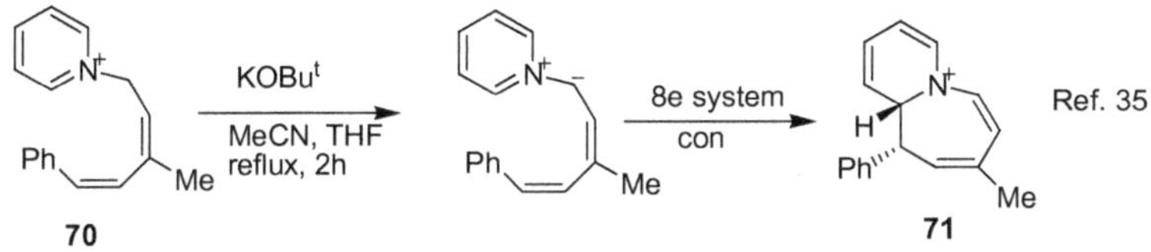

Hydrobenzamide **72** on treatment with phenyl lithium gives a pentadienyl anion, which on electrocyclization gives dihydroimidazole, amarine **73** [36].

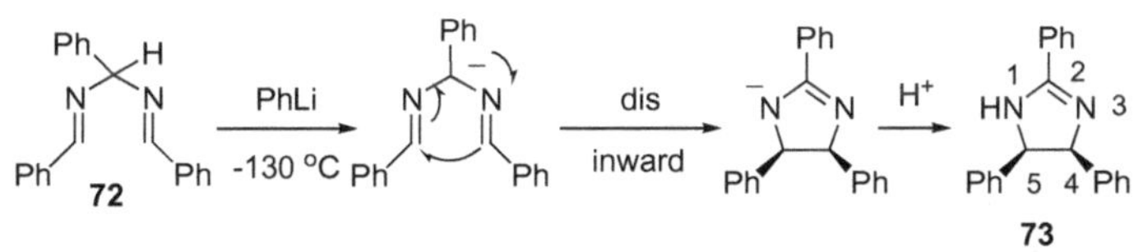

Similarly, heterocyclic compound **74** in the presence of a strong base gives **75** by disrotatory reaction [36].

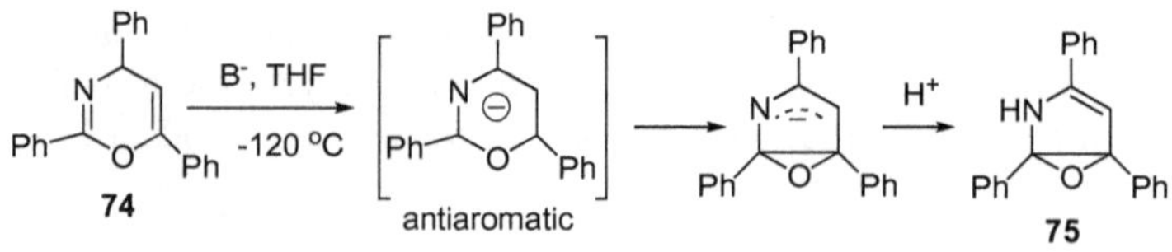

2.6 Problems

2.6.1. Predict the structure, including stereochemistry, of the product for each of the following reactions.

a. 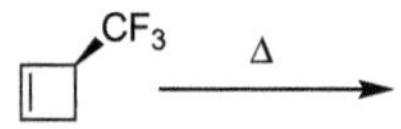

b. 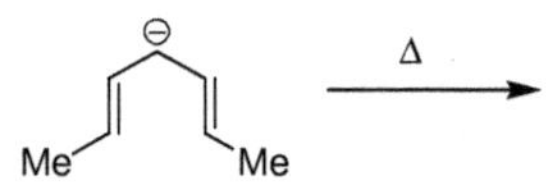

c. 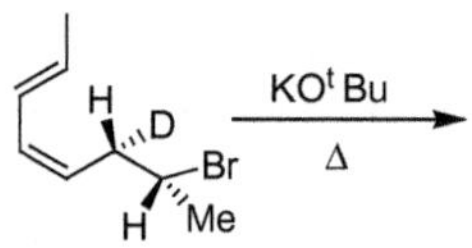

d. 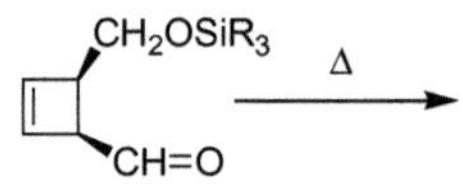

e. 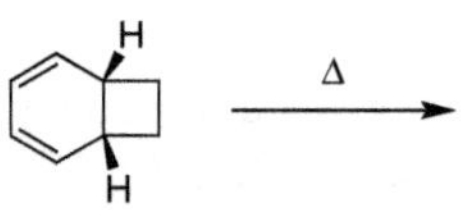

f. 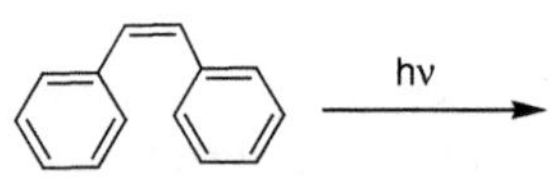

g. 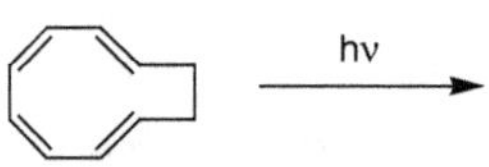

h. 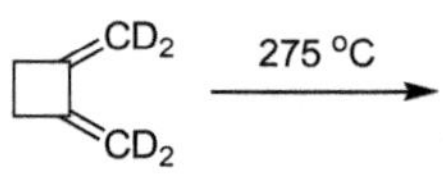

i. 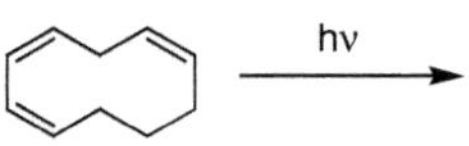

j. 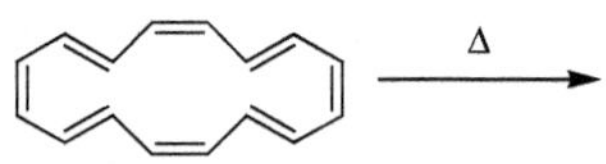

k. 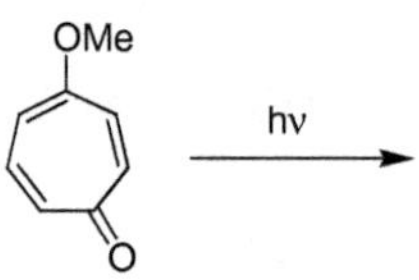

2.6.2 Rationalize the following reactions.

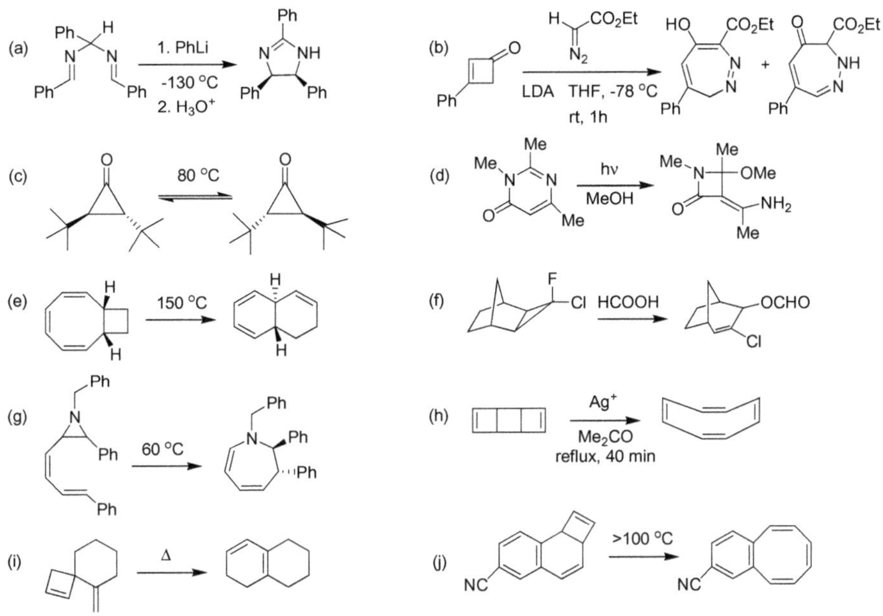

2.6.3. Offer a mechanistic explanation for each of the following observations.

a. It has been found that both compound A and B undergo ring opening about 10^4 times faster than C in the presence of lithium di-t-butylamide.

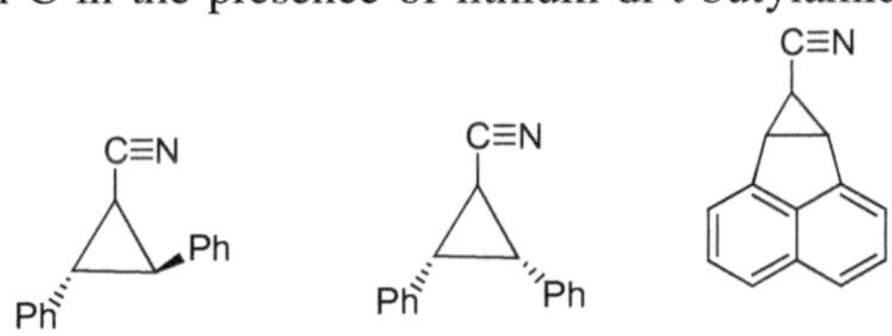

A B C

b. 2-Vinylcyclopropanols undergo facile rearrangement to give cyclopent-3-enols.

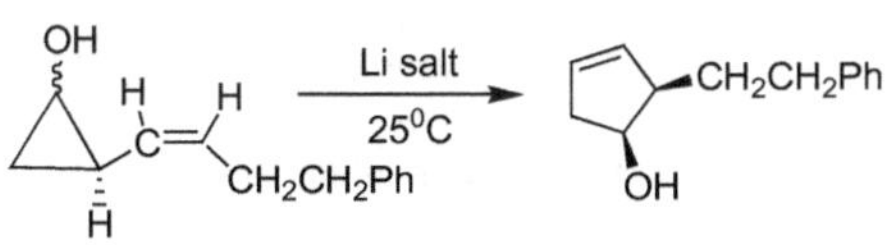

c.

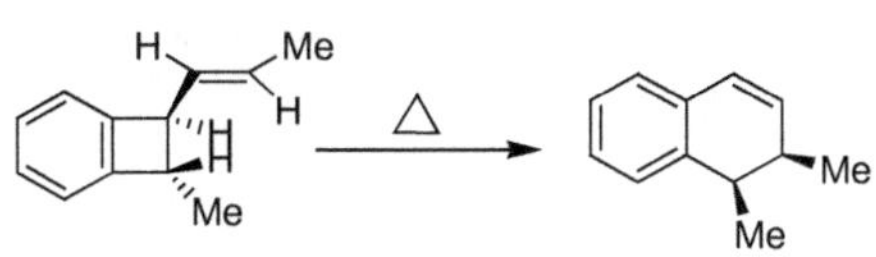

d. Compound D undergoes solvolysis readily with AcOH at 125 °C, whereas compound E remains unchanged on prolonged heating with AcOH at 210 °C.

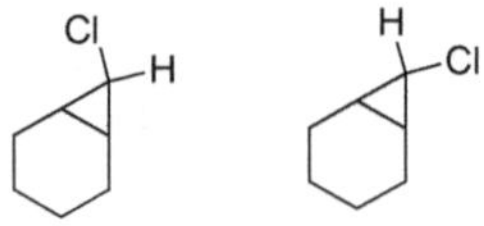

D E

2.7 Further Reading

1. Marvell EN (1980) Thermal electrocyclic reactions. Academic Press, New York
2. Durst T, Breau L (1991) Cyclobutane ring opening. In: Trost BM, Fleming I (eds) Comprehensive organic synthesis. Pergamon, Oxford (Chapter 6)
3. Marchand AP, Lehr RE (1977) Pericyclic reactions, vols I & II. Academic Press, New York
4. Carey FA, Sundberg RJ (2007) Advanced organic chemistry, Parts A & B. Springer, New York

References

1. Winter REK (1965) Tetrahedron Lett 1207
2. Brauman JI, Archie WC Jr (1972) J Am Chem Soc 94:4262
3. Marvell EN, Caple G, Schatz B (1965) Tetrahedron Lett 385
4. Vogel E, Grimme W, Dinne E (1965) Tetrahedron Lett 391
5. Woodward RB, Hoffmann R (1965) J Am Chem Soc 87:395
6. Fukui K, Fujimoto H (1968) In: Thygarajan BS (ed) Mechanisms of molecular migrations, vol 2. Interscience, New York, p. 113; Longuet-Higgins HC, Abrahamson EW (1965) J Am Chem Soc 87:2045
7. Schumate KM, Neuman PN, Fonken GJ (1965) J Am Chem Soc 87:3996
8. Bronton R, Frey HM, Montaque DC, Stevens IDR (1966) Trans Faraday Soc 62:659
9. Buda AB, Wang Y, Houk KN (1989) J Org Chem 54:2264; Hayes R, Ingham S, Saengchantrara ST, Walker TW (1991) Tetrahedron Lett 32:2953
10. Criegee R, Seebach D, Winter RE, Boerretzen B, Brune H (1963) Chem Ber 98:2339; Maier G, Bothur A (1998) Eur J Org Chem 2063
11. Binns F, Hayes R, Ingham S, Saengchantara ST, Turner RW, Wallace TW (1992) Tetrahedron 48:515
12. Dolbier WR Jr, Koroniak H, Houk KN, Sheu C (1996) Acc Chem Res 29:471
13. Jefford CW, Boschung AF, Rimboult CG (1974) Tetrahedron Lett 3387
14. Rondon NG, Houk KN (1985) J Am Chem Soc 107:2099
15. Hamura T, Miyamoto M, Imura K, Matsumoto T, Suzuki K (2002) Org Lett 4:1675
16. Morrison RT, Boyd RN (1989) Organic chemistry, 5th edn. Prentice-Hall, New Delhi; De Puy CH (1968) Acc Chem Res 1:33
17. Goldstein MJ, Leight RS (1977) J Am Chem Soc 99:8112

18. Maier G (1967) Angew Chem Int Ed Engl 6:402
19. von Zezschwitz P, Petry F, de Meijre A (2001) Chem Eur J 7:4035
20. Eberbath W (1975) Chem Ber 108:1052; Mosamune S, Dubby N (1972) Acc Chem Res 5:272
21. Fleming I (2002) Pericyclic Reactions. Oxford University Press, New York
22. Jefford CW, Delay F (1975) J Am Chem Soc 97:2272
23. Staley SW, Henry TJ (1971) J Am Chem Soc 93:1292
24. Huisgen R, Dahmen A, Huber H (1969) Tetrahedron Lett 1461; (1967) J Am Chem Soc 89:7130; Dahmen A, Huisgen R (1969) Tetrahedron Lett 1465
25. Masamune S, Takeda S, Scidner RT (1969) J Am Chem Soc 91:7769; Anastassiou A, Cellura RP (1969) Chem Commun 1521
26. Zhang M, DiRico Kj, Kirchhoff MM, Phillips KM, Cuff LM, Johnson RP (1993) J Am Chem Soc 115:12167
27. de Puy CH (1968) Acc Chem Res 1:33
28. Schleyer PVR, Sliwinski WF, van Dine GW, Schollkopf U, Paust J, Fellenberger K (1972) J Am Chem Soc 94(125):132
29. Whitham GH, Wright M (1971) J Chem Soc C 883
30. Reese CB, Shaw A (1970) J Am Chem Soc 92:2566
31. Woodward RB (1969) In: Aromaticity, chemical society special publication no. 21, p. 217; Habermas KL, Denmark SE, Jones TK (1994) Org React 45:1
32. Bates RB, Gosselink DW, Kaczynski JA (1967) Tetrahedron Lett 199, 205; Bates RB, McCombs DA (1969) Tetrahedron Lett 977
33. Williams DR, Reeves JT, Nag PP, Pitcock WH, Baik MH (2006) J Am Chem Soc 128:12339
34. Cantrell TS, Shechter H (1967) J Am Chem Soc 89:5868
35. Marx K, Eberbach W (1997) Tetrahedron 53:14687
36. Hunter DH, Sim SK (1969) J Am Chem Soc 91: 6202; ibid (1972) Can J Chem 50:669

Chapter 3
Cycloaddition Reactions

3.1 Introduction

Cycloaddition reactions are the most useful pericyclic reactions in organic synthesis. These are the reactions of two π systems to form ring compounds by the breaking of two π bonds and making of two σ bonds in a concerted process. The reverse of cycloaddition reactions are known as retrocycloaddition reactions. Both cycloadditions and cycloreversions proceed through cyclic transition states in which continuous flow of electrons occur among the reacting molecules. These reactions are classified according to the number of π electrons involved in each reacting molecules. The major classes are $[\pi^2+\pi^2]$, $[\pi^4+\pi^2]$, $[\pi^6+\pi^2]$, $[\pi^8+\pi^2]$, and $[\pi^6+\pi^4]$. These are simply known as [2+2]-, [4+2]-, [6+2]-, [8+2]-, and [6+4]-cycloaddition reactions and are illustrated below.

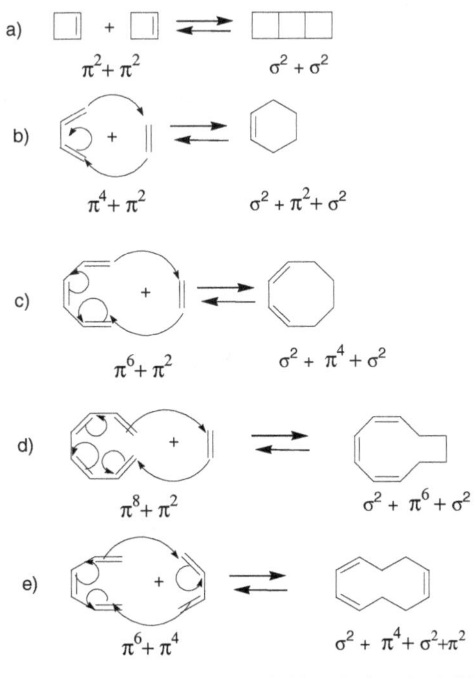

B. Dinda, *Essentials of Pericyclic and Photochemical Reactions*, Lecture Notes in Chemistry 93, DOI 10.1007/978-3-319-45934-9_3

3.2 [2+2]-Cycloaddition Reactions

3.2.1 Overview of Thermal and Photochemical [2+2]-Cycloaddition Reactions

Thermal [2+2]-cycloaddition reactions are less common, but photochemical [2+2]-cycloaddition reactions are very common. This fact can be explained by analyzing these cycloaddition reactions using Woodward–Hoffmann selection rules. In frontier orbital approach, the thermal reaction of two ethene molecules (one is HOMO and other is LUMO) is orbital symmetry forbidden process for its suprafacial–suprafacial $[\pi^2s+\pi^2s]$-cycloaddition, but a suprafacial–antarafacial $[\pi^2s+\pi^2a]$-cycloaddition reaction is symmetry allowed process (Fig. 3.1). It signifies that the cycloaddition of one two-π electron system with another two-π electron system will be a thermally allowed process when one set of orbitals is reacting in a suprafacial mode and other set in an antarafacial mode ("s" means suprafacial and "a" means antarafacial). Thermal $[\pi^2s+\pi^2a]$-reactions usually occur in the additions of alkenes to ketenes, when alkene is in the ground state and ketene in the excited state [1] (Fig. 3.2).

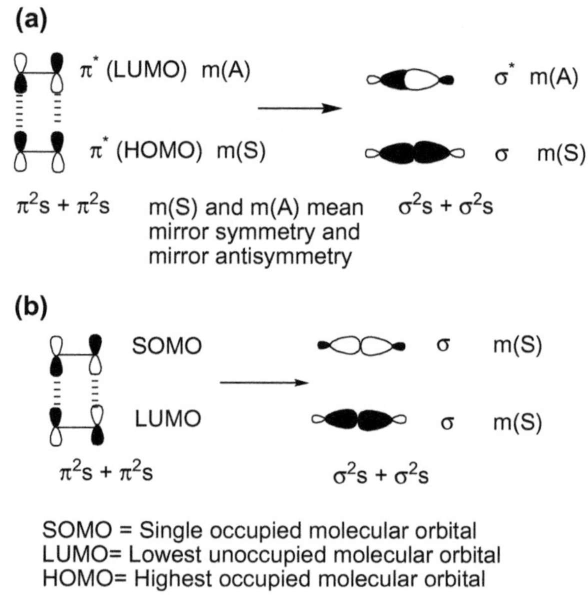

Fig. 3.1 Frontier orbital interactions of **a** thermally forbidden $[\pi^2s+\pi^2s]$-cycloaddition reaction, **b** photochemically allowed $[\pi^2s+\pi^2s]$-reaction of alkenes

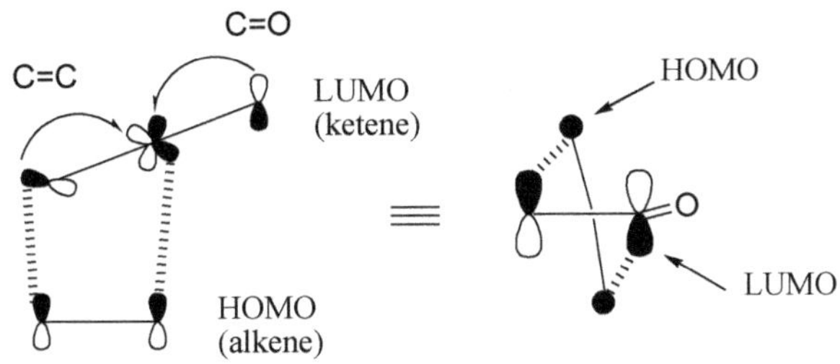

Fig. 3.2 Frontier orbital interactions of thermally allowed antarafacial interaction of a ketene (LUMO) and an olefin (HOMO)

3.2.2 Applications of [2+2]-Cycloaddition Reactions

Cycloaddition of ethoxy ketene **1** with *cis*- and *trans*-2-butene is a concerted process and gives stereospecific products **2** and **3**, respectively [2].

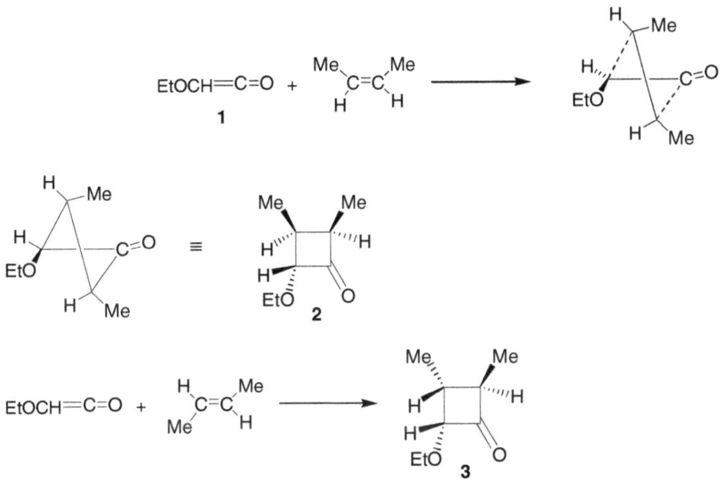

Similarly, ethylketene **4** and dimethylketene **5** react with cyclopentadiene to give **6** and **7**, respectively.

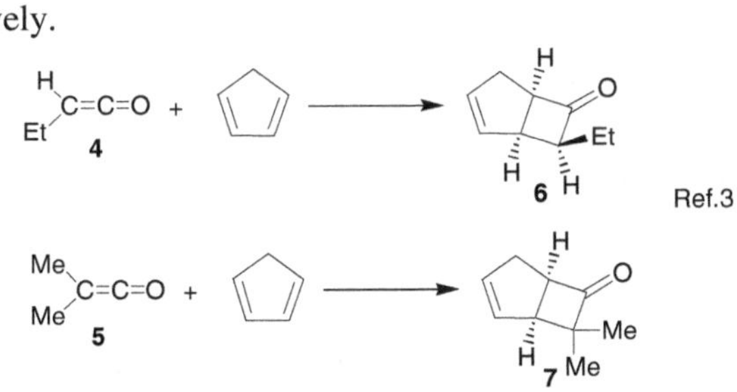

Ketene **8** reacts with ethoxy acetylene to give **9** [3] and phenyl methyl ketene **10** reacts with Z-alkene **11** to give **12** [4].

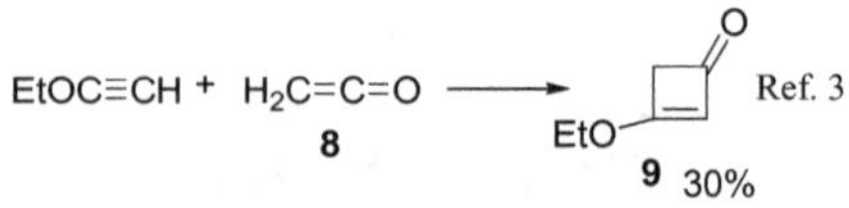

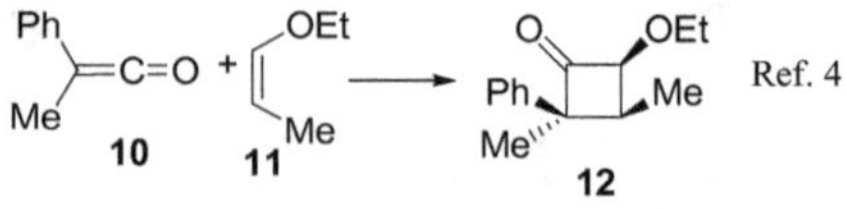

Ethoxyethylene reacts with diphenylketene **13** to give cycloadduct **14** in high yield [5].

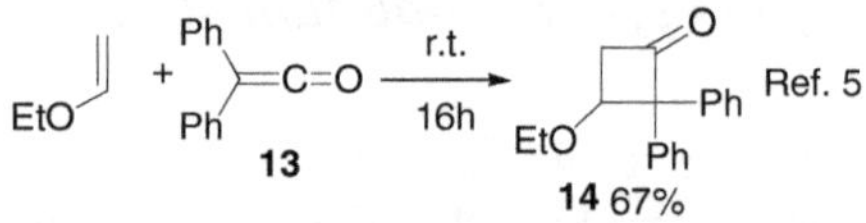

Ketenes also undergo [$\pi^2+\pi^2$]-cycloadditions with ketones, in the presence of Lewis acids as catalysts, e.g., **8** reacts with ketone **15** to give **16** and with **17** to give **18** in high yields [6].

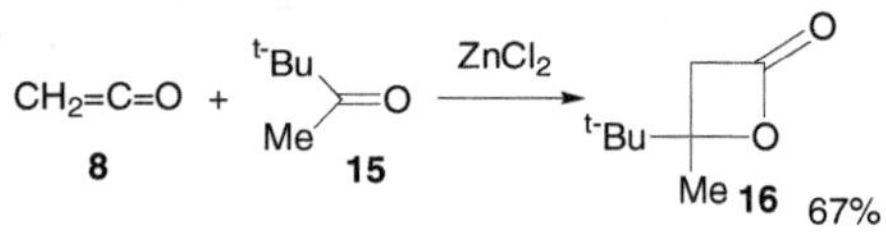

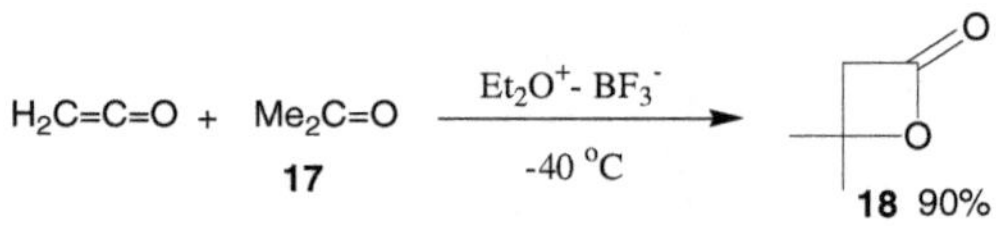

Intramolecular ketene cycloadditions are also observed in compounds having both ketene and alkene functionalities in appropriate orientations. For example, **19** gives **20** [6] and **21** gives **22** [7].

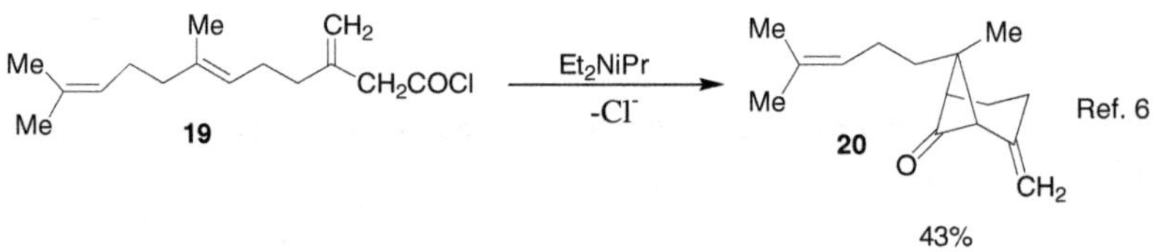

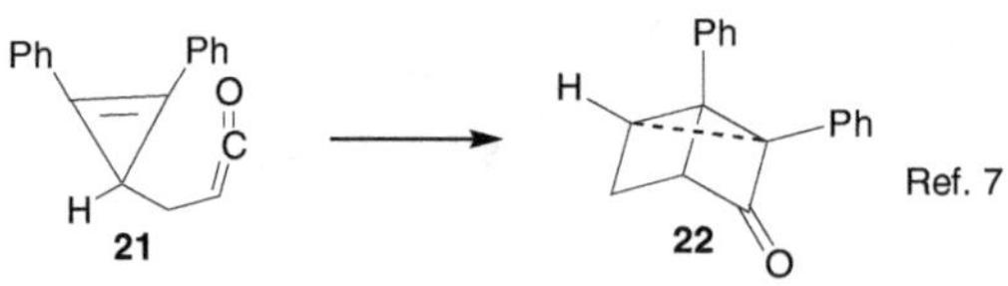

Several non-concerted [2+2]-cycloadditions have been reported, where diradical or zwitterionic intermediates are produced. The following examples are illustrative:

1. 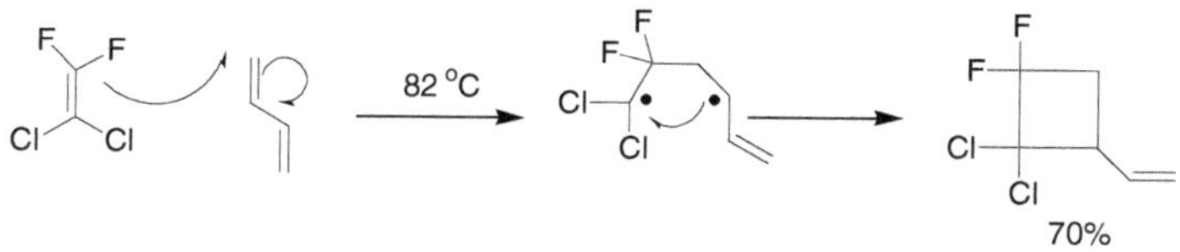

2. 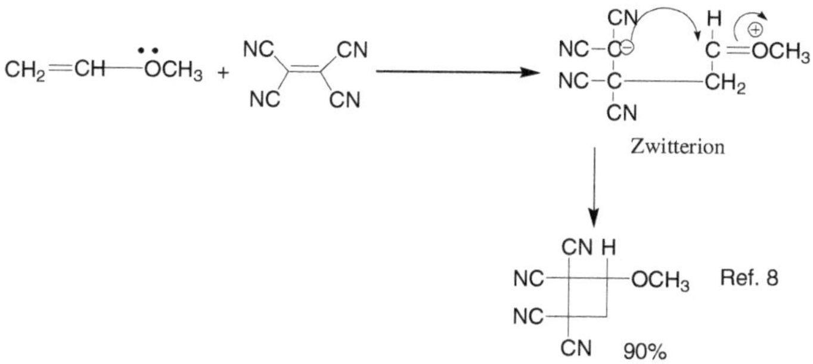

3. 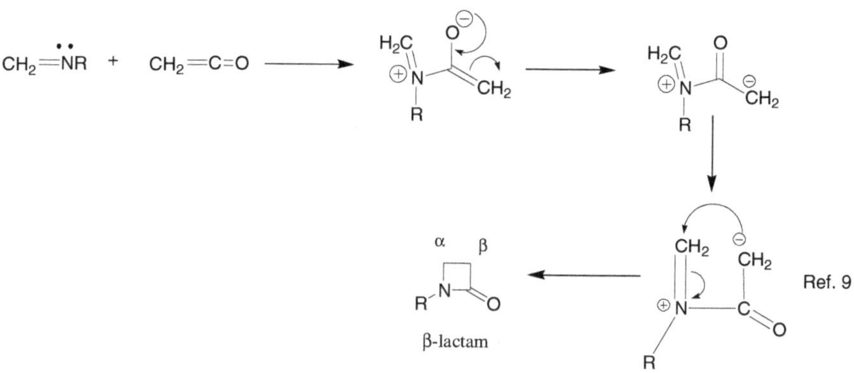

4. 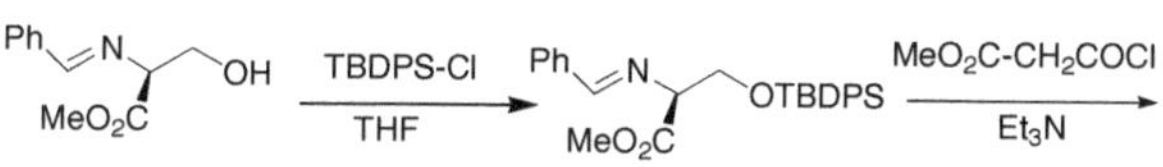

TBDPS= tert-butyldiphenylsilyl

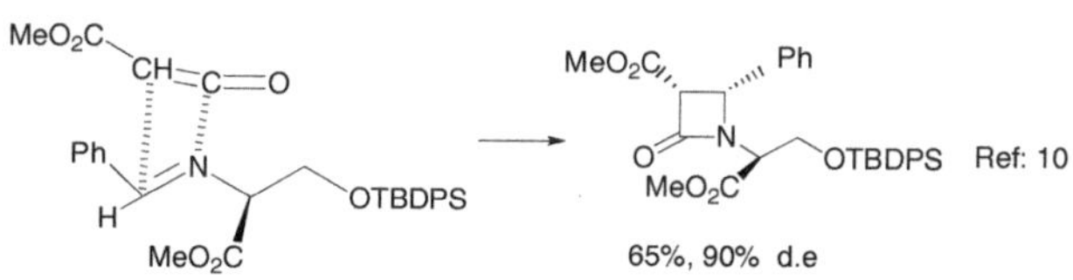

Photochemical [2+2]-cycloaddition in suprafacial mode is symmetry allowed process and occurs in dimerization of alkenes, intermolecular additions of alkenes, and intramolecular cycloadditions of dienes and alkynes. The following examples are illustrative:

(a) Dimerization of alkenes

1. 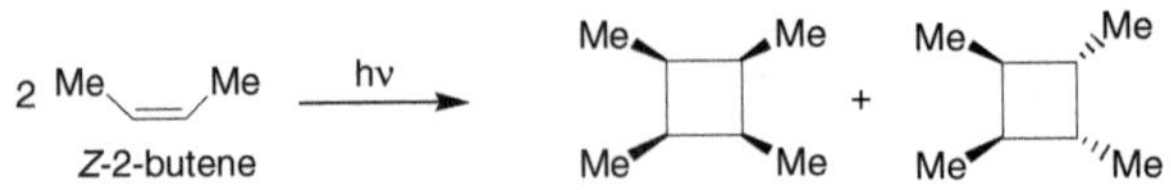 Ref: 11

2. 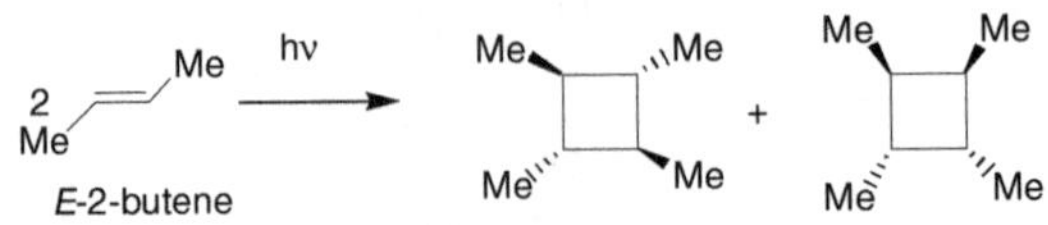

(b) Intermolecular additions of alkenes

3. 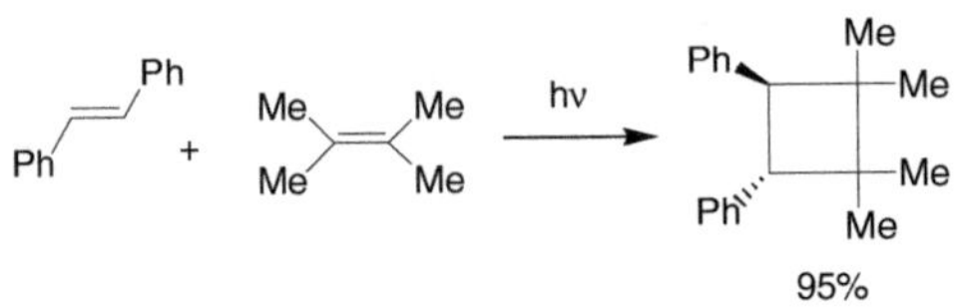 Ref: 12

(c) Intramolecular addition of dienes

4. 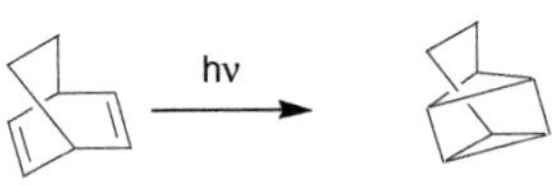 Ref: 13

5. 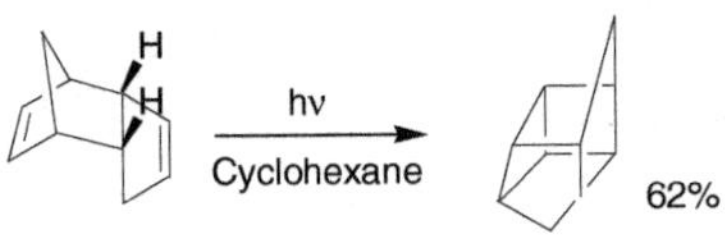 Ref. 14

6. 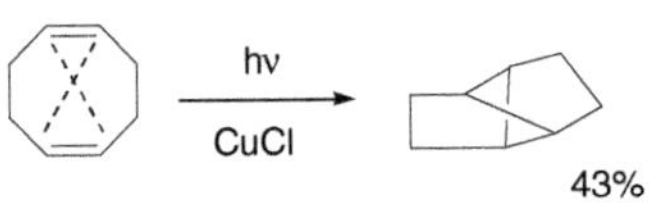 Ref. 15

7. 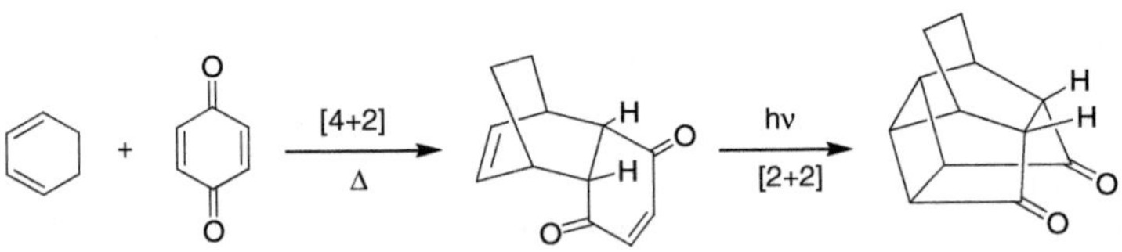 Ref. 15

(d) Intramolecular cycloaddition of enone and compounds bearing enolate and yne functions

8. 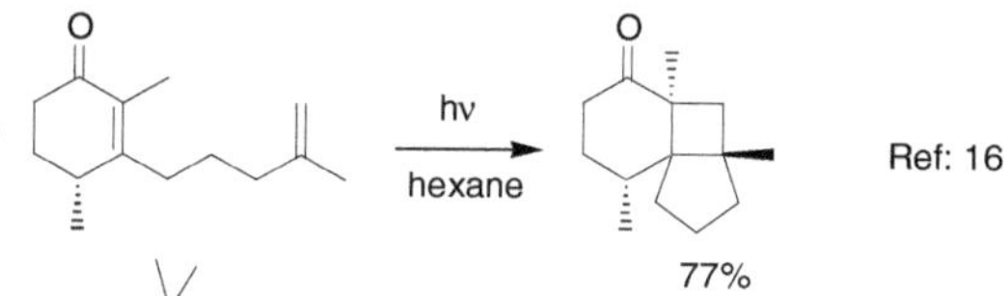

9. 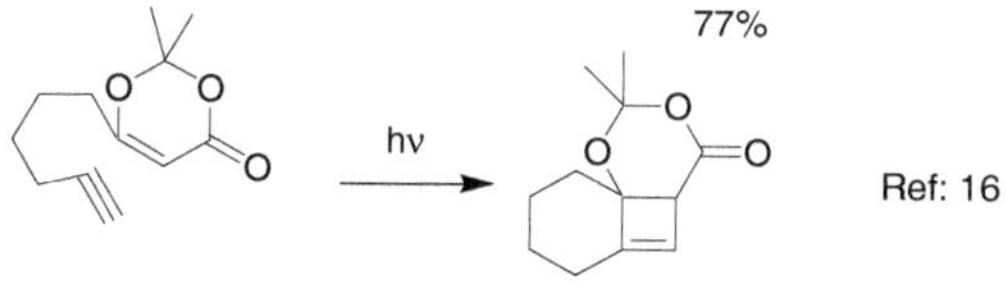

Non-stereospecific photochemical [2+2]-cycloadditions occur in the dimerization of phenyl cyclohexene **23** in the presence of a sensitizer to produce **24** and **25** [17], and in reactions of Z/E-2 butene with cyclohexenone **26** to give **27** and **28** [18] through the formation of intermediate diradicals. The photoaddition of cyclohexene to an enolised form of 1,3 diketone **29** gives **30** in a concerted process via the formation of an unstable cycloadduct [18].

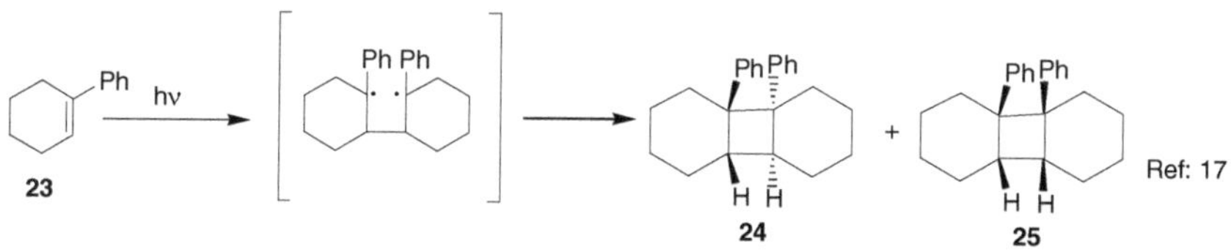

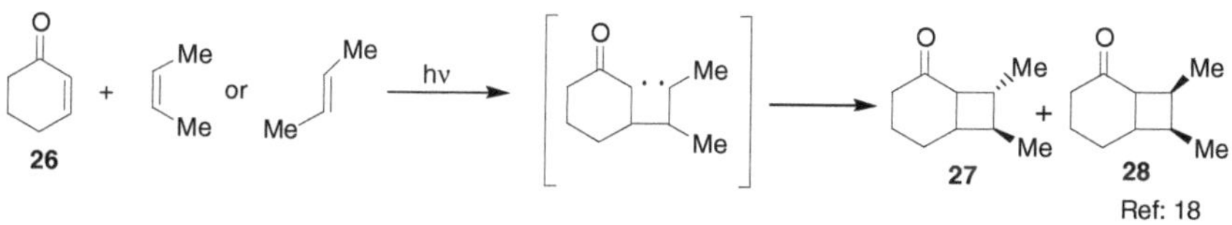

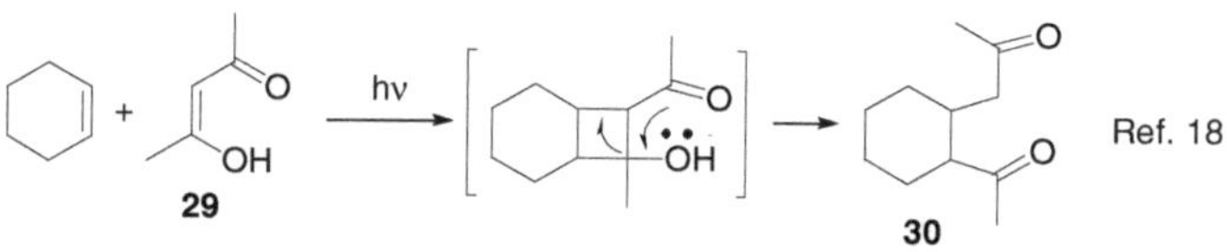

3.3 [4+2]-Cycloaddition Reactions

The Diels–Alder reactions and 1,3-dipolar cycloaddition reactions are known as [4 +2]-cycloaddition reactions because four electrons from diene or 1,3-dipole, and two electrons from the dienophile or dipolarophile are involved in these reactions. The 1,3-dipolar cycloaddition reactions are also called [3+2]-cycloaddition reactions because three atoms of dipolar compound and two atoms of dipolarophile are involved in the cyclization process.

3.3.1 *The Diels–Alder Reactions*

3.3.1.1 Overview of the Diels–Alder Reaction

The most important type of thermal [4+2]-cycloaddition reactions is known as the *Diels–Alder reaction*, as this reaction was discovered by Otto Paul Herman Diels and Kurt Alder in 1928 [19]. It may be noted that both of them awarded the Nobel Prize in Chemistry in 1950 for their contributions on the development of the Diels–Alder reaction. These reactions are defined as the concerted [4+2]-cycloaddition reactions of conjugated dienes with an alkene or alkyne. The alkene or alkyne is known as dienophile. Hence, these reactions are described as $[\pi^4+\pi^2]$-cycloaddition reactions. These reactions are carried out by heating the compounds alone or in an inert solvent or in the presence of a Lewis acid. An alkene or alkyne having electron-withdrawing substituent acts as an effective dienophile. These reactions proceed stereospecifically to *syn*-addition with respect to both diene and dienophile.

The following cycloaddition reactions of butadiene, furan, and cyclopentadiene are illustrative examples:

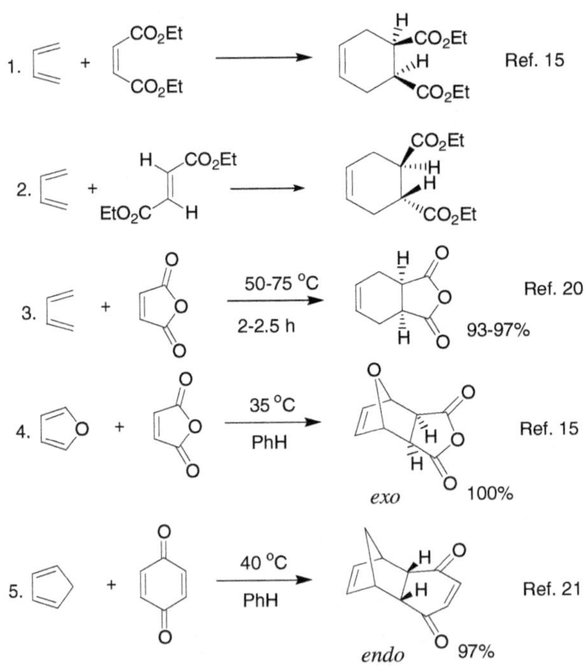

These reactions are broadly classified as normal and inverse electron demand Diels–Alder reactions. In normal electron demand Diels–Alder reactions (NED D–AR), diene system acts as electron-rich HOMO system and dienophile as electron-seeking LUMO system, while in inverse electron demand Diels–Alder reactions (IED D–AR), diene system serves as LUMO and dienophile as HOMO [22]. For example, reaction of cyclopentadiene **31** with methyl acrylate gives **32** in NED D–AR [23] and of ethoxyethylene with β-(2-cyclohexenone)-ethyl acrylate **33** gives **34** in NED D–AR [24].

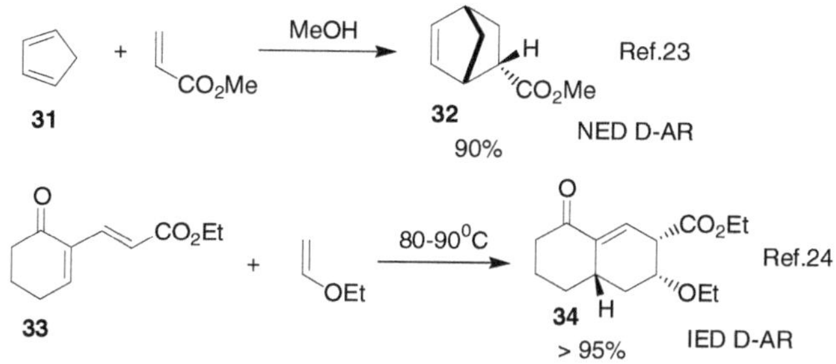

3.3.1.2 Regioselectivity

Regioselectivity of Diels–Alder reaction depends on the position of substituents in both diene and dienophile. Usually four types of D–A reactions are observed (Scheme 3.1)

This regioselectivity of Diels–Alder reactions can be interpreted on the basis of electron density (orbital coefficient) at C-4 of the diene and C-2 of the dienophile. For dienes with electron-releasing groups at C-1, HOMO has its largest electron density (largest HOMO orbital coefficient) at C-4. For dienophiles with electron-accepting substituents, C-2 has minimum electron density (largest LUMO orbital coefficient) in their LUMO. Hence, the strongest frontier orbital interaction occurs between C-4 of the diene and C-2 of the dienophile in a normal electron demand D–A reaction to give '*ortho*'-like product. For example, reaction of **35** with **36** gives cycloadduct **37** [5].

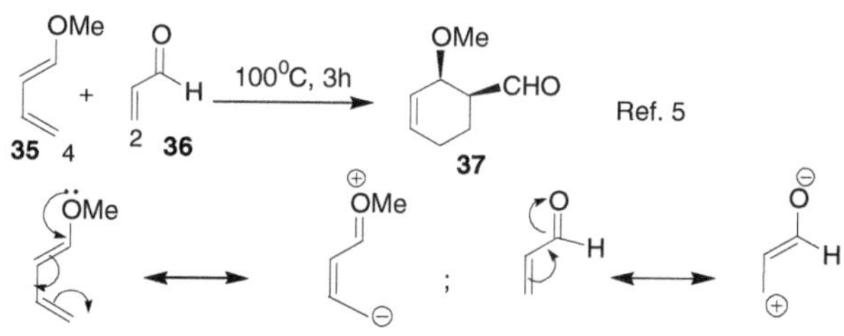

Similarly, in other type of normal Diels–Alder reactions, dienes with electron-releasing groups at C-2 have the largest electron density at C-1 in their HOMO. Therefore, the strongest frontier orbital interaction occurs between C-1 of

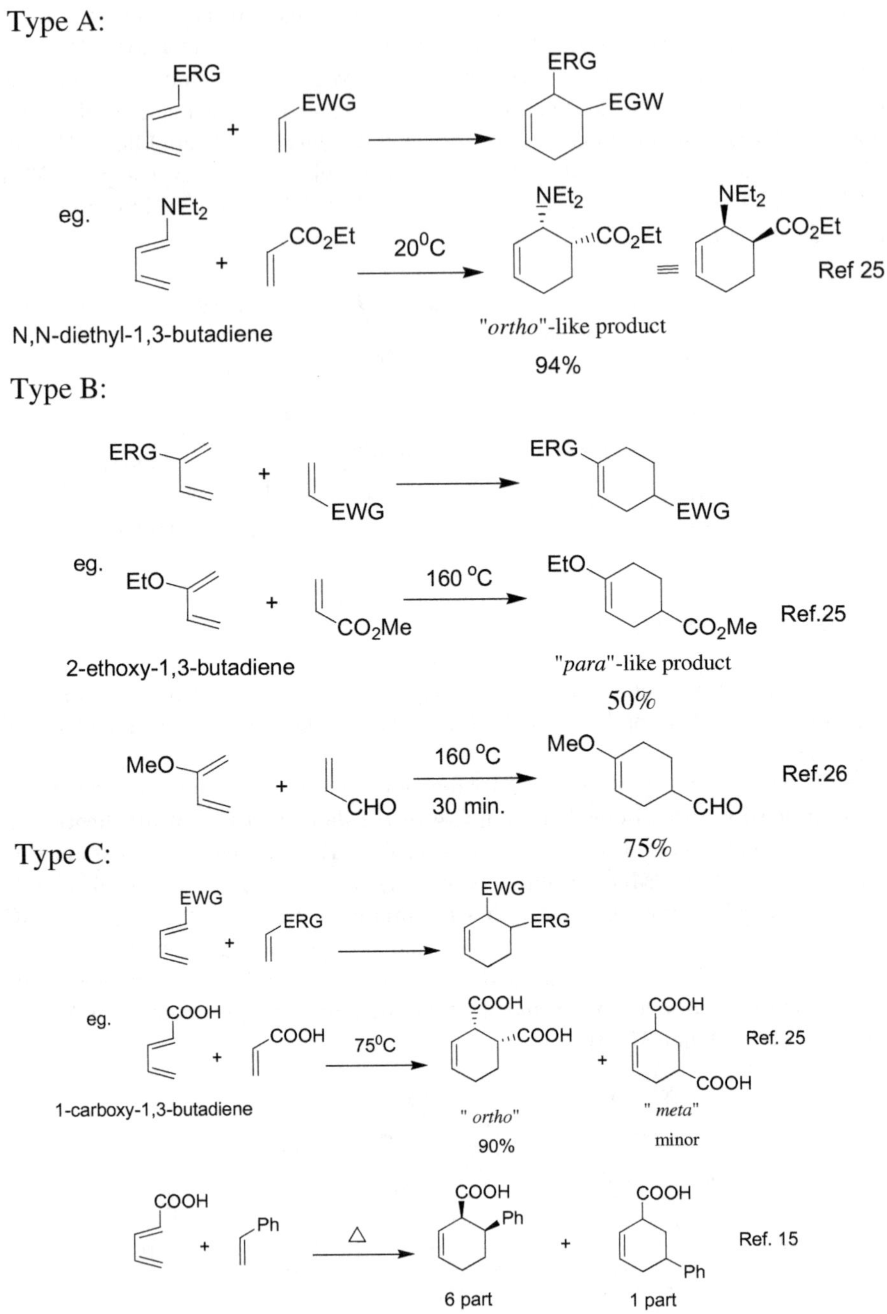

Scheme 3.1 Regioselectivity of Diels–Alder reaction

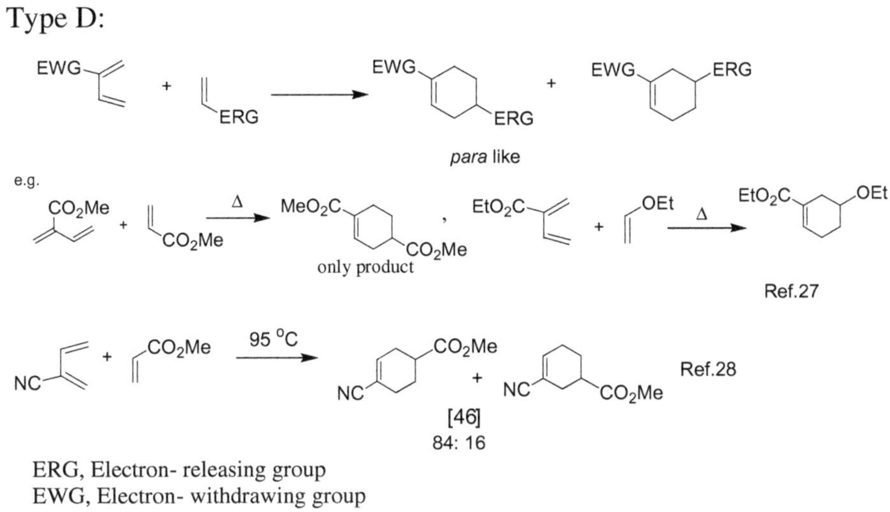

Scheme 3.1 (continued)

the diene and C-2 of the dienophile to give '*para*'-like product as major product, e.g., reaction of **38** with **39** gives **40** and **41** as major and minor product, respectively [29].

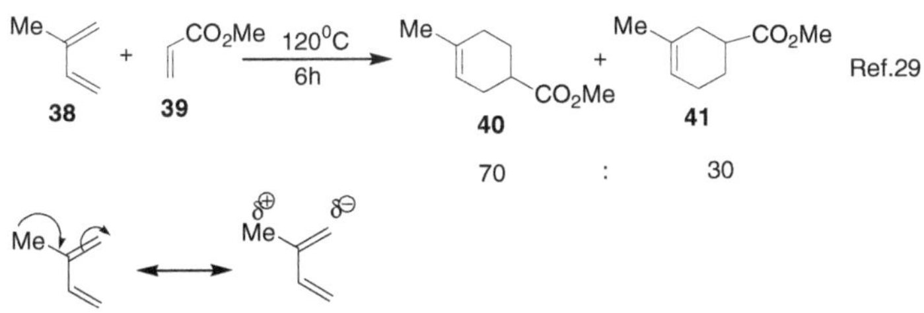

The regioselectivity of inverse electron demand Diels–Alder reactions can be rationalized on the basis of orbital coefficient rather than partial positive charge that is expected at C-4 of the diene. This is because the positive charge at C-4 of the diene will be repelled by the positive charge of the β-carbon of the dienophile (acrylic acid). In such cases, the orbital coefficients of C-4 of the diene and C-2 of the dienophile were comparable to that of normal electron demand Diels–Alder reactions. The LUMO of the diene and HOMO of the dienophile are in lower energy levels (Fig. 3.3).

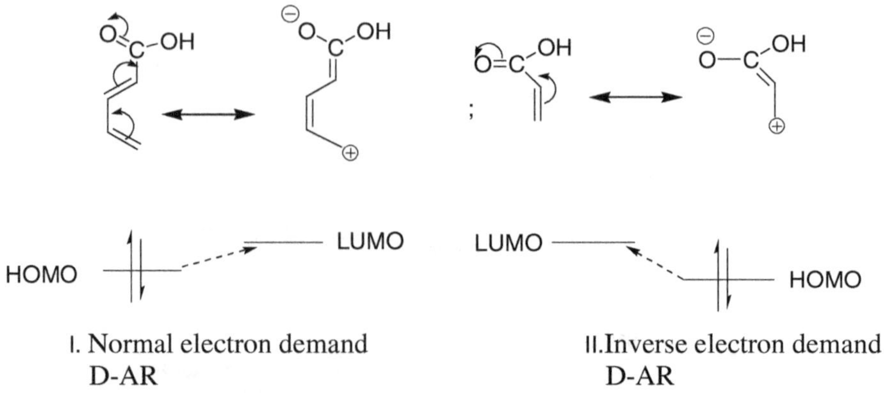

Fig. 3.3 Frontier orbital interactions in Diels–Alder reactions

Hence, frontier orbital interaction takes place preferentially at these positions to give "*ortho*"-like product as major product [30].

3.3.1.3 Stereochemistry

Woodward–Hoffmann selection rules predict the allowedness of thermal and photochemical [4+2]-cycloaddition reactions (Table 3.1).

The Woodward–Hoffmann rules for cycloaddition reactions can be explained from frontier orbital interactions, orbital correlation diagram and aromatic transition state approaches.

(a) *Explanation from the FMO approach*

For a favorable TS structure of low energy for a Diels–Alder cycloaddition reaction, the HOMO of diene and LUMO of dienophile or LUMO of diene and HOMO of dienophile will approach face to face from the same side for maximum orbital interactions (Fig. 3.4).

(b) *Explanation from the orbital correlation diagram*

Let us consider the orbital symmetry properties of the reactants and products for the Diels–Alder cycloaddition reaction of butadiene and ethylene into cyclohexene (Fig. 3.5). The addition of the diene and dienophile takes place face to face, where diene assumes *S-cis*-conformation. The reactants, TS and product maintain symmetry (plane of symmetry) among their orbitals during the course of cycloaddition. To understand this fact, an orbital correlation diagram [31] is constructed by arranging the orbitals with respect to their energy content, or correlation lines are

Table 3.1 Woodward–Hoffmann rules for [m+n]-cycloaddition reactions

[m+n]πe	Number of nodes	Aromaticity	Thermal	Photochemical
4n	0	Antiaromatic	Forbidden	Allowed (supra/supra antara/antara)
4n	1	Aromatic	Allowed (supra/antara antara/supra)	Forbidden
4n+2	0	Aromatic	Allowed (supra/supra antara/antara)	Forbidden
4n+2	1	Antiaromatic	Forbidden	Allowed (supra/antara antara/supra)

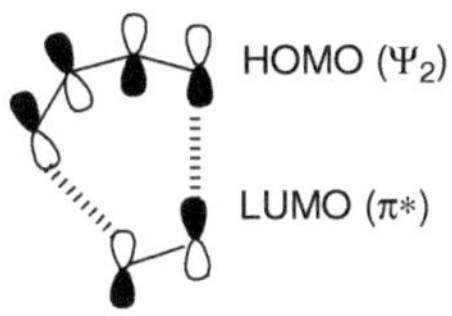

(a) Orbital symmetry allowed TS ($\pi^4{}_s + \pi^2{}_s$) for NED D-AR

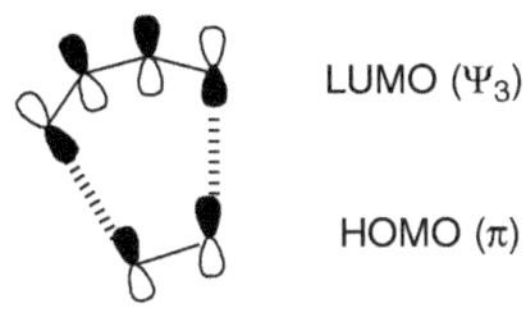

(b)Orbital symmetry allowed TS ($\pi^4{}_s + \pi^2{}_s$) for IED D-AR

(c)Orbital symmetry forbidden TS ($\pi^4{}_a + \pi^2{}_s$)

Fig. 3.4 Orbital interactions of HOMO of diene and LUMO of dienophile and vice versa in a Diels–Alder reaction

drawn as in Fig. 3.6. From this diagram, it is reflected that thermal concerted reaction between butadiene and ethylene is allowed process, because ψ_2 and π orbitals of butadiene and ethylene are correlated with σ_1 and π orbitals of cyclohexene in the ground state.

(c) *Transition state stability*

Now, we may consider the TS structure for a Diels–Alder cycloaddition reaction. The Huckel TS structure of zero node is aromatic in nature. Thus, the reaction takes place in suprafacial mode following the Huckel topology, whereas in Mobius

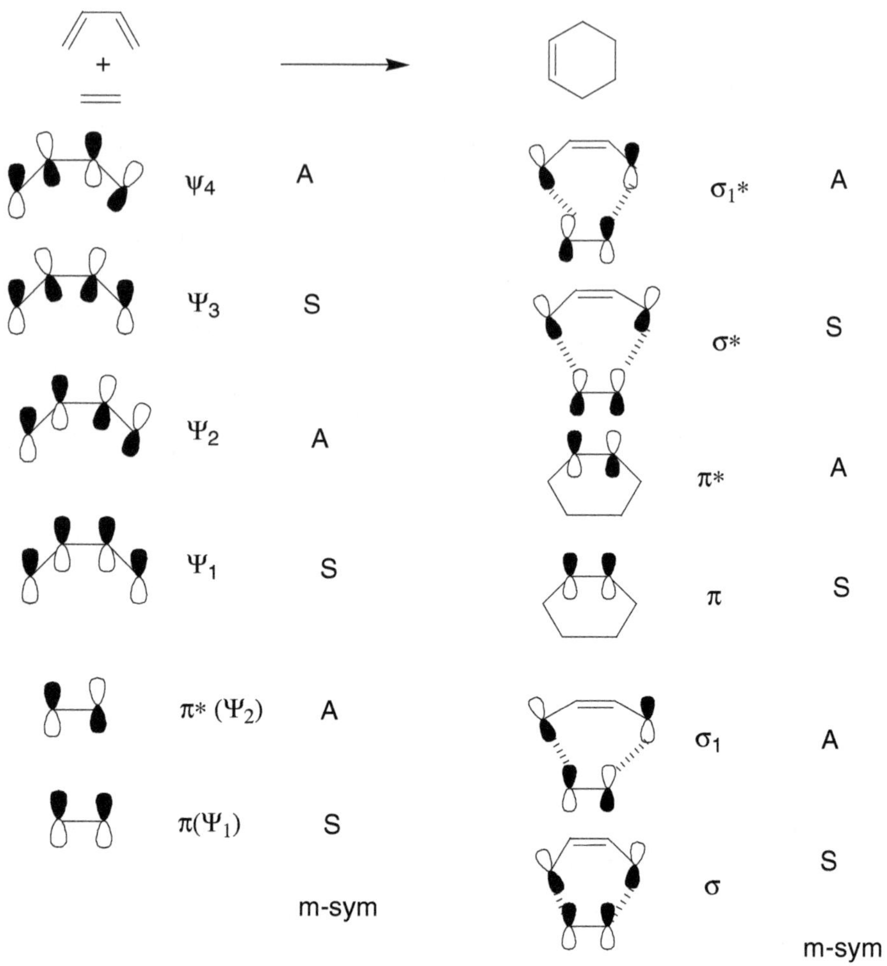

Fig. 3.5 Symmetry properties of butadiene, ethylene, and cyclohexene orbitals with respect to plane of symmetry. *m-sym* means mirror, *S* means symmetric, and *A* means antisymmetric

topology, addition to opposite faces of the π system is required and is unfavorable for its antiaromatic nature (Fig. 3.7) [32].

3.3.1.4 Stereochemical Features of the Products

The D–A reaction of a cyclic diene with a cyclic dienophile may give two stereoisomeric products, *endo*- and *exo*-products depending on the conformation of the transition states. The product in which the unsaturated substituents of the dienophile are *cis* to the double bond of the newly formed cyclohexene ring is called

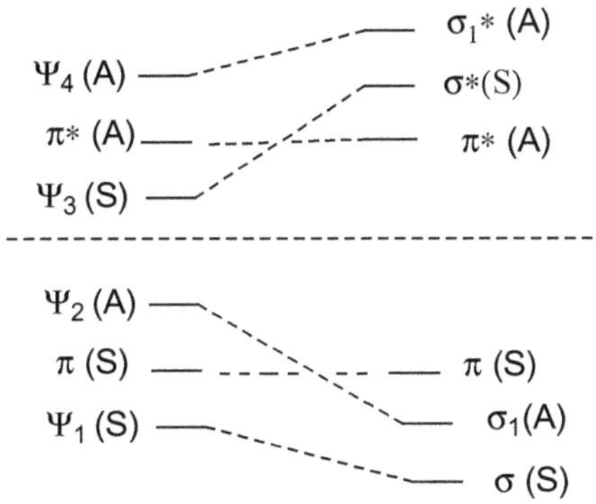

Fig. 3.6 Symmetry correlation diagram for ethylene, butadiene, and cyclohexene orbitals

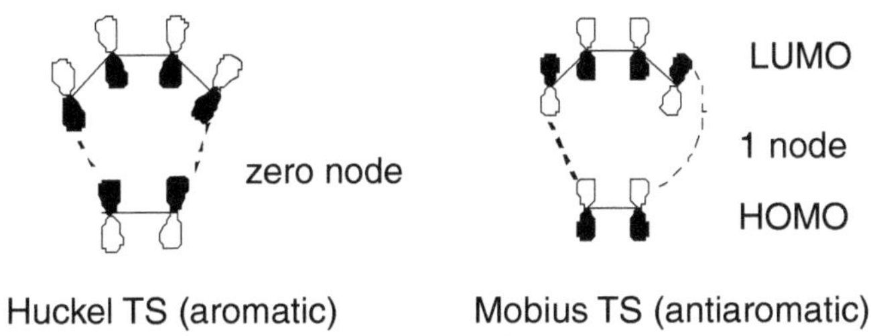

Fig. 3.7 The orbitals set for supra-, supra-[π^4+π^2]-cycloaddition in Huckel and Mobius TSs

the *endo*-product, whereas in the product, where these substituents are *trans* to the double bond is called the *exo*-product. For example, furan **42** on reaction with succinimide **43** gives *endo*- and *exo*-products **44** and **45**, respectively, at different reaction conditions.

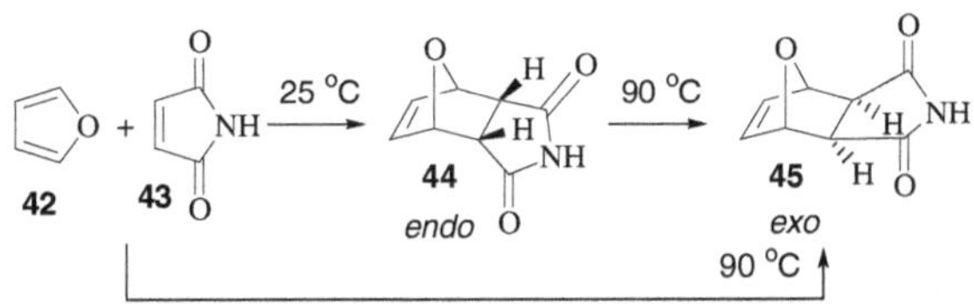

The D–A reactions of acyclic dienes with acyclic dienophiles also give *endo*- and *exo*-products. For example, D–A reaction of 1-deuterio-1,3-pentadiene **46** with *trans*-3-penten-2-one **47** gives *endo* **48** and *exo* **49** products as shown.

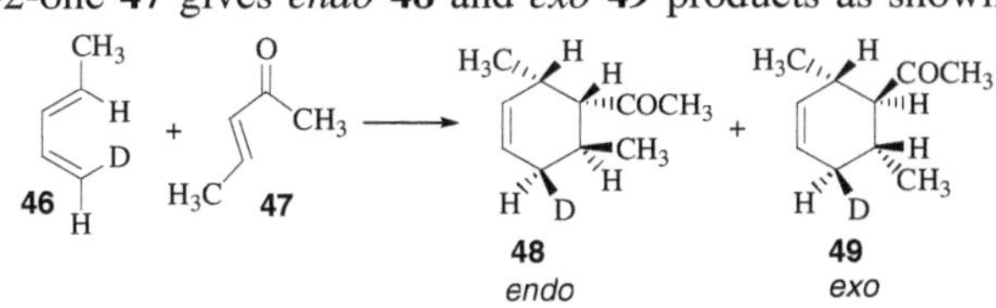

The relative orientations of the diene and dienophile in a favorable TS for a Diels–Alder reaction is predicted by *Alder's endo-rule* [33]. The *Alder's endo-rule* states that for Diels–Alder reactions of substituted butadiene derivatives with dienophiles having an electron-withdrawing substituent, kinetically controlled *endo*-TS will be preferred over *exo*-TS because of secondary orbital interactions of the electron-withdrawing substituent with the butadiene π system. The *endo*-TS has lower activation energy than that of *exo*-TS. The product derived from *endo*-TS is called kinetically controlled product and the product derived from *exo*-TS is called thermodynamically controlled product. Frequently a mixture of both stereoisomers is formed and sometimes the thermodynamically controlled *exo*-product predominates. It has been observed that reaction of butadiene with maleic anhydride using deuterium-labeled butadiene gives 85 % of the *endo*-product **50** from *endo*-TS [33]. The reaction of cyclopentadiene with maleic anhydride also gives 97.5 % *endo*-product **51**. The secondary orbital interactions in preferred *endo*-TS are shown in Fig. 3.8.

For reaction of cyclopentadiene with acyclic dienophile-like methyl acrylate, the *endo*- and *exo*-TS would be

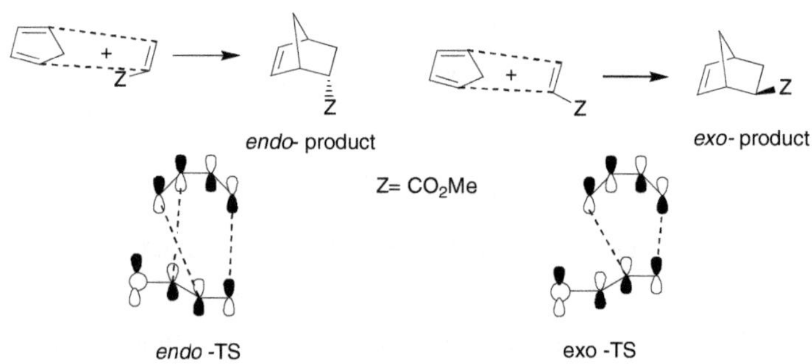

3.3.1.5 Substituent Effects on the Reaction Rates

(a) *Steric effects of substituents*

Diels–Alder reactions are sensitive to steric effects. The presence of bulky substituents on the dienophile or on the diene hinders the approach of the components to each other and decreases the rate of reaction. For example, 1-*tert*-butyl butadiene is about 20 times less reactive than butadiene toward maleic anhydride [34].

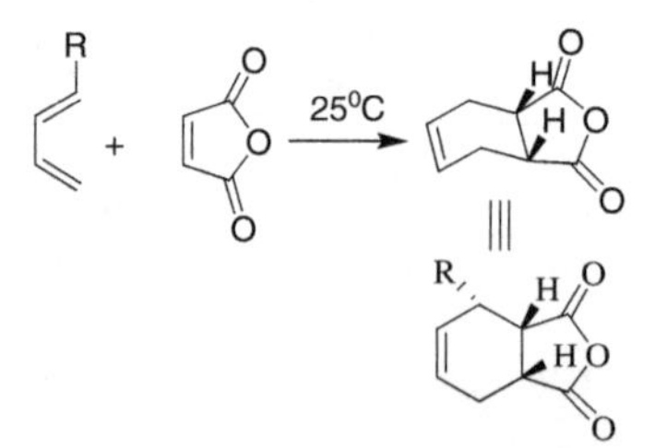

R	K_{rel}
H	1
$C(Me)_3$	< 0.05

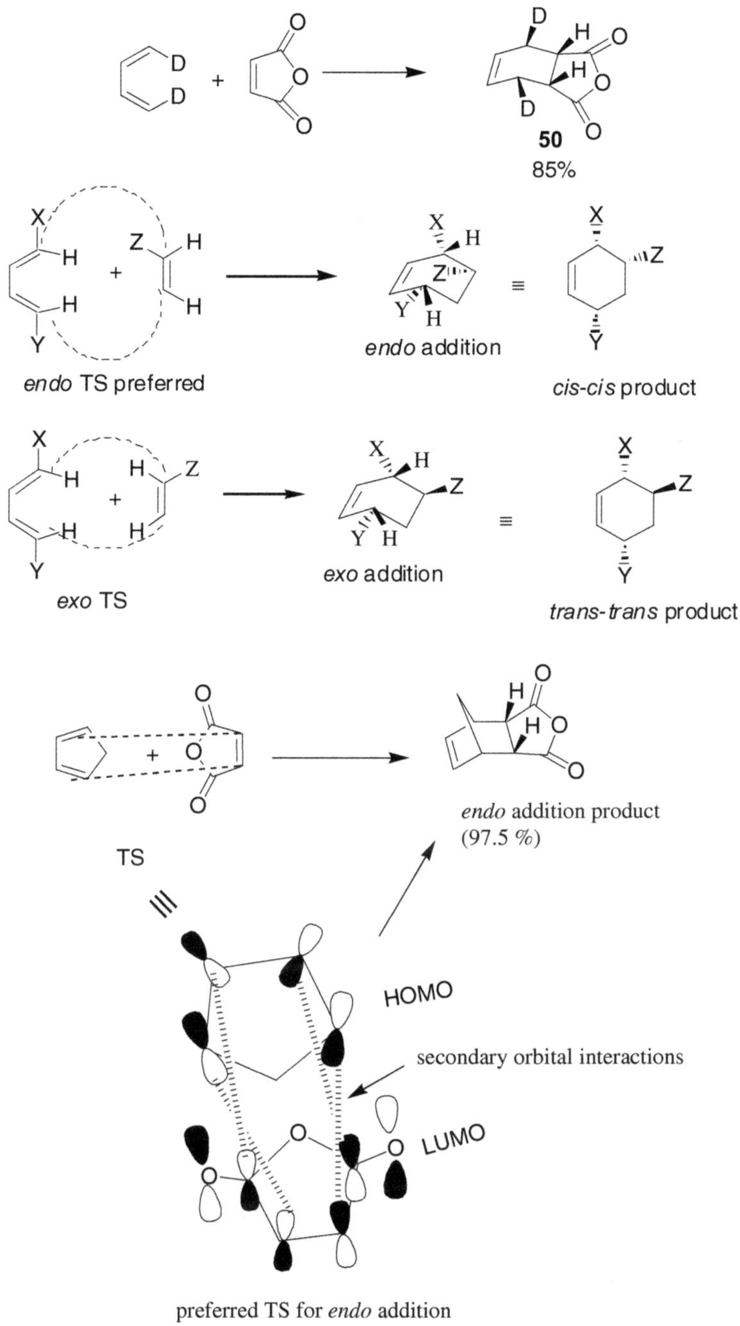

Fig. 3.8 The orbital interactions in *endo*- and *exo*-transition states (TSs) in a Diels–Alder reaction

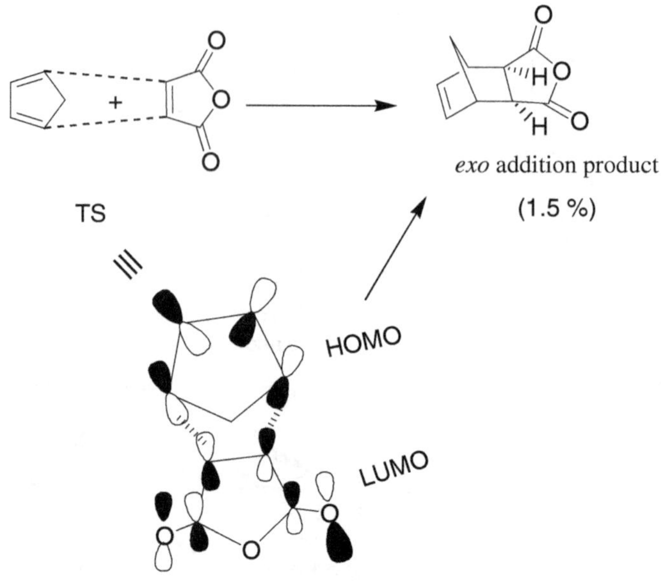

less preferred TS for *exo* addition

Fig. 3.8 (continued)

2,3-Dimethylbutadiene reacts with maleic anhydride about ten times faster than butadiene due to electron-releasing effect of the methyl group.

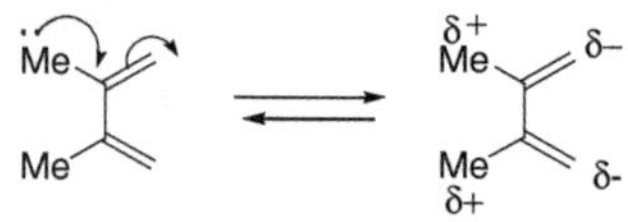

2-*tert*-Butyl-1,3-butadiene is 27 times more reactive than butadiene toward maleic anhydride because the *tert*-butyl group favors the *s-cis* conformation due to high 1,3-steric interaction in the *s-trans* conformation [35].

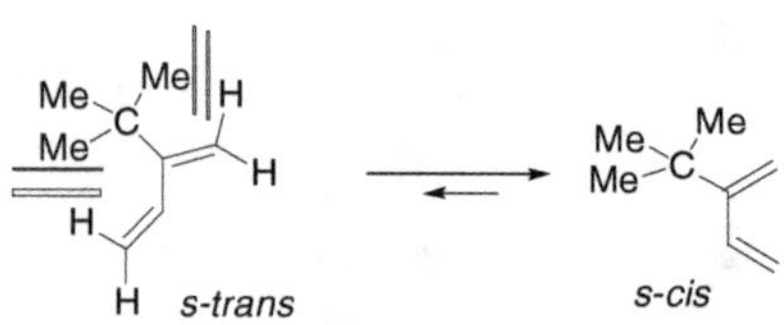

The presence of two bulky substituents at C-1 position of a diene prevents the adoption of *s-cis* conformation of the diene and decreases the reaction rate. For example, 4-methyl-1,3-pentadiene is about 1000 times less reactive than *trans*-1,3-pentadiene toward tetracyanoethene [36].

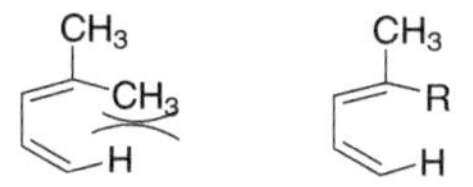

R	K_{rel}
H	1
CH_3	10^{-3}

Usually cyclic dienes such as cyclopentadiene and *ortho*-quinodimethane are more reactive than open-chain dienes because of their *s-cis* conformations. The rates of DA reactions are also affected by the nature of dienes and dienophiles.

(b) *Electronic effects of substituents*

In a Diels–Alder (D–A) reaction, TS is formed by a process of charge transfer. The electron-rich reactant diene in normal electron demand D–A(NED D–A) reaction and dienophile in inverse electron demand (IEDD–A reaction) acts as an electron donor component (nucleophile) and the electron-poor reactant (dienophile in NED D–A reaction and diene in IEDD–A reaction) acts electron acceptor component (electrophile). The greater the extent of charge transfer, strongest is the interaction between the reactant components and faster is the rate of the reaction. The reactivity of 1,3-butadiene increases with increasing the electron acceptor capacity of the dienophile. Tetracyanoethylene is a very strong dienophile having global electrophilic power, Δω, 4.91 eV (where ω is the global electrophilic parameter) and is highly reactive toward 1,3-butadiene. The global electrophilic powers of some dienophiles are provided in Table 3.2 [37].

Similarly, the reactivity of some dienes in Diels–Alder reactions increases by increasing the electron donor ability of the dienes. The relative reactivity of some substituted butadienes with maleic anhydride is provided in Table 3.3 [38].

This electronic effect of the substituents in a Diels–Alder reaction can be explained from the FMO theory with respect to the energy gap between the energy levels of HOMO of the diene and LUMO of the dienophile for a normal electron demand D–A reaction. The stronger the electron donor ability of diene and the greater the electron acceptor ability of dienophile, the closer will be the energy gap between the energy levels of HOMO of diene and LUMO of dienophile, and thus the activation energy of the reaction will be lowered and the reaction rate will be faster (Fig. 3.9).

Table 3.2 Global electrophilicity of some dienophiles in D–A reactions with 1,3-butadiene (Δω = 1.05 eV)

Dienophile	Δω
Tetracyanoethylene	4.91
1,1-Dicyanoethylene	1.77
Acrolein	0.79
Acrylonitrile	0.69
Ethylene	−0.32

Table 3.3 Relative rates of reactivity of some substituted butadienes in D–A reactions with maleic anhydride

Diene	Relative rate
Cyclopentadiene	1350
1-methoxy-1,3-butadiene	12.4
2-phenyl-1,3-butadiene	8.8
1-phenyl-1,3-butadiene	1.65
1-Methyl-1,3-butadiene	3.3
2-Methyl-1,3-butadiene	2.3

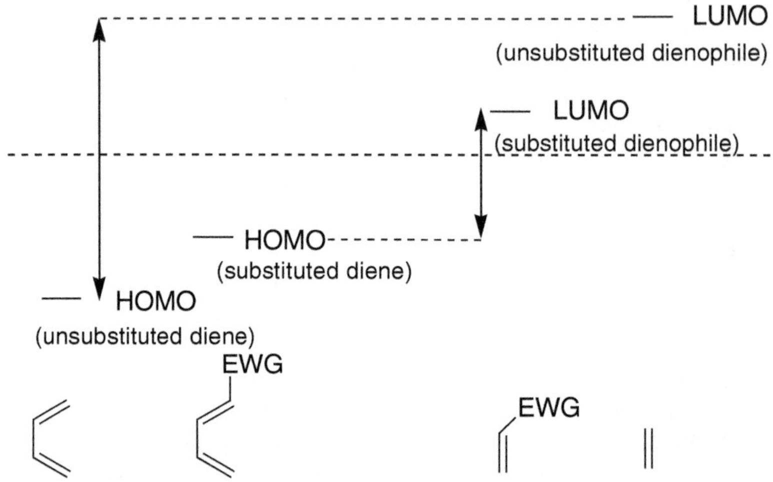

Fig. 3.9 The figure illustrates the HOMO–LUMO energy gap in terms of FMO theory on the reactivity of diene and dienophile in normal electron demand Diels–Alder reaction. The narrower the gap the higher will be the TS stability and faster will be the reactivity

3.3.1.6 The Dienes and Dienophiles

Different kinds of diene and dienophile are used in the Diels–Alder reactions. Dienes and dienophiles with a heteroatom such as N, O, or S in their π systems are known as heterodienes and heterodienophiles, and their cycloaddition reactions are called the *hetero-Diels–Alder reactions*. Some highly reactive dienes and dienophiles used in Diels–Alder reactions are listed in Table 3.4.

Table 3.4 Representative dienes and dienophiles used in Diels–Alder reactions

A. ***Dienes***

1. Butadiene, 

2. 1-Substituted butadiene  R= Me, OMe, Ph, COOH

3. 2-Substituted butadiene 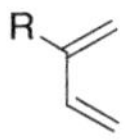 R= Me, OR, CN, COOR, $OSiMe_3$

4. 1,4-Diacetoxy butadiene 

5. 1,2-Dimethylene cyclohexane 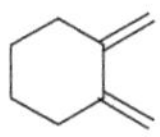

6. α-Pyrones 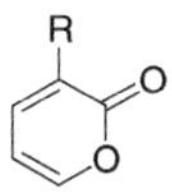 R= OH, CO_2Me

7. 1- Vinyl cyclohexene 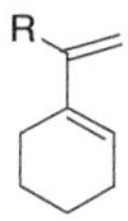

8. *ortho*-Quinodimethane 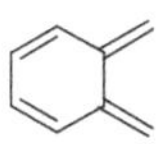

9. Cyclopentadiene 

10. 1,3-Cyclohediene 

11. Furan 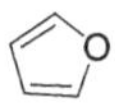

B. ***Heterodienes***

12. Acrolein 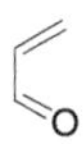

13. 1-Azabutadiene 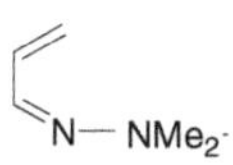

14. 2-Azabutadiene 

C. ***Dienophiles***

1. Maleic anhydride 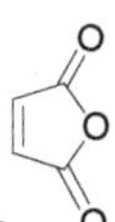

2. Benzoquinone 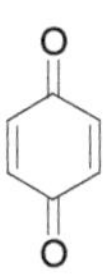

(continued)

Table 3.4 (continued)

3.	α,β-Unsaturated aldehydes, ketones, esters, nitriles and nitro compounds	$RCH{=}CH{-}Z$ Z= CHO, COR, CO_2R, CN, NO_2 R= H, Me, Ph, CO_2Me
4.	2-Bromo 2-cyclobutenone	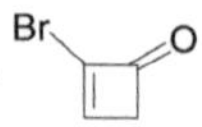
5.	Alkyl and aryl vinyl sulfones	$CH_2{=}CH{-}SO_2R$ R=Me, Et, Ph
6.	α,β-Unsaturated phosphonates	$RCH{=}CH{-}P(O)(OR)_2$
7.	Tetracyanoethylene	$(CN)_2C{=}C(CN)_2$
8.	Esters of acetylene dicarboxylic acids	$RO_2C{-}C{\equiv}C{-}CO_2R$
9.	Dlbenzoyl ethylene	$Ph{-}C(O){-}CH{=}CH{-}C(O)Ph$
10.	Dicyanoacetylene	$NC{-}C{\equiv}C{-}CN$
11.	Ethynyl sulfones	$HC{\equiv}C{-}SO_2R$ R = Ph, *p*-tolyl
D.	***Heterodieneophiles***	
12.	Esters of azodicarboxylic acid	$RO_2C{-}N{=}N{-}CO_2R$; R= Me, Et
13.	4-Phenyl 1,2,4- triazoline-3,5-dione	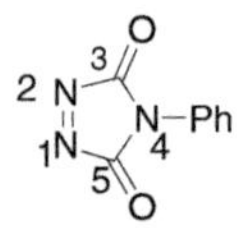
14.	Imino urethanes	$CH_2{=}N{-}CO_2R$; R = Me, Et
15.	Nitrosobenzene	$Ph{-}N{=}O$

Some Diels–Alder reactions with less commonly used dienes or dienophiles are illustrated:

1. 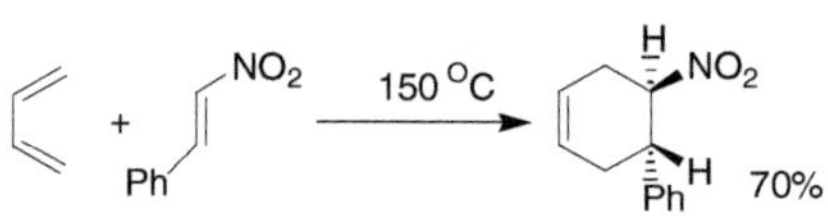 Ref. 15

2. 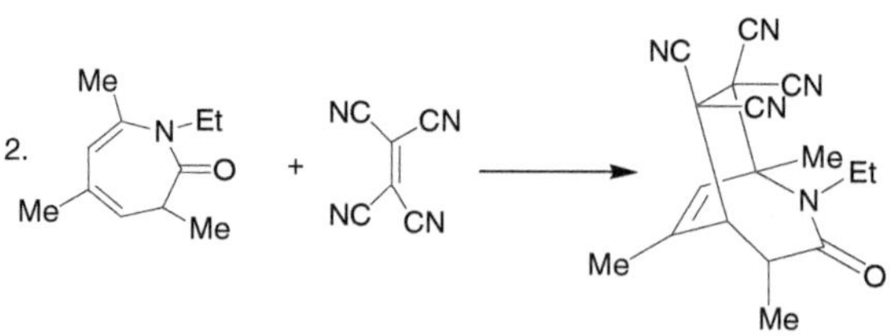 Ref. 39

3. 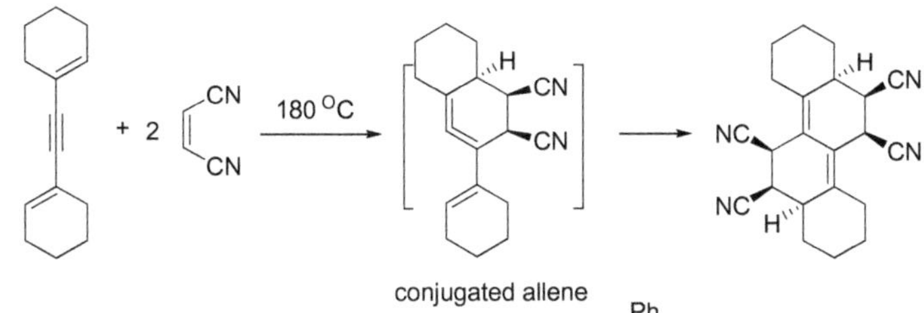

Ref. 40

4. 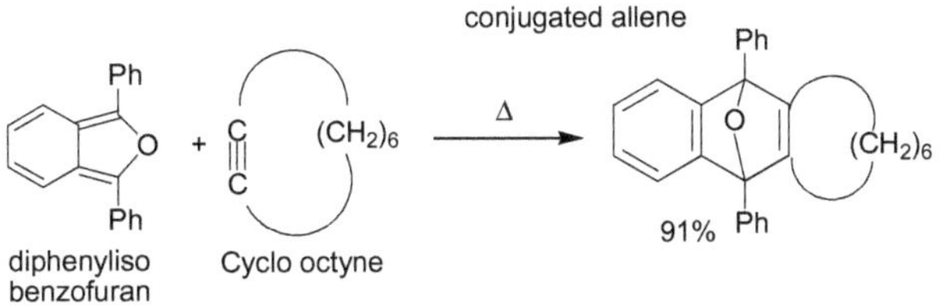

Ref. 41

5. 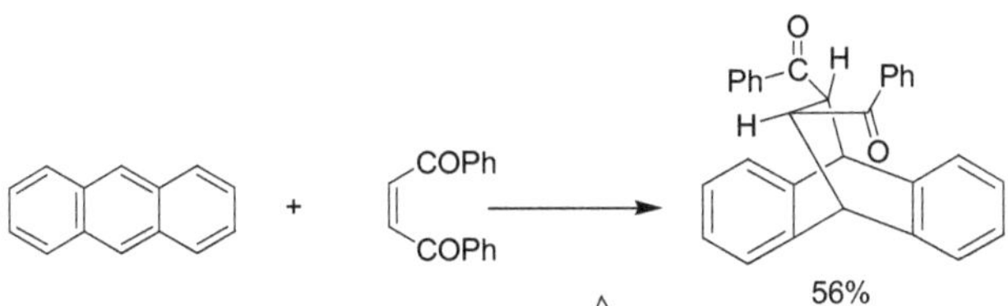

Ref. 42

6. 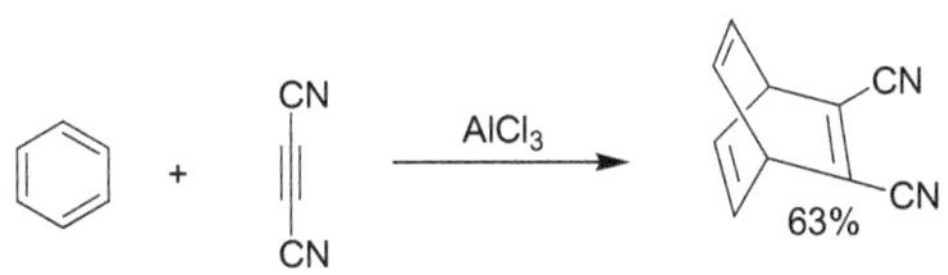

Ref. 43

7. 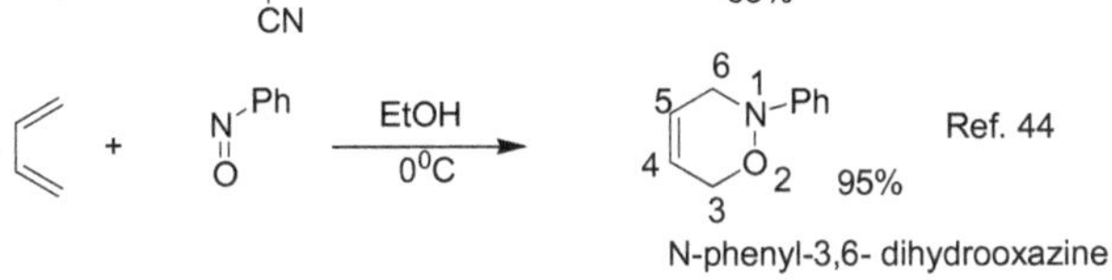

Ref. 44

3.3.1.7 Lewis Acid-Catalyzed Diels–Alder Reactions

Lewis acids such as $ZnCl_2$, $SnCl_4$, $AlCl_3$, derivatives of $AlCl_3$, Me_2AlCl, and Et_2AlCl act as effective catalysts to accelerate the rates of Diels–Alder reactions by increasing the electron-withdrawing capacity of the dienophiles via the formation of Lewis acid complex. For example, the reaction of 2-methyl-1,3-butadiene **38** with methyl acrylate takes place at room temperature and in the shorter time (3 h) in the presence of $AlCl_3$ compared to uncatalyzed reaction, which occurs on heating at 120 °C for 6 h [29].

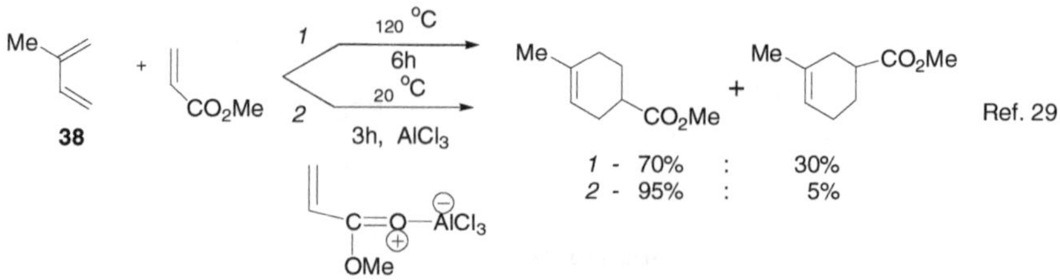

Lewis acid complex of the dienophile

Similarly, the cycloaddition reaction of cyclopentadiene with 2-pyridyl styryl ketone **52** gives **53** in the presence of $Cu(NO_3)_2$ takes place much faster than the uncatalyzed reaction [45].

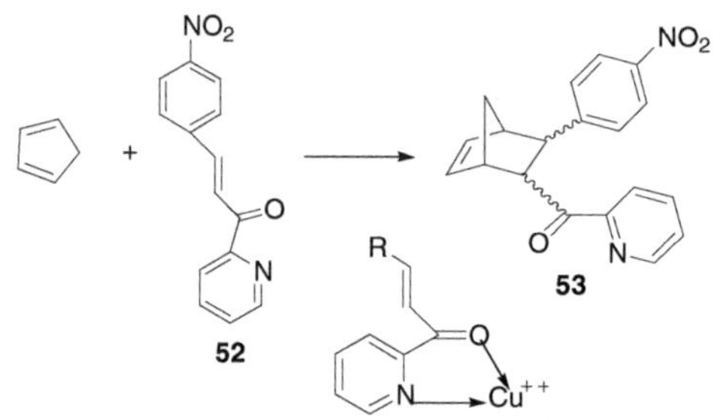

Solvent	Rel. Rate
CH_3CN	1
H_2O + 0.01M $Cu(NO_3)_2$	250,000

Ref. 45

Lewis acid complex of the dienophile

Anthracene reacts with methyl fumarate **54** at room temperature in the presence of $AlCl_3$, whereas without catalyst, the reaction occurs at high temperature, 101 °C, and in longer time (2–3 days) [46].

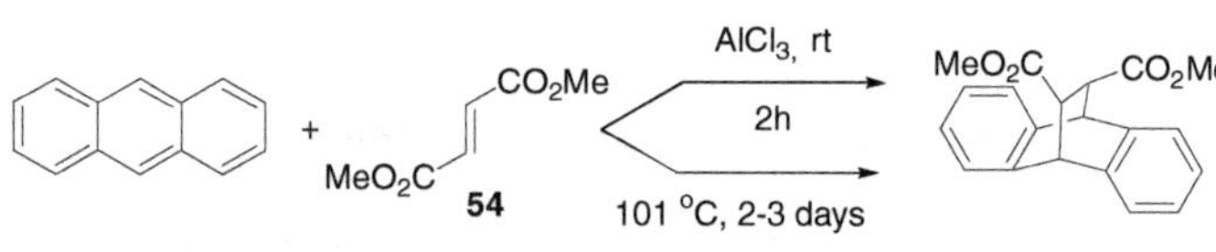

3.3.1.8 Applications of Neutral Dienes and Dienophiles in Diels–Alder Reactions

Diels–Alder reactions using neutral dienes and dienophiles have been utilized in the synthesis of various types of organic compounds. For example, in the synthesis of steroids, the angular methyl group may be introduced by the reaction of 1,3-butadiene with 2-methoxy-5-methylbenzoquinone **55**. 5,6-Double bond of the quinone **55** is more reactive as dienophile because the electron donating effect of methoxy group at C-2 position weakens the dienophilic character of 2,3-double bond by the delocalisation with C-4 carbonyl group [47].

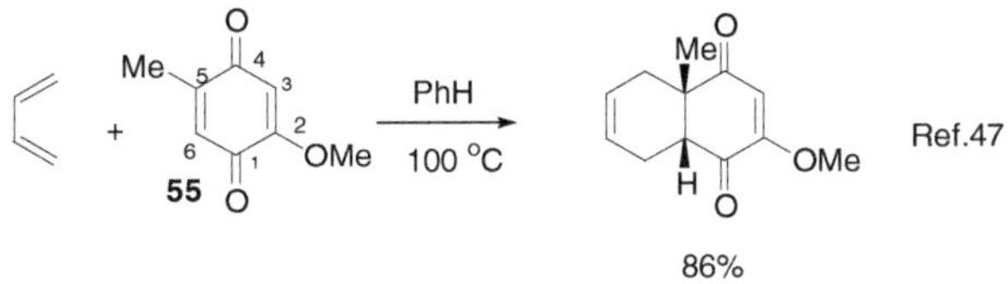

Dienophiles such as nitroethene **56**, α-chloroacrylonitrile **57,** and vinyl sulfoxide **58** can be used as ketene equivalent in the synthesis of organic compounds to improve the yield of the products. The cycloadducts of cyclopentadiene derivatives obtained from these dienophiles are used in the synthesis of prostaglandins.

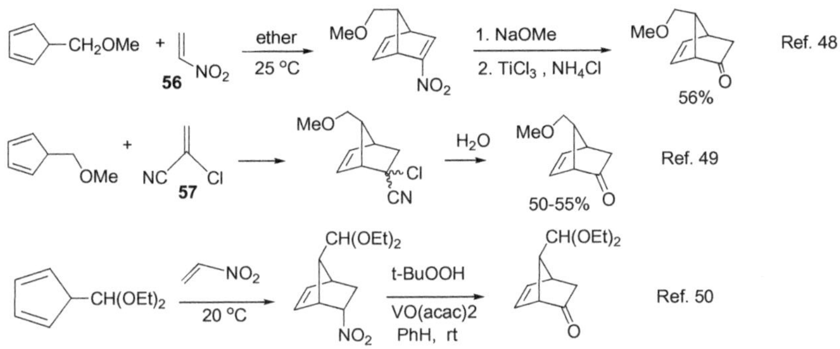

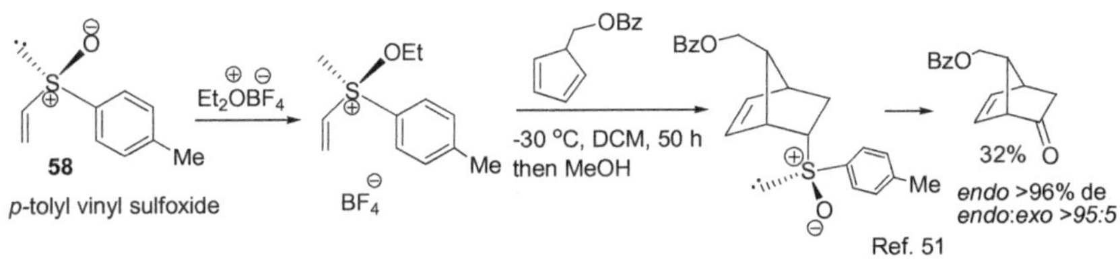

Dienophiles, phenyl vinyl sulfone **59,** and ethynyl sulfone **60** in Diels–Alder reactions can serve as potential ethene and ethyne equivalents, respectively.

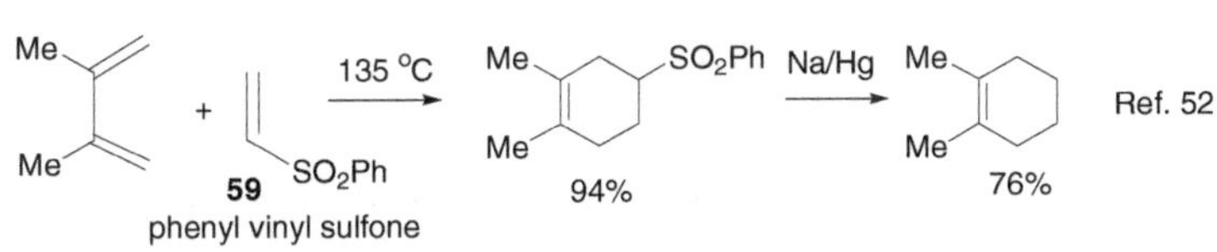

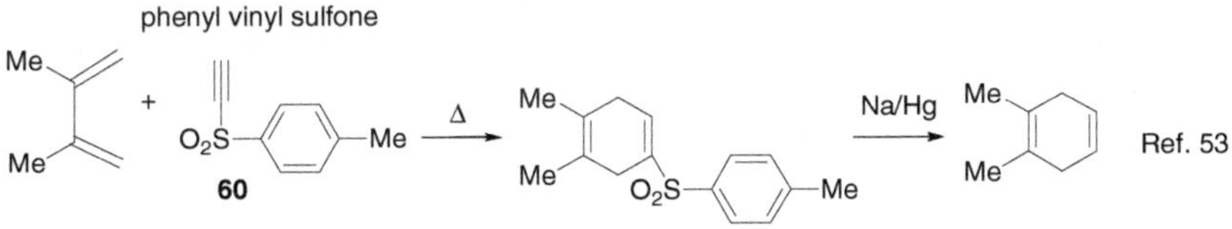

Similarly, dienophile, vinyl triphenyl phosphonium bromide **61** is used as an allene equivalent.

Furan, thiophene **62**, fulvene **63**, and aromatic hydrocarbons **64** are used as efficient dienes in the synthesis of various heterocycles and carbocycles.

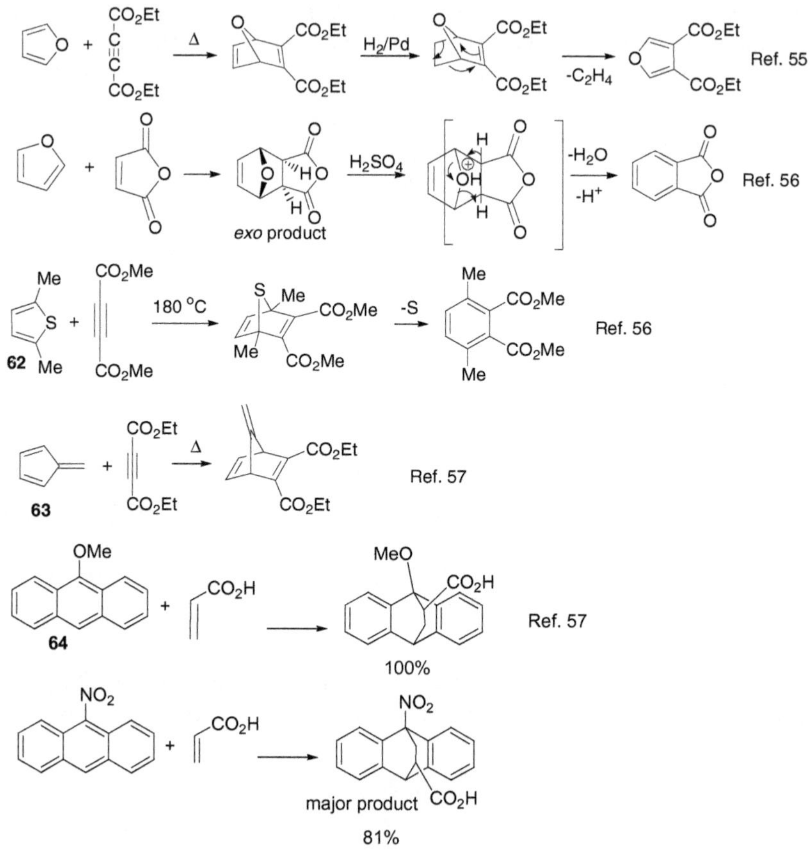

Cyclopentadiene is a highly reactive diene and reacts with both strong and weak dienophiles, cyclopropene **65** and chiral allenes, **66** and **67** to give cycloadducts in high yields.

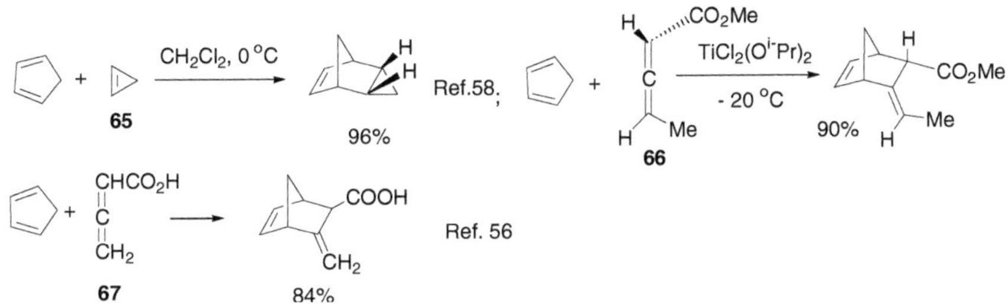

Cyclobutene **68** and benzocyclobutenes **69** and **70** are used for generation of dienes in situ.

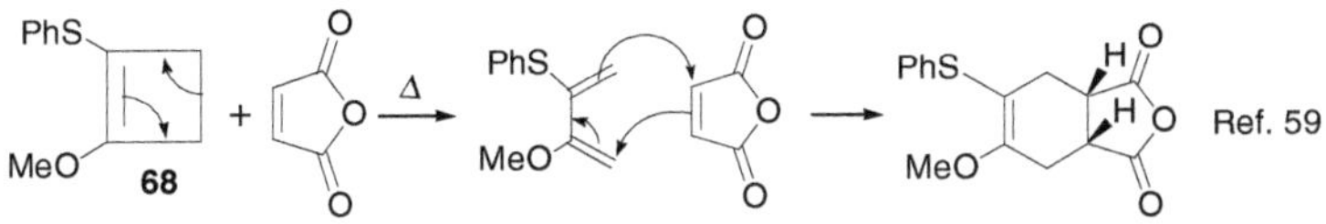

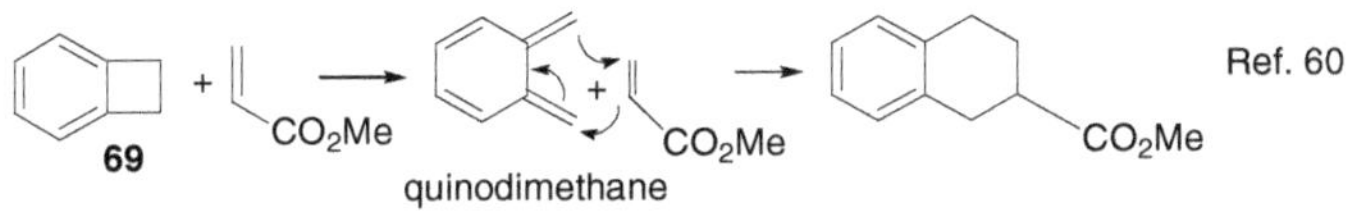

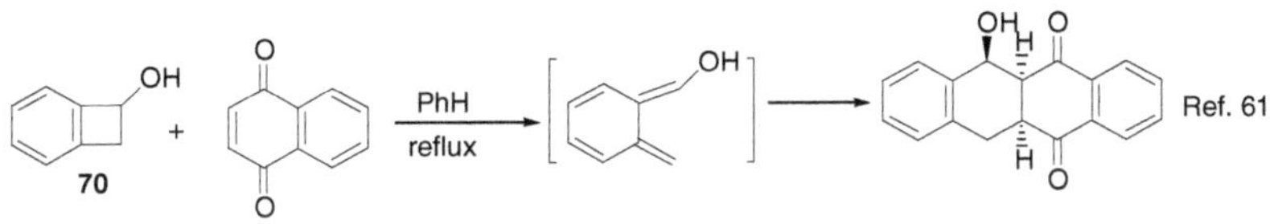

Quinodimethane may also be generated from cheletropic elimination of cycloadduct **71**.

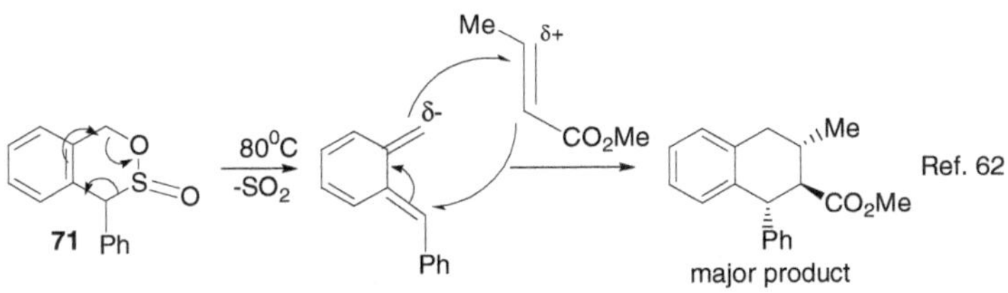

Diphenlyisobenzofuran **72** can be used as highly reactive diene for reaction with electron-rich alkene [63].

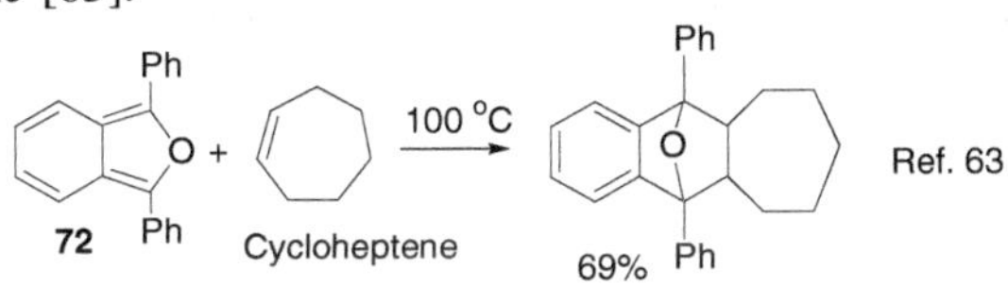

Butadiene and its derivatives such as 1,3-pentadiene **73** and α-pyrone **74** and 1,3-cylohexadiene **75** are used as effective dienes.

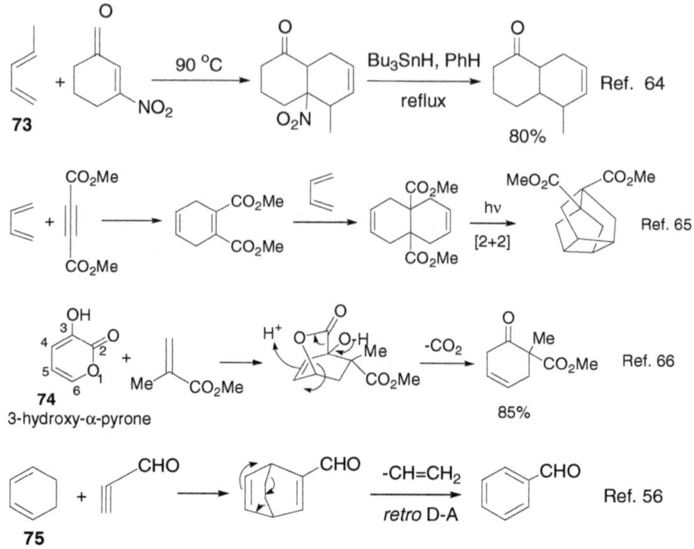

1-Methoxy-3-trimethylsilyloxy-1,3-butadiene known as Danishefsky's diene **76** is used for the synthesis of cyclic α,β-unsaturated ketone.

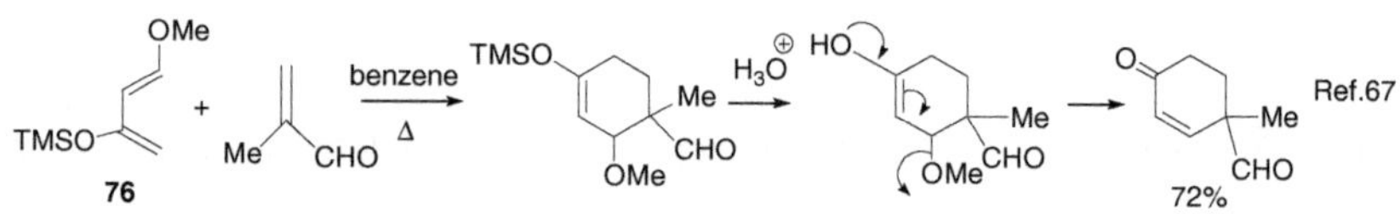

1,1-Dimethoxy-3-trimethylsilyloxy-1,3-butadiene **77** is used as a diene for synthesis of resorcinol derivative **78**.

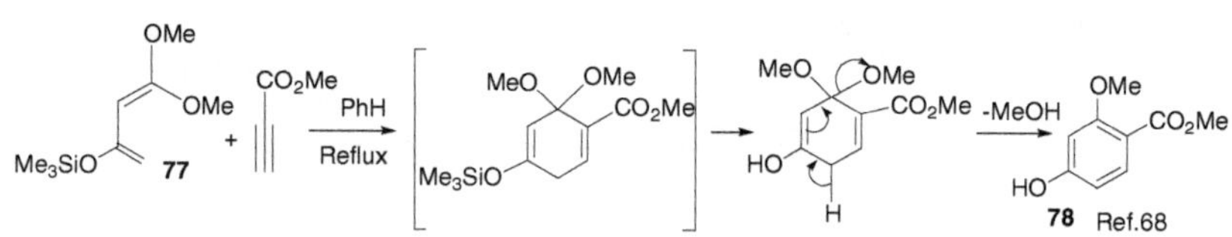

Vinylcycloalkene **79** is used as a diene in the synthesis of cyclobutane derivative **80** [69].

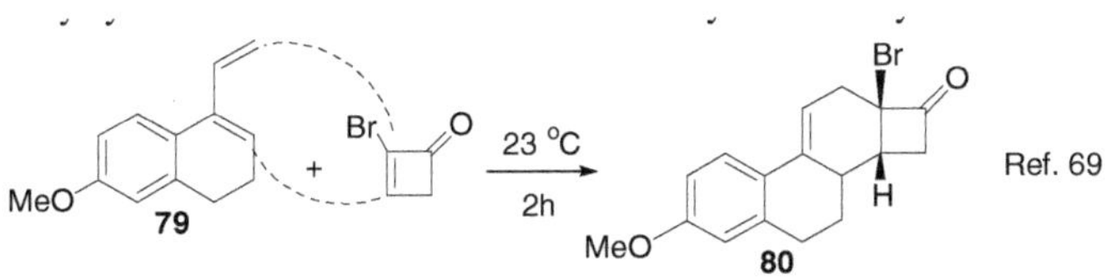

Several heterodienes and heterodienophiles are used in hetero-Diels–Alder reactions in the synthesis of heterocyclic compounds. For example, aza-butadiene **81** and **81a** and oxazoles **82** and **83** are used as heterodienes for synthesis of heterocycles.

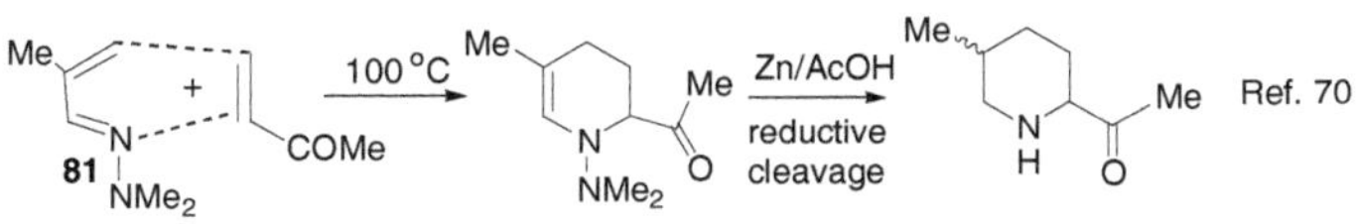

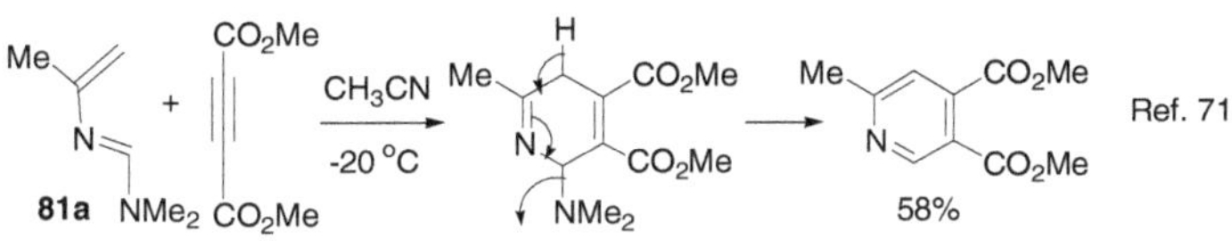

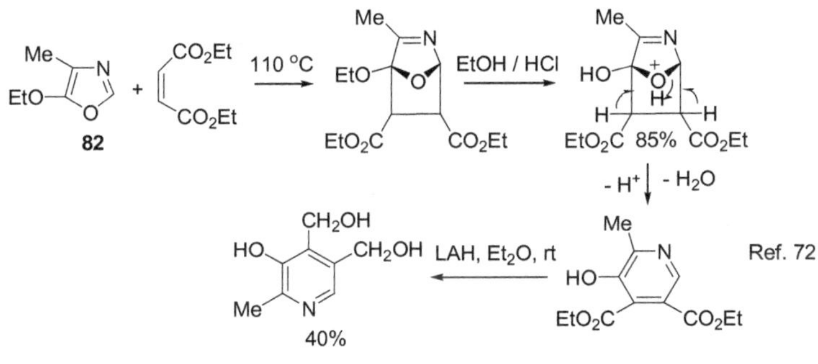

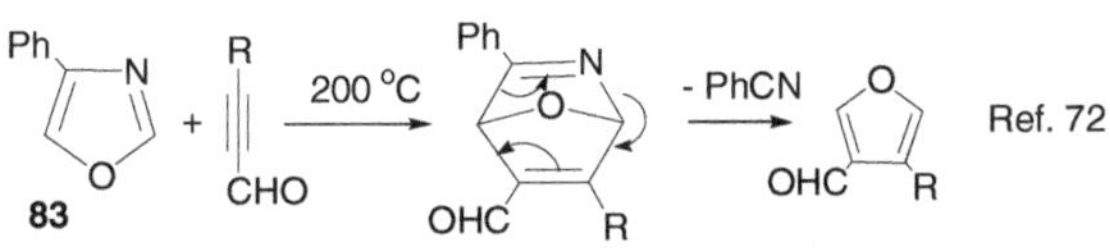

Nitroso compound **84**, imine **85**, azodicarboxylate **86**, and carbonyl compound **87** are used as heterodienophiles in Diels–Alder reactions for the synthesis of heterocycles.

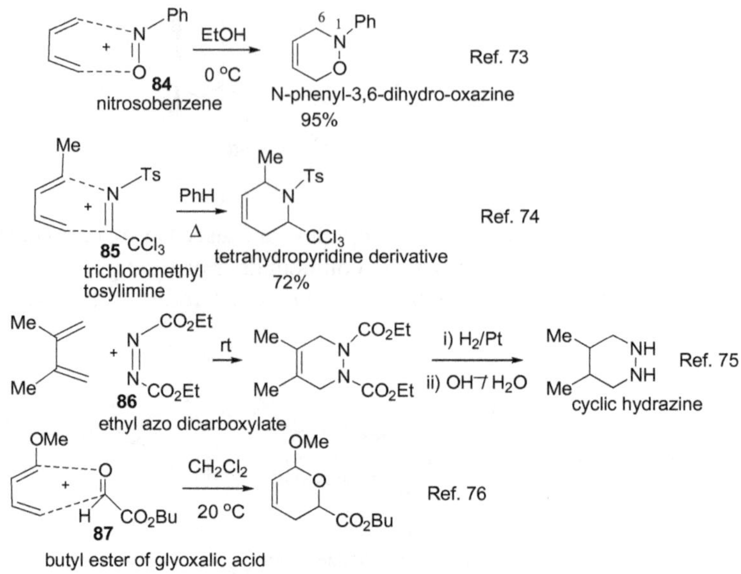

3.3.1.9 Applications of Ionic Dienes and Dienophiles in Diels–Alder Reactions

Several allyl cations can serve as dienophiles and allyl anions and pentadienyl cations as dienes in Diels–Alder cycloadditions; for example, cycloaddition of 2-methyl allyl cation **88** with cyclopentadiene.

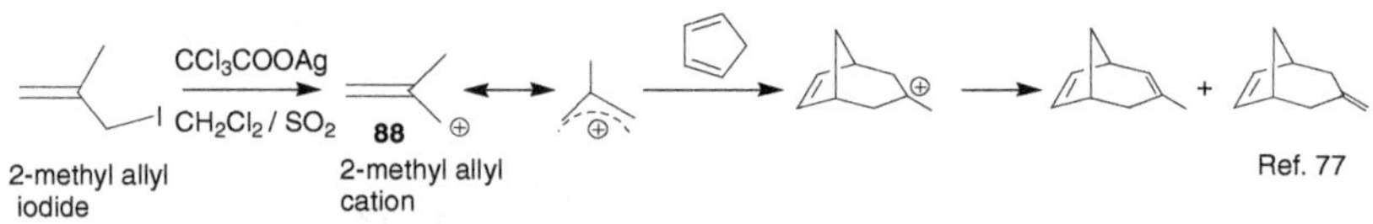

Trimethylsilyloxy-substituted allyl halides in the presence of silver perchlorate in nitromethane generate allyl cations **89**, which react efficiently with cyclopentadiene.

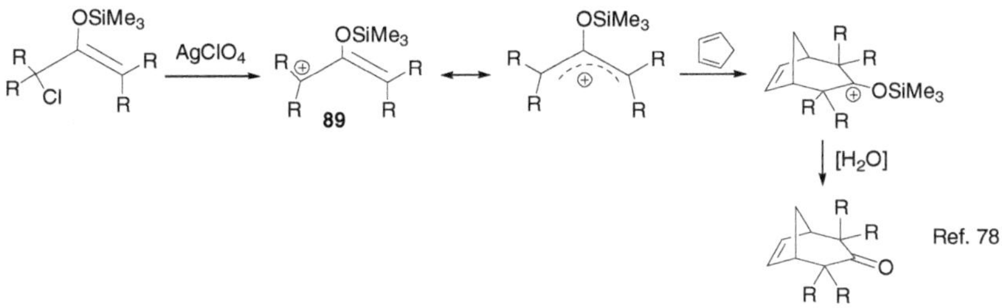

Enamine of 2-chloro cyclohexanone in the presence of $AgBF_4$ generates allyl cation **90**, which serves as reactive dienophile for synthesis of tricyclic ketone **91** [5].

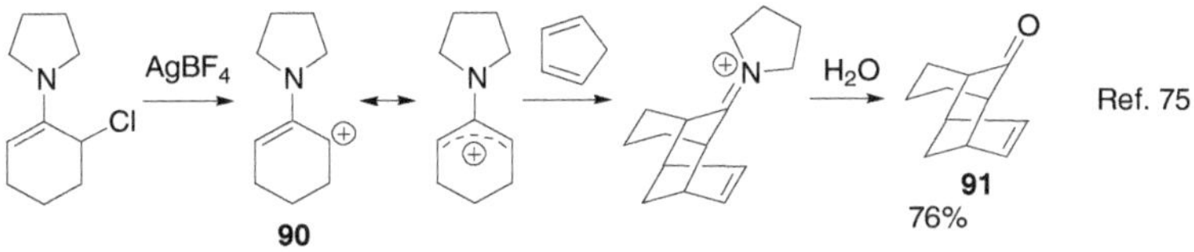

2,2-Dimethylcyclopropanone remains in equilibrium with zwitterionic oxyallyl cation **92** in situ and undergoes D–A reaction with furan. Similarly, allene oxide **92a** generated from silyl epoxide gives cyclopropanone and oxyallyl cation, which is trapped as furan adduct [79].

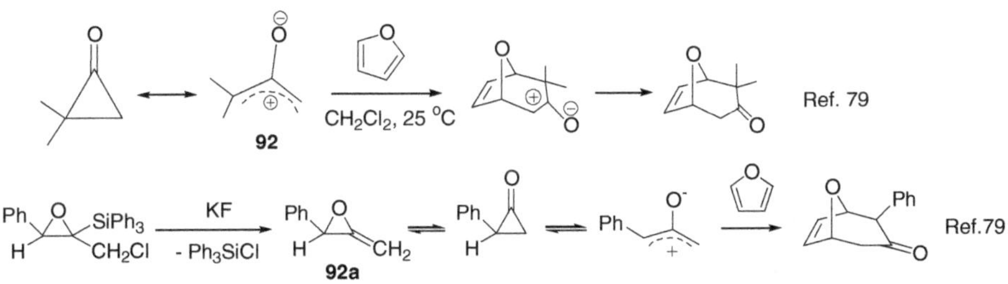

α-Methyl styrene in the presence of a strong base generates allyl anion **93** in situ, which undergoes D–A cycloaddition with an alkene [5].

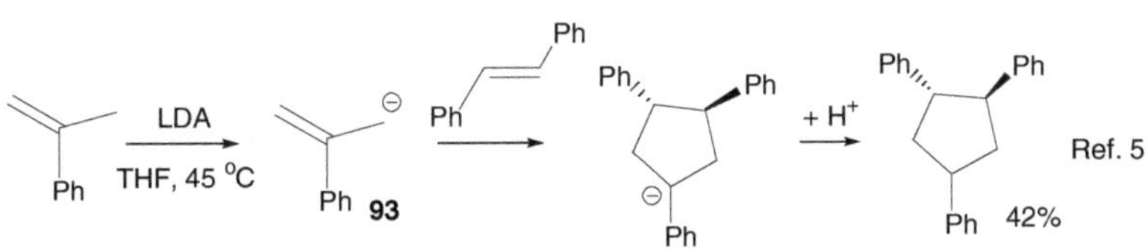

2-Aza allyl anion **94** serves as a diene in D–A reaction [80].

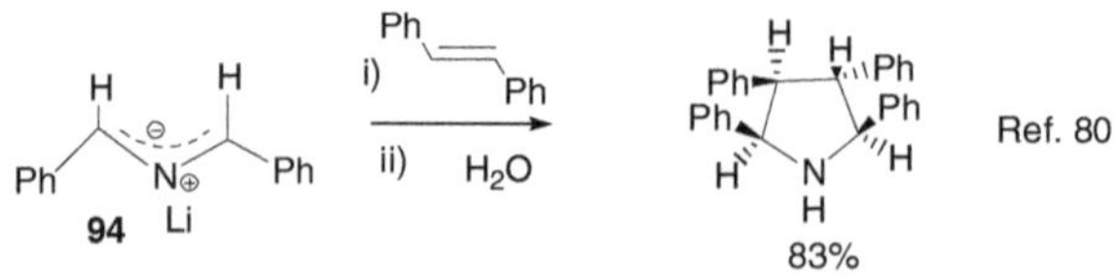

3.3.1.10 Enantioselective Diels–Alder Reactions

In Diels–Alder (D–A) reactions, racemic products are obtained from enantiomeric diene or dienophile. D–A reactions are widely applied in the synthesis of bioactive asymmetric natural products and hence enantio- and diastereoselectivities of the D–A reactions are very much needed to get the desired products as major products. Several approaches have been developed in the last three decades in this respect by the use of different Lewis acid catalysts as chiral auxiliary or asymmetric catalysts. Catalytic D–A reactions have twofold benefits. On one side, it provides high enantiomeric/ diastereomeric excess of product and on the other side, it affords high yield of products by reducing the activation energy of the transition states (TSs). In NED D–A reactions, it lowers the LUMO energy of the dienophile and in IED D–A reactions, it lowers the energy of LUMO of the diene so that the reaction occurs at low temperature with ease (Fig. 3.10). At higher temperature the stereoselectivity of the D–A reactions is lost.

The choice of Lewis acid is very important in D–A reactions as in few cases it leads to non-concerted stepwise process. The calculation of energy of the TSs by DFT methods is very useful in the study of steric, stereoelectronic and chelating

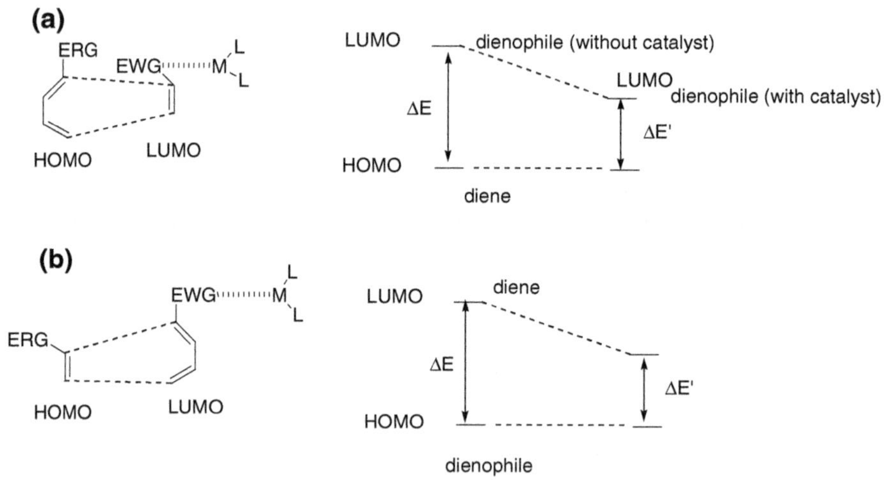

Fig. 3.10 **a** LUMO energy of dienophile is lowered by Lewis acid catalyst in NED D–A reactions and **b** LUMO energy of diene is lowered by Lewis acid catalyst in IED D–A reactions

interactions of the Lewis acid catalyst with dienophile or diene, as the case arises. To achieve diastereoselectivity, a chiral auxiliary is installed in the reaction by the use of chiral Lewis acid catalyst or chiral esters or amides of acrylic acids [81]. The latter method is better because the chiral auxiliary can be recovered by hydrolysis of the purified adduct.

Enantioselective D–A Reactions Using Chiral Auxiliary

Enantioselective D–A reactions of chiral esters and amides of acrylic acid can be achieved using achiral Lewis acid such as $TiCl_4$. After the reaction, enantiomeric pure carboxylic acid can be recovered on hydrolysis. For example, the reaction of acrylic acid with cyclopentadiene using chiral α-hydroxy ethyl propionate **95a** as chiral auxiliary in the presence of $TiCl_4$ gives only one enantiomeric product in large excess (93 %) [82]. The chiral auxiliary **95a** reacts with dienophile to produce a chiral ester **95**, which participates in the reaction with cyclopentadiene.

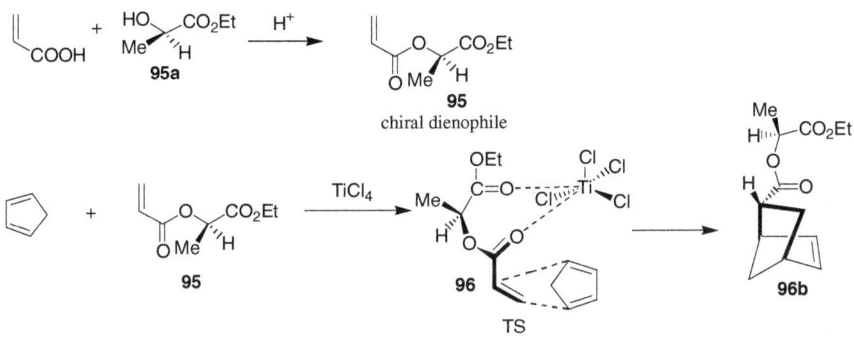

The β-chloride ligand of $TiCl_4$ shields the top face of the dienophile in the TS **96** and only bottom face of the dienophile is able to react with the diene to produce mostly one enantiomeric product **96b**, which on hydrolysis affords the desired product **96a**.

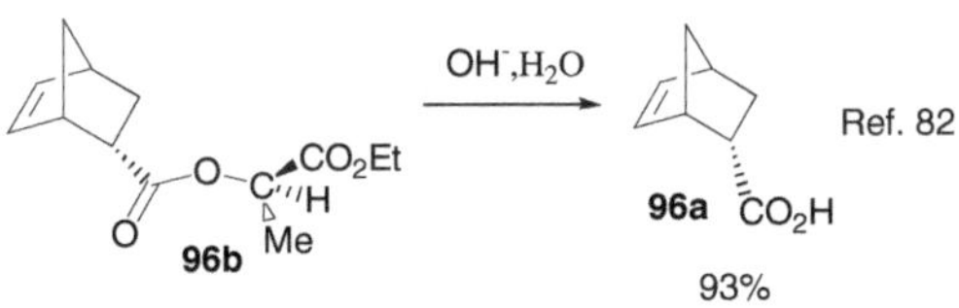

Similarly, 2,4-dihydroxy-3,3-dimethylbutyrolactone known as D-(-)-pantolactone **97** is used as chiral auxiliary along with $TiCl_4$ in several D–A reactions. For example, the reaction of 2,3-dimethylbutadiene with α-cyanocinnamic acid affords the product of 92 % diastereomeric excess.

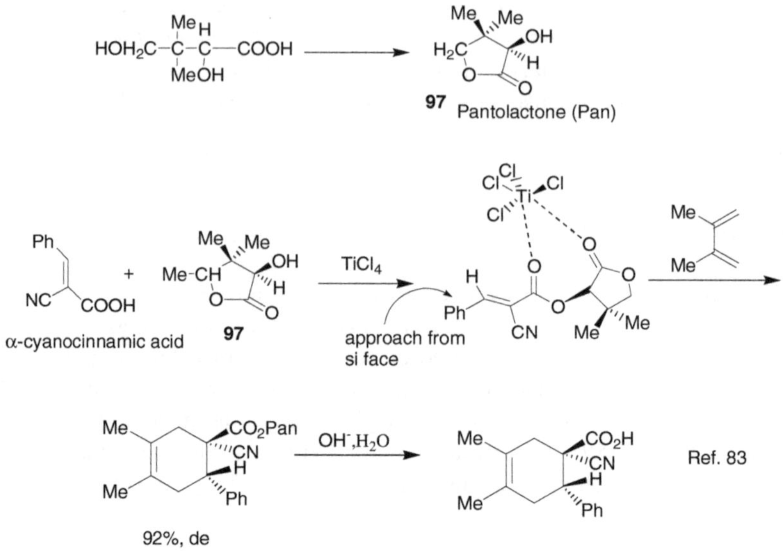

Ref. 83

Enantioselective D–A Reactions Using Chiral Catalysts

Different types of chiral catalysts including nonmetal and metal complexes have been introduced in enantioselective D–A reactions.

Among nonmetal chiral complexes, chiral oxazaborolidines have been found effective in many D–A reactions. The adduct obtained from the D–A reactions of 5-benzyloxymethyl-1,3-cyclopentadiene **98a** with α-bromoacrolein in the presence of catalyst **98** (S-tryptophan-derived oxazaborolidine) is an important intermediate in the synthesis of prostaglandins. The aldehyde group of the dienophile is bound to the catalyst by coordination with boron by Lewis interaction and the Lewis complex is stabilized by H-bonding. The upper face of the aldehyde is shielded by indole moiety of the catalyst. The benzyloxymethyl substituent of the cyclopentadiene produces a steric differentiation on the two faces of cyclopentadiene ring resulting the approach of the diene preferably from one face.

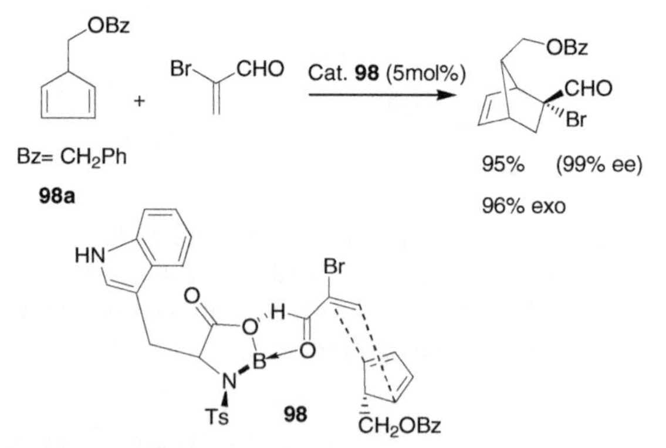

Ref.84

TS of the Lewis complex with catalyst **98**

Chiral *bis*-oxazolines known as BOX–Cu^{2+} complexes are efficient catalysts for asymmetric hetero D–A reactions. For example, BOX–Cu (II) complex **99** catalyzes the reaction of pentane-2,3-dione with Danishefsky's diene **76** efficiently with 76 % yield of the product and of 97.8 % ee [85].

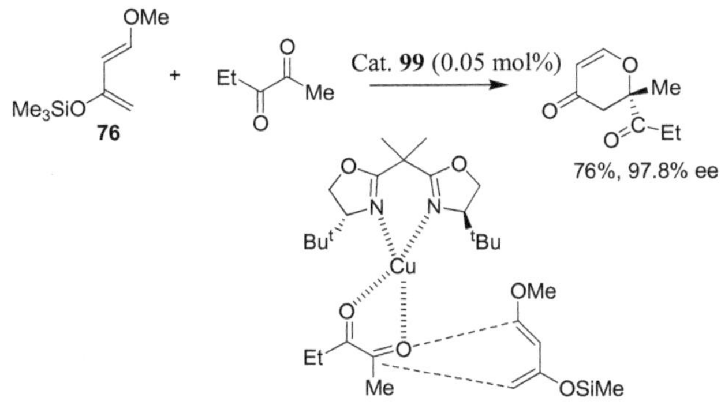

TS of BOX-Cu(II) complex (Cat. **99**)

Binaphthol (BINOL)–Ti complexes **100** are found to be effective catalysts in NED D–A reactions. It catalyzes the reaction of 5-hydroxynaphthoquinone with butadienyl acetate in 86 % yield and 96 % ee of *endo*-product. This intermediate is useful in the synthesis of anthracycline antibiotics [86].

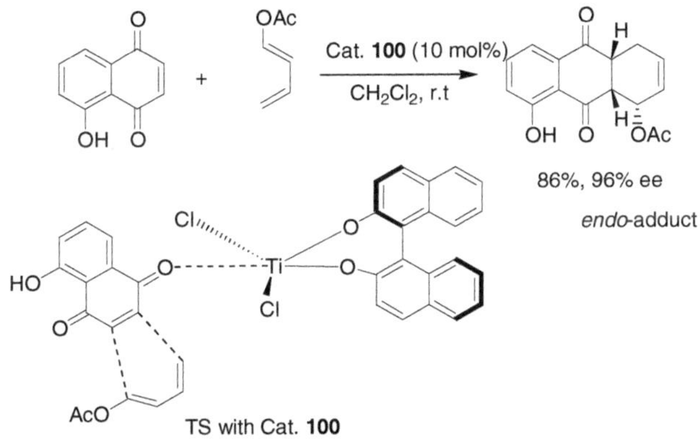

TS with Cat. **100**

R(+)-BINOL–Yb triflate complex **101** catalyzes the D–A reaction of cyclopentadiene with 3(2-butenoyl)-1,3-oxazolidin-2-one **102** in the presence of some additives, 3-acetyl-1,3-oxazolidin-2-one, MS 4A and *cis*-1,2,6-trimethylpiperidine at room temperature to yield enantiomeric excess of *endo*-adduct (2*S*, 3*R*) (93 % ee) with *endo–exo* ratio of 96.5: 3.5, whereas this chiral Yb triflate complex prepared from $Yb(OTf)_3$, R(+) binaphthol, *cis*-1,2,6-trimethylpiperidine, MS 4A, and 3-phenylacetylacetone affords *endo*-adduct (81 % ee) of other configuration (2*R*, 3*S*). Possibly 3-phenylacetyl acetone reverses the enantiofacial selectivity of the dienophile [87].

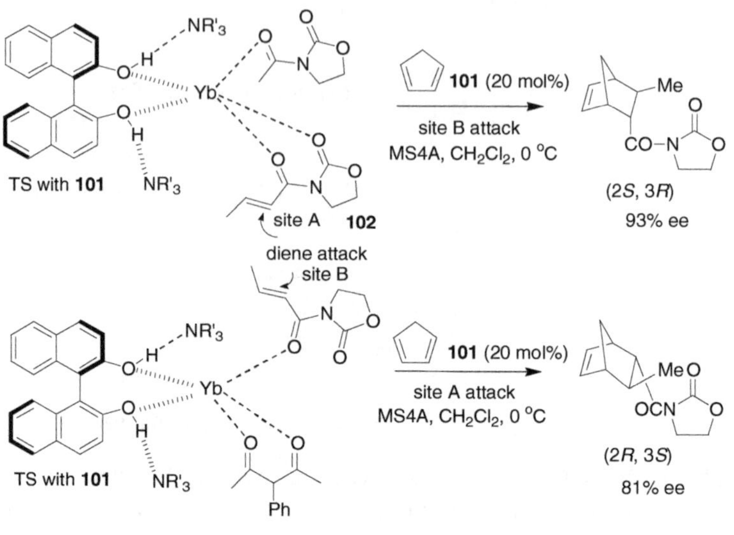

Organocatalysts such as TADDOLs (α, α, α, α-tetraaryl-1,3-dioxolane-4,5-dimethanols) are effective in asymmetric oxa-D–A reactions. For example, TADDOL **103** catalyzes the HDA reaction of less reactive Danishefsky's diene with benzaldehyde to afford corresponding hydropyranone in moderate yield [88].

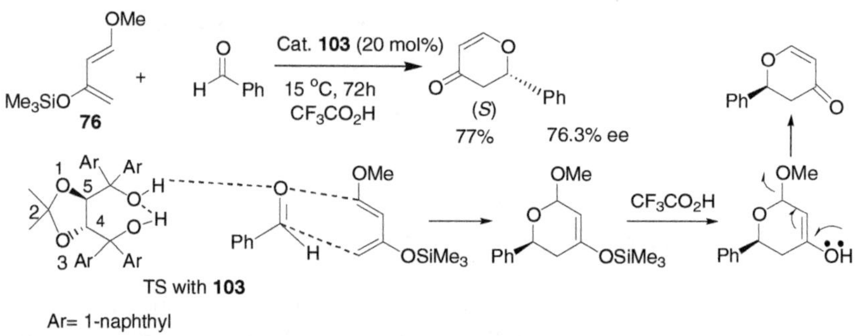

Asymmetric aza-Diels–Alder reaction of chiral imines with Danishefsky's diene in chiral ionic liquids have been reported [89]. These reactions occur at room temperature under green chemistry conditions without using Lewis acid catalyst and organic solvent. For example, the reaction of Danishefsky's diene **76** with imine **104** in the presence of chiral ionic liquid (IL) **105**, gives pyridone derivative in moderate diastereoselectivity [89]. In the TS, the chiral liquid **105** binds the diene and the dienophile in such a way that the diene approaches from top face of dienophile.

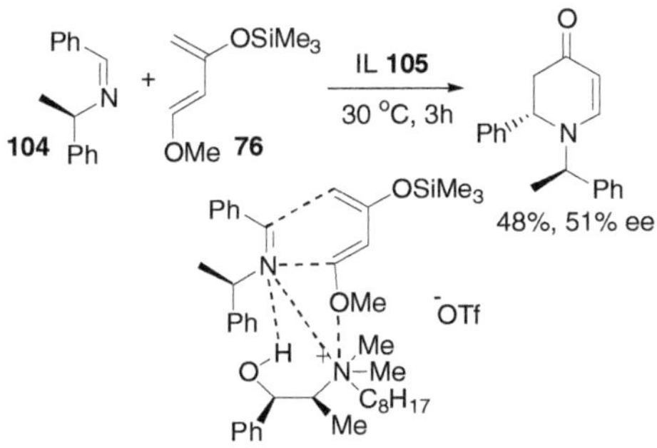

TS with **105**

3.3.1.11 Intramolecular Diels–Alder Reactions

Intramolecular Diels Alder (IMDA) reactions are extensively applied in the synthesis of polycyclic compounds. Most of these reactions require high temperatures to occur and hence these are catalyzed by Lewis acids to occur at ordinary temperatures. These reactions are classified into two types according to the connectivity of alkyl chain to the diene part at C-1 and C-2. Type-1 is very common.

Type 1:

Type 2:

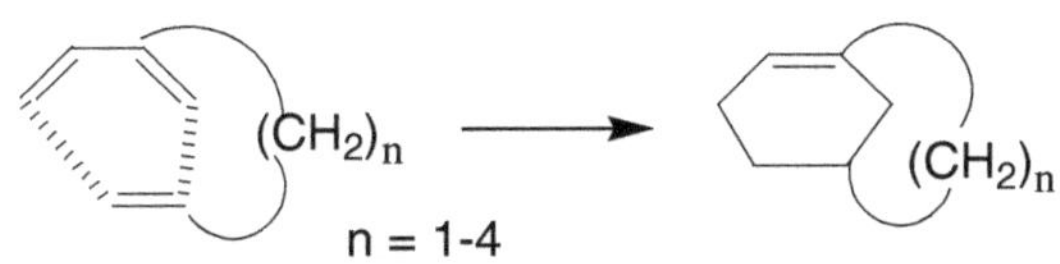

Study of the synchronicity, i.e., the formation of two sigma bonds to the same extent of the reactions indicated that the formation of a bicyclo-[4.3.0]-non-2-ene system ($n = 3$) is preferred kinetically than a bicyclo-[4.4.0]-dec-2-ene for both *cis*- and *trans*-ring junctions. Steric and torsional strains are the important factors to control the TS of the reactions. Usually a *cis*-ring junction is favored for $n = 1-3$ and a *trans*-ring junction is favored for $n = 4$. The following examples are illustrative:

1. 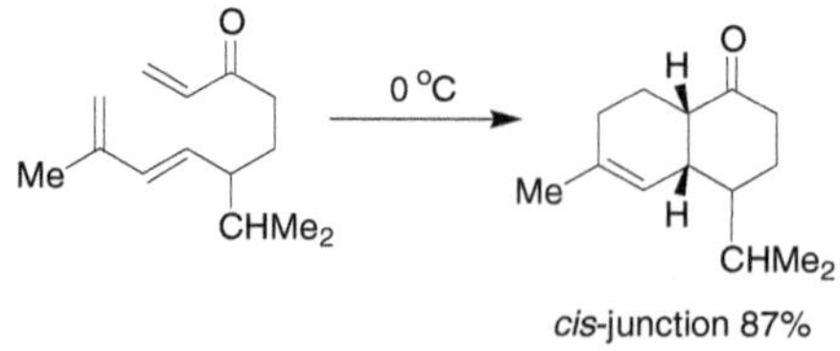

Ref. 90

2. 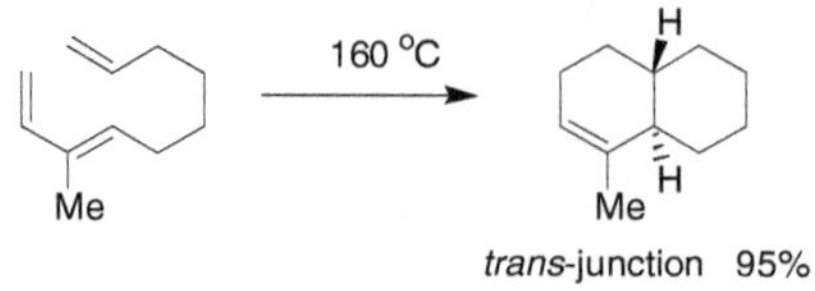

Ref. 91

3. 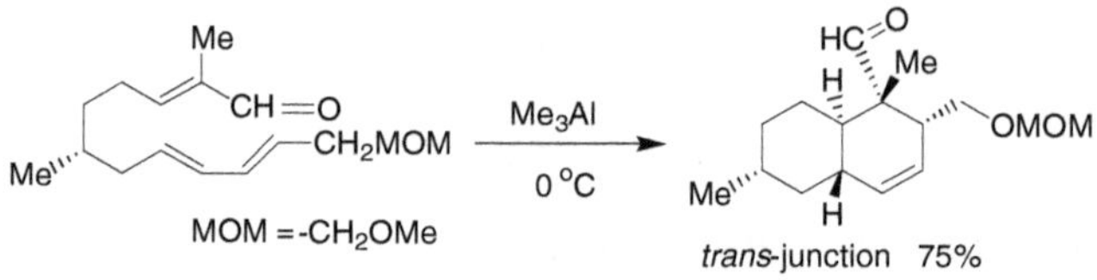

Ref. 92

4. 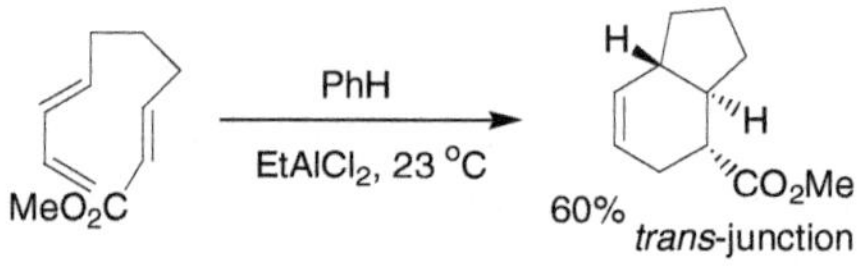

Ref. 93

5. 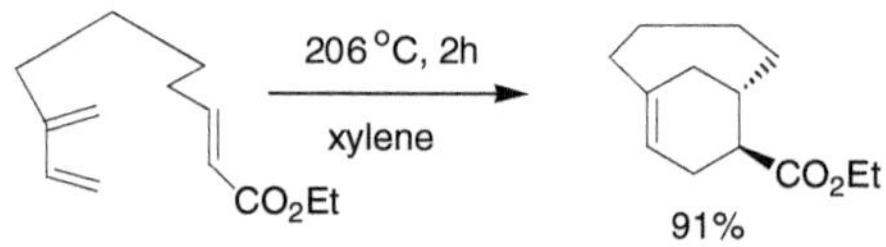

Ref. 94

6. 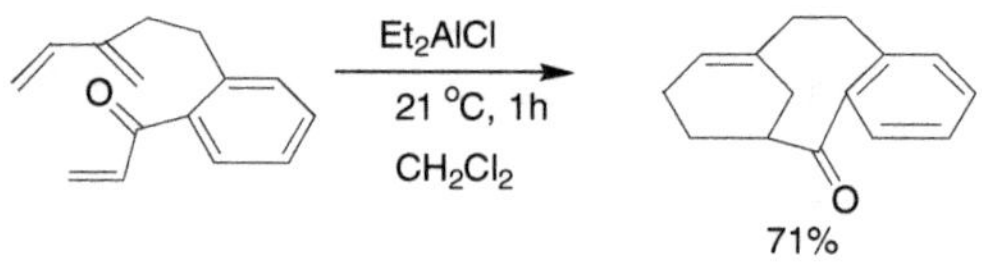

Ref. 95

The geometry of the TS of these reactions is helpful to understand the stereochemistry of the ring junction. In entry 1, the preferred TS is *endo*-boat, which gives the *cis*-fused product.

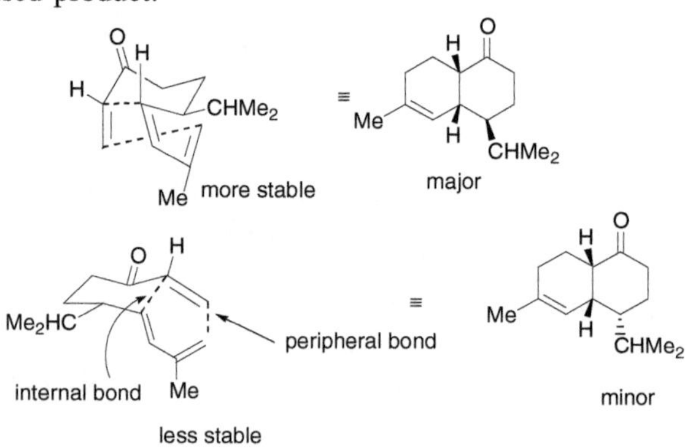

In entry 2, preferred TS is *exo*-chair, which gives the product of *trans*-ring junction.

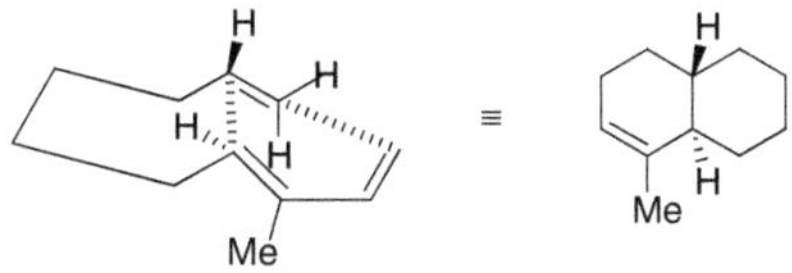

In entry 3, preferred TS is *endo*-chair, which gives *trans*-adduct.

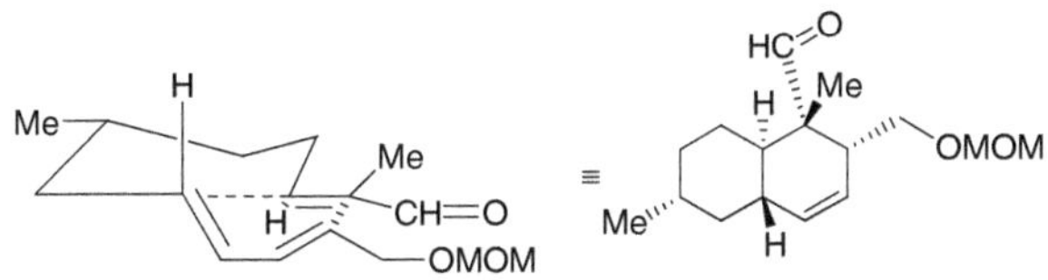

In entry 4, preferred TS is *endo*-boat, which gives *trans*-adduct.

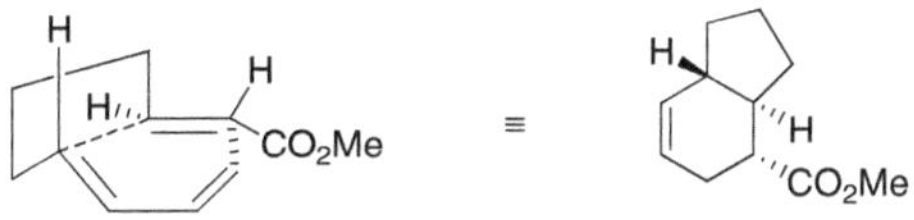

In entry 5, preferred TS is *exo*-chair and favors *trans*-ring junction.

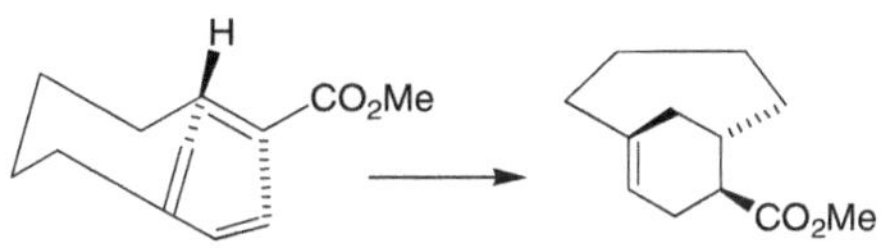

In entry 6, preferred TS is *exo*, and gives *exo*-product.

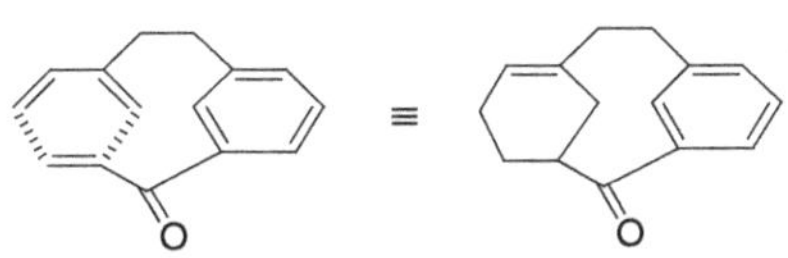

Lewis acid-catalyzed intramolecular D–A reactions prefer *endo*-TS to yield kinetically controlled *endo*-products in preference. For example, triene **106** gives mainly the *endo*-product **107** in the presence of Lewis catalyst, Et_2AlCl [96].

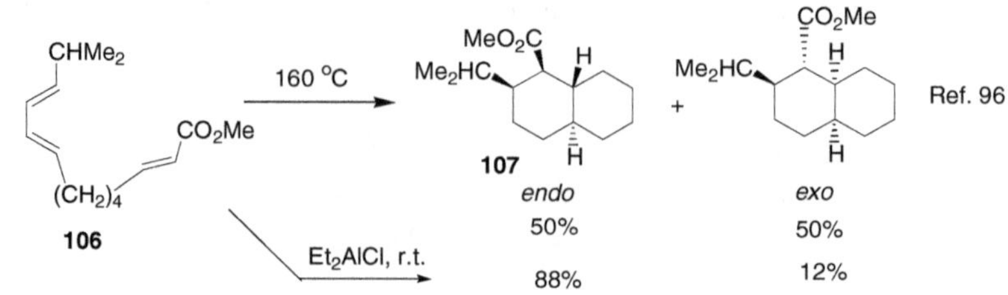

IMDA reaction of benzocyclobutane **108** having a vinyl cyclopentane substituent is useful for construction of steroidal skeleton **109** [97]. Benzocyclobutane generates an *ortho*-quinodimethane in situ.

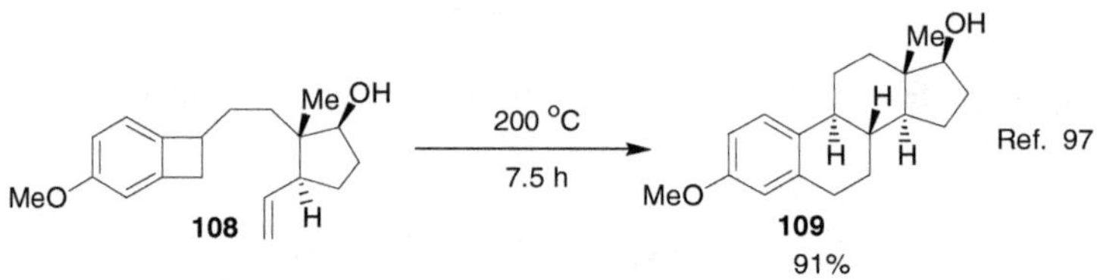

Other typical examples are:

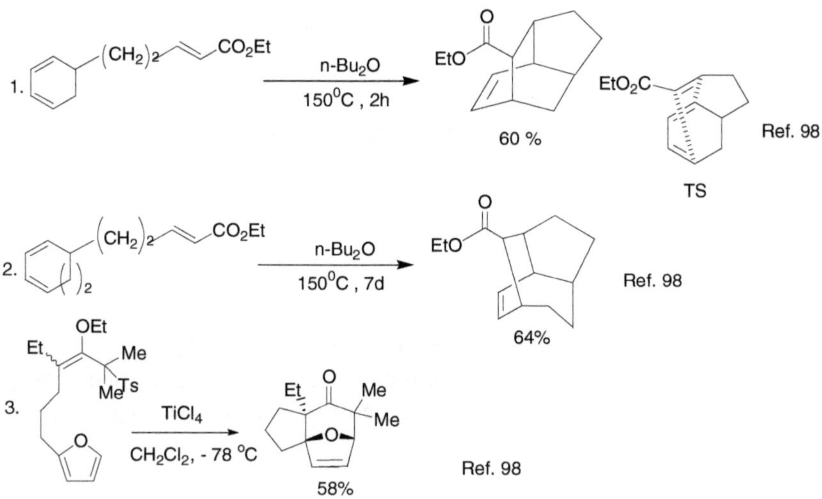

3.3.1.12 The *Retro*-Diels–Alder Reactions

The reverse reactions of Diels–Alder reactions for thermal dissociations of cycloadducts in to dienes and dienophiles at higher temperatures or in the presence of Lewis acid or base are known as the *retro-Diels–Alder (rDA) reactions*. These reactions in most cases proceed in a concerted process. These reactions are often used for separation of diene or dienophile from their mixture with other compounds. Proper selection of conditions of these reactions provides new dienes and dienophiles, which are important synthons for synthesis of several bioactive natural products and organic molecules of complex structures. For example, the D–A adduct of 4-phenyl oxazole **110** with methyl acetylene dicarboxylate, on *retro*-D–A reaction gives new compounds, benzonitrile, and furan 3,4-dicarboxylic acid methyl ester **111** [65].

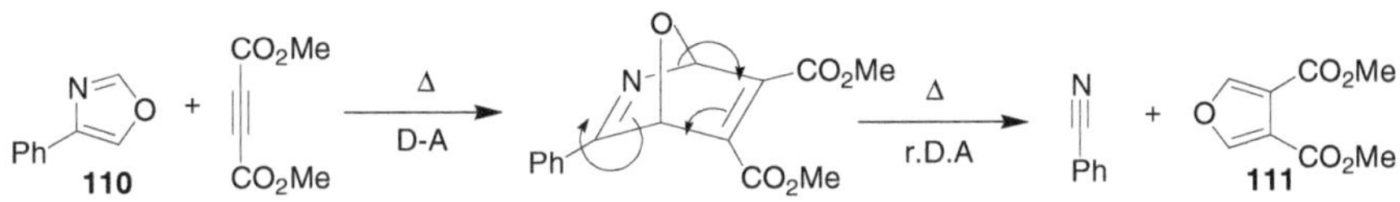

Similarly, 1,3-dienic δ-sultone **112** can be used for the synthesis of highly substituted aromatic compound **113** by a domino DA/rDA process [99].

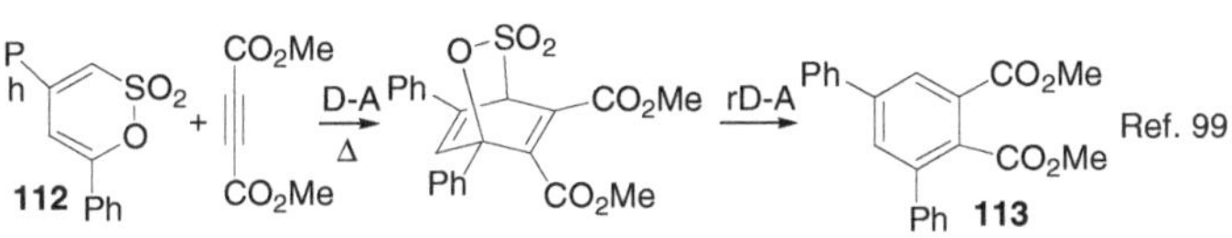

These reactions are very useful in the synthesis of highly substituted aromatic, heterocyclic and bicyclic compounds. The following examples are illustrative:

1. 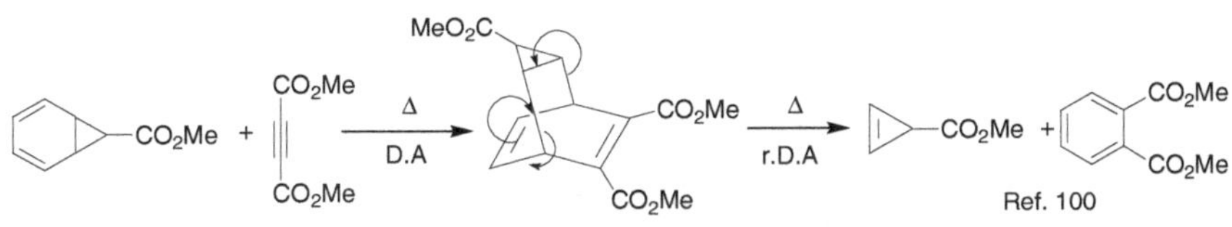

2. 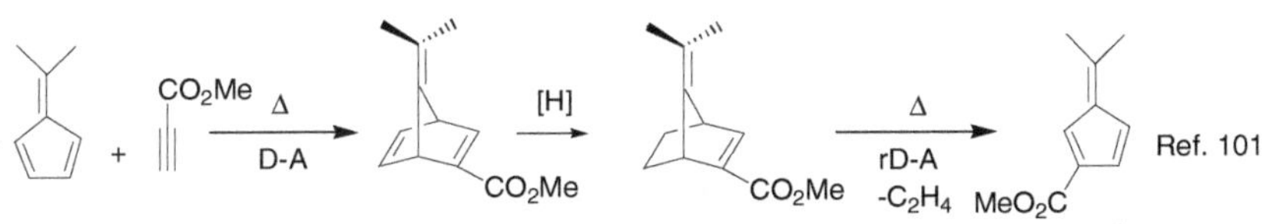

3. 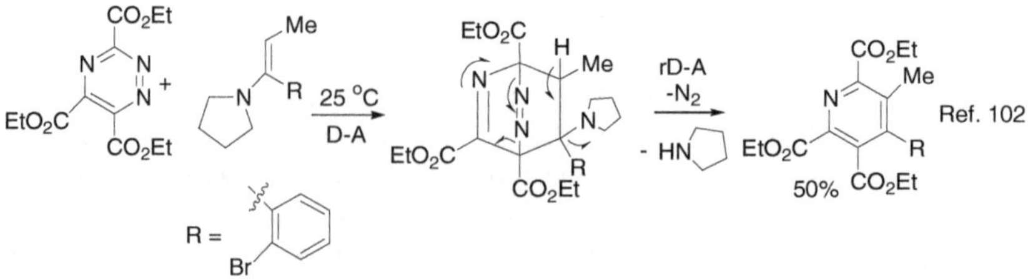

4. 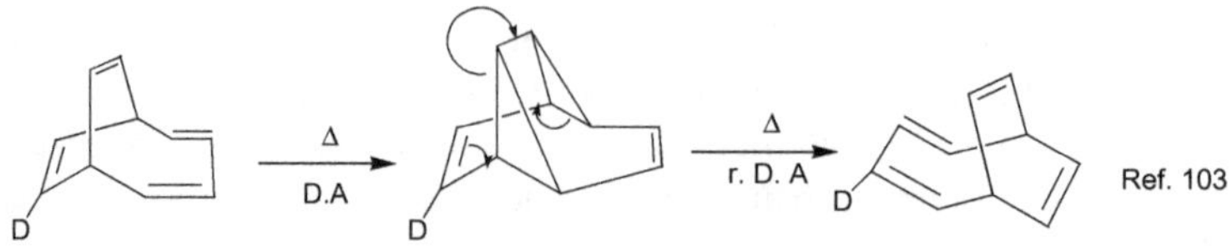

5. 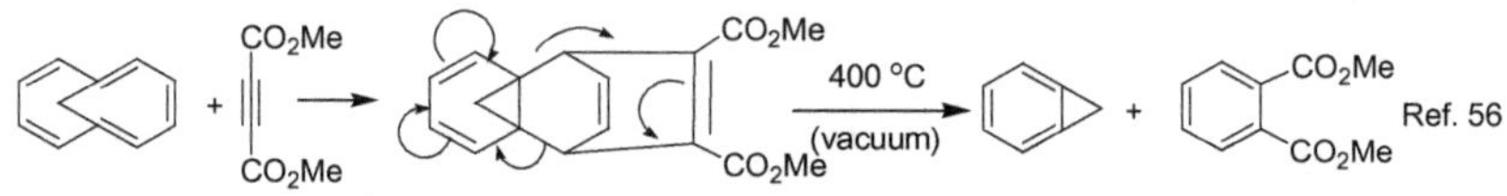

3.3.1.13 1,3-Dipolar Cycloaddition Reactions

Overview of 1,3-Dipolar Cycloaddition Reactions

The cycloaddition reactions of 1,3-dipolar compounds with alkenes or alkynes or heteroatom containing double or triple bonds are known as 1,3-Dipolar cycloaddition reactions, or simply 3-DPCA reactions. These are another type of concerted $[\pi_s^4+\pi_s^2]$-cycloaddition reactions, analogous to D–A reactions [104]. 1,3-Dipolar compounds are known as 1,3-dipoles and alkenes or alkynes are known as dipolarophiles. These reactions are represented as

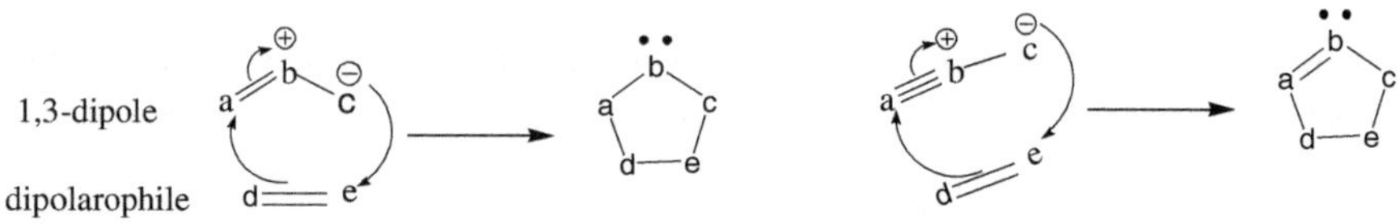

1,3-Dipoles are a class of organic molecules, analogous to allyl or propargyl anions having a 4π elctron system, while dipolarophiles are alkenes or alkynes with an electron-withdrawing or electron-releasing group. Other multiple-bonded groups such as carbonyl, azo, imino, and nitroso groups can also serve as dipolarophiles. Some common 1,3-dipoles with their resonating structures are listed in Table 3.5.

Methods for Generation of 1,3-Dipolar Compounds

Most of the 1,3-dipoles are short lived and are generated in situ. The common methods for generation of 1,3-dipolar compounds in situ are:

Table 3.5 List of common 1, 3-dipoles with resonating structures

a. Allylic anion type

i) Nitrogen atom in the middle

Nitrones

Azo methine imines

Azo methine ylides

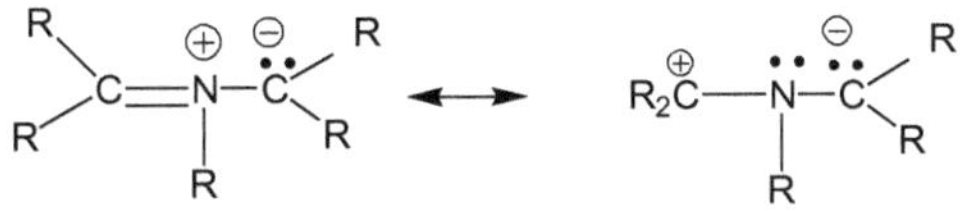

Azimines

ii) Oxygen atom in the middle

Ozone

Carbonyl ylides

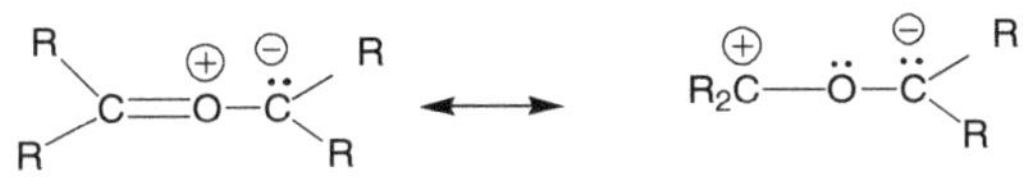

Nitroso imines

iii) Propargyl anion type

Nitrile oxides

Nitrile imines

Nitrile ylides

Diazoalkanes

Azides

Nitrous oxide

Nitrile oxides:

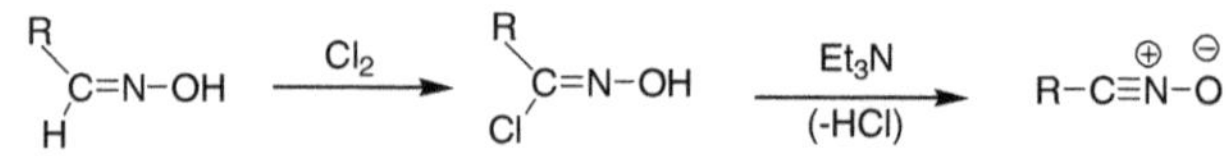

Nitrile ylides:

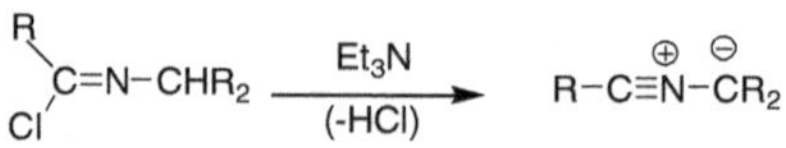

Nitrones:

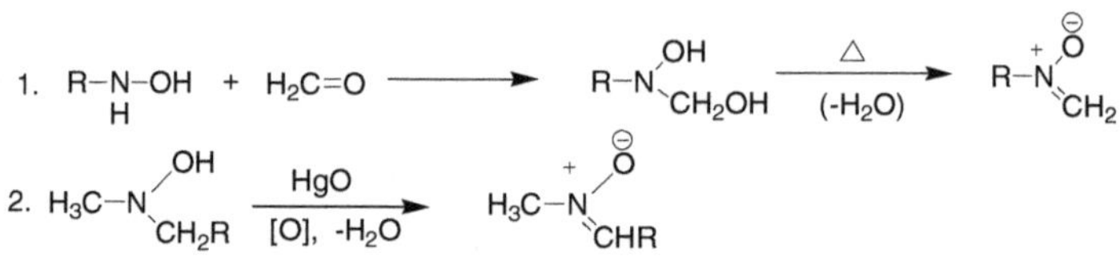

Nitrile oxides:

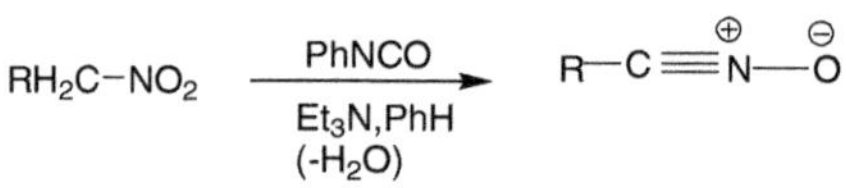

Regioselectivity and Stereoselectivity of 1,3-Dipolar Cycloaddition Reactions

Most of the 1,3-DPCA reactions are highly stereospecific with respect to dipolarophiles and give *syn*-addition products. For example, diazomethane reacts with *cis*- and *trans*-2-methyl-methyl-2-butenoate **114** and **115** separately to afford *cis*- and *trans*-product **116** and **117**, respectively, with more than 99.9 % stereospecificity [105].

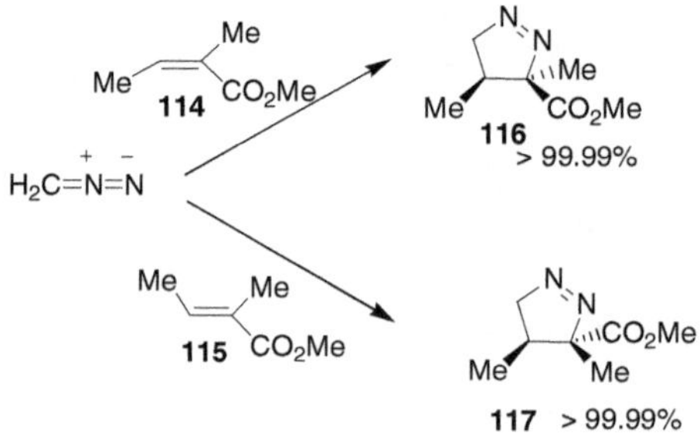

Ref. 105

While the reaction of phenyl diazoalkane with unsymmetrical dipolarophile α-methyl-methyl maleate **118** gives two diastereoisomeric products **119** and **120** from the *endo-* and *exo-*transition states [106].

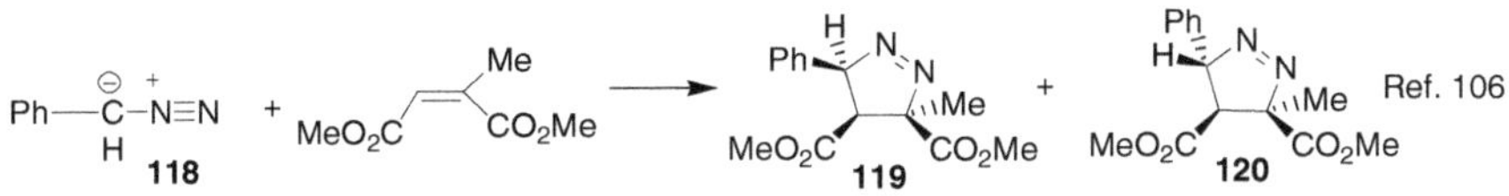

118 119 120

When both 1,3-dipole and dipolarophile are unsymmetrical, two products are possible. The formation of major product can be predicted by consideration of their TSs. The most stable TS will provide the major product. The stability of the TS is controlled by both electronic and steric factors. Therefore, the regioselectivity of a 1,3-DPCA reaction is determined by the steric and electronic properties of the substituents attached to 1,3-dipole and dipolarophile. The FMO theory may also be applied to analyze the regioselectivity of 1,3-DPCA reaction [107]. A relatively stronger donor–acceptor interaction between HOMO and LUMO and lowest dipole moment favors the TS. The HOMO and LUMO of a 1,3-dipole are similar to that of a diene in a Diels–Alder reaction. The interactions of HOMO or LUMO of a dipole with a LUMO or HOMO of a dipolarophile depend on their electron donor and electron acceptor property. The orbital interactions of HOMO and LUMO of dipole and dipolarophile are shown in Fig. 3.11.

To understand the strong interaction between HOMO and LUMO, the knowledge about the orbital coefficients of the frontier orbitals of the 1,3-dipoles and dipolarophiles is essential. The orbital coefficients of the common dipolarophiles at C(1) and C(2) carbons depend on the nature of the substituents.

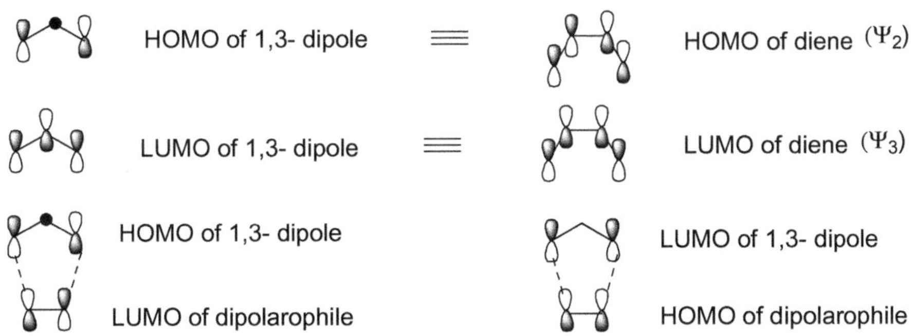

Fig. 3.11 Frontier orbital interactions in a 1,3-dipolar cycloaddition reaction

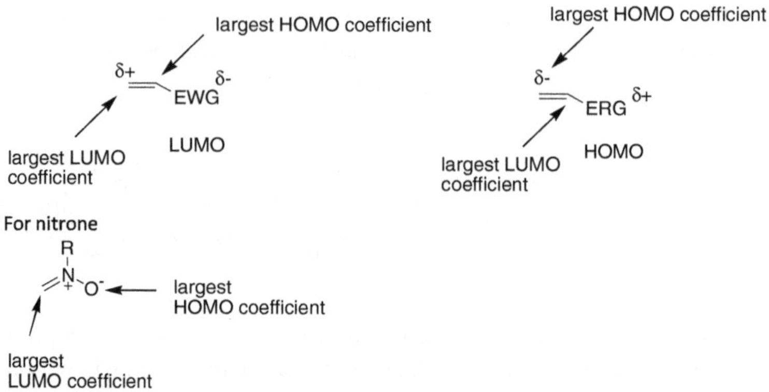

The reported orbital coefficients of some 1,3-dipoles in their HOMO and LUMO states are given in Fig. 3.12 [108].

The HOMO or LUMO of the 1,3-dipole will interact through the atom having highest orbital coefficient to LUMO or HOMO of the dipolarophile. For example, in

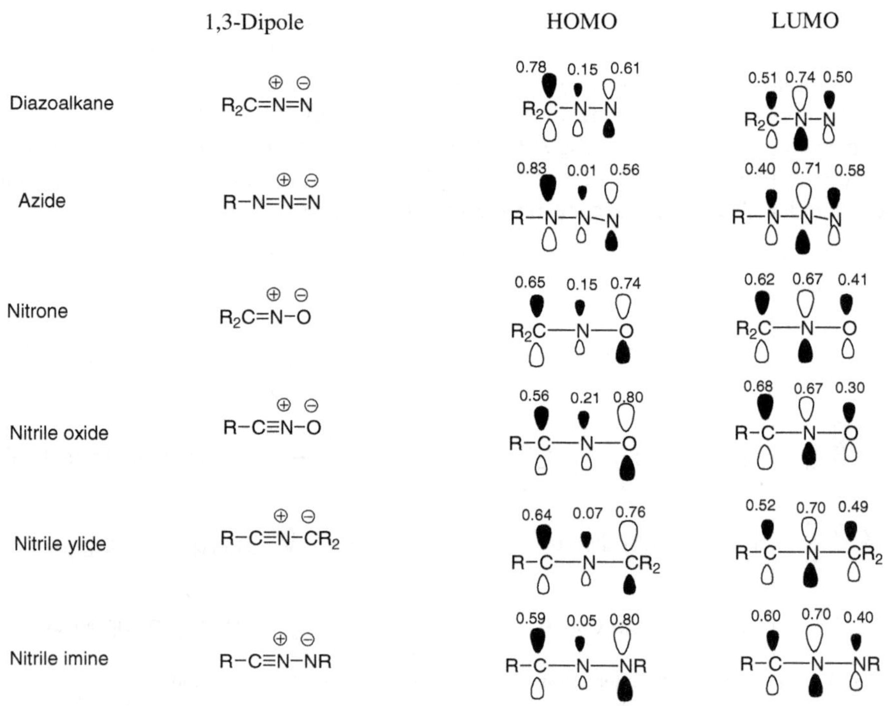

Fig. 3.12 Orbital coefficients of the HOMO and LUMO of some 1, 3-dipoles. Adapted with permission from (Houk et al. 1973 J Am Chem Soc, 95:7287). Copyright (1973) American Chemical Society

the reaction of phenyl azide **121** with methyl acrylate, azide is HOMO and dipolarophile is LUMO. The dominant interaction gives **122** as major product [109]:

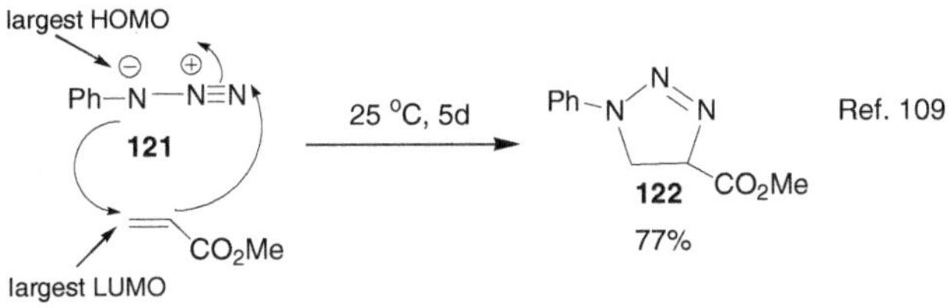

In the reaction of p-nitrophenyl azide **123** with ethoxyethylene, the azide is LUMO and dipolarophile is HOMO. Hence, their dominant interaction gives **124** as only product [109].

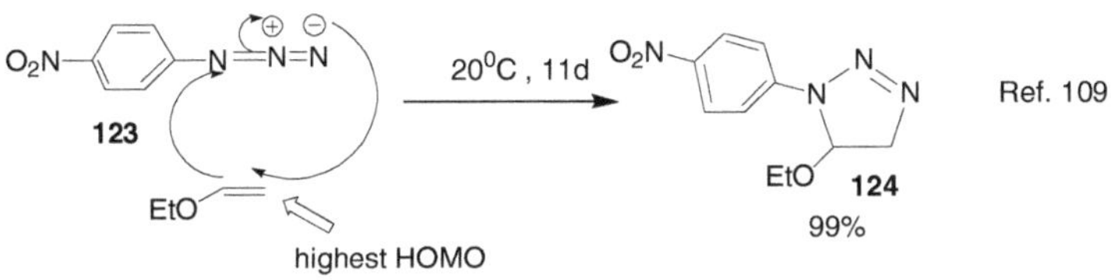

Similarly, in the reaction of C-phenyl-N-methylnitrone **125** with methyl acrylate, nitrone is HOMO and methyl acrylate is LUMO. Hence, their dominant interaction gives **126** because steric factor prevents the formation of 4-substituted isomer as per electronic factor [5].

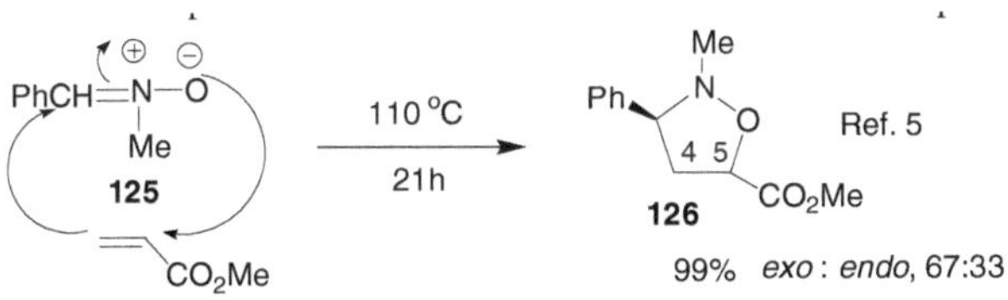

The reaction of nitrone **125** with nitro ethene, bearing strong electron-withdrawing group, gives only 4-substituted products **125a** and **125b** as per FMO-controlled electronic factor [108].

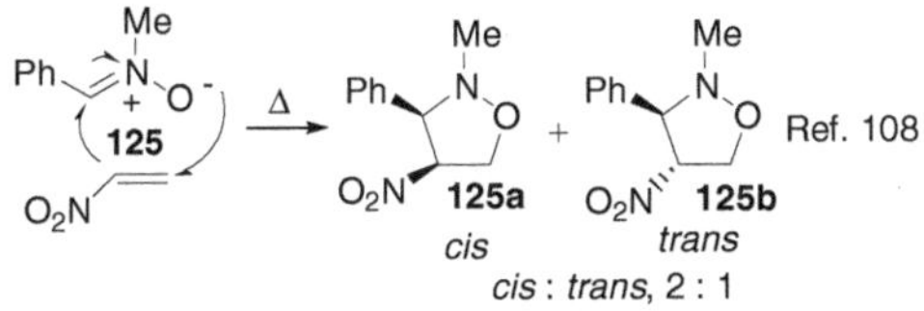

The cycloaddition reaction rate of nitrone with α,β-unsaturated aldehydes in the presence of a pinhole Lewis catalyst is enhanced dramatically and gives only electronically controlled cycloadducts, isoxazolidine-4-carboxaldehydes. For example, C,N-diphenyl nitrone **127** reacts with acrolein **127** at room temperature to give a 20:80 mixture of regioisomeric cycloadducts, 2,3-diphenyl-isoxazolidine-4-carboxaldehyde **128** and -5-caboxaldehyde **129** in only 5 % of total yield. When the same reaction is carried out at 0 °C in the presence of a catalytic amount (10 mol%) of aluminum–tris (2,6-diphenyl phenoxide) (ATDP) **130**, a single regioisomer (**128**:**129**, 99:1) is obtained, but in a poor diastereoselectivity (*endo*:*exo*, 77:23). On the other hand, crotonaldehyde **131**, a 1,2-disubstituted alkene, reacts with **127** under both non-catalyzed and ATDP-catalyzed conditions to give only the electronically controlled product **132**. No **132a** was found. It indicates that steric factor is minimized in case of 1,2-disubstituted alkenes [110].

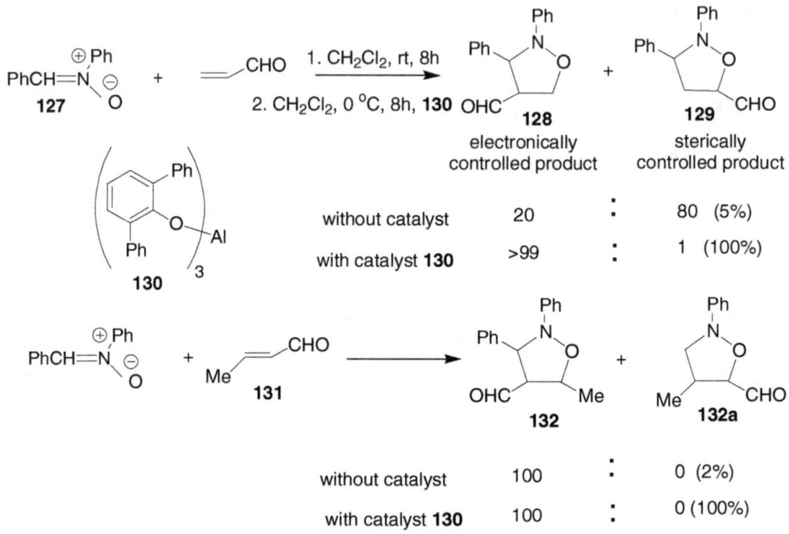

The use of an auxiliary of α,β-unsaturated compound in the cycloaddition reaction of acyclic nitrone improves the yield and diastereoselectivity of product. Moreover, the use of bulky catalyst favors the *exo*-selectivity of the product. For example, N-crotonoyl succinimide **133** reacts with C,N-diphenyl nitrone **127** in toluene at rt to give exclusively *endo*-product **135**, whereas in the presence of 5 mol% of $TiCl_2$-TADDOLate **134** in toluene gives exclusively *exo*-product **136** with high (73 %) ee. The X-ray study of the TS indicates that the nitrone approaches the alkene from an Re face to give the *exo*-product [111].

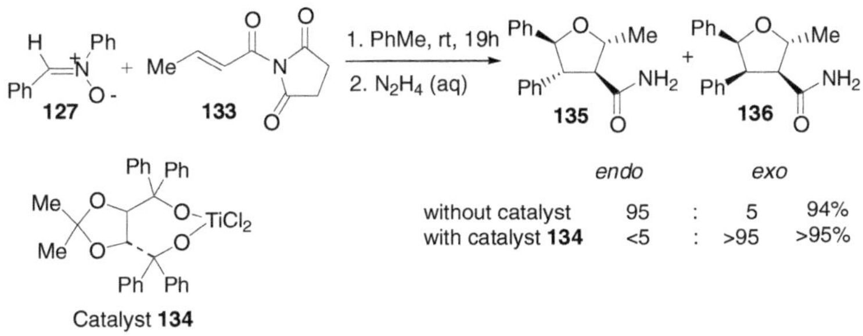

The cycloaddition of C-phenyl-N-methyl nitrone with a dipolarophile having an *ortho*-hydroxy group is completely controlled by intermolecular hydrogen bonding between dipolar compound and dipolarophile leading to an *E-endo* transition state to give exclusively *cis*-cycloadduct.

For example, reaction of C-phenyl-N-methyl nitrone **125** with *ortho*-hydroxystyrene **137** gives only *cis*-adduct (100 %) in 92 % yield [112].

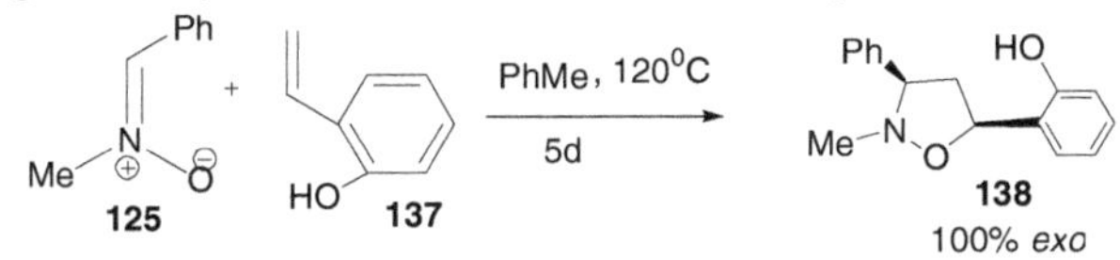

The reaction proceeds through an *E-endo*-TS in which intermolecular, H-bonding and secondary orbital interactions make the TS tighter and gives only *cis*-product from the interaction of LUMO nitrone and HOMO dipolarophile (Fig. 3.13).

In the case of styrene, such rigidity of the TS is not occurred and hence both *cis*- and *trans*-adducts are formed from *E-endo*- and *Z-endo*-TS, respectively [112].

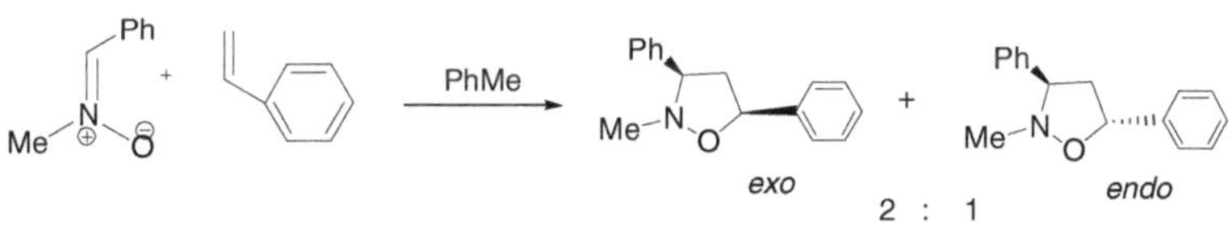

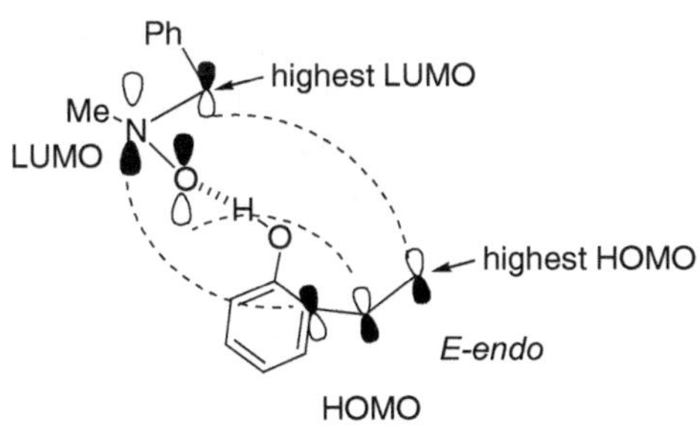

Fig. 3.13 The orbital interactions of HOMO and LUMO in the TS in the reaction of nitrone **125** with *ortho*-hydroxyl styrene **137**

Nitrone reacts with vinylethers, vinylalkyls and vinylnitrile to give major 5-substituted *exo*-products. The following examples are illustrative:

1. 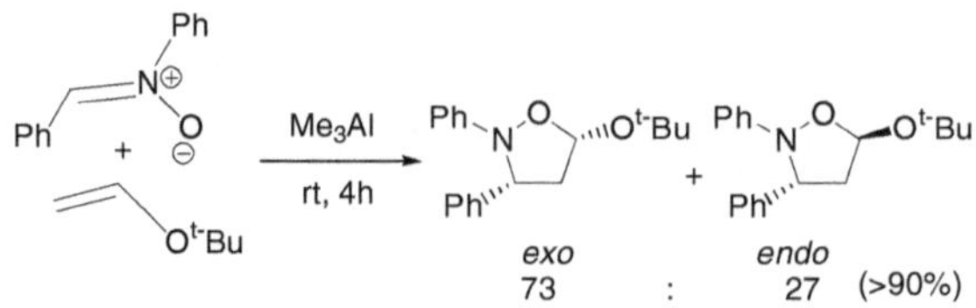

Ref. 113

2. 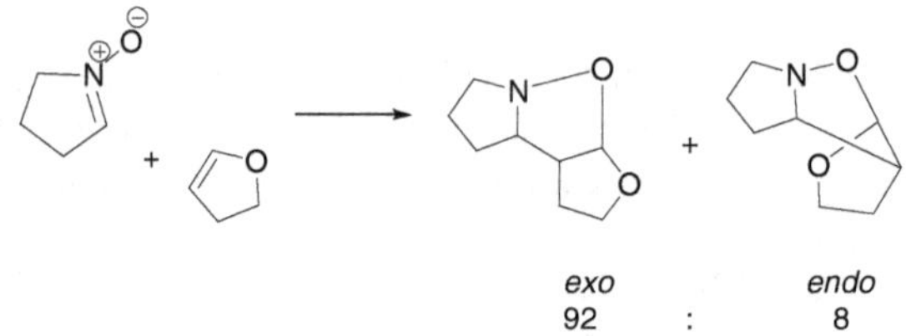

Ref. 5

3. 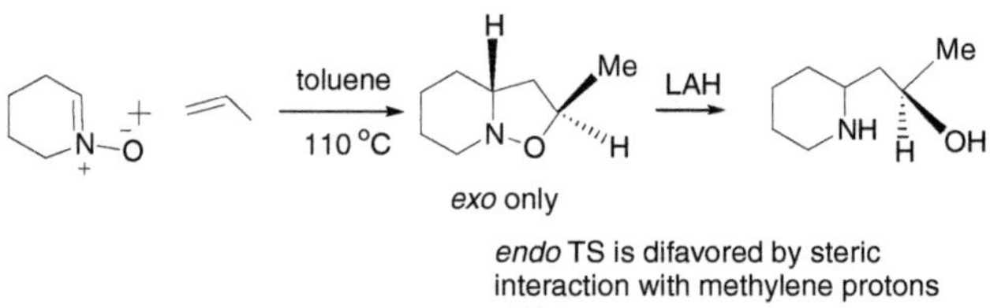

Ref. 114

4. 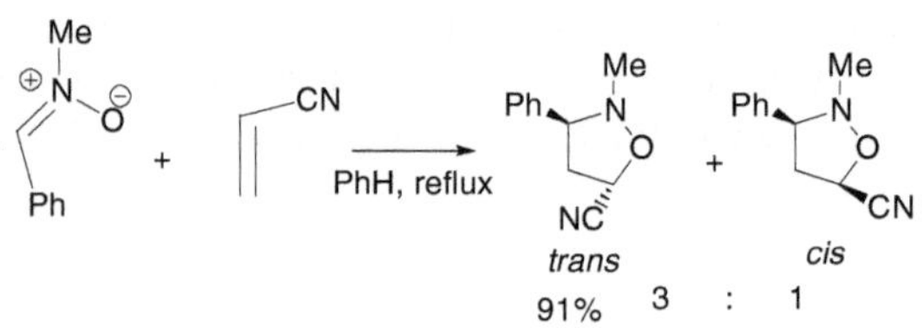

Ref. 115

Synthetic Applications of Intermolecular 1,3-Dipolar Cycloaddition Reactions

1,3-DPCA reactions are very useful for the synthesis of five-membered heterocyclic compounds. Sometimes the reaction products undergo hydrogen shifts to afford stable heterocyclic molecules. Some of these reactions are illustrated to highlight the yields and regioselectivity of the products.

Azides are used for the synthesis of triazoles, e.g.,

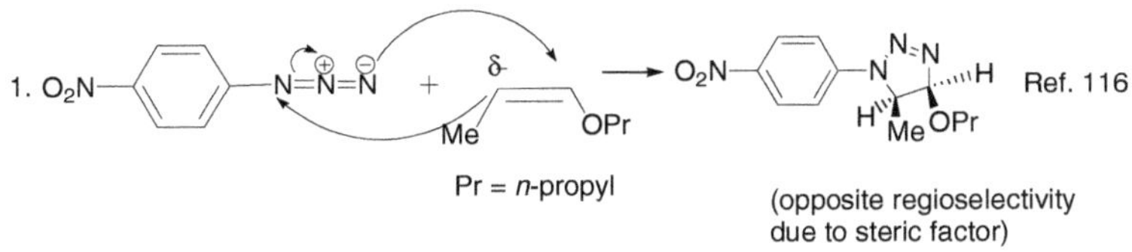

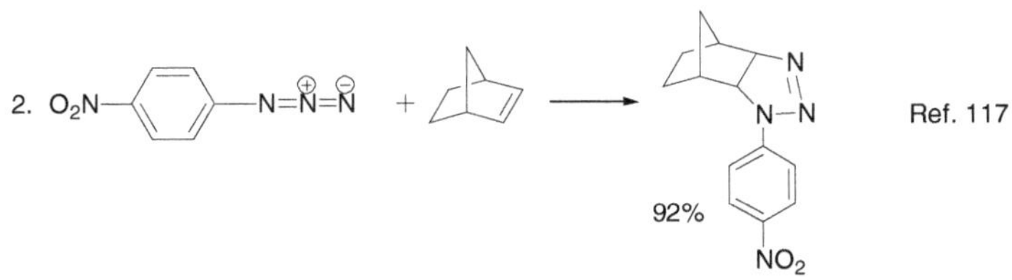

Nitrile oxides are used for synthesis of isoxazoles, e.g.,

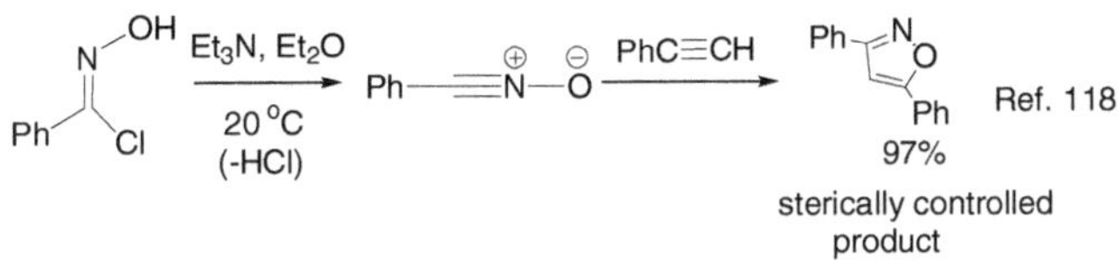

Diazoalkanes are used in the synthesis of pyrazolines.

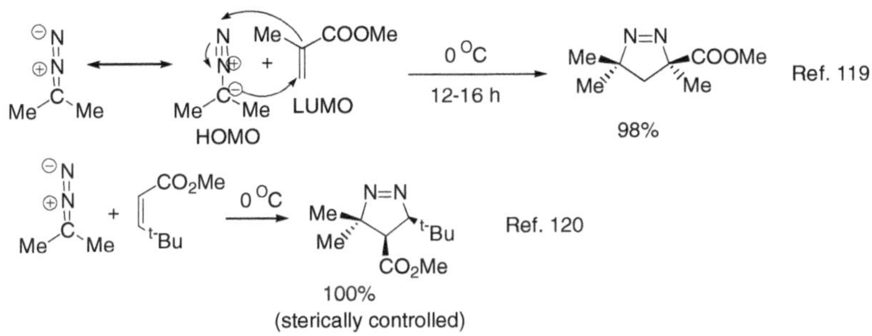

Nitrile ylides generated in situ are used for the synthesis of Δ^1-pyrrolines.

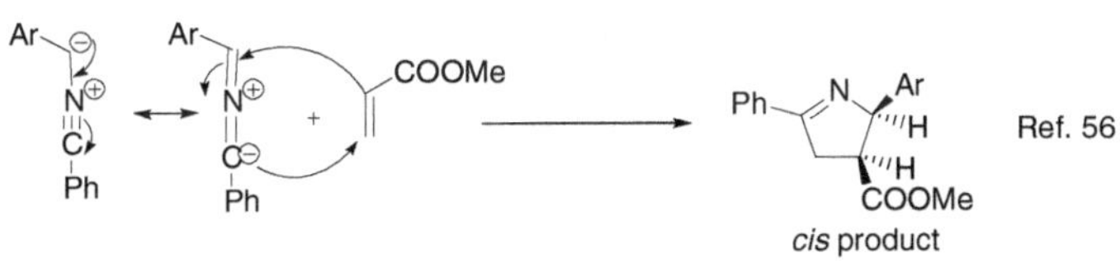

Azomethine ylides generated in situ are used in the synthesis of pyrrolidines by trapping with dipolarophiles.

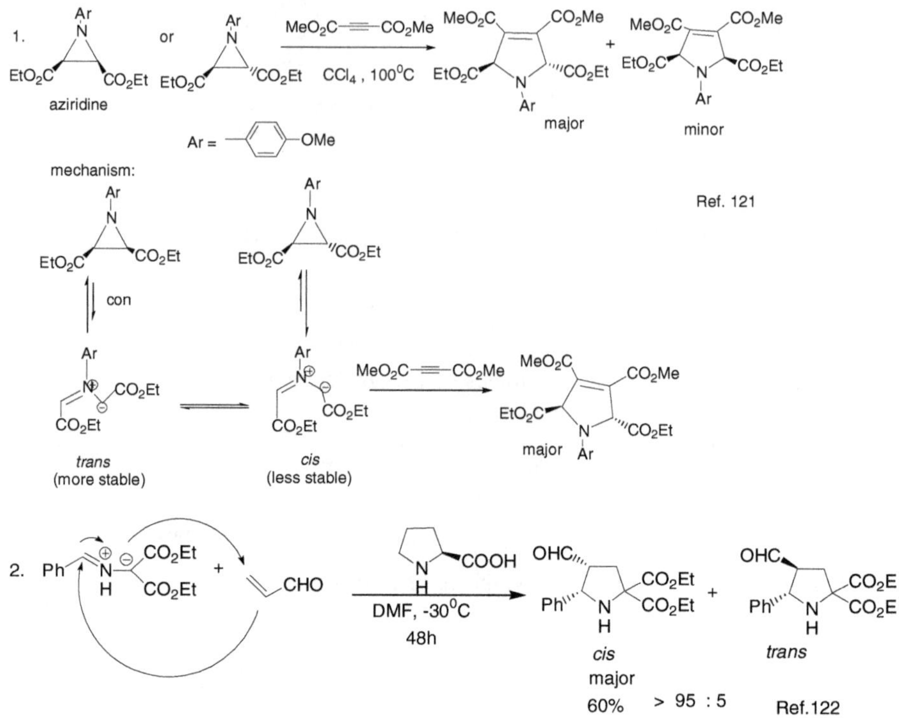

Nitrones are used in the synthesis of isoxazoles. For example,

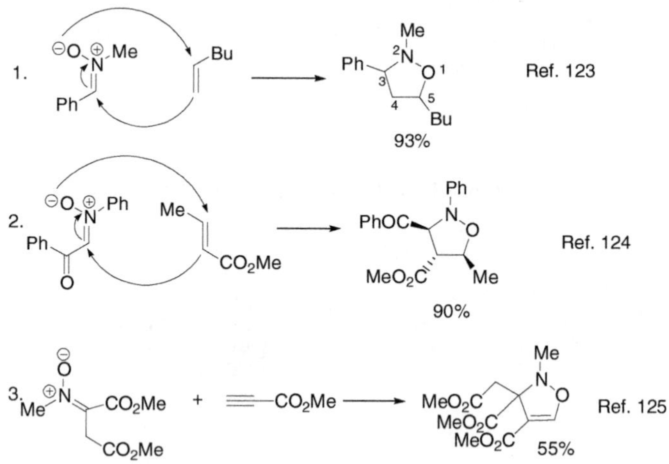

Carbonyl ylides generated by reversible ring opening of epoxides undergo cycloadditions with electrophilic alkenes. For example, tetracyanoethylene oxide generates a carbonyl ylide, which is trapped as tetrahydrofuran.

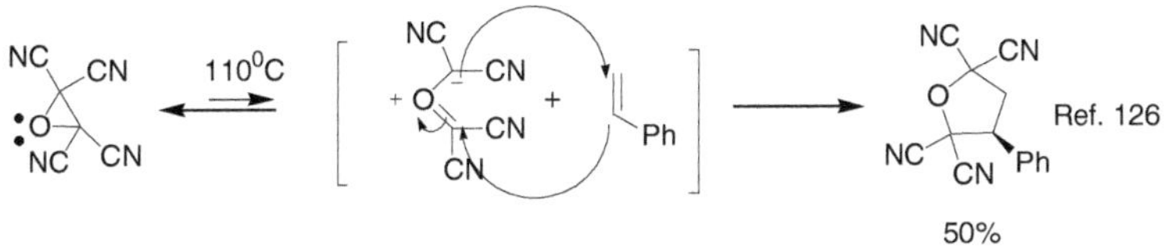

Carbonyl ylide generated by Cu (1) chloride gives regioselective product with benzaldehyde, which is governed by dipole character of 1,3-dipole and dipolarophile [5].

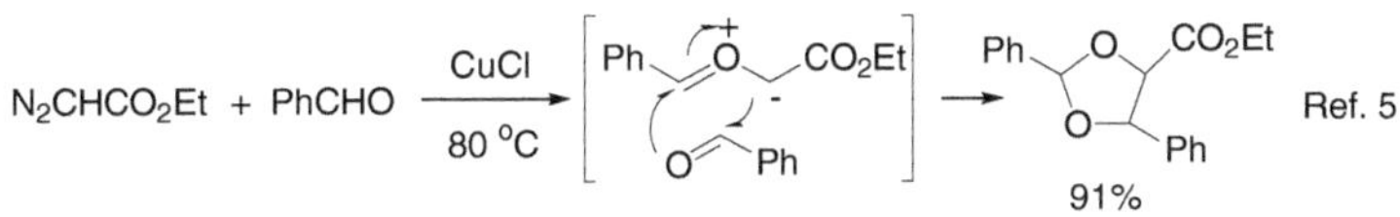

Carbonyl ylides generated by the transition metal-catalyzed decomposition of diazo compounds are trapped with dipolarophiles [127].

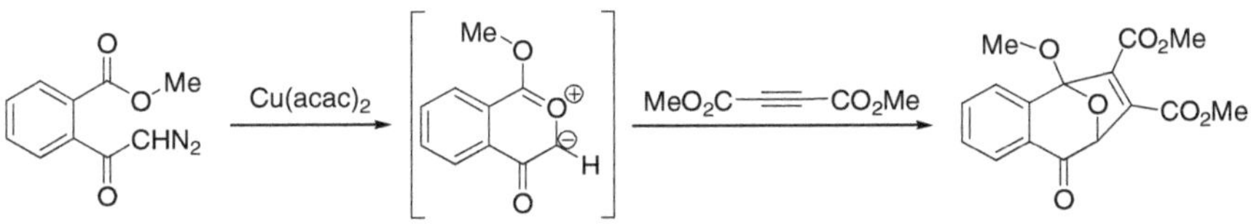

Similarly, carbonyl ylide generated from the diazo compound in the presence of rhodium (II) acetate reacts with dimethyl acetylene dicarboxylate to give dihydrofuran derivative [127].

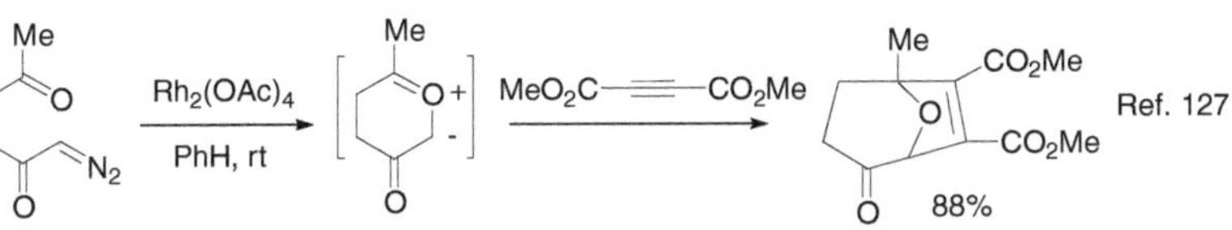

3.3.1.14 Synthetic Applications of Intramolecular 1,3-Dipolar Cycloaddition Reactions

Several intramolecular 1,3-DPCA reactions are used in the synthesis of heterocyclic compounds. The following reactions are illustrative:

A. Intramolecular additions of nitrones

1. 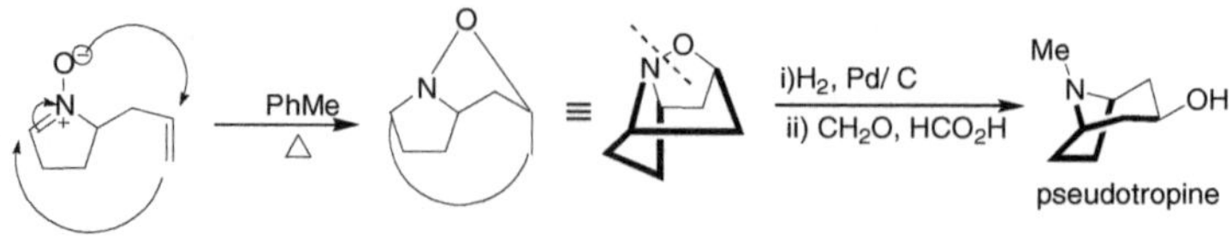
Ref. 128

2. 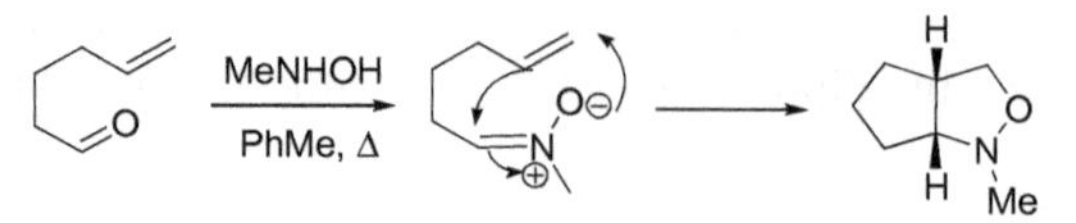
Ref. 129

3. 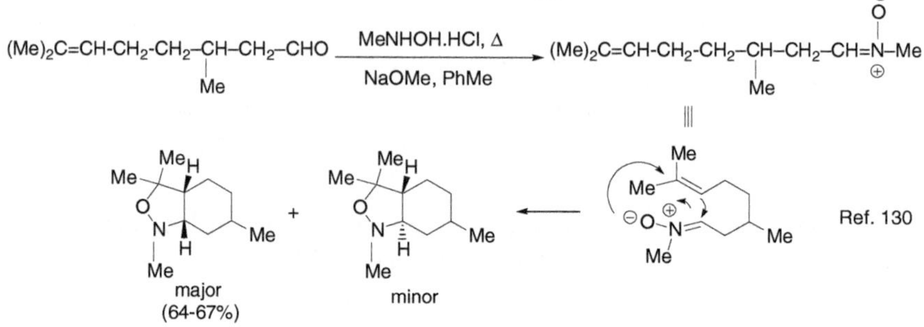
Ref. 130

4. 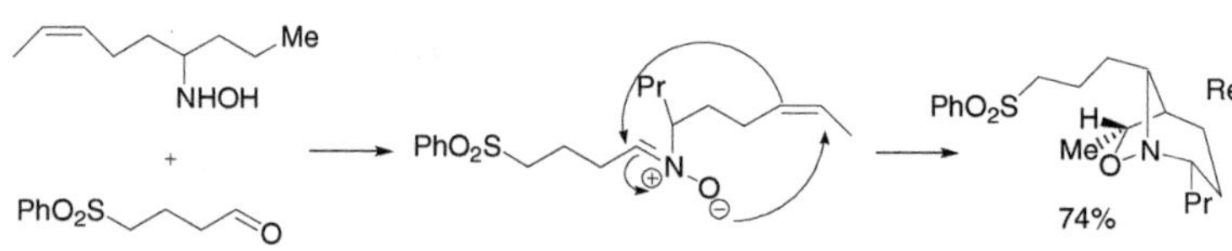
Ref. 131

B. Intramolecular addition of azomethine ylides

5. 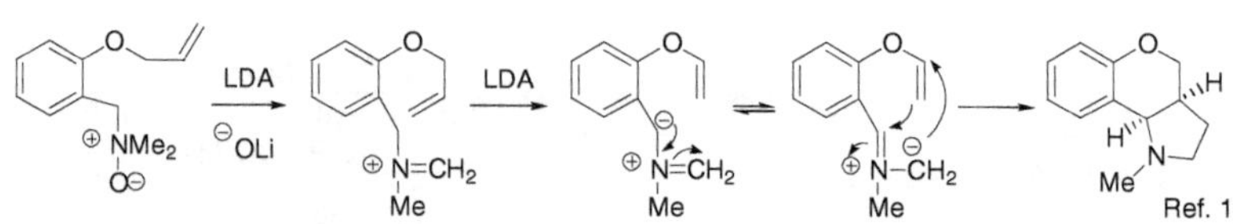
Ref. 132

6. 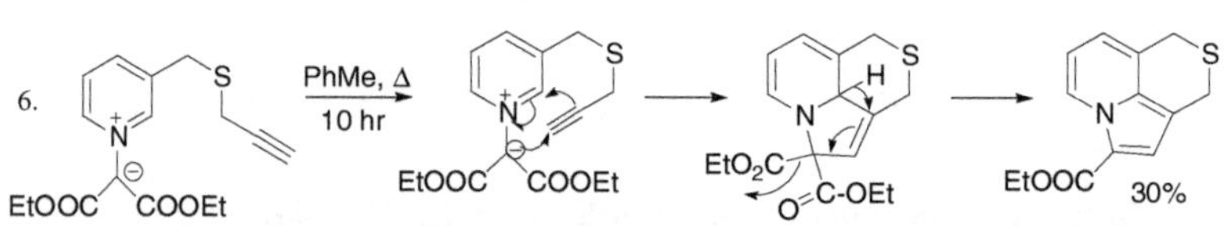
Ref. 133

7. 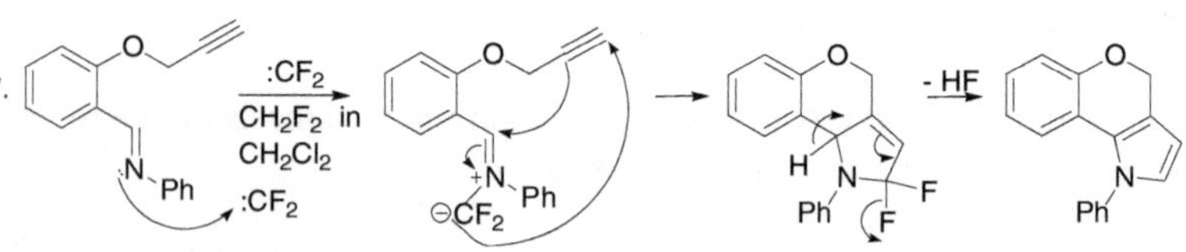
Ref. 134

C. Intramolecular addition of nitrile ylide

8. 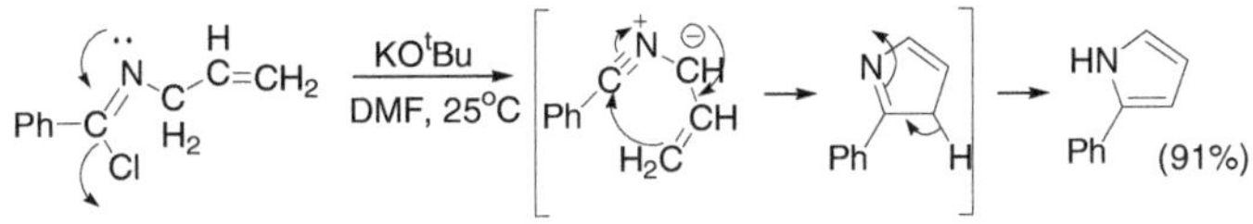
Ref. 135

D. Intramolecular addition of nitrile oxides

9. 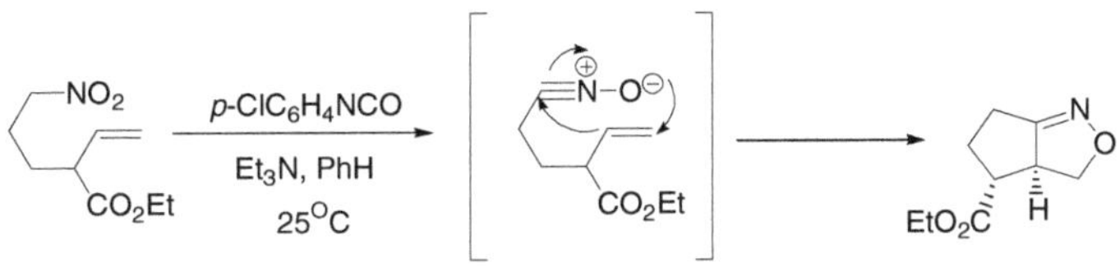
Ref. 136

10. 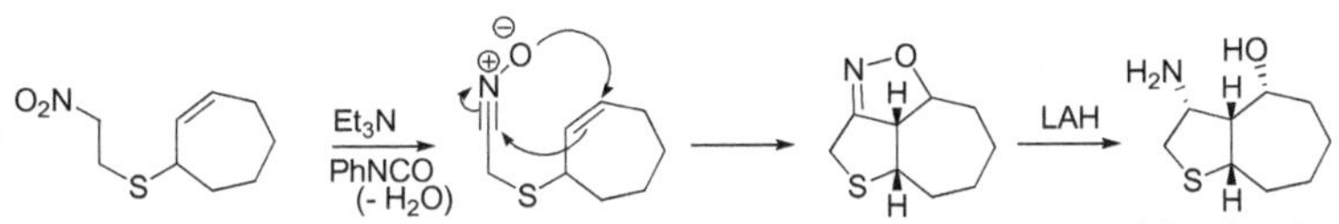

Ref. 137

11. 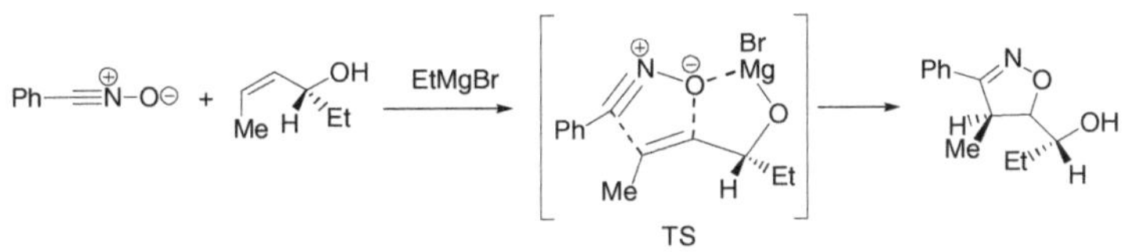
Ref. 138

3.4 Cycloaddition Reactions of More Than Six Electrons Systems: [4+4]-, [6+6]-, [6+4]-, [8+2]-, [12+2]-, and [14 +2]-Cycloadditions

Some cycloaddition reactions of more than six π electron systems have been reported. Thermal suprafacial $\left[\pi^4_S + \pi^4_S\right]$, $\left[\pi^6_S + \pi^2_S\right]$ and $\left[\pi^6_S + \pi^6_S\right]$-cycloadditions are forbidden according to Woodward–Hoffmann rules. These cycloadditions are photochemically allowed processes. Thermal antarafacial $\left[\pi^4_S + \pi^4 a\right]$ addition is possible, but is rare. The following examples are illustrative for [4+4]- and [6+6]--cycloadditions:

1. 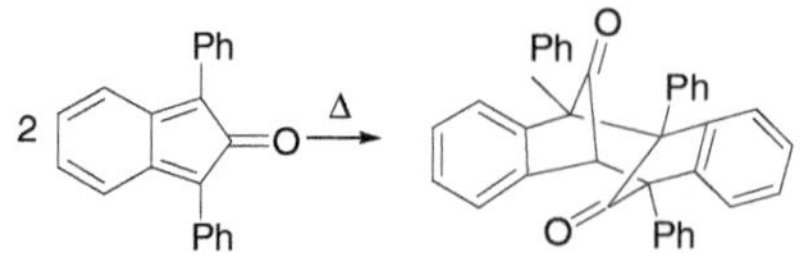

1,3-diphenylisoindenone
Ref. 139

2. 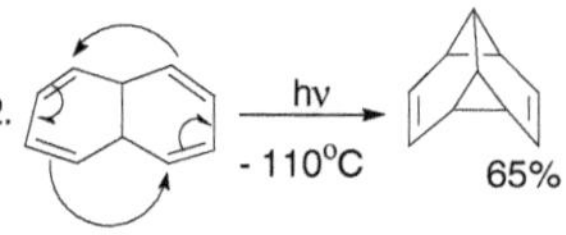

9,10-dihydro-naphthalene
Ref. 5

3. 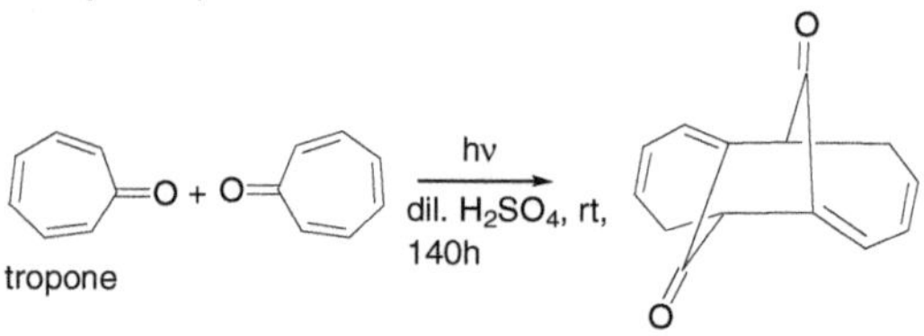

tropone
Ref. 140

Thermal [6+4]- and [12+2]-cycloadditions are allowed processes. Among them, [6+4]-additions are common. These additions are favored than D–A cycloaddition reaction because of the largest coefficients of HOMO and LUMO. The following examples are illustrative:

1. 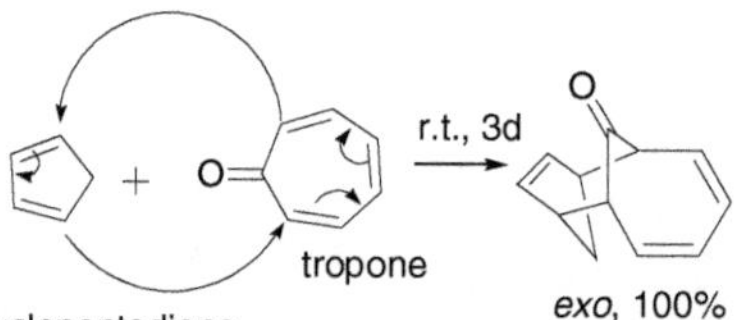

cyclopentadiene
Ref. 141
exo, 100%
(repulsive seconary orbital interaction destabilizes *endo* TS)

2. 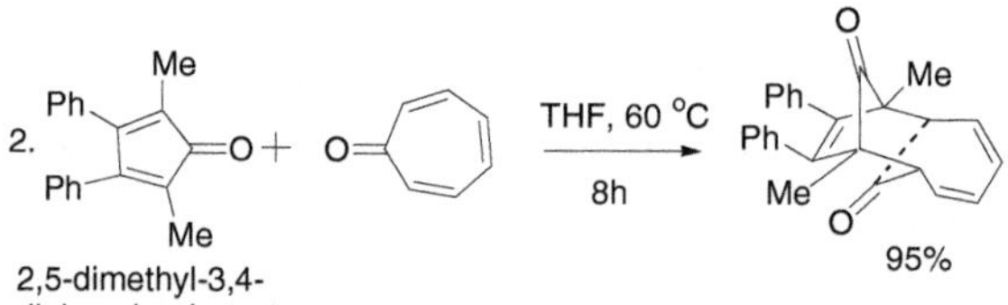

2,5-dimethyl-3,4-diphenylcyclopentadienone
Ref. 142

3. 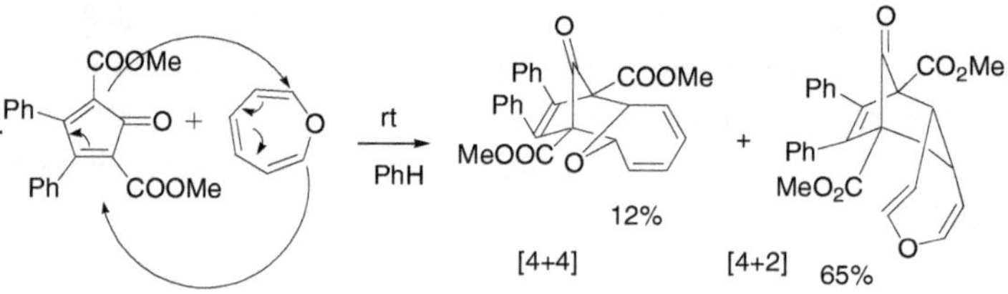

Ref. 143

4. 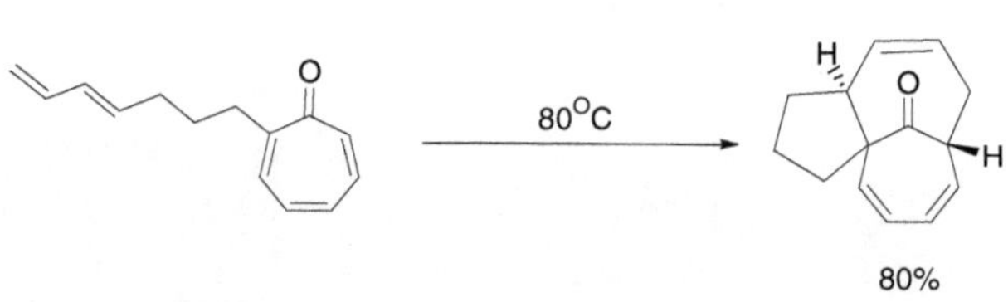

80%
Ref. 144

N-Ethoxycarbonylazepine **139** dimerizes to **140** on heating, where one molecule acts as a 6π system and other as a 4π electron component.

Ref. 145

1,3-Pentadiene reacts with tropone to give [6+4]-cycloadduct **141**.

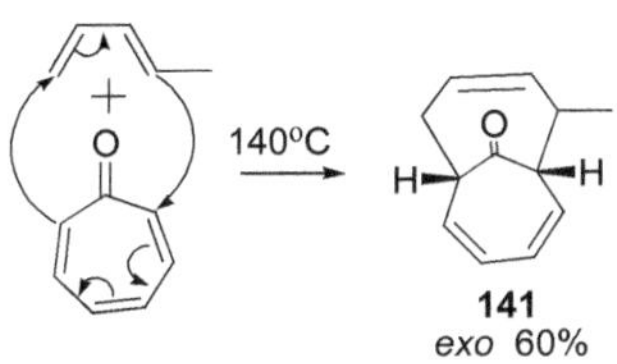

Ref. 146

Conjugated 8π electron systems undergo thermal [8+2]-cycloaddition reactions with 2π electron system. The following examples are illustrative:

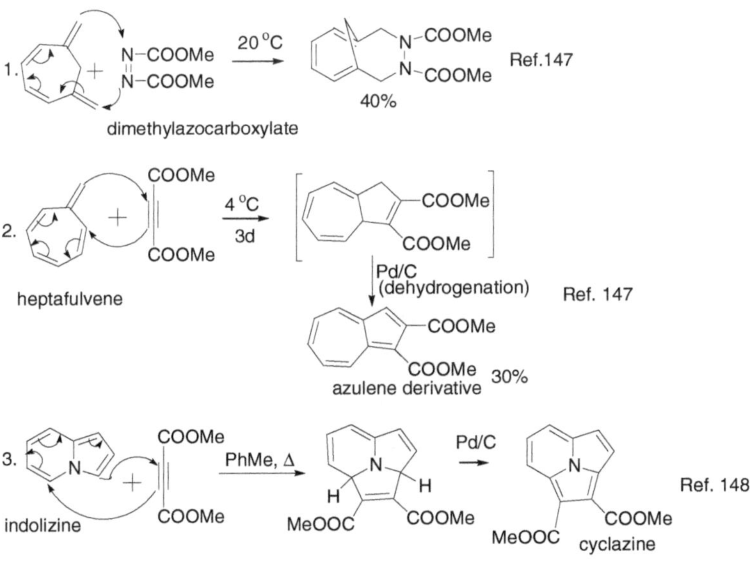

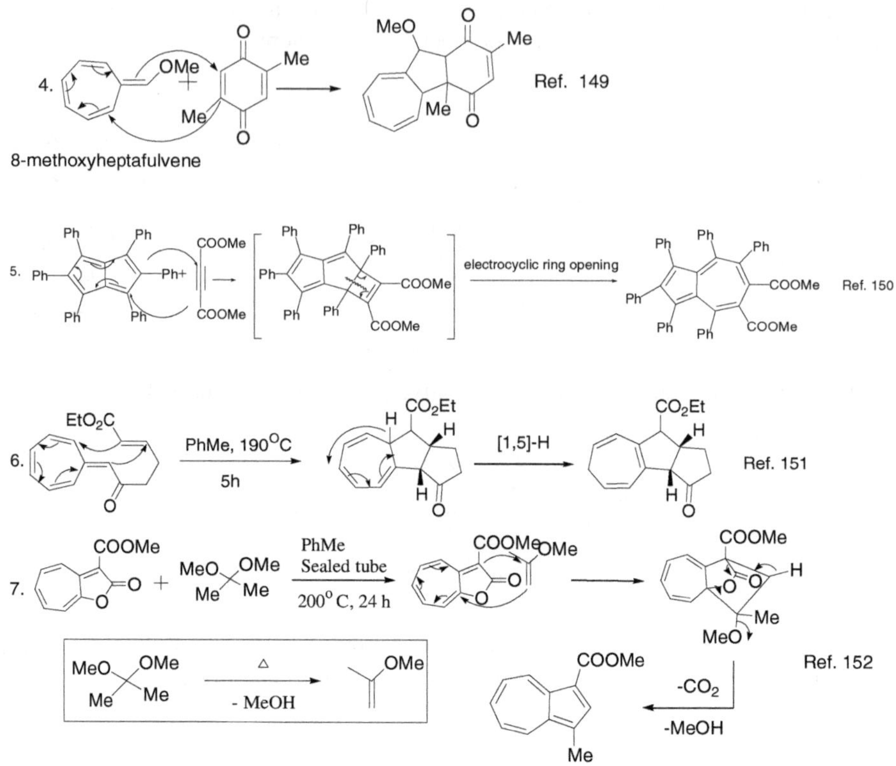

Thermal suprafacial [12+2]-cycloaddition is an allowed process, e.g., **142** undergoes cycloaddition with tetracyanoethylene suprafacially to give **143**.

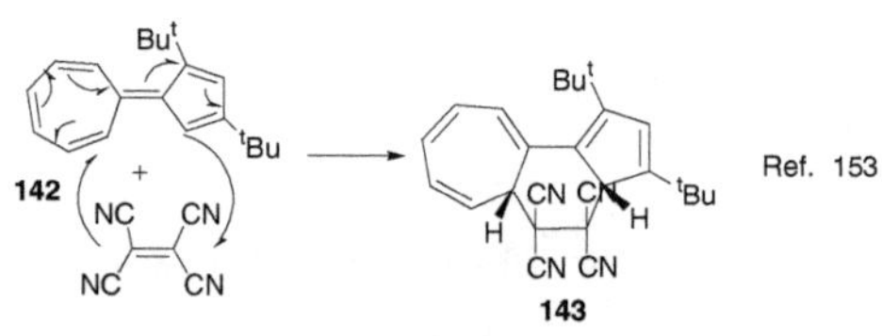

Thermal [14+2]-cycloaddition of heptafulvene **144** with tetracyanoethylene takes place in an antarafacial manner, in which the heptafulvene acts as an antara component.

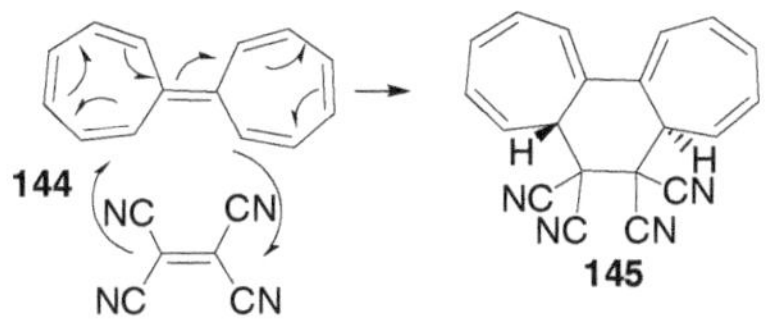

Ref.154

3.5 Cheletropic Reactions

3.5.1 *Overview of Cheletropic Reactions*

The cycloaddition of a conjugated π system to an electrophilic molecule by the formation of two new σ bonds to an atom of the electrophile in a concerted manner is known as cheletropic addition reaction and its reverse process in which two σ bonds are broken from the same atom of the adduct is known as cheletropic elimination reaction. In cheletropic elimination, the driving force is often from the entropic benefit of gaseous elimination of N_2, CO, and SO_2. For example, the cycloaddition of 1,3-butadiene and its derivatives with SO_2 and of alkene with a carbene are cheletropic addition reactions.

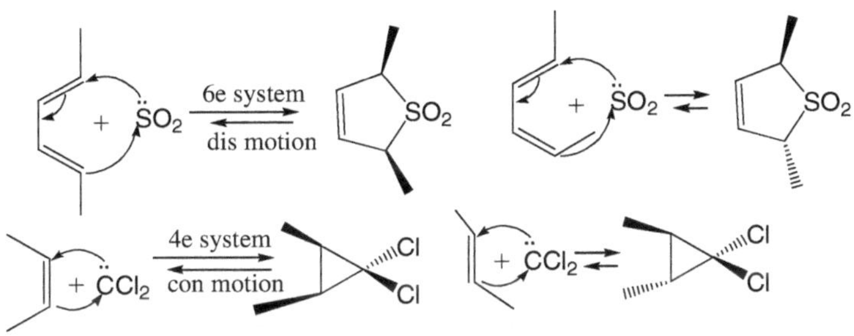

The mechanism of this cycloaddition can be explained by FMO theory, in which one component acts as a HOMO and other as LUMO in a favorable low-energy TS to afford a stereoselective product. The Woodward–Hoffmann rules for electrocyclic reactions are also applied to this cycloaddition reaction. The reaction of an alkene with a carbene is considered as a 4n electron process and of a conjugated diene with an electrophilic molecule as a 4n+2 electron process. Therefore, for thermal reaction of 4n electron process, conrotatory motion of the substituents from the termini of the π system will favor a low-energy TS to afford the product and in photochemical process, the reverse disrotatory mode of motion will be the favored path. Similarly, for a 4n + 2 electron process, disrotatory mode is a symmetry allowed process in thermal reaction and conrotatory mode for its photochemical reaction.

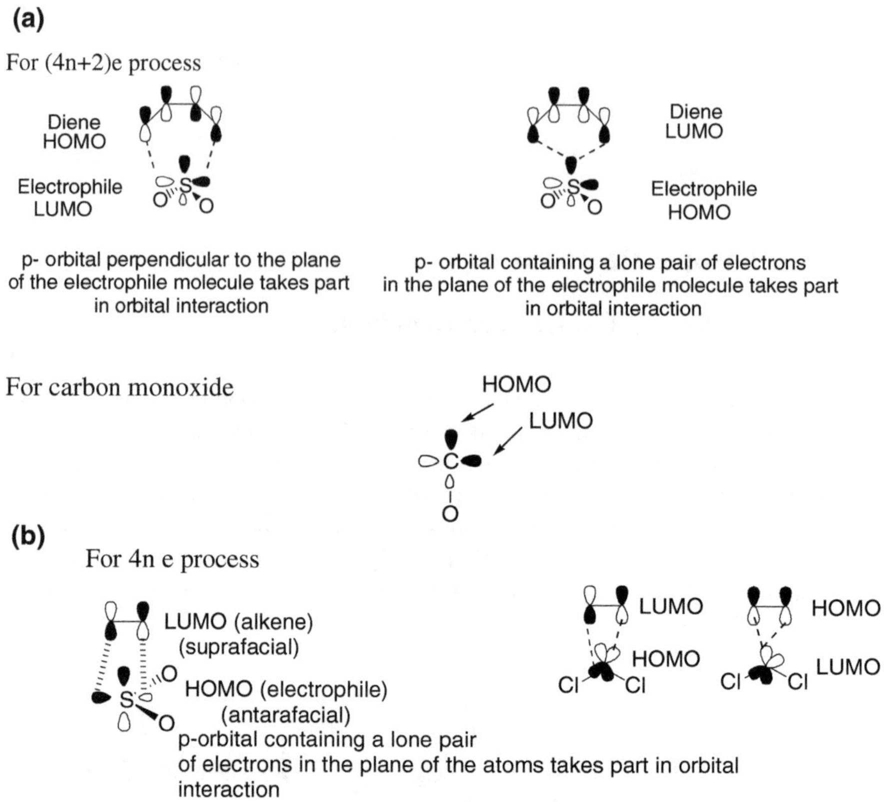

Fig. 3.14 Orbital interactions in the TS for cheletropic addition reactions in (4n+2) and 4n electron systems

The orbital symmetry allowed HOMO and LUMO interaction for the reactions of 4n+2, and 4n processes can be explained as follows (Fig. 3.14)

3.5.2 Applications of Cheletropic Reactions

The cheletropic elimination reactions are widely used for generation of relatively unstable dienes in situ for Diels–Alder reactions. For example, extrusion of nitrogen from diazene **146**, and sulfur dioxide from 2,5-dihydrothiophene-1,1-dioxide **147** and 2,7-dihydrothiepin-1,1-dioxide **148** generate diene and triene for Diels–Alder reactions.

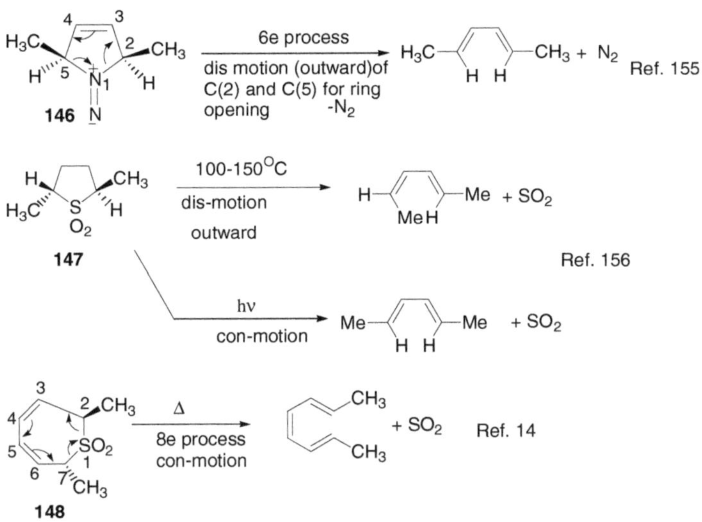

Extrusion of carbon monoxide from Diels–Alder adducts **149** and **150** gives benzene and cyclopentanone derivatives [157, 158].

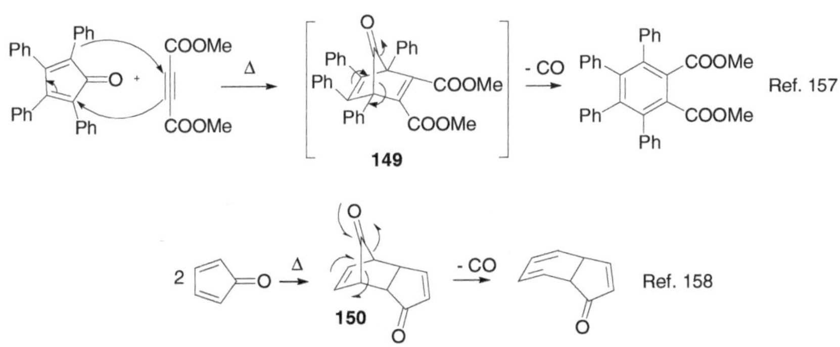

Cheletropic eliminations of SO_2 from cycloadducts **151–155** generate dienes and triene, which are usedin Diels–Alder and electrocyclic reactions.

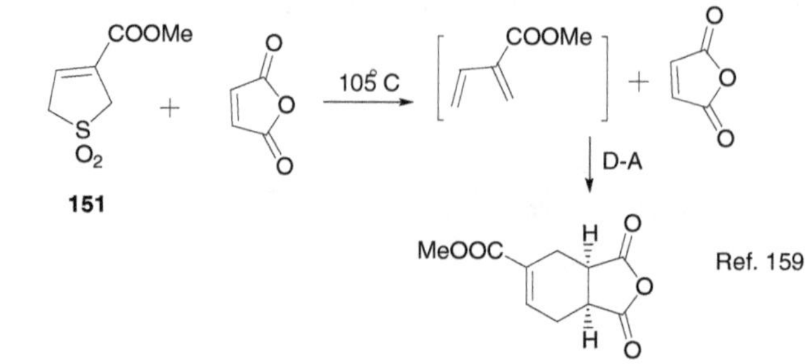

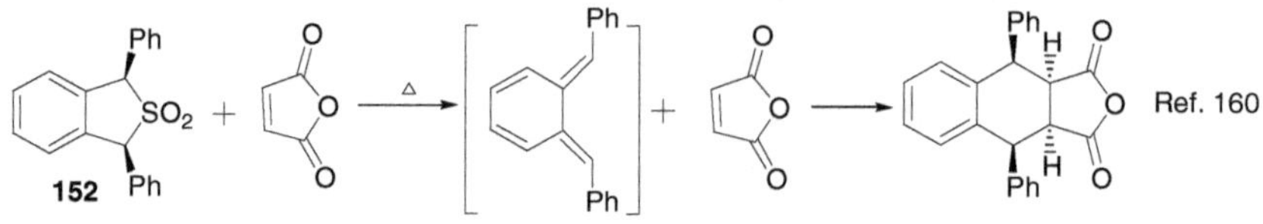

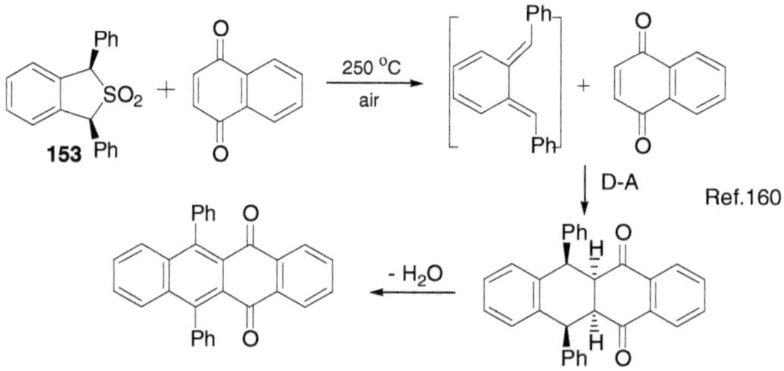

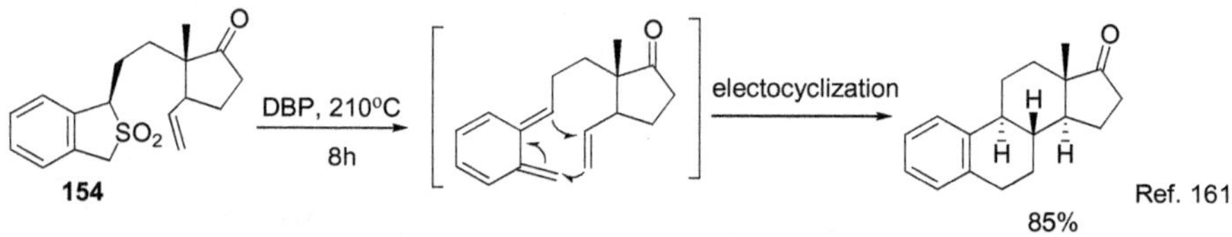

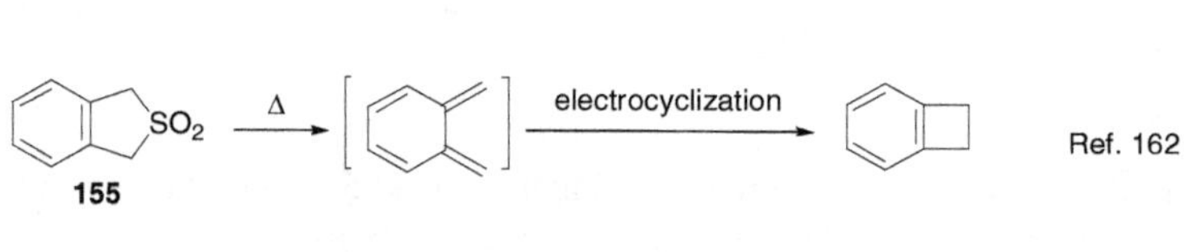

Cheletropic addition of SO_2 to a polyene depends on the geometry of the intrusion adduct. For example, cheletropic addition of SO_2 to 1,3,5-cyclooctatriene **156** gives cycloadduct **157** using 4π e system in a disrotatory motion process [163].

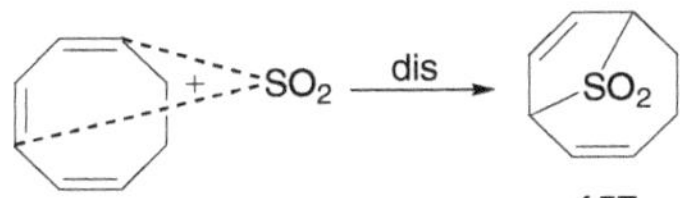

156 157

Similarly, the reverse process of extrusion is more facile for 6e process. For example, the adducts **158** and **159**, on heating above 300 °C undergo extrusion process; the rate of **158** is about 60,000 times faster than **159** [163].

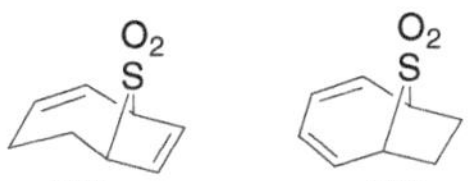

158 159

Cheletropic addition of SO_2 to divinylallenes takes place preferably at the most substituted vinyl allene and at the *E*-site of the vinyl groups. For example, divinylallene **160** having both *E*-and *Z*-vinyl units selectively undergoes addition at the *E*-site to give **161** [164].

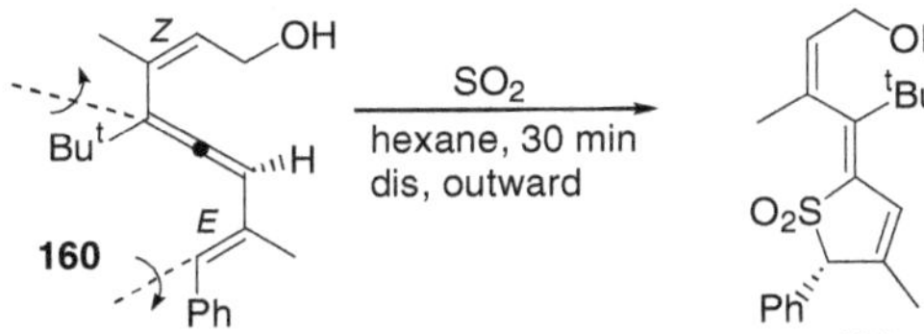

160 161 79% Ref. 164

3.6 Problems

3.6.1 Suggest a mechanism for each of the following transformations. More than one pericyclic step may be involved in each case.

(a)
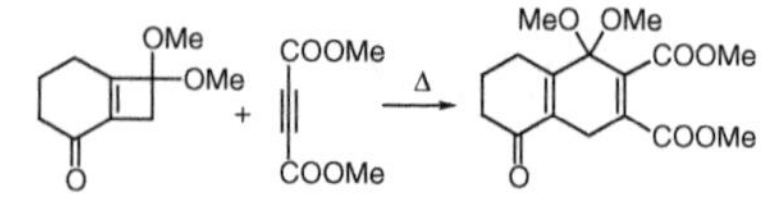

(b)
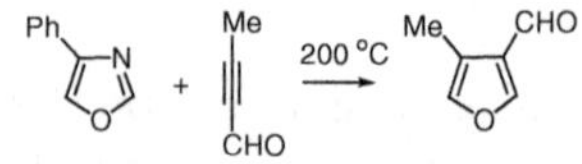

(c)
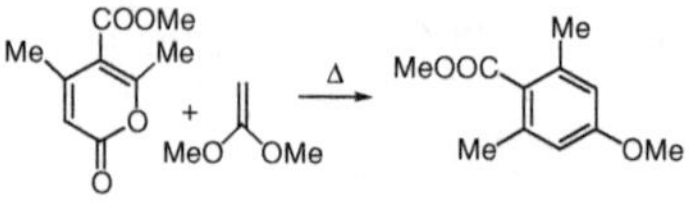

(d)
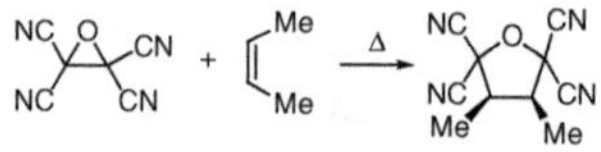

(e)
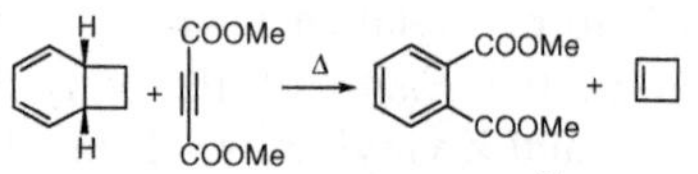

(f)
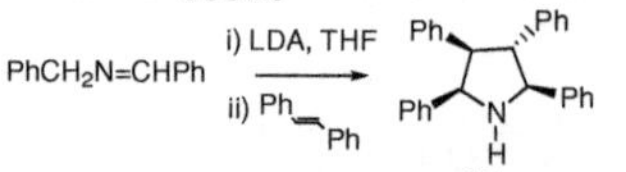

(g)
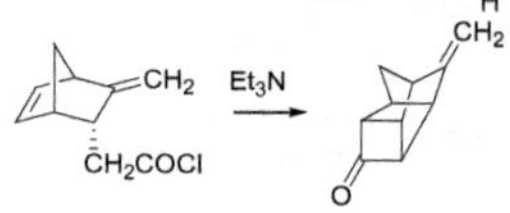

(h)
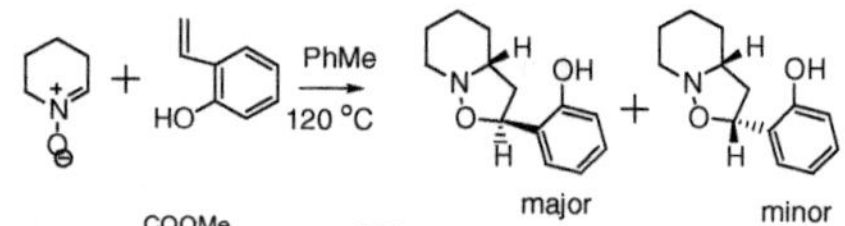

(i)
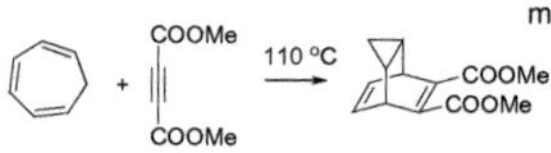

(j)
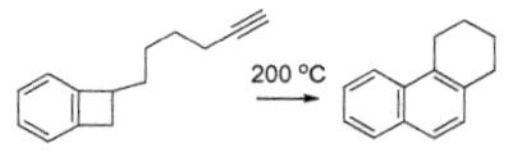

(k)
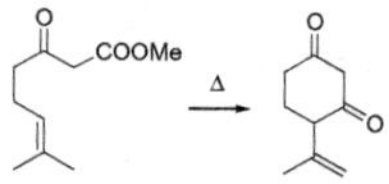

(l)
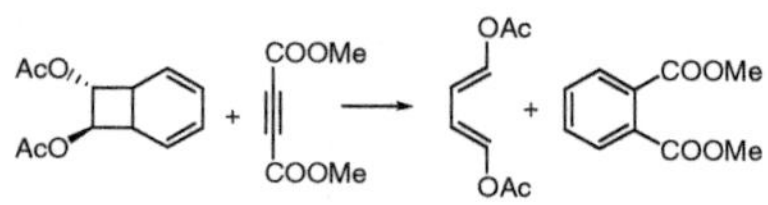

(m)
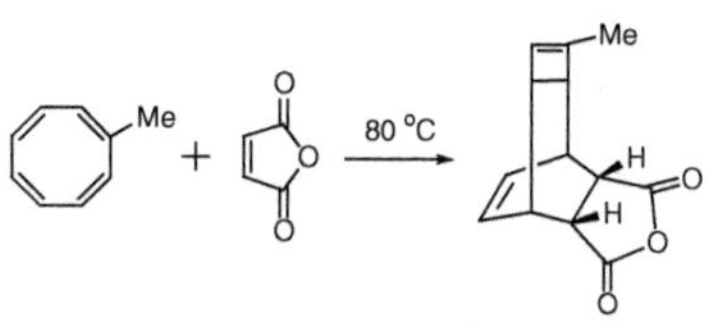

(n)
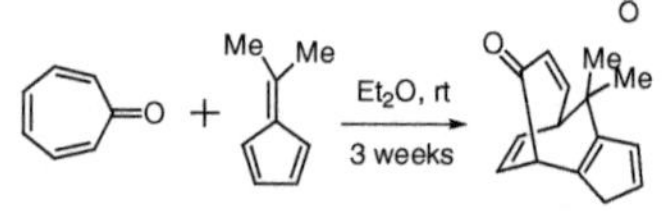

3.6.2 Predict the structure of expected product including stereochemistry in each reaction and indicated the basis of your prediction

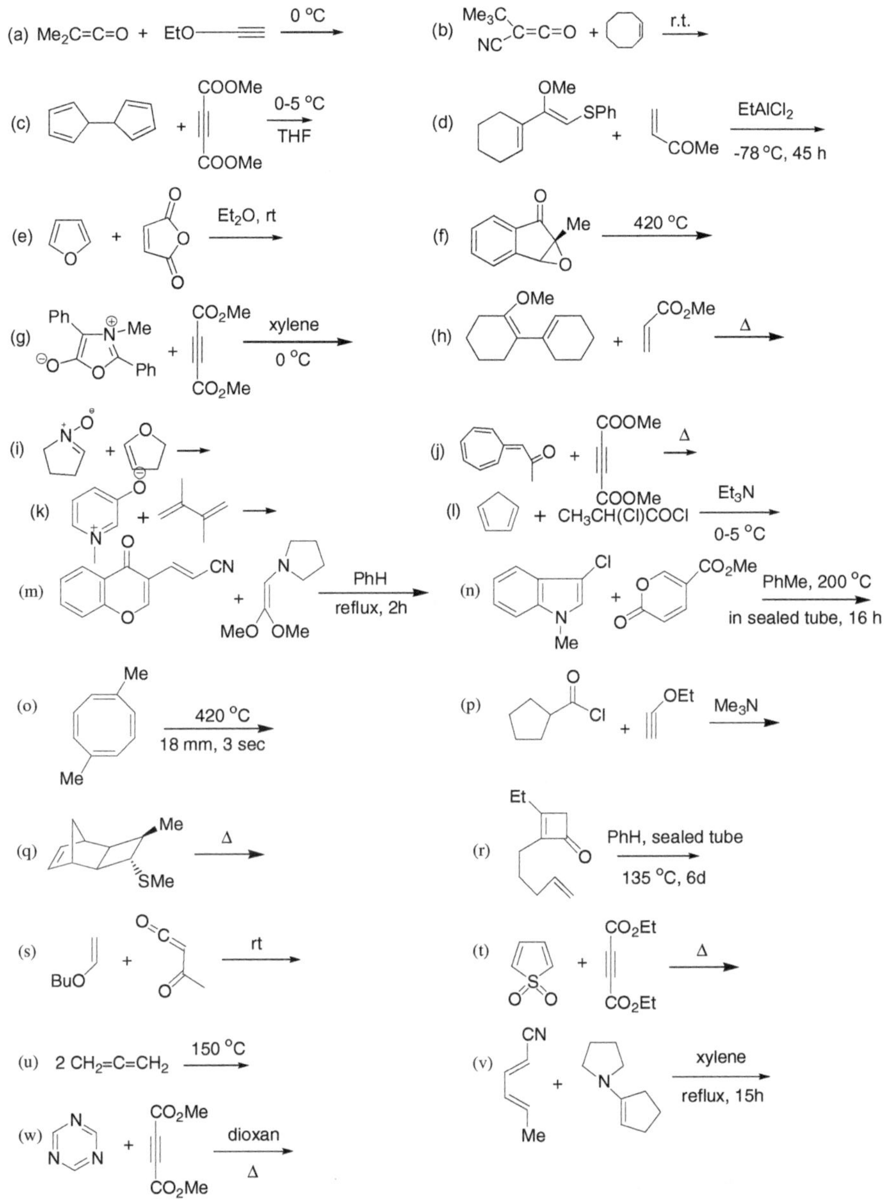

3.6.3 Predict the product(s) in the following reaction sequences and justify your answer.

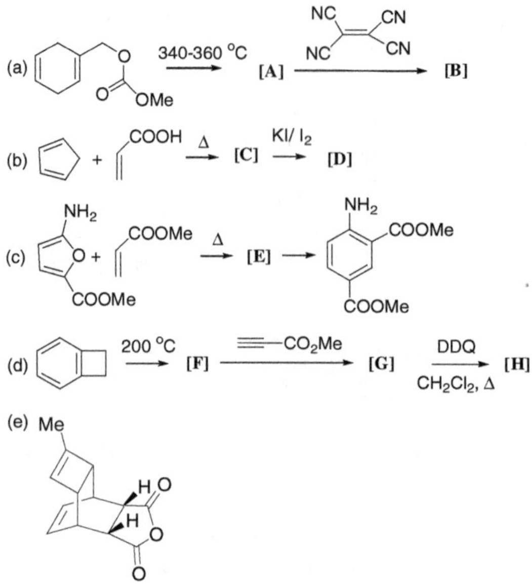

3.6.4 Outline the synthesis of the following compounds

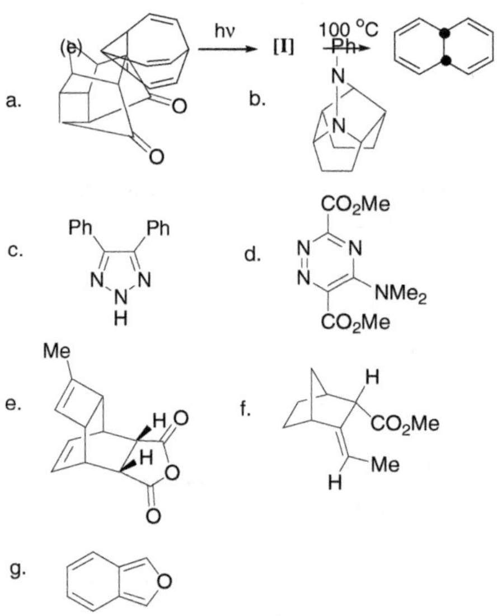

3.7 Further Reading

1. Padwa A (1984) 1,3-dipolar cycloaddition chemistry. Wiley, New York.
2. Taber DF (1984) Intramolecular Diels–Alder and alder ene reactions. Springer, Berlin.

3. Hamer J (1967) 1,4-cycloaddition reactions: the Diels–Alder reactions in heterocyclic syntheses. Academic Press, New York.
4. Marchand AP, Lehr RE (1977) Pericyclic reactions, vols 1 & 2. Academic Press, New York.
5. Gothelf KV, Jorgensen KA (1998) Asymmetric 1,3-dipolar cycloaddition reactions. Chem Rev 98:863.

References

1. Pasto DJ (1979) J Am Chem Soc 101:37; Brady WT, Roe R, Jr (1971). J Am Chem Soc 93:1662
2. DoMinh T, Strausz OP (1970) J Am Chem Soc 92:1766
3. Rey M, Roberts SM, Dreiding AS, Roussel A, Vanlierde H, Toppet S, Ghosez l (1982) Helv Chim Acta 65:703; Wasserman HH, Piper JU, Dehmlow EV (1973) J Org Chem 38:1451
4. Rey M, Roberts S, Dieffenbacher A (1970) Helv Chim Acta 53:417
5. Fleming I (2002) Pericyclic reactions. Oxford University Press, New York
6. Corey EJ, Desai MC (1985) Tetrahedron Lett 26:3535; Metzger C, Borrmann D, Wegler R (1967) Chem Ber 100:1817
7. Masamune S, Fukumoto K (1965) Tetrahedron Lett 4647
8. Williams JK, Wiley DW, Mckusick BC (1962) J Am Chem Soc 84:2210
9. Palomo C, Aizpurua JM, Ganboa I, Oiarbide M (1999) Eur J Org Chem 3223
10. Farina V, Hauck SI, Walker DG (1992) Synlett 761
11. Yamasaki H, Cvetanovic RJ (1969) J Am Chem Soc 91:520
12. Chapman OL, Adams WR (1968) J Am Chem Soc 99:2333
13. Dauben WG, Cargill RL (1961) Tetrahedron 15:197
14. Norman ROC, Coxon JM (1993) Principles of organic synthesis. Chapman & Hall, Oxford
15. Srinivasan P (1971) Org Photochem Synth 1:101
16. Oppolzer W (1982) Acc Chem Res 15:135; Winkler JD, McLaughlin EC (2005) Org Lett 7:227
17. Unett DJ, Caldwell RA, Hrncir DC (1996) J Am Chem Soc 118:1682
18. Corey EJ, Bass JD, Le Mahieu R, Mitra RB (1964) J Am Chem Soc 86:5570; de Mayo P (1971) Acc Chem Res 4:41
19. Diels O, Alder K (1928) Justus Liebigs Ann Chem 460:98
20. Cope C, Herrichy EC (1963) Org Syn IV:890
21. Wasserman A (1935) J Chem Soc 1511
22. Sustmann R (1974) Pure Appl Chem 40:569; (1971) Tetrahedron Lett 2717
23. Berson JA, Hamlet Z, Mueller WA (1962) J Am Chem Soc 84:297
24. Bodwell GJ, Pi Z (1997) Tetrahedron Lett 38:309
25. Sauer J (1967) Angew Chem Int Ed Engl 6:16
26. De Graw JI, Goodman L, Baker BR (1961) J Org Chem 26:1156
27. Spino C, Crawford J, Cui Y, Gugelchuk M (1998) J Chem Soc. Perkin Trans 2:1499
28. Inukai T, Kojima T (1971) J Org Chem 36:924
29. Inukai T, Kojima T (1966) J Org Chem 31:1121
30. Houk KN (1975) Acc Chem Res 8:361; (1979) Topics Current Chem 79:1; Sustmann R, Schubert R (1972) Angew Chem Int Ed Engl 11:840
31. Longuet-Higgins HC, Abrahamson EW (1965) J Am Chem Soc 87:2045
32. Zimmerman HE (1971) Acc Chem Res 4:272
33. Stephenson lM, Smith DE, Current SP (1982) J Org Chem 47:4170
34. Craig D, Shipman JJ, Fowler RB (1961) J Am Chem Soc 83:2885

35. Backer HJ (1939) Rec Trav Chim Pays-Bas 58:643
36. Stewart CA Jr (1963) J Org Chem 28:3320
37. Domingo LR, Jose Aurell M, Perez P, Contreras R (2002) Tetrahedron 58:4417
38. Rucker C, Lang D, Sauer J, Friege H, Sustmann R (1980) Chem Ber 113:1663
39. Paquette LA (1964) J Org Chem 29:3447
40. Ionescu D, Silverton JV, Dickinson LC, Miller B (1996) Tetrahedron Lett 37:1559
41. Wittig G (1962) Angew Chem Int Ed 1:415
42. Dewar MJS, de Llano D (1969) 91;789
43. Ciganek E (1967) Tetrahedron Lett 3321
44. Hamer J, Ahmad M (1967) In: Hamer J (ed) 1,4-cycloaddition reactions. Academic Press, New York
45. Domingo LR, Andres J, Alves CN (2002) Eur J Org Chem 2557
46. Atterton JCC, Jones S (2003) Tetrahedron 59:9039
47. Woodward RB, Sondheimer F, Taub D, Heusler K, Mc Lamore WM (1952) J Am Chem Soc 74:4223
48. Ranganathan S, Ranganathan D, Mehrotra AK (1974) J Am Chem Soc 96:5261
49. Corey EJ, Weinshenker NM, Schaaf TK, Huber W (1969) J Am Chem Soc 91:5675
50. Bartlett PA, Green FR, Webb TR (1977) Tetrahedron Lett 331; Ranganathan D, Rao CB, Ranganathan S, Mehrotra AK, Iyengar R (1980) J Org Chem 45:1185
51. Ronan B, Kagan HB (1992) Tetrahedron Asymm 3:115
52. Paquette LA, Moerck RE, Harirchian B, Magnus PD (1978) J Am Chem Soc 100:1597
53. Davis AP, Whitham GH (1980) J Chem Soc Chem Commun 639
54. Girard C, Bloch R (1982) Tetrahedron Lett 23:3683
55. Carruthers W (1993) Some modern methods of organic synthesis. Cambridge University Press, New Delhi
56. Ciganek E (1967) Tetrahedron Lett 3321
57. Fringuelli F, Tatiedri A (1990) Dienes in diels alder reactions. Wiley, New York; Alston PV, Ottenbrite RM, Newby J (1979) J Org Chem 44:4939
58. Deem ML (1972) Synthesis 675; Oppolzer W, Chapuis C (1983) Tetrahedron Lett 24:4665
59. Trost BM, Bridges AJ (1976) J Am Chem Soc 98:5017
60. Oppolzer W (1977) Angew Chem Int Ed Engl 16:10
61. Arnold BJ, Sammes PG, Wallace TW (1974) J Chem Soc Perkin Trans 1(409):415
62. Mann J, Piper SE (1982) J Chem Soc, Chem Commun 430
63. Wittig G, Burger TF (1966) 0). Justus Liebigs Ann Chem 632:85
64. Ono N, Miyaki H, Kaji A (1982) Chem Commun 33
65. Singh J (2009) Photochemistry and pericyclic reactions 3rd edn. New Academic Science
66. Ireland RE, Daub JP (1983) J Org Chem 48:1303
67. Danishefsky S, Kitahara T (1974) J Am Chem Soc 96:7807
68. Danishefsky S, Singh RK, Gammill RB (1978) J Org Chem 43:379
69. Ross AG, Townshed SD, Danishefsky S (2013) J Org Chem 78:204
70. Serckx Poncin B, Hesbain- Prizque AM, Ghosez L (1982) Tetrahedron Lett 23:3261
71. Demoulin A, Gorissen H, Hesbain- Frisque AM, Ghosez L (1975) J Am Chem Soc 97:4409
72. Harris EE, Firestone RA, Pfister K, Boettcher RR, Cross FJ, Currie R (1962) J Org Chem 27:2705
73. Hamer J, Ahmad M (1967) In: Harmer J (ed) 1,4 Cycloaddition reactions. Academic Press, New York; Mc Clure KM, Danishefsky SJ (1991) J Org Chem 56:850
74. Weinreb SM, Staib RR (1982) Tetrahedron 39:3087
75. Gillis BT (1967) In: Hamer J (ed) 1,4 cycloaddition reactions. Academic Press, New York
76. Jurczak J, Zamojski A (1972) Tetrahedron 28:1505
77. Hoffmann HMR, Joy DR, Suter AK (1968) J Chem Soc B 57; Hoffmann HMR, Joy DR (1968) J Chem Soc B 1182
78. Shimizu N, Tanaka M, Tsuno Y (1982) J Am Chem Soc 104:1330

79. Turro NJ, Adelson SS, Williams JR, Darling TR, Hammond WB (1969) J Am Chem Soc 91:2283; Elderson SS, Turro NJ (1970) J Am Chem Soc 92:2770; Chan CH, Li MP, Mychajlowskij W, Harp DN (1974) Tetrahedron Lett 35
80. Kauffmann T (1974) Angew Chem Int Ed 13:627
81. Kagan HB, Riant O (1992) Chem Rev 92:1007; Oppolzer W (1984) Angew Chem Int Ed Engl 23:876
82. Poll T, Helmchen G, Bauer B (1984) Tetrahedron Lett 2191
83. Avenoza A, Cativiela C, Mayoral JA, Peregrina JM (1992) Tetrahedron Asymmetry 3:913; Cativiela C, Mayoral JA, Avenoza A, Peregrina JM, Lahoz FJ, Gimeno S (1992) J Org Chem 57:4664
84. Corey EJ, Loh TP (1991) J Am Chem Soc 113:8966
85. Yao S, Johannsen M, Audrain H, Hazell RG, Jorgensen KA (1998) J Am Chem Soc 120:8599
86. Mikami K, Motoyama Y, Terada M (1994) J Am Chem Soc 116:2812
87. Kobayashi S (1998) Pure Appl Chem 70:1019
88. Zhang X, Du H, Wang Z, Wu YD, Ding K (2006) J Org Chem 71:2862
89. Pegot B, Buu ONV, Gori D, Vo-Thanh G (2006) Beilstein J Org Chem 2:18
90. Taber DF, Gunn BP (1979) J Am Chem Soc 101:3992; Kitahara T, Kurata H, Matsuoka T, Mori K (1985) Tetrahedron 41:5475
91. Wilson SR, Mao DT (1978) J Am Chem Soc 100:6289
92. Yuki K, Shindo M, Shishido K (2001) Tetrahedron Lett 42:2517
93. Roush WR, Gillis HR, Ko AI (1982) J Am Chem Soc 104:2269
94. Shea KJ, Wise S, Burke LD, Davis PD, Gilman JW, Greeley AC (1982) J Am Chem Soc 104:5708
95. Shea KJ, Gilman JW (1983) Tetrahedron Lett 24:657
96. Roush WR, Gillis HR (1982) J Org Chem 47:4825
97. Gricco PA, Takigawa T, Schilling WJ (1980) J Org Chem 45:2247
98. Yeh MCP, Tsao WC, Liu CJ (2005) J Chin Chem Soc 52:2247; Harmata M (2001) Acc Chem Res 34:595
99. Gaitzsch J, Rogachev VO, Metz P, Yusubov MS, Filimonov VD, Kataeva O (2009) J Sulfur Chem 30:4
100. Alder K, Muders R, Krane W, Wirtz P (1959) Ann 627:58
101. Starck F, Jones PG, Herges R (1998) Eur J Org Chem 2533
102. Boger DL, Duff SR, Panck JS, Yasuda M (1985) J Org Chem 50:5782
103. Jones M Jr, Reich SD, Scott LT (1970) J Am Chem Soc 92:3118
104. Kadaba PK (1969) Tetrahedron 25:3053; Huisgen R, Szeimies G, Mobius L (1967) Chem Ber 100:2494
105. Bihlmaier W, Geittner J, Huisgen R, Reissig HU (1978) Heterocycles 10:147
106. Huisgen R, Eberhard P (1971) Tetrahedron Lett 4343
107. Sustmann R, Trill H (1972) Angew Chem Int Ed Engl 11:838
108. Houk KN, Sims J, Duke BE, Jr, Strozier RW, George JK (1973) J Am Chem Soc 95:7287; Sims J, Houk KN (1973) J Am Chem Soc 95:5798
109. Huisgen R, Szeimies G, Mobius L (1967) Chem Ber 100:2494
110. Kanemasa S, Ueno N, Shirahase M (2002) Tetrahedron Lett 43:657
111. Jensen KB, Gothelf KV, Hazell RG, Jorgensen KA (1997) J Org Chem 62:2471
112. Chiacchio U, Casuscelli F, Corsaro A, Rescifina A, Romeo G, Uccella N (1994) Tetrahedron 50:6671
113. Simonsen KB, Bayon P, Hazel RG, Gothelf KV, Jorgensen KA (1999) J Am Chem Soc 121, 3845
114. Kozikowski AP, Ghosh AK (1982) J Am Chem Soc 104:5788
115. Huisen R, Hauck H, Grashey R, Seidl H (1968) Chem Ber 101:2568
116. Huisgen R, Szeimies G (1965) Chem Ber 98:1153
117. Scheiner P, Schomaker JH, Deming S, Libbey WJ, Nowack GP (1965) J Am Chem Soc 87:306

118. Curran DP (1982) J Am Chem Soc 104:4024
119. Huisgen R (1963) Angew Chem Int Ed 2:633
120. Padwa A (1983) 1,3-dipolar cycloaddition chemistry, general heterocyclic chemistry series 1. Wiley, New York, pp 141–145
121. Huisgen R, Scheer W, Szeimies G, Huber H (1966) Tetrahedron Lett 397
122. Ugarriza I, Uria U, Reyes E, Carillo L, Vicario JL (2015) Asymmetric Catal 2:7
123. Huisgen R, Hauck H, Grashey R, Scidl H (1968) Chem Ber 101:2568
124. Joucla M (1973) Tetrahedron 29:2315
125. Winterfeldt E, Krohn W, Stracke H (1969) Chem Ber 102:2346
126. Huisgen R (1977) Angew Chem Int Ed Engl 16:572
127. Toyoda J, Ibata T, Tamura H, Ogawa K, Nishino T, Takebayashi M (1985) Bull Chem Soc Jpn 58: 2212; Ibata T, Toyada J, Sawada M, Tanaka T (1986) J Chem Soc, Chem Commun 1266
128. Tufariello JJ, Mullen GB, Tegeler JJ, Trybulski EJ, Wong SC, Ali SA (1979) J Am Chem Soc 101:2435
129. Tufariello JJ (1979) Acc Chem Res 12:396
130. Le Bel NA, Hwang D (1978) Org Synth 58:106
131. LeBel NA, Balasubramanian N (1989) J Am Chem Soc 111:3363
132. Roussi G (1990) Heterocycles 31:1445
133. Seitz G, Tegethoff R (1993) Arch Pharm 326:135
134. Novikov MS, Khlebnikov AF, Besedina OV, Kostikov RR (2001) Tetrahedron Lett 42:533
135. Engel E, Steglich W (1978) Angew Chem 90:719; ibid(1978) Angew Chem Int Ed Engl 17:676
136. Kozikowski AP, Stein PD (1982) J Am Chem Soc 104:4023
137. Confalone PN, Lollar ED, Pizzolato G (1978) J Am Chem Soc 100:6291
138. Kanemasa S (2002) Synlett 1371
139. Holland JM, Jones DW (1969) Chem Commun 587
140. Mukai T, Tezuka T, Akasaki Y (1966) J Am Chem Soc 88:5025
141. Mukherjee D, Watts CR, Houk KN (1978) J Org Chem 43:817
142. Houk KN, Woodward RB (1970) J Am Chem Soc 92:4145
143. Ban T, Wakita Y, Kanematsu K (1980) J Am Chem Soc 102:5416
144. Funk RL, Bolton GL (1986) J Am Chem Soc 108:4655
145. Paquette LA, Barrett JH (1966) J Am Chem Soc 88:2590
146. Rigby JW, Ateeq HS, Charles NR, Cuisiat SV, Ferguson MD (1993) J Am Chem Soc 115:1382
147. Doering WVE, Wiley DW (1960) Tetrahedron 11:183
148. Galbraith A, Small T, Barnes RA, Boekelheide V (1961) J Am Chem Soc 83:453
149. Bindl J, Burgemeister T, Daub J (1985) Justus Liebigs Ann Chem 1346
150. Le Goff E (1962) J Am Chem Soc 84:3975
151. Liu CY, Houk KN (1987) Tetrahedron Lett 28:1367
152. Pham W, Weissleder R, Tung CH (2002) Tetrahedron Lett 43:19
153. Prinzbach H, Knoefel H (1969) Angew Chem Int Ed Engl 8:881
154. Woodward RB, Hoffmann R (1971) The conservation of orbital symmetry. Academic Press, New York, p 85
155. Lemal DM, Mc Gregor SD (1966) J Am Chem Soc 88:1335
156. Mock WL (1966) J Am Chem Soc 88:2857; Mc Gregor SD, Lemal DM (1966) J Am Chem Soc 88:2858
157. Ogliaruso MA, Romanelli MG, Becker EI (1965) Chem Rev 65:261
158. Allen CFH, Van Allen JA (1950) J Am Chem Soc 72:5165
159. Mc Intosh JM, Sieler RA (1978) J Org Chem 43:4431
160. Cava MP, Mitchell MJ, Deana AA (1960) J Org Chem 25:1481
161. Nicolaou KC, Barnette WE, Ma P (1980) J Org Chem 45:1463
162. Cava MP (1959) J Am Chem Soc 81:4266
163. Isaac NS, Laila AAR (1976) Tetrahedron Lett 17:715
164. Souto JA, Lopez CS, Faza ON, Alvarez R, de Lera AR (2005) Org Lett 7:1565

Chapter 4
Sigmatropic Rearrangements

4.1 Introduction

Sigmatropic rearrangements are another class of pericyclic reactions which are governed by the orbital symmetry. Woodward and Hoffmann coined the term 'sigmatropic shifts' because one sigma-bonded atom or group is shifted from its allylic position in these rearrangements [1, 2]. This rearrangement involves the shift of one sigma-bonded atom or group from its allylic type position to the distant end of the conjugated π system followed by simultaneous shift of π electrons. The rearrangement is described by the order [i, j], where i specifies the number of atom or atoms shifted and j specifies the number of atoms in the π system that are directly involved in the bonding changes. If the sigma-bonded atom or group shifts from one surface of the conjugated system and arrives at the other end of the same surface, the rearrangement is described as suprafacial and is an orbital symmetry allowed pathway for the total number of [4n+2] electrons [1]. Alternatively, if it leaves from one surface of the π system and arrives at the opposite surface, the rearrangement is called antarafacial, and this is an orbital symmetry allowed pathway when the total number of electrons in the process is 4n, where n is a positive integer. The topological properties of the interacting orbitals dictate the facility of most of the sigmatropic rearrangements and their stereochemistry. The sigmatropic rearrangements are classified on the basis of their orders. The major types of sigmatropic rearrangements are illustrated by general examples in Scheme 4.1.

B. Dinda, *Essentials of Pericyclic and Photochemical Reactions*,
Lecture Notes in Chemistry 93, DOI 10.1007/978-3-319-45934-9_4

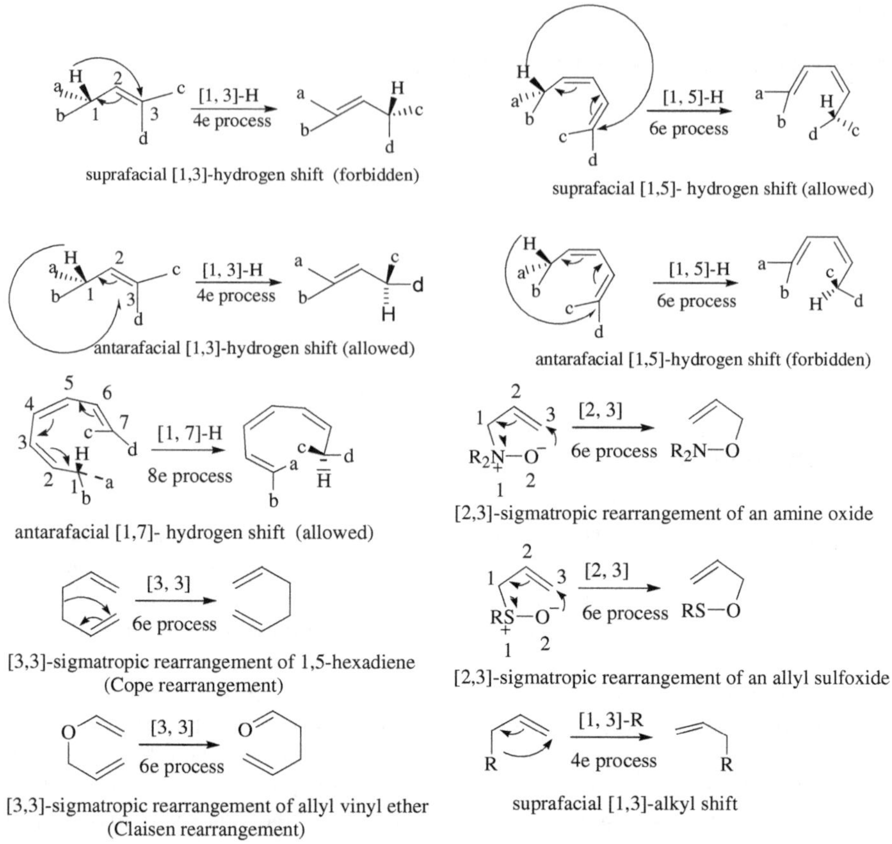

Scheme 4.1 Major types of sigmatropic rearrangements

4.2 Orbital Symmetry Basis for Allowed and Forbidden Sigmatropic Rearrangements and Their Stereochemistry

4.2.1 Orbital Symmetry Analysis of [1,3]-, [1,5]-, and [1,7]-Sigmatropic Shifts of Hydrogen and Alkyl Groups

The orbital symmetry study of sigmatropic reactions will help us to understand the stereochemistry of these reactions. An FMO analysis of this process indicates the interaction between the frontier orbitals of the π system and orbitals of the migrating atom or atoms. Let us consider the simplest case of 1,3-sigmatropic shift of hydrogen. In this case, π system is an allyl radical and the migrating atom is hydrogen. Their frontier orbitals are allyl ψ_2 and hydrogen 1s for thermal reaction

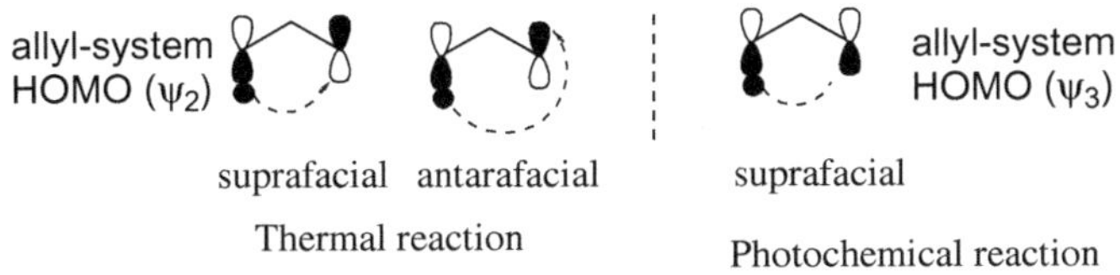

Fig. 4.1 Orbital interactions in thermal and photochemical reactions of [1,3]-sigmatropic hydrogen shift

and allyl orbital ψ_3 in photochemical reaction. The interactions between these orbitals are depicted in Fig. 4.1.

It is seen from the figure that in thermal reaction, a bonding interaction (in the same phase) can be maintained only in the antarafacial mode of shift. Therefore, thermal 1,3-suprafacial shift of hydrogen is forbidden from orbital symmetry considerations. The antarafacial shift is orbital symmetry allowed process and will be a concerted process. Photochemically, [1,3]-suprafacial shift of hydrogen is a symmetry allowed process because the bonding interaction takes place in the same phase of allyl group [1, 2].

A similar FMO approach for 1,5-shift of hydrogen may be considered for study of symmetry property of the process. In a thermal reaction of [1,5]-suprafacialsigmatropic hydrogen shift, the interaction of 1s orbital of hydrogen with pentadienyl radical ψ_3 orbital takes place maintaining the orbital symmetry [1, 2].

In thermal reaction, bonding interaction is maintained in the suprafacial mode of 1,5-shift and hence this process is symmetry allowed, while the antarafacial shift is symmetry forbidden. The suprafacial shift also corresponds to a favorable six-electron Huckel-type transition state in thermal reaction, whereas Huckel-type TS for suprafacial [1,3]-sigmatropic hydrogen shift is antiaromatic and is a forbidden process (Fig. 4.2) [1, 2]. Photochemically, [1,5]-hydrogen shift in the suprafacial mode is a symmetry forbidden process, but antarafacial shift is a symmetry allowed process (Fig. 4.3).

Analysis of a 1,7-hydrogen shift process indicates that the suprafacial hydrogen shift is symmetry forbidden in a thermal reaction. Photochemically, [1,7] suprafacial shift of hydrogen is symmetry allowed (Fig. 4.4) [1, 2].

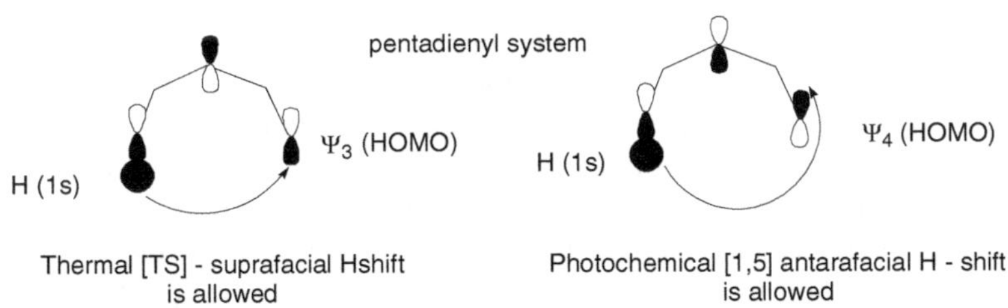

Fig. 4.2 Orbital interactions in thermal and photochemical reactions of [1,5]-sigmatropic hydrogen shift

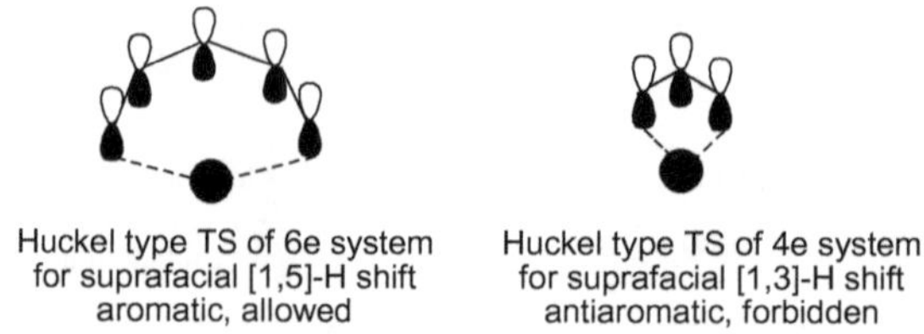

Fig. 4.3 Orbital interactions in Huckel-type TSs for thermal [1,5]-, and [1,3]-sigmatropic hydrogen shifts

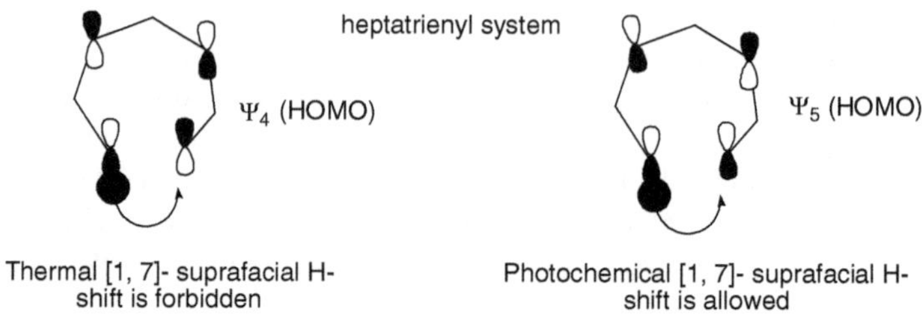

Fig. 4.4 Suprafacial orbital interactions in thermal and photochemical reactions of [1,7]-sigmatropic hydrogen shift

When an allyl group migrates, an additional stereochemical feature arises because of p-orbital of migrating carbon of alkyl group. In this case, the thermal [1,3]-suprafacial alkyl shift is an allowed process with inversion of configuration of migrating alkyl carbon and the thermal suprafacial [1,5]-alkyl shift is also allowed process with retention of configuration of alkyl carbon (Fig. 4.5) [1, 2].

1,7-Alkyl shift is analogous to 1,3-alkyl shift as both are 4ne processes. Hence 1,7-suprafacial alkyl shift with inversion of configuration is symmetry allowed process [1, 2].

4.2.2 *Orbital Symmetry Analysis of [3,3]- and [2,3]-Sigmatropic Rearrangements*

In [3,3]-sigmatropic rearrangements, the suprafacial orbital interactions of two allyl radicals take place. Suprafacial interactions of the orbitals of two allyl radicals produce both chair-like and boat-like TS (depicted below). Both these TSs are thermally allowed, but in most cases preferred TS is chair like (Fig. 4.6) [3]. [3,3]-sigmatropic rearrangements of 1,5-hexadienes known as the Cope rearrangements and of allyl vinyl ethers or allyl ethers of phenols known as the Claisen rearrangements have both chain-like and boat-like TS. The major product is derived from chair-like TS having larger substituent in the pseudoequatorial position.

1,5-suprafacial alkyl shift with
retention of configuration is allowed
6e process, zero node, Huckel topology
aromatic

1,5-suprafacial alkyl shift with
inversion of configuration is forbidden
6e process, one node, Mobius topology
antiaromatic

Fig. 4.5 Orbital interactions in the TSs of thermal reactions of [1,3]- and [1,5]-sigmatropic suprafacial alkyl shifts

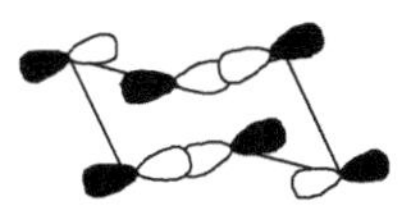

Chair- like TS, 6e process
Huckel topology
aromatic

Boat- like TS, 6e process
Huckel topology
aromatic

Fig. 4.6 Suprafacial orbital interactions in chair- and boat-like TSs in thermal [3,3]-sigmatropic rearrangements

Fig. 4.7 Suprafacial orbital interactions in the TS (Huckel type) of [2,3]-sigmatropic rearrangements

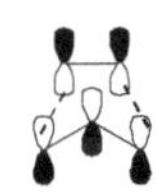

Huckel type TS, 6e, aromatic

The TS is stabilized by hydrogen bonding and the reaction rate increases with increasing solvent polarity. Water is a favorable solvent in many cases.

In [2,3]-sigmatropic rearrangement, the interactions between the orbitals of allylic radical and migrating group take place in a suprafacial manner to produce a TS of Huckel-type topology consisting of 6e. TS is aromatic in nature and symmetry allowed process (Fig. 4.7).

Table 4.1 Woodward-Hoffmann rules for sigmatropic rearrangements

a. Order [*i*, *j*], *i* = 1	Thermal allowed	Photochemical allowed
4*n*	Supra/inversion, antara/retention	Supra/retention, antara/inversion
4*n*+2	Supra/retention, antara/inversion	Supra/inversion, antara/retention
b. Order [*i*, *j*], *i* > 1		
4*n*	Supra/antara	Supra/supra, antara/antara
4*n*+2	Supra/supra, antara/antara	Supra/antara

Analysis of these sigmatropic processes using Woodward–Hoffmann rules leads to selection rules for sigmatropic rearrangements (Table 4.1) [1, 2]. For sigmatropic rearrangements of order [i, j], where i > 1, the suprafacial or antarafacial nature of migration for both the components should be specified.

4.3 [1,3]-, [1,5]-, and [1,7]-Sigmatropic Hydrogen and Alkyl Shifts and Their Applications

4.3.1 [1,3]-Sigmatropic Hydrogen and Alkyl Shifts

Thermal [1,3]-suprafacial hydrogen shift is orbital symmetry forbidden process, but [1,3]-suprafacial alkyl shift is symmetry allowed process with inversion of configuration of migrating alkyl carbon. For example, the thermal rearrangement of bicyclo-[3.2.0]-heptene 1 to bicyclo-[2.2.1]-heptene **2** [4].

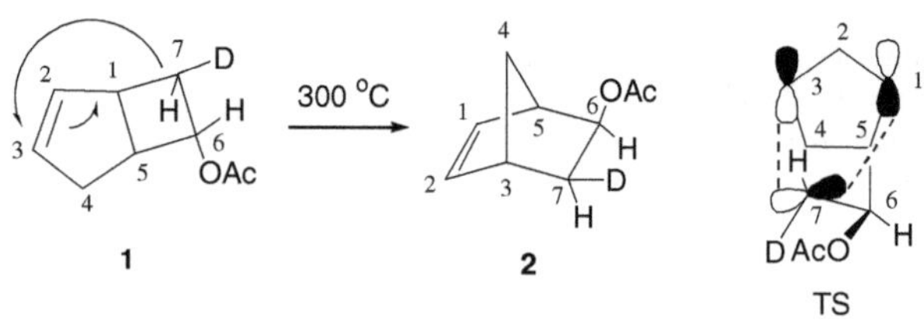

In compound **1**, deuterium is *trans* to the acetoxy group, while in the product **2**, it is *cis*. This indicates that the inversion of configuration at C-7 occurs during this rearrangement *via* the TS.

Similarly, bicyclo[2.1.1]-hexenes **3** give bicyclo[3.1.0]-hexenes **4** on heating through a favored TS [5].

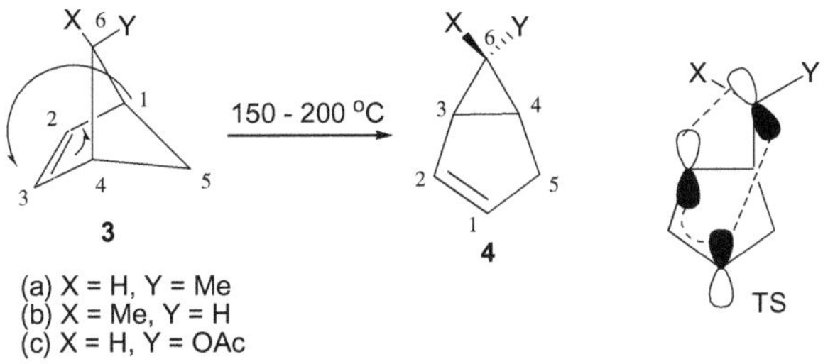

Although inversion of configuration of alkyl carbon is symmetry allowed process in thermal [1,3]-suprafacial alkyl shift, the presence of strong electron-withdrawing or electron-donating substituents results in the retention of configuration. For example, thermal rearrangement of **5** to **6** proceeds with about 95 % retention of configuration [6]. Possibly, the presence of strong EWG lowers the energy level of LUMO and of ERG raises the energy level of HOMO causing closure of HOMO and LUMO energy levels, respectively, to facilitate this conversion at ease.

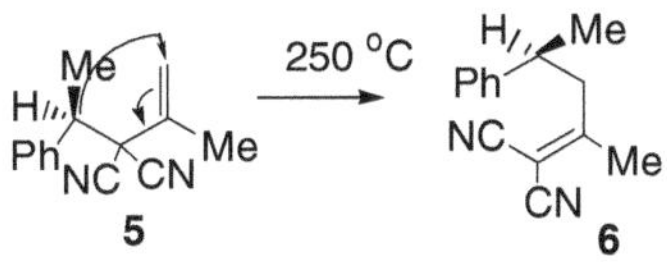

1,3-Alkyl shift also occurs in cyclopropane derivative **6a**.

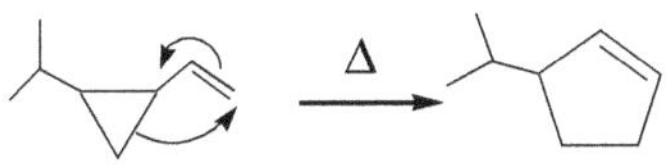

6a

Photochemically, [1,3]-alkyl shift is symmetry allowed process with retention of configuration of alkyl carbon. The photochemical rearrangement of **7** to **8** is an illustrative example [7].

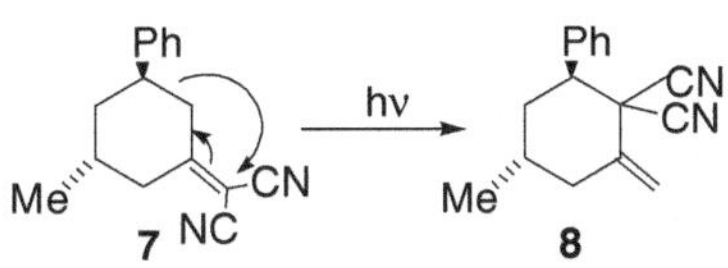

4.3.2 [1,5]-Sigmatropic Hydrogen and Alkyl Shifts

Thermal suprafacial [1,5]-hydrogen shift is orbital symmetry allowed process. For example, diene **9** is rearranged to stable diene **10** on heating [8].

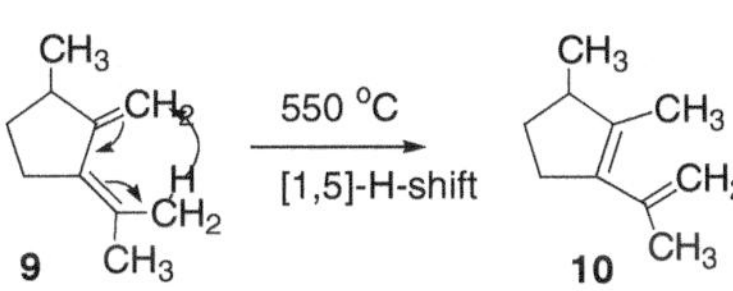

In cyclic systems, [1,5]-hydrogen shift is preferred than [1,5]-alkyl shift or [1,5]-deuterium shift. For example, [1,5]-H-shift of diene **11** gives **12**, and of indenes **13** and **14** gives **13a** and **14a**, respectively [9, 10].

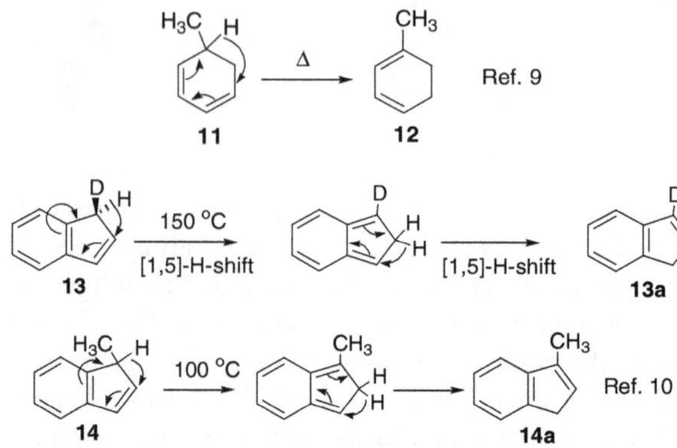

The suprafacial nature of thermal [1,5]-hydrogen and deuterium shifts is indicated in the rearrangement of chiral diene **15** into its stereoisomers **16** to **18** [11].

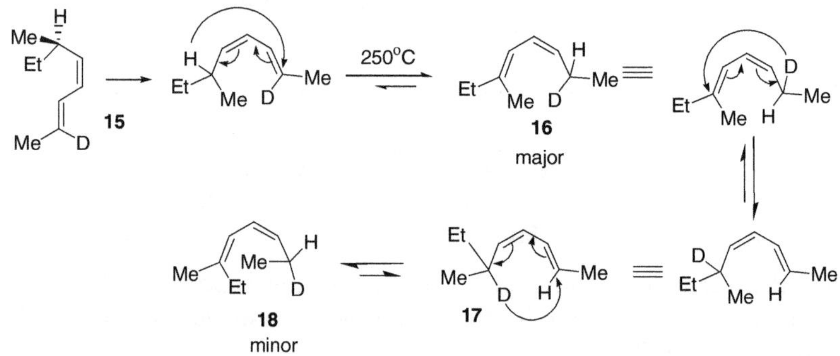

Thermal [1,5]--hydrogen shift occurs in cyclobutane **19** and arylallene **21** to give **20** and **22**, respectively. The product **20** undergoes electrocyclic ring opening on heating at slightly higher temperature to give **20a** [12].

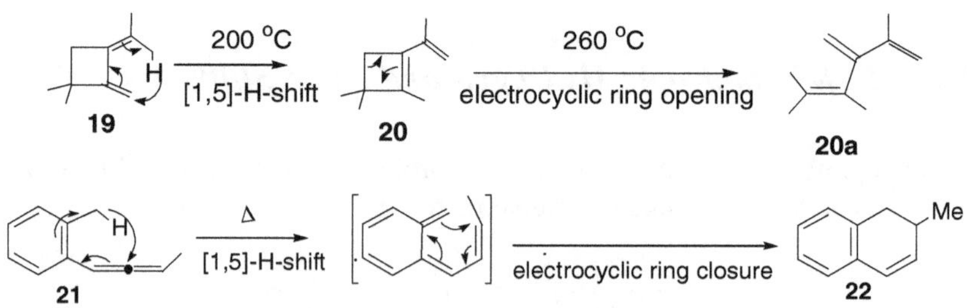

2-Methylcyclohepta-1,3-diene **23** at room temperature gives a mixture of isomeric methylcycloheptadienes by 1,5-hydrogen shifts [13].

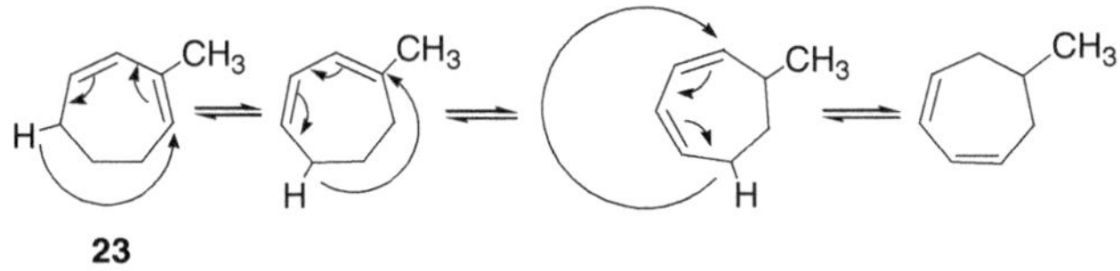

Similarly, 5-methyl-1,3-cyclopentadiene **24** rapidly rearranges at room temperature to yield a mixture of 1-methyl-, 2-methyl-, and 5-methyl-substituted products by [1,5]-hydrogen shifts [14].

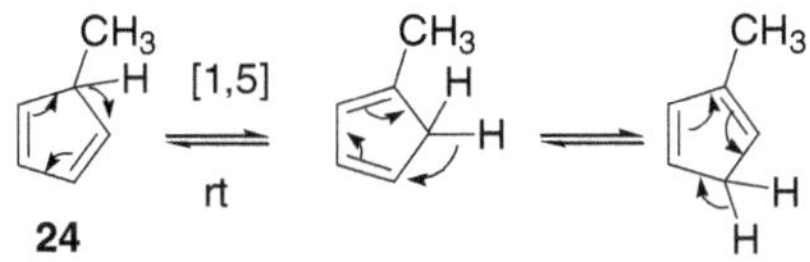

Both [1,5]-alkyl and hydrogen shifts take place in the thermal conversion of the spiro-dienes, 6,9-dimethylspiro-[4,4]-nona-1,3-diene **25** to **26** and of spiro-[4.2]-heptadiene **27** to **28** [15].

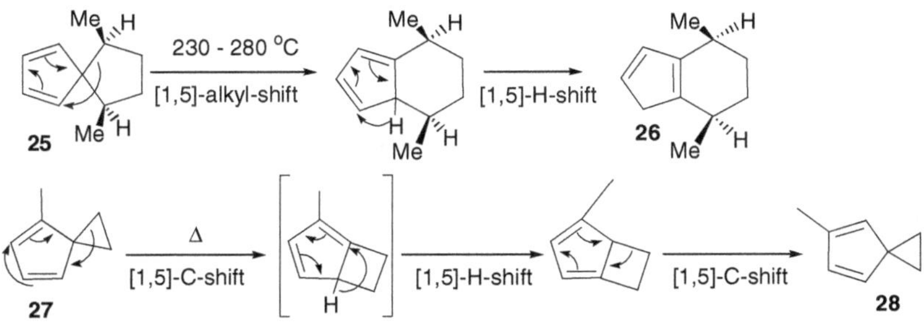

In [1,5]-shifts, the migratory aptitude of carbonyl group is greater than methyl group in indene skeleton **29** [16].

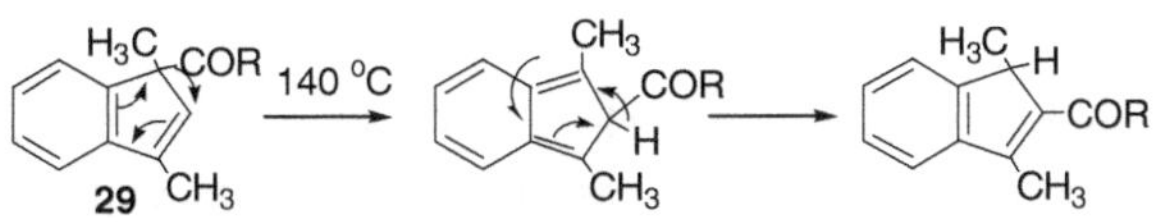

Thermal conversion of cycloheptatriene **30** to **31** and **32** takes place by [1,5]-alkyl shift followed by electrocyclic ring opening of norcaradienes. This phenomenon is known as valence tautomerism of norcaradiene [17].

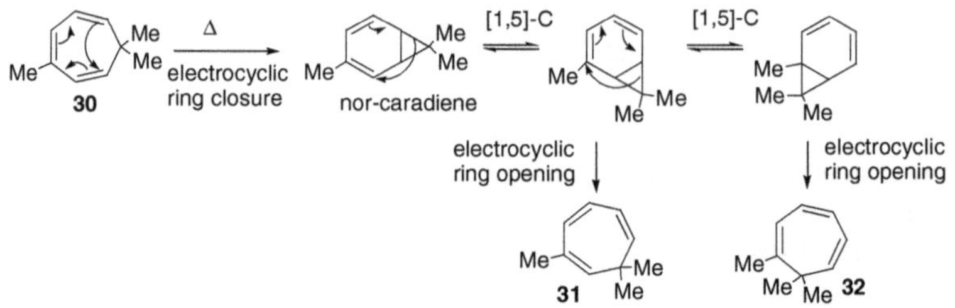

Thermal [1,5]--hydrogen shift also occurs in the conversion of cyclic enone **33** to **33a**.

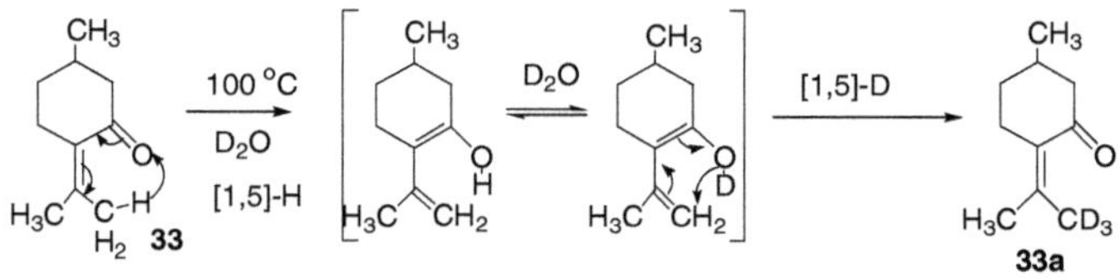

7-Methyl-cycloheptatriene **34** on heating undergoes slow [1,5]-suprafacial hydrogen shift rather that [1,7]--antarafacial H-shift to yield a mixture of methyl-substituted isomers [9].

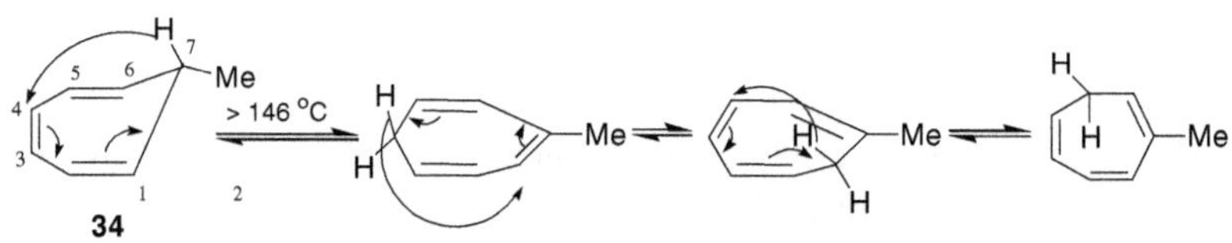

Thermal [1,5]-sigmatropic hydrogen shift also occurs in vinyl allenes **35** to **37** to give products, which in many cases undergo spontaneous [1,7]-sigmatropic hydrogen shifts to give thermodynamically stable products of linearly conjugated system [18, 19].

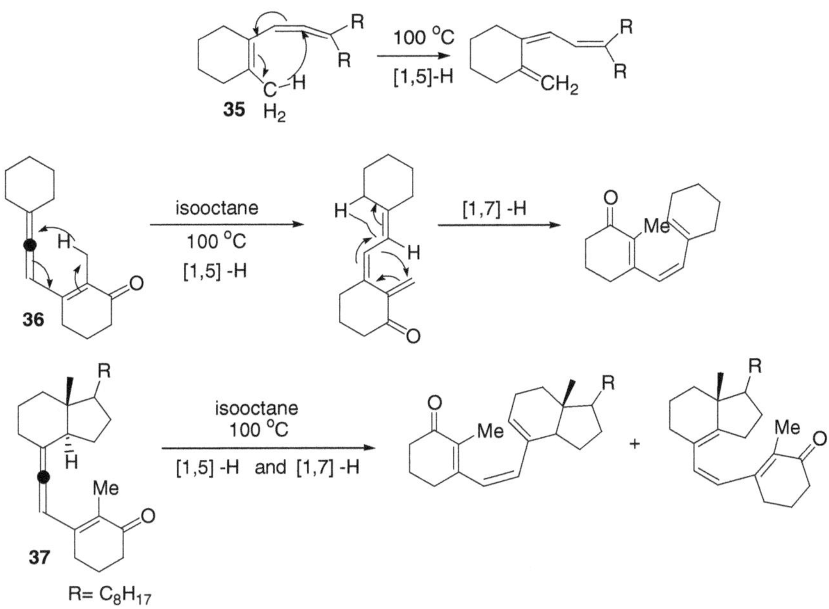

Photochemical [1,5]-hydrogen shift is rare as it is possible in antarafacial path. Photochemical conversion of the diene **38** to **39** illustrates the case [20].

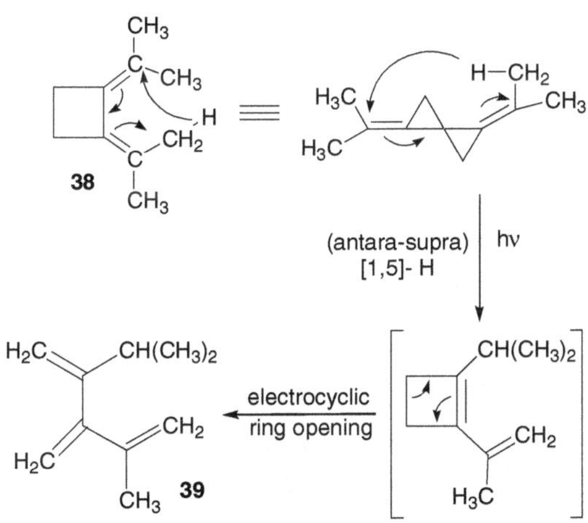

Similar to [1,5]-suprafacial H-shift, [1,6]-suprafacial H-shift occurs in cyclooctadienyl cation **40**, which is generated from cyclooctadienyl epoxide **41** [21].

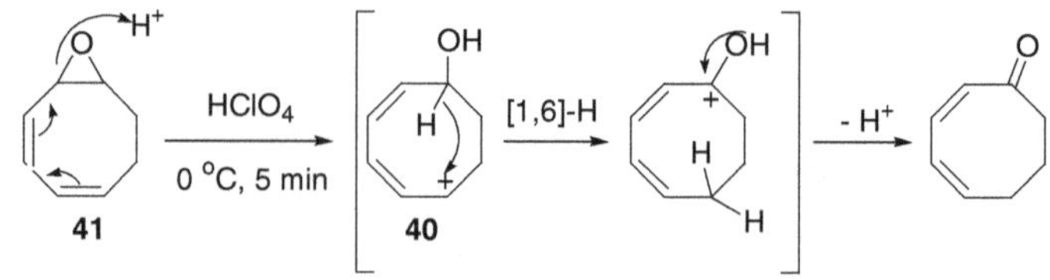

4.3.3 [1,7]-Sigmatropic Hydrogen and Alkyl Shifts

[1,7]-hydrogen shift occurs in antarafacial manner in a flexible π system. This type of hydrogen shift is common when one part of the π system is in the ring and other part in open chain. The most important reaction of this shift is the thermal conversion of precalciferol (previtamin D_3, **42**) to calciferol (vitamin D_3, **43**) in our body [22]. For this reason, it is advisable to expose our bodies to sunlight sometimes in a day for synthesis of vitamin D_3 in natural process.

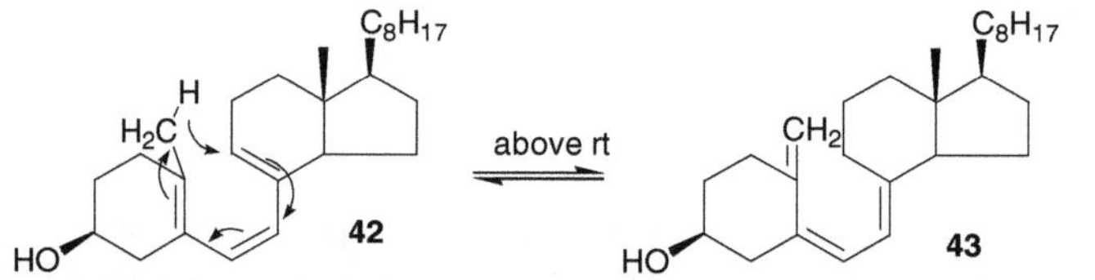

Thermal conversion of *ortho*-butadienyl phenol **44** to **45** and of **46** to **47** involves [1,7]-D and [1,7]-H-shift, respectively [23].

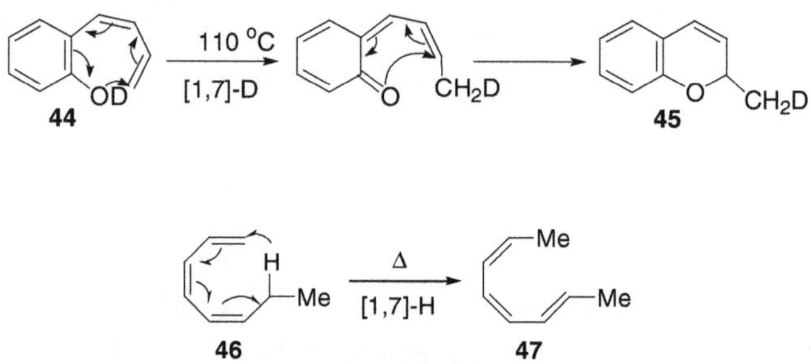

A similar [1,7]-antarafacial hydrogen and deuterium shifts are observed in the rearrangement of the steroid **48** on heating [24].

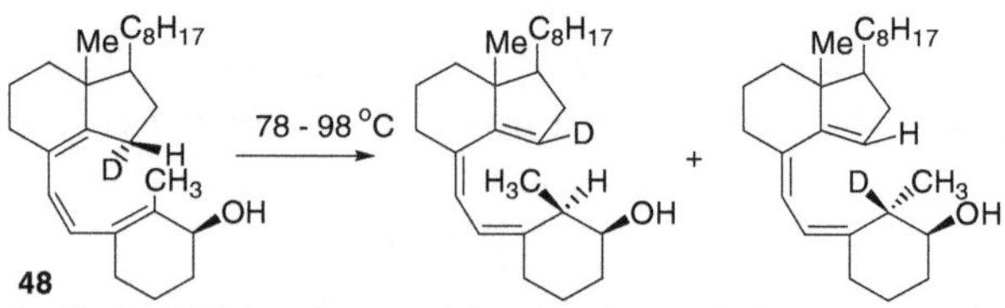

[1,7]-Suprafacial alkyl shift is observed in the thermal interconversion of bicyclo [6.1.0]-nonatriene **49** into its isomers. The interconversion occurs much rapidly and

hence it is difficult to detect the endo- and exo-substituents. This rearrangement is known as Walk rearrangement [25].

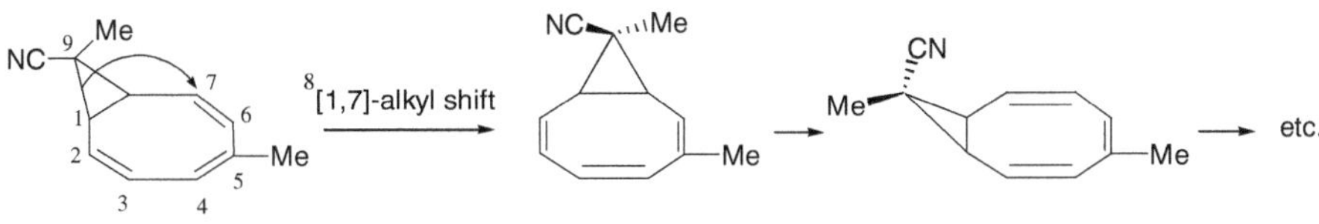

49

Photochemical [1,7]-suprafacial alkyl and hydrogen shifts have been observed in the conversion of cycloheptatriene derivative **50** into its isomers [26].

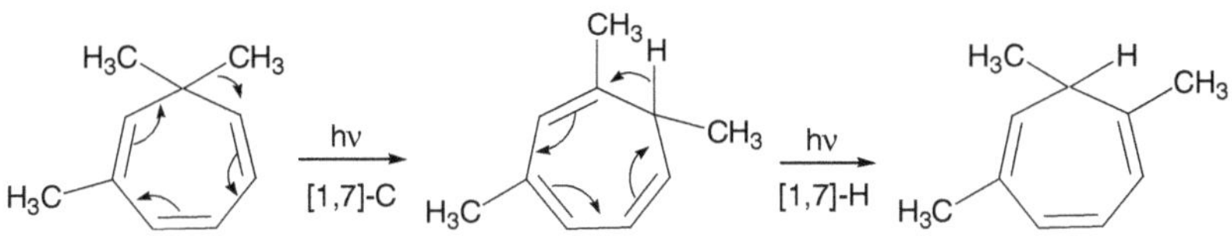

50

4.4 [3,3]-Sigmatropic Rearrangements

4.4.1 The Cope Rearrangements

The Cope rearrangement is the most important category of sigmatropic rearrangements from synthetic point of view. The thermal [3,3]-sigmatropic rearrangements of 1,5-hexadienes are called the *Cope rearrangements*. These rearrangements are reversible in nature. The Cope rearrangement proceeds through a chair-like or boat-like transition state. Usually, a chair-like transition state with minimum steric interactions between the substituents provides the major product. For a chair-like TS of a trans-3,4-disubstituted 1,5-hexadiene, the major product would be either, *E,Z*- or *Z,E*-diene. For example, diene **51** on heating gives **53** and **54** as major and minor product, respectively, through chair-like TSs **52a** and **52b** [27]. Enantiomerically pure compound **51** gives an optically pure product of >95 % ee [27].

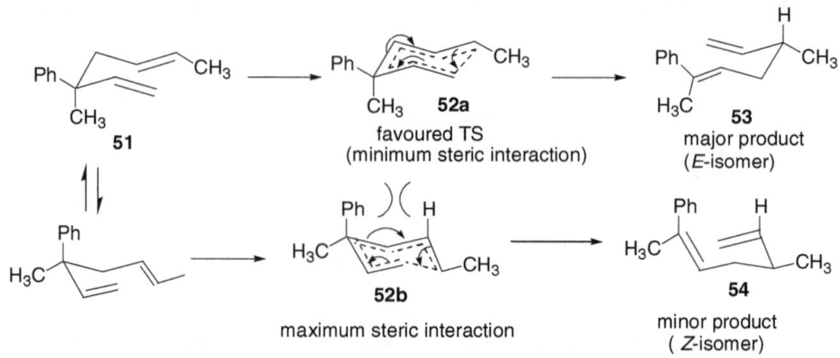

Conjugated substituents at C-1 (or C-6) and C-3 (or C-4) of 1,5-dienes accelerate the rates of the reactions, since the substituents at these positions weaken the bonds being broken in the reactions [28]. Donor substituents at C-2 and C-3 also accelerate the reaction [29]. The following examples support these facts.

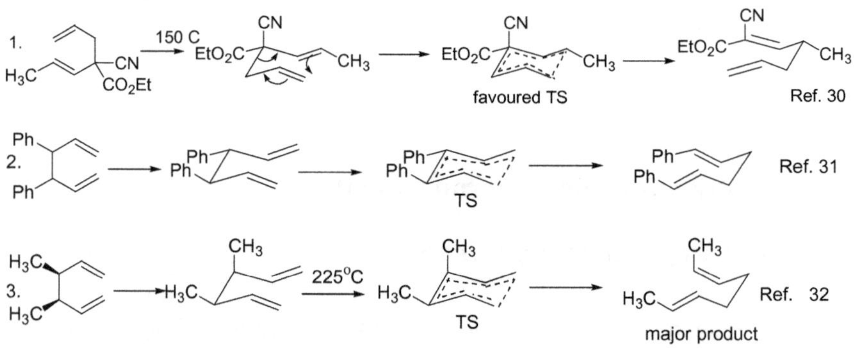

The reversible nature of Cope rearrangement is supported by the fact that optically active compound **55** undergoes racemization on heating by reversible Cope process [33].

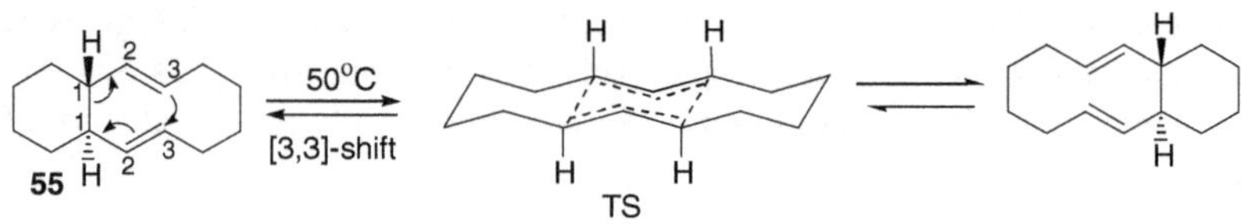

Bicyclic compounds **56** and **57** on Cope rearrangement undergo skeletal rearrangements to give **58** and **58a**, respectively [34].

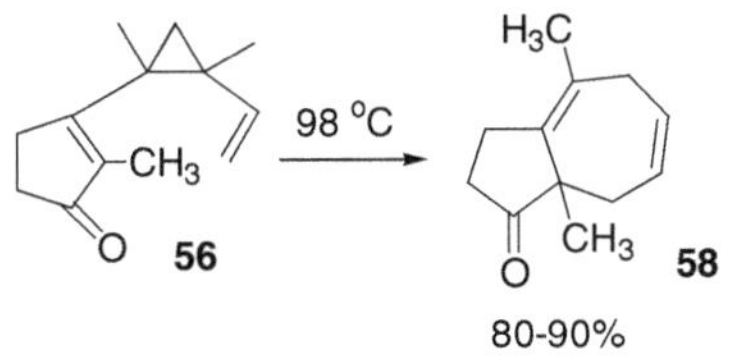

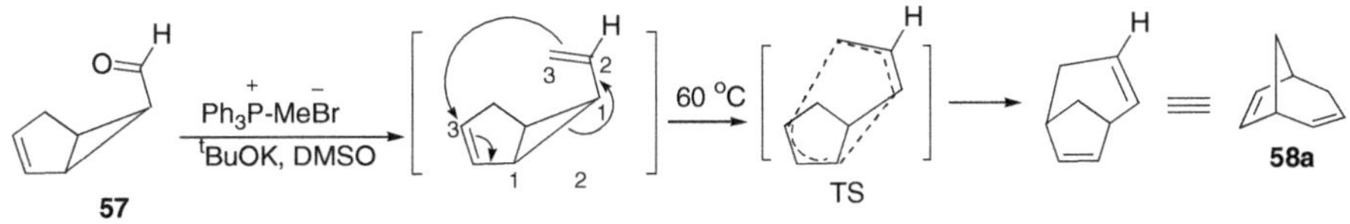

Analysis of the product ratio from chair and boat TS geometry from a Cope rearrangement of deuterated 1,5-hexadiene indicated that the boat TS is about 6 kcal/mol less stable than the chair TS [35]. It is reflected in the Cope rearrangement of cyclic dienes **59** and **60**. Comparison of their reaction rates showed that diene **59** reacted faster by a factor of 18,000. This fact can be rationalized by considering their TS. Compound **59** reacts through a chair-like TS while **60** through a boat-like TS. The chair-like TS has lower activation energy and hence **59** reacts much faster [36].

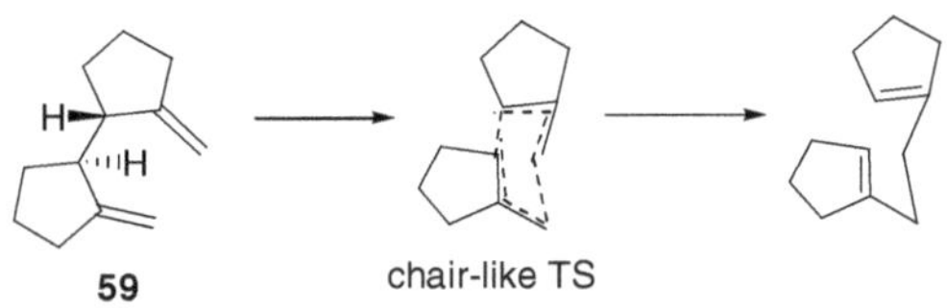

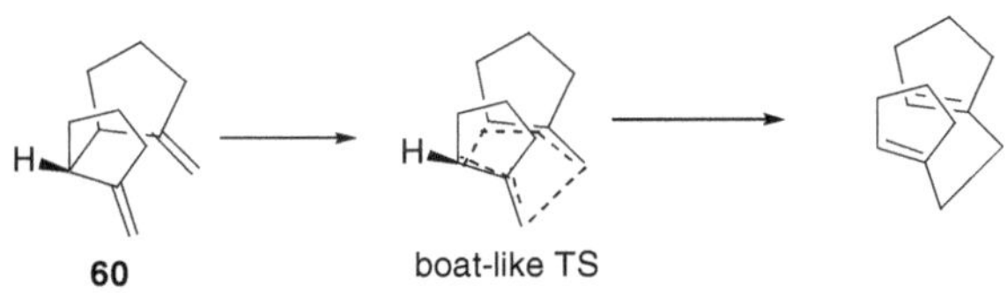

When the C-3 and C-4 carbons of 1,5-dienes are connected to a cyclopropane ring, the reaction rates are accelerated due to favorable interactions of the diene termini resulting in the reduction of the enthalpy of activation. For example, the conversion of *cis*-divinylcyclopropane **61** to 1,4-cycloheptadiene **62** occurs readily at temperatures below −40 °C [37].

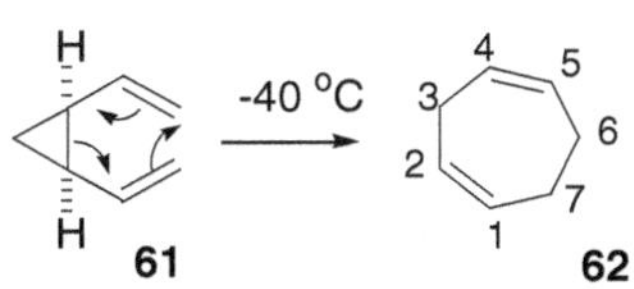

This reaction occurs at much lower temperature to relieve the ring strain. The *cis*-geometry of the vinyl substituents plays one of the key roles to accelerate this reaction. This is substantiated by the fact that the reactions of *trans*-

divinylcyclopropane **63** and vinylcyclopropane **64** take place at much higher temperatures in a non-concerted process [38].

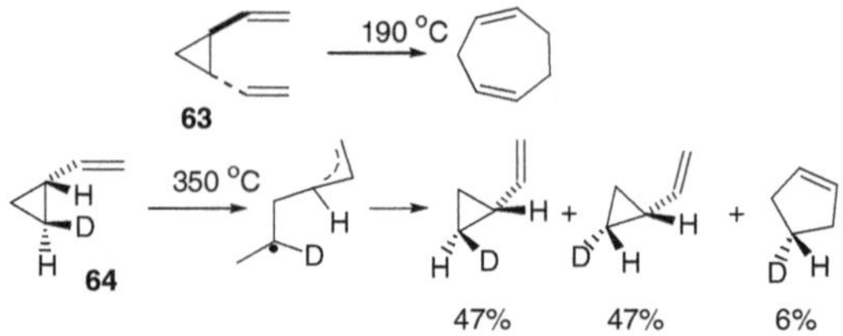

The ring strain is another important factor to increase the reaction rate of *cis*-divinylcyclopropane. For instance, the rearrangements of *cis*-vinyloxirane **65** and *cis*-divinylthiirane **66** require relatively higher temperatures because of lesser degree of ring strain in their rings [39]. Similarly, the change of ring size from three-membered to four- and five-membered requires higher temperature for the reaction. Thus, the rearrangement of *cis*-divinylcyclobutane **67** occurs at 120 °C and of *cis*-divinylcyclopentane does not occur even at 250 °C [40].

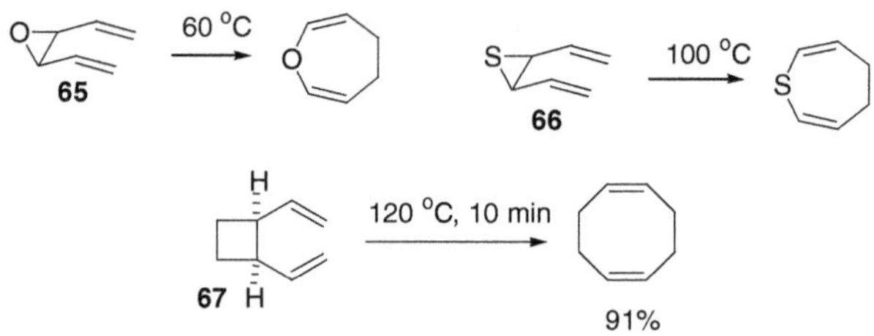

Divinylcyclopropane rearrangements take place with even greater ease if the vinyl groups are incorporated in another ring. This condition favors the entropy of activation of the reaction to be less negative. This is found in the degenerate rearrangement of homotropilidene **68** [41]. A degenerate rearrangement is a reaction when the product of the rearrangement is structurally identical to the starting material in terms of nature and types of bond order. The occurrence of a dynamic equilibrium in the reaction of homotropilidene is evident from the NMR study of the reaction. At low temperature, the rate of interconversion is slow and the NMR

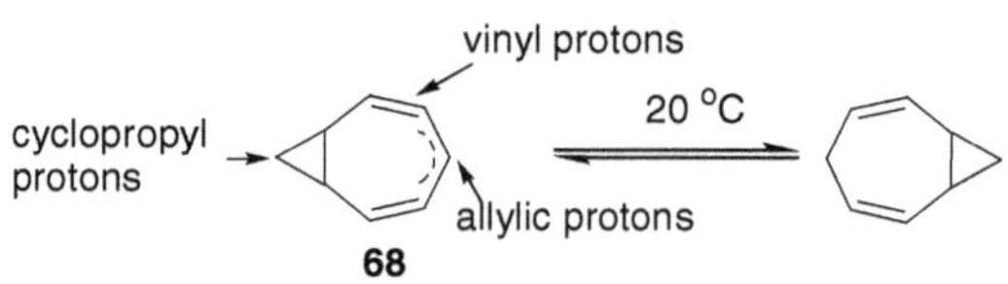

spectrum showed the presence of four vinyl protons, two allylic protons and four cyclopropyl protons. When the temperature is raised, the rate of rearrangement increases and the NMR spectrum recorded the signals of two vinyl protons and signals of other two vinyl protons coalesce with two cyclopropyl protons and the signals of two allylic protons coalesce with two cyclopropyl protons. This indicates that the sets of protons undergoing rapid interchange with one another show an averaged signal.

Several degenerate Cope rearrangements are known. One of the most interesting cases is of bullvalene **69** [42]. At 10 °C, its ^{1}H-NMR spectrum showed a single signal at δ_H 4.22 ppm, indicating the 'fluxional' nature of its molecule and identical environment of all the carbons. The first-order rate constant of the reaction is $3.4 \times 10^{3}\ s^{-1}$ at 25 °C with ΔG of 12.6 kcal/mol. The rearrangement of bullvalene **69** is shown by the change of environment of labeled carbons.

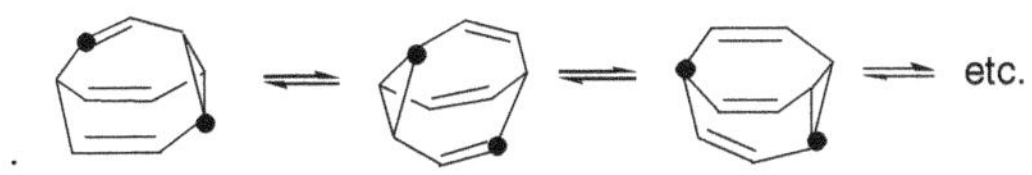

69

• indicates labelled carbon

Among other degenerate rearrangements, the rearrangement of barbaralane **70** [43] and semibullvalene **71** [44] are important. The free energy of activation ΔG for their rearrangements is 7.6 kcal/mol at 25 °C and 5.5 kcal/mol at −143 °C, respectively. It indicates that degenerate rearrangement of **71** is much more rapid than the conformational inversion of cyclohexane.

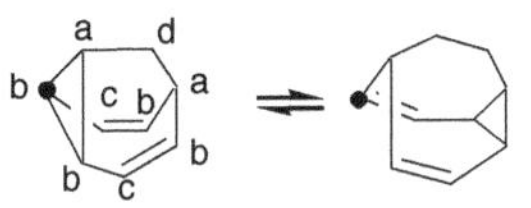

70

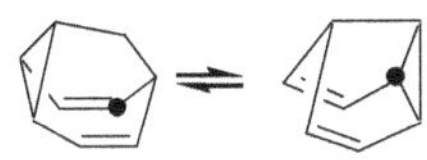

71

4.4.2 *The Oxy-Cope and the Anionic Oxy-Cope Rearrangements*

The presence of a hydroxyl group at C-3 position of the 1,5-diene system drives the rearrangement into the less stable enol product, which is converted into stable carbonyl product. This version of this Cope rearrangement is termed the *oxy-Cope rearrangement* [45]. A particular advantage of this process is that it proceeds under mild conditions, so that many sensitive functional groups are tolerated and is useful for construction of complex organic molecules. The simplest case is the conversion of 3-hydroxy-1,5-hexadiene to 5-hexenal [46].

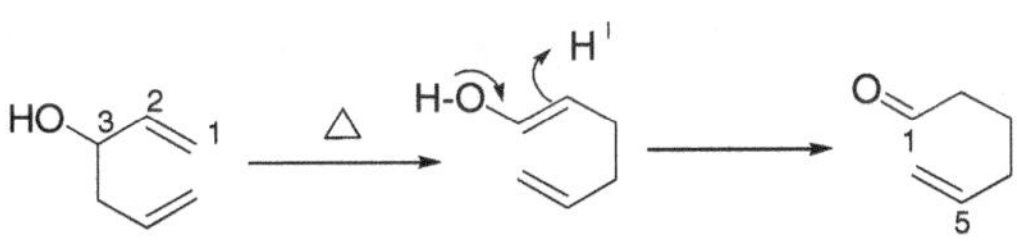

If the hydroxyl group at C-3 position of the 1,5-diene is converted into the respective potassium alkoxide, the rearrangement is known as the *anionic oxy-Cope rearrangement*. The conversion of C-3 hydroxyl group into an alkoxide ion accelerates the reaction rate by a factor of 10^{17} [47]. The following examples [48–53] of this reaction are illustrative:

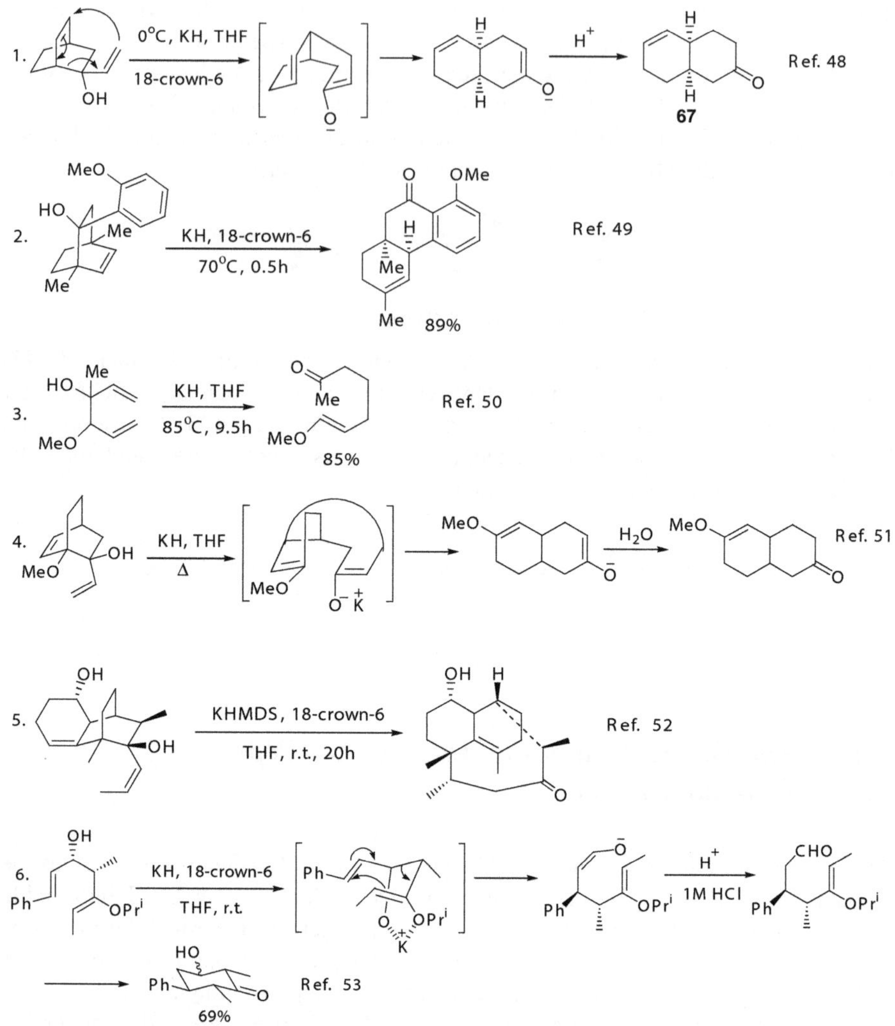

Acyclic 1,5-diene-3-ols **72** and **73** undergo anionic oxy-Cope rearrangements to give major products from equatorial orientations of the oxyanions.

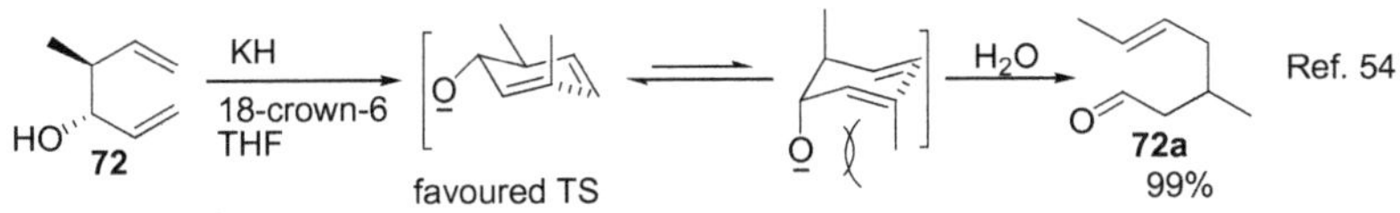

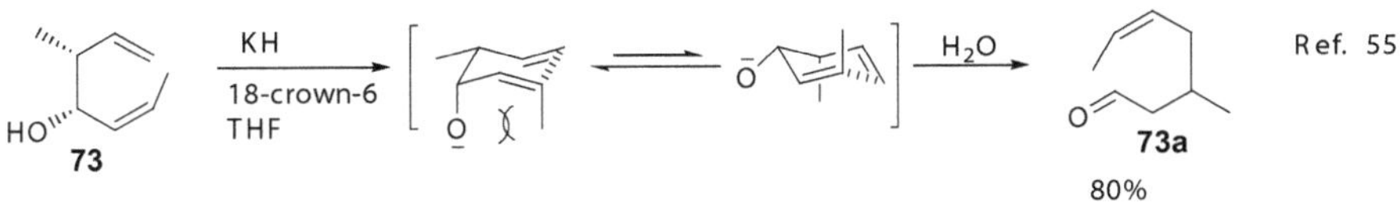

The reaction rate of an anionic oxy-Cope rearrangement is accelerated by the presence of an additional unsaturation function on the terminal position. This additional unsaturation function probably stabilizes the transition state and helps the reaction to occur at ease. For example, bicyclic allyl alcohol **74** on rearrangement gives bicyclo-[5.3.1]-undecenone **75** in 88 % yield, while under the same conditions, isomeric alcohol **76** gives **77** in 25 % yield [56].

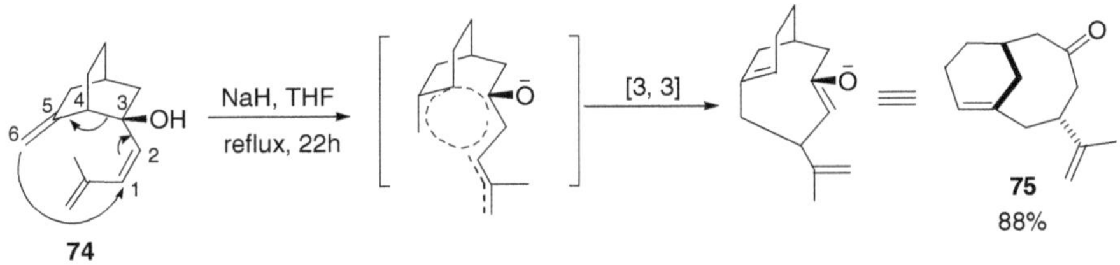

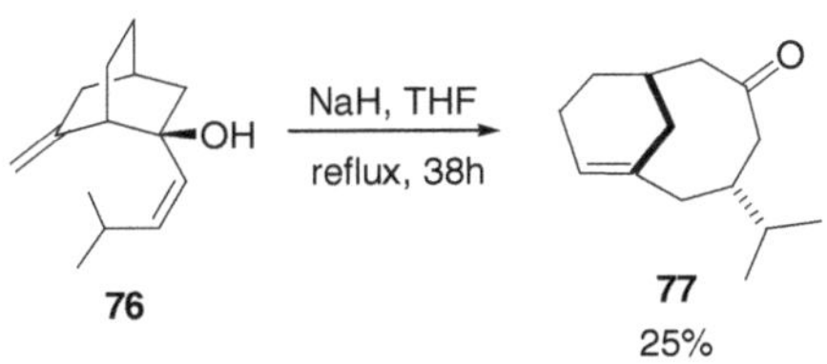

Base-catalyzed allenic oxy-Cope rearrangement of **78** to **78a** is useful for synthesis of bicyclic compound [57].

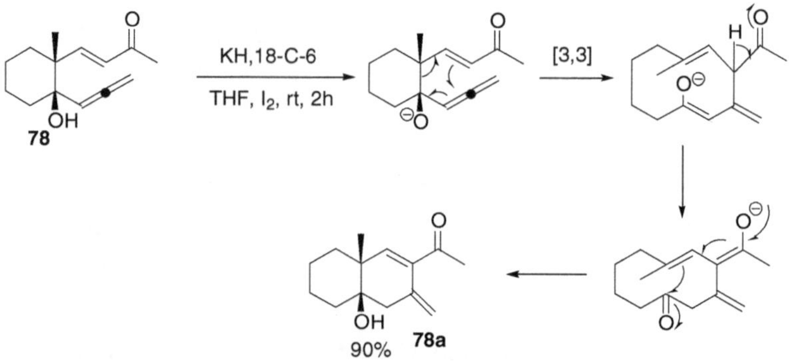

The products from the anionic oxy-Cope reactions of norbornenyl derivatives **79** and **80** depend on the steric demands of the oxyanions. In both cases, the orientation of the oxyanion favors chair-like **TS** structures for *exo*-bond formation to give the major aldehyde **81** along with minor aldehyde **82** [58].

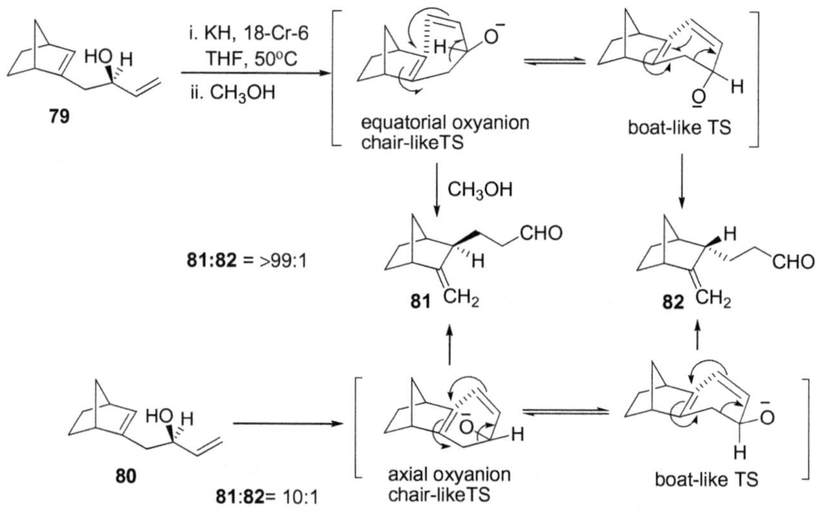

4.4.3 The Amino- and Aza-Cope Rearrangements

3-Amino group in the 1,5-diene system also accelerates the Cope rearrangement. This version of the Cope rearrangement is known as the *amino-Cope rearrangement*. The products of the reactions are useful in the synthesis of unsaturated aldehydes. The following examples are illustrative [59, 60]:

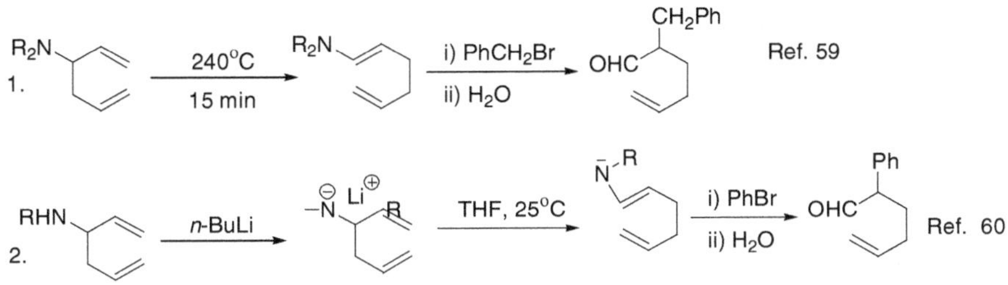

When nitrogen atom belongs to a part of 1,5-diene skeleton, the Cope rearrangement is known as the *aza-Cope rearrangement*. The conversion of the mesylate **83** to **84** is an example of 2-aza-Cope rearrangement [61]. Similarly, the reaction of pyridine 3-aldehyde with N-methyl-2-hydroxy-2-methylbutenamine gives an iminium salt as an intermediate **85**, which undergoes 2-aza-Cope rearrangement to give acetylnicotine derivative in high yield [62].

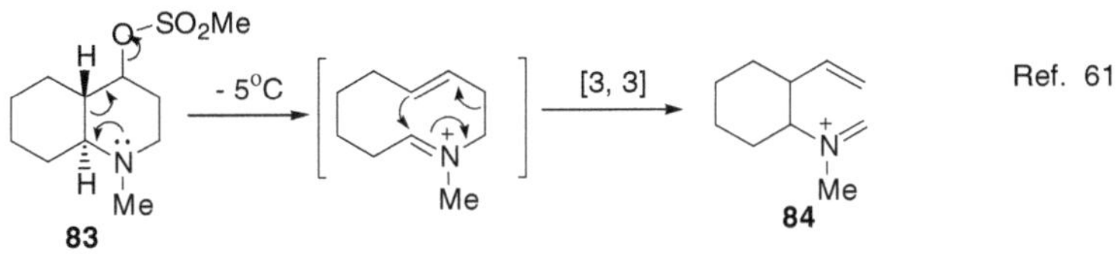

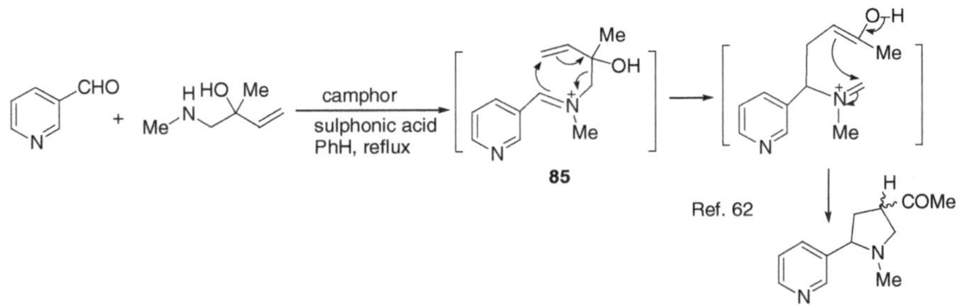

The aza-Cope rearrangement of cyclopentane derivative **86** gives a bicyclic pyrrolidine **87** in which original ring is expanded by one carbon atom [63].

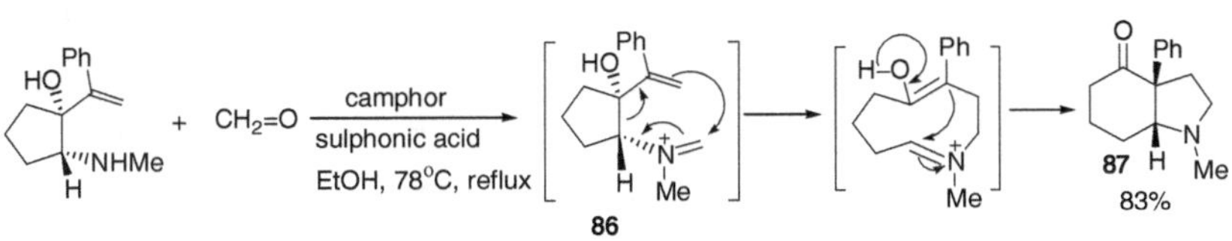

When cyclic ketone was used to generate an iminium ion, the yield of the reaction becomes low due to unfavorable steric strain in chair-like **TS**; for example, the synthesis of 1-aza-spiro-[4,5]-decane ring system **88** from cyclohexanone via iminium ion **88a** [64].

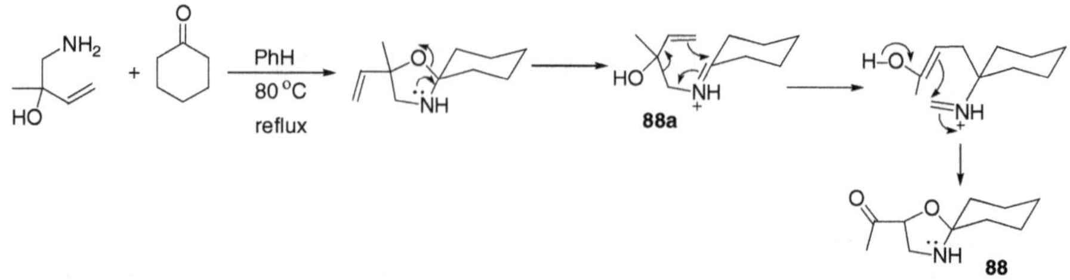

The rearrangements of 4-hydroxy-2-aza-1,5-dienes **89** to **89a** in anionic forms are known as the anionic-4-oxy-2-aza-Cope rearrangements [65].

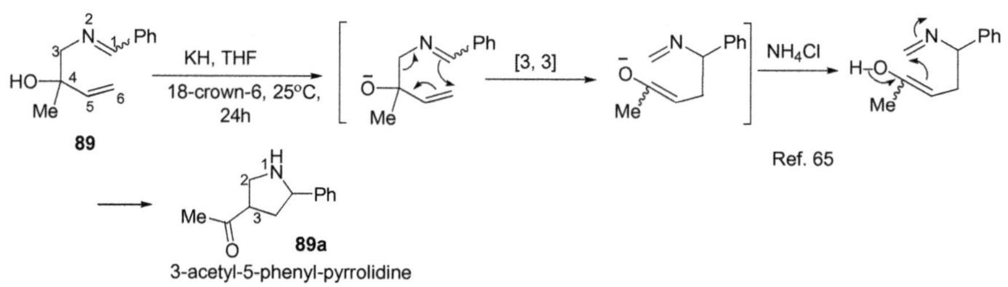

The Cope rearrangements of 1-aza-1,5-dienes are known as the *1-aza Cope rearrangements* (1-ACR). For example, N-acylimine **90** obtained from flash vacuum pyrolysis (FVP) of N-acyl hydroxylamine derivative gives pyridine derivative **91** by 1-ACR [66].

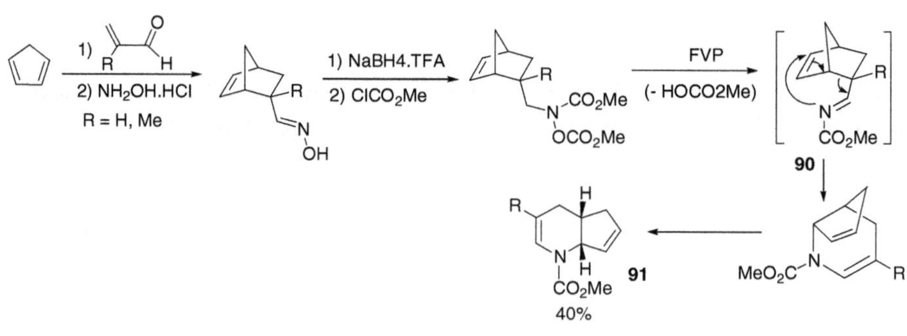

The [3,3]-sigmatropic rearrangement of 4-nitroso-1-butene **92** and nitrosobicyclo-[2.2.2]-octene **92a** is known as the *1,2-oxaza-Cope rearrangements* [67].

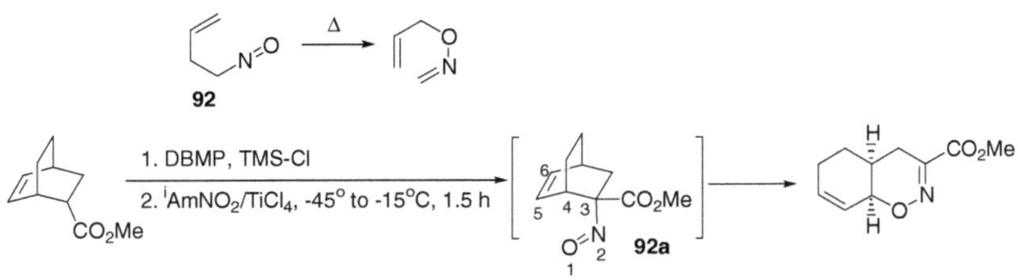

4.4.4 The Claisen Rearrangements and Their Modified Versions: The Carroll, Eschenmoser, Ireland, Johnson, Gosteli, Bellus, and Enzymatic Claisen Rearrangements

Thermal [3,3]-sigmatropic rearrangements of allyl aryl ethers and allyl vinyl ethers are known as the *Claisen rearrangements* [68]. These reactions are sensitive to solvent polarity and the rates of the reactions are increased by increasing the solvent polarity [69]. The simplest examples are the thermal conversion of allyl phenyl ether to *ortho*-allyl phenol and of allyl vinyl ether to 4-pentenal [70].

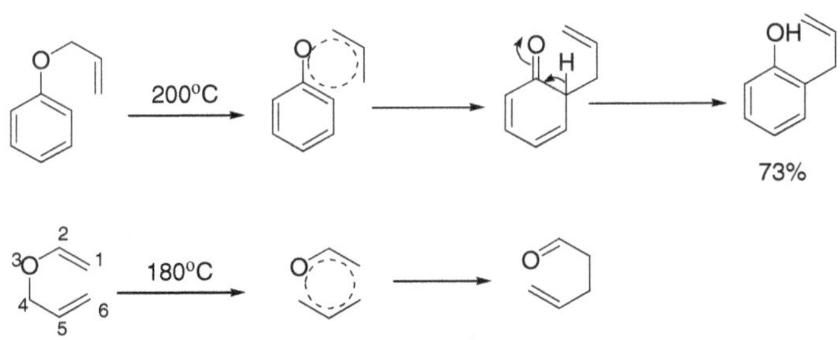

Allyl vinyl ethers **93** and **94** are generated from the reaction of allyl alcohols with alkyl vinyl ether in the presence of $Hg(OAc)_2$ [71].

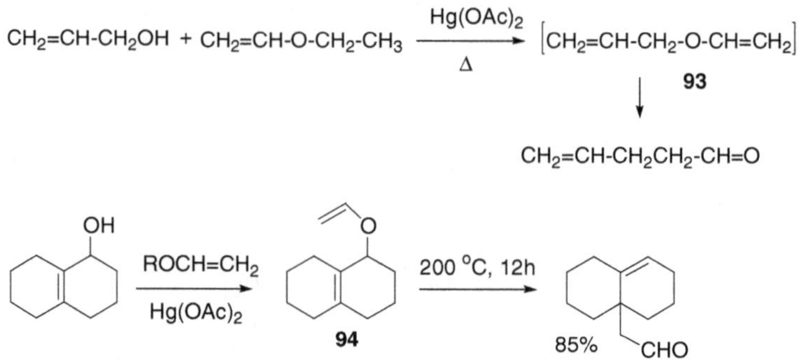

The Claisen rearrangement is intramolecular in nature. It was confirmed by a crossover experiment in which two aromatic allyl ethers **95** and **96** were heated together and found to yield same products **97** and **98** as when they were heated separately. No crossover products **99** and **100** were found [72].

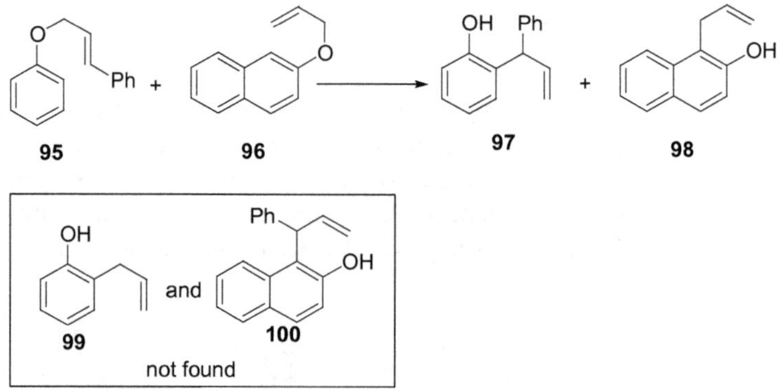

The intramolecular mechanism of the Claisen rearrangement was also verified by the use of ^{14}C-labeled allyl phenyl ether **101** [73].

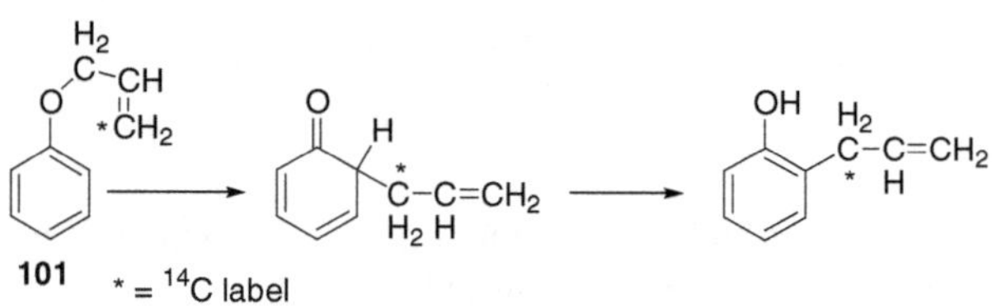

The major product of the Claisen rearrangement is derived from a preferred chair-like transition state in which the larger substituent occupies the pseudoequatorial position. For example, in the Claisen rearrangement of **102**, the major product **103** was obtained [74].

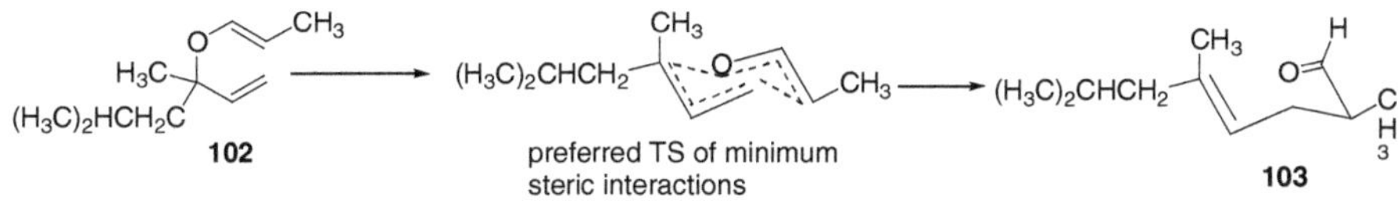

In the rearrangement of aryl allyl ethers, when both the *ortho*-positions of the aromatic ring are substituted, the migrating allyl group will shift to the *para*-position. This rearrangement is known as the *para*-Claisen rearrangement. For example, the rearrangement of the aromatic ether **104** gives the major product **105** [75].

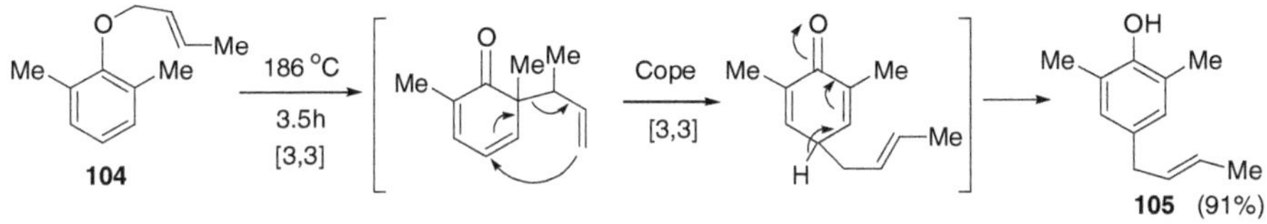

The products of Claisen rearrangements undergo further rearrangements to yield rearranged products, these rearrangements are known as the abnormal Claisen rearrangements [76]. For example, vinyl ether **106** gives **107**.

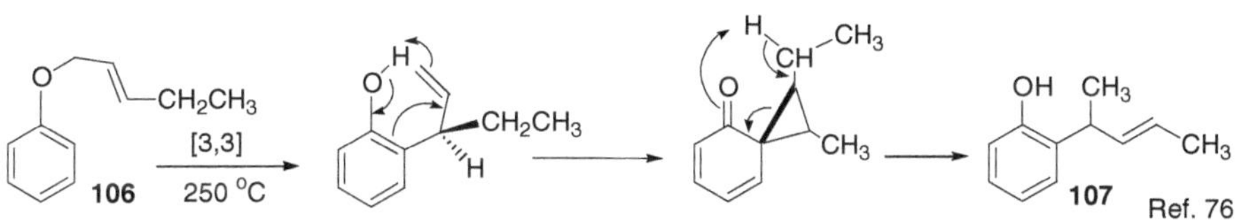

Similarly, the phenyl allyl ether **108** gives abnormal product **109** on heating. But in the presence of dimethylaniline gives normal Claisen rearrangement product **110**.

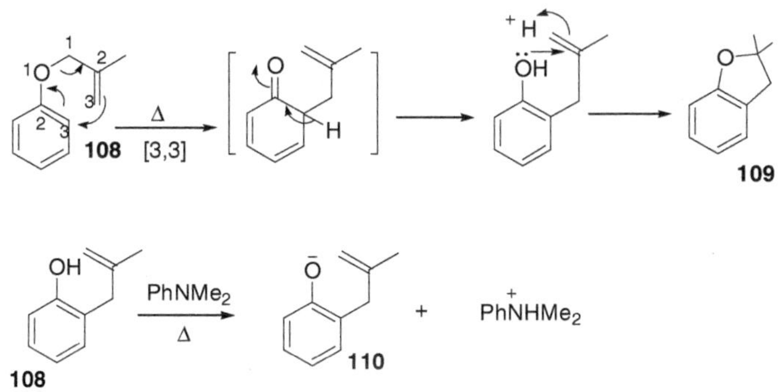

When an allyl aryl ether **111** contains a vinyl group at the *ortho*-position of aryl group, the Claisen product undergoes further Cope rearrangement and [1,5]-H-shift to give the product **111a** [76].

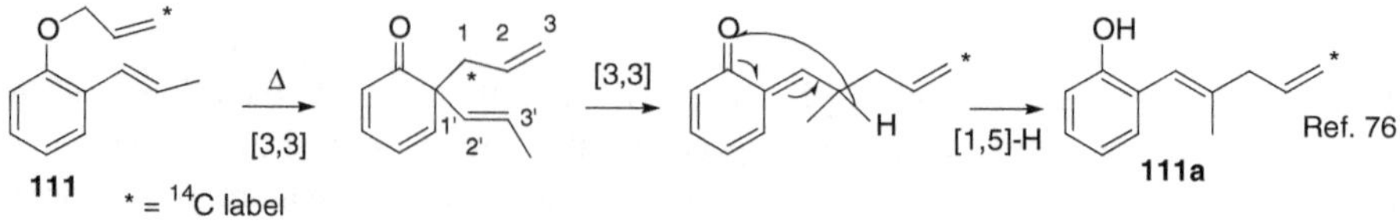

Several modifications of Claisen rearrangement were developed to increase the versatilities of this rearrangement in the synthesis of different classes of organic compounds. For instance, thermal rearrangement of allylic β-keto esters **112** to **112a** are known as the Carroll–Claisen rearrangements [77].

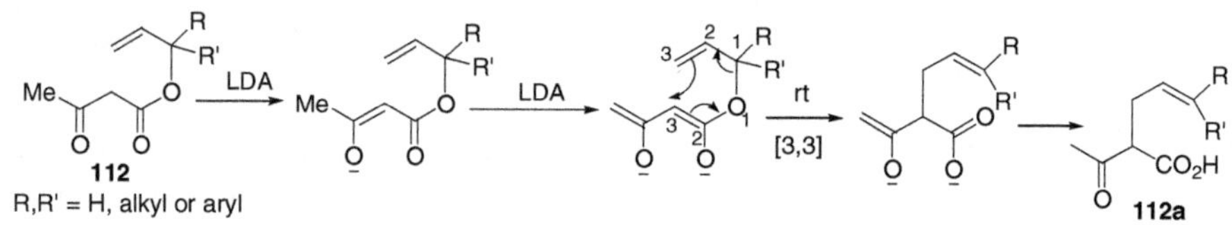

The Claisen rearrangements of amide acetals of allyl or crotyl alcohols are known as the Eschenmoser–Claisen rearrangements [78]. For example, *E*- and Z-isomers of **113** give **113a** and **113b** as major product, respectively [78].

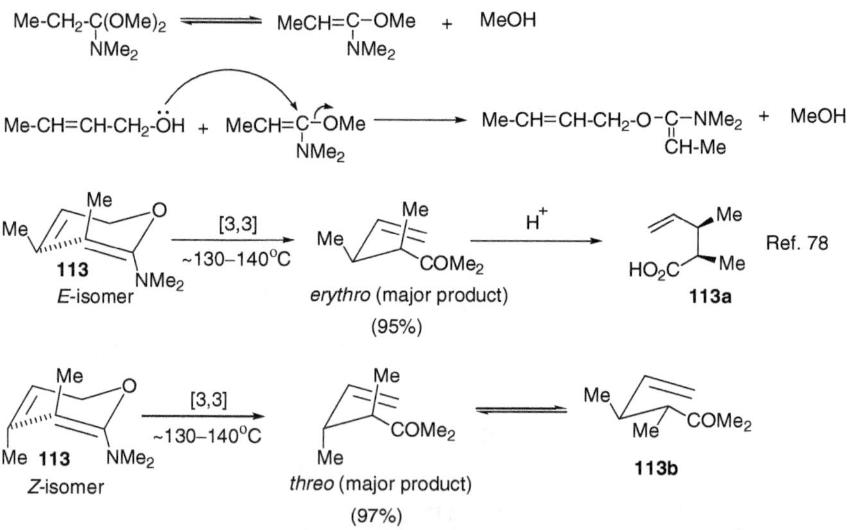

This version of Claisen rearrangement is not suitable for compounds having acid-sensitive functional groups.

Trimethylsilyl derivatives of enol esters of allylic alcohols **114** undergo [3,3] sigmatropic rearrangements on heating below 100 °C. This version of Claisen rearrangement is known as the Ireland–Claisen rearrangement. The major product of the reaction is derived from the geometry of the silyl derivative, which depends on the condition of solvent used for the preparation of lithium enolate with LDA. When THF is used, *Z*-lithium enolate gives the major product **114a**, whereas use of 23 % HMPA (hexamethyl phosphoramide)—THF, gives the thermodynamic controlled *E*-lithium enolate-derived product **114b** as major product [79].

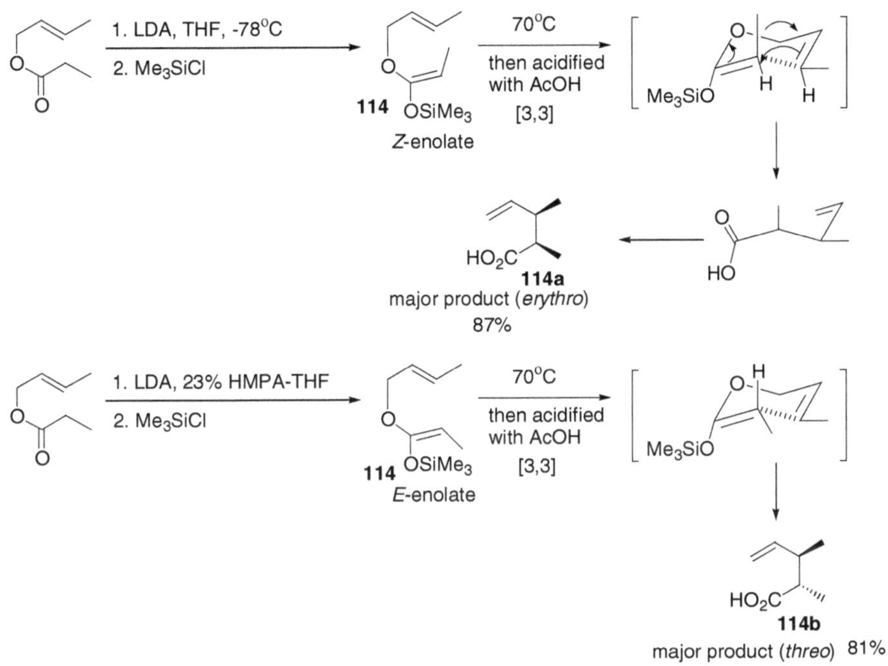

Aromatic ketone **115** containing allyloxy group at the α-position undergoes Ireland–Claisen rearrangement [80] to give **115a**.

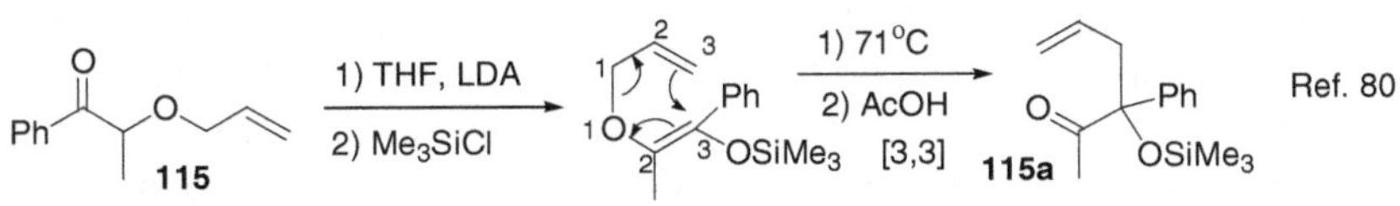

TBS (*tert*-butyldimethylsilyl) and TES (triethylsilyl) were also used instead of TMS for preparation of silyl derivatives.

The rearrangements of orthoesters of allyl alcohols are known as the Johnson–Claisen rearrangements. This modification of Claisen rearrangement improves the yield of the reaction and transfers the chirality of the alcohols in high levels and allows the introduction of carboalkoxyalkyl group. For example, orthoester **116** of (2*R*,3*E*)-3-penten-2-ol gives ethyl ester of (3*R*,4*E*)-3-methyl-4 hexenoic acid **117** in 90 % optical yield [81] and methyl orthoester **118** gives **119** [82].

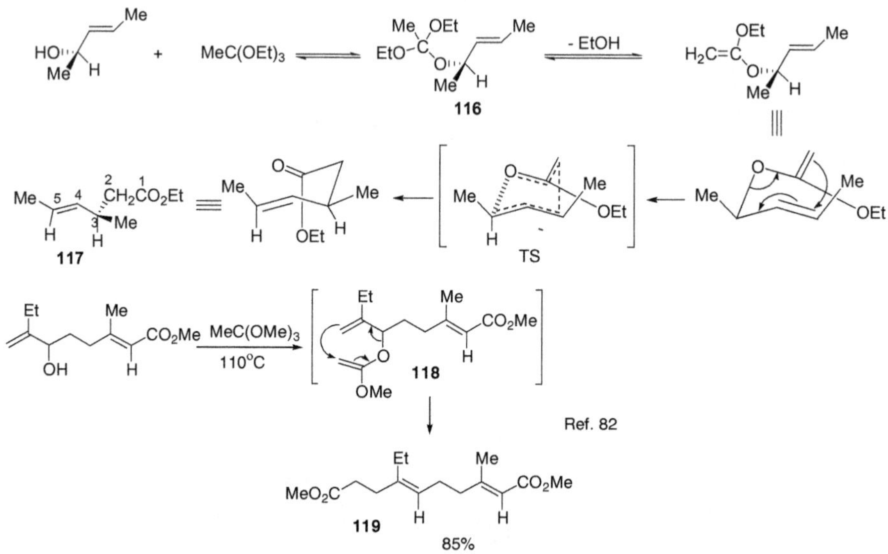

The [3,3]-sigmatropic rearrangements of 2-alkoxycarbonyl-substituted allyl vinyl ethers **120** are known as the Gosteli–Claisen rearrangements [83].

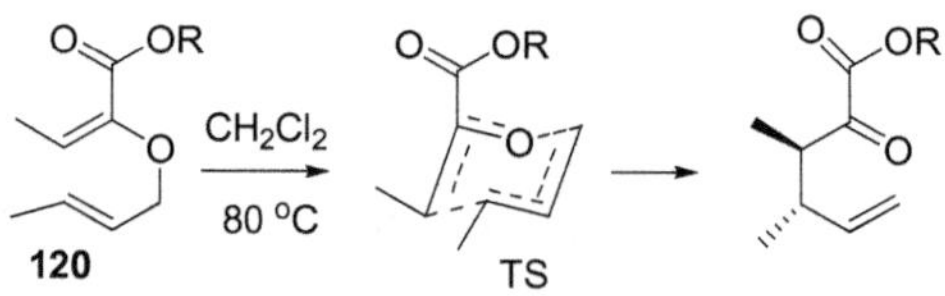

The catalyzed version of this rearrangement is useful for the synthesis of medium and large-sized carbocycles [84]. For example, the enantioselective synthesis of carbocyclic natural product, (−)-9,10-dihydroecklonialactone B **121** was done successfully by catalytic asymmetric Claisen rearrangement of a Gosteli-type allyl vinyl ether **122** in the presence of (S,S)-Cu (box)-catalyst A to produce a chiral α-ketoester **123**, as a building block unit [85].

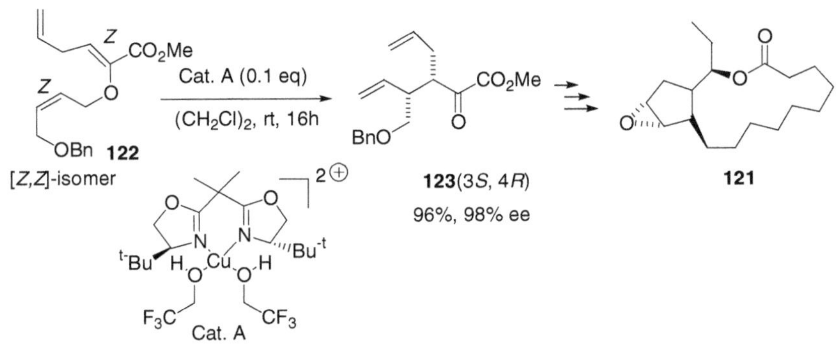

The ketene-Claisen rearrangement of a 1,3-dipolar allyl vinyl ether **124** is known as the Bellus–Claisen rearrangement [86].

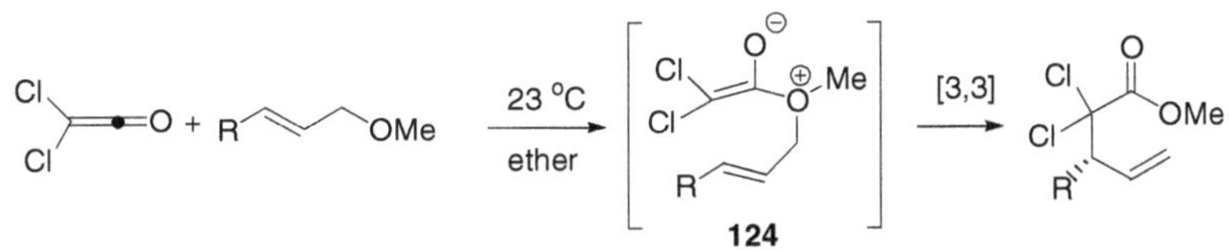

Subsequently, this version of Claisen rearrangement was extended to tertiary allyl amines and acyl chlorides [87]. Using Lewis acid catalyst, excellent stereoselectivity of the product was found. For example, 3,3-disubstituted allyl morpholine **125** with propionyl chloride in the presence of $TiCl_4$ gives highly stereoselective *syn* product **126** [87]. This method is useful for the synthesis of α,β-disubstituted-γ,σ-unsaturated carbonyl compounds.

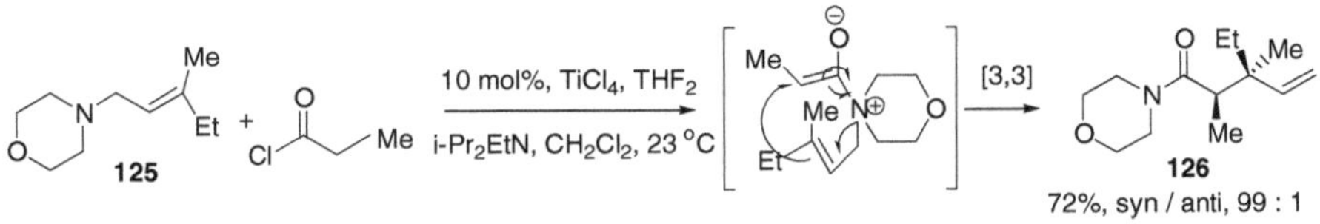

In the rearrangement of allyl vinyl ethers in conformationally rigid cyclohexane system **127**, the major product **128** is derived from the preferred axial bond formation in the TS [88].

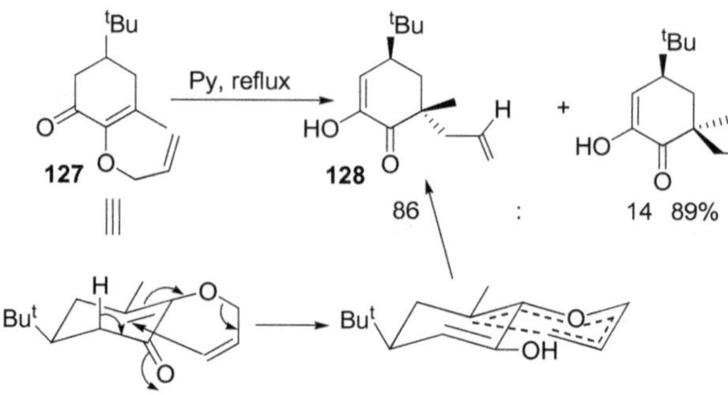

Ref. 88

Enzymatic Claisen rearrangement is observed in the rearrangement of chorismate **129** into prephenate **130** [89].

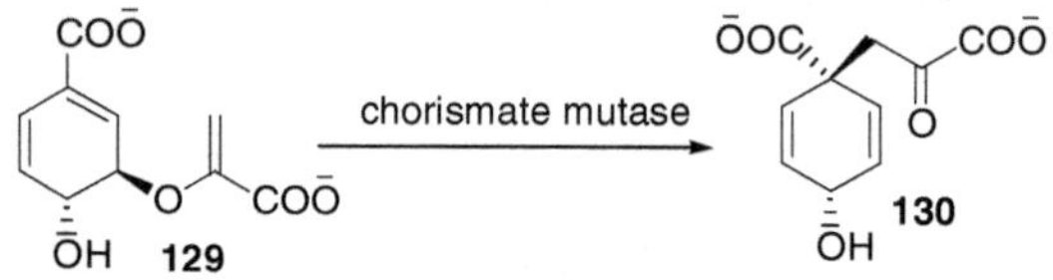

In the Claisen rearrangements of chiral substrates, the chirality is maintained in the products. The following examples are illustrative [90, 91]:

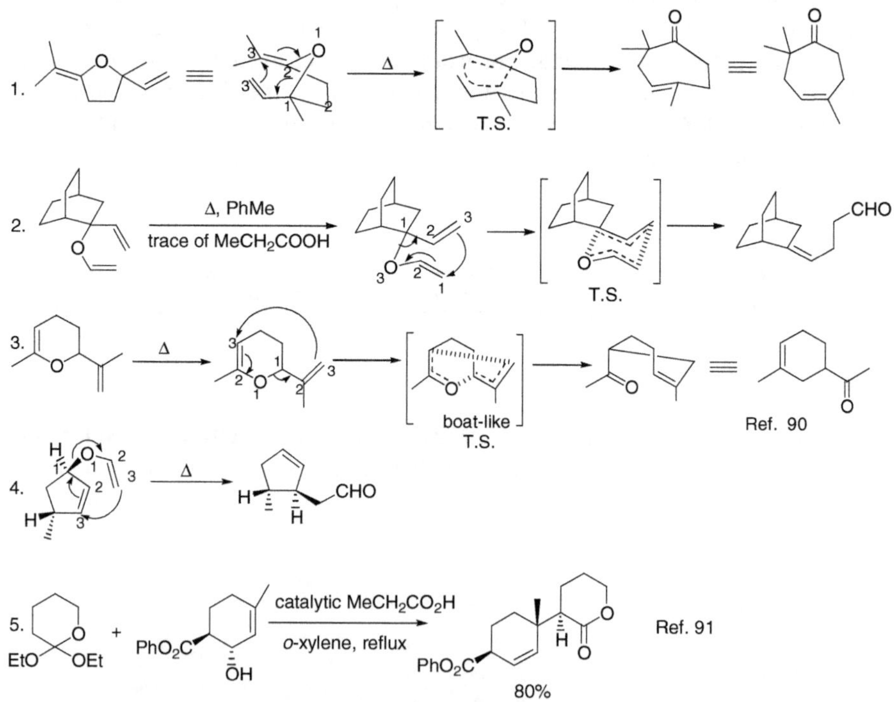

Ref. 90

Ref. 91

The Claisen rearrangements are extended to allyl and propargyl esters **131** and **132** in the presence of catalysts [92, 93].

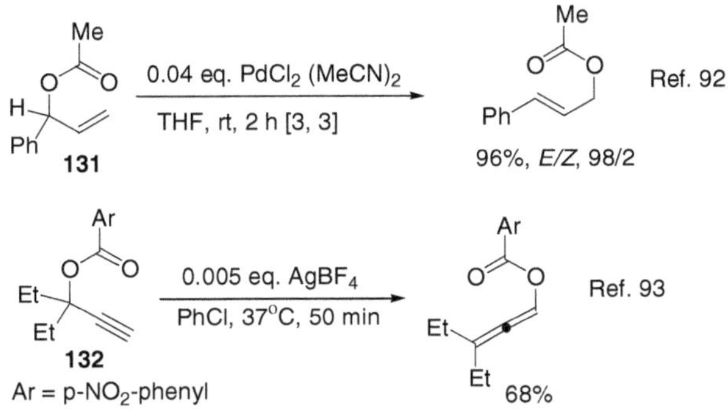

The Claisen rearrangements of aryl propargyl ethers **133** are used in the synthesis of flavonoids [94].

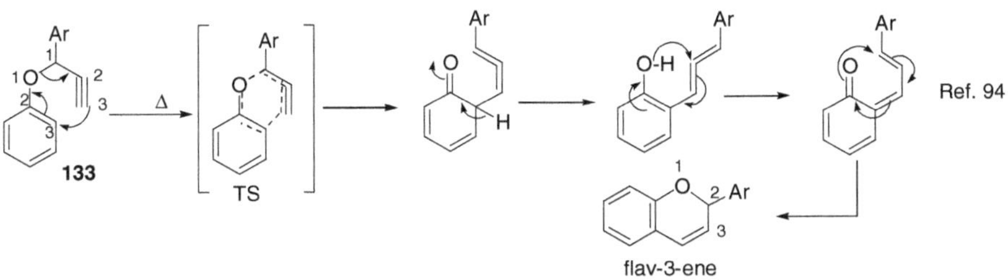

4.4.5 *The Thio- and Aza-Claisen Rearrangements*

4.4.5.1 The Thio-Claisen Rearrangements

[3,3]-Sigmatroic rearrangements of phenyl allyl sulfides or vinyl allyl sulfides are known as the *thio-Claisen rearrangements*. For example, **134** gives **135** [95].

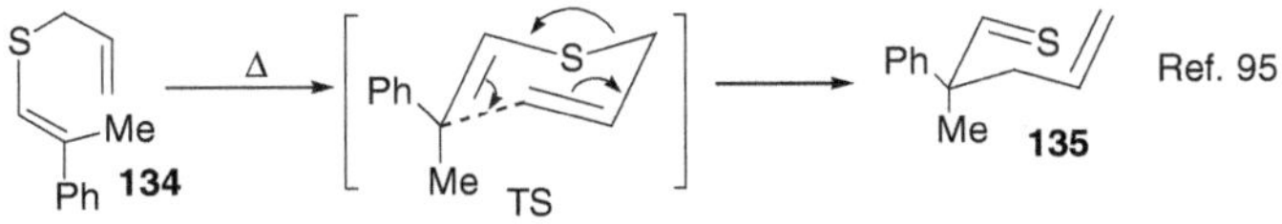

Thio-Claisen rearrangements (TCRs) of vinyl allyl sulfides **136** are useful in the synthesis of unsaturated aldehydes [96].

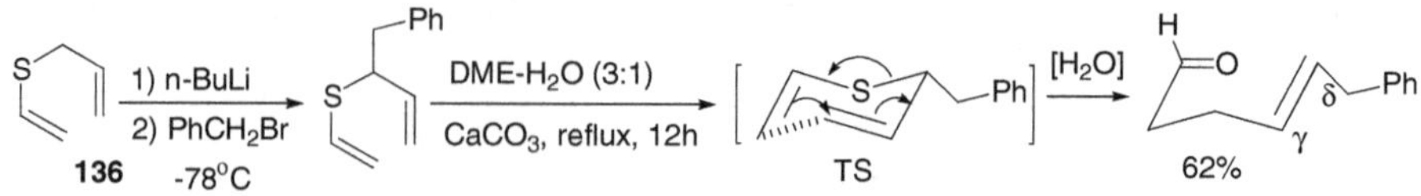

TCR provides an efficient synthetic route of several sulfur heterocycles. The following examples are illustrative [97–100]:

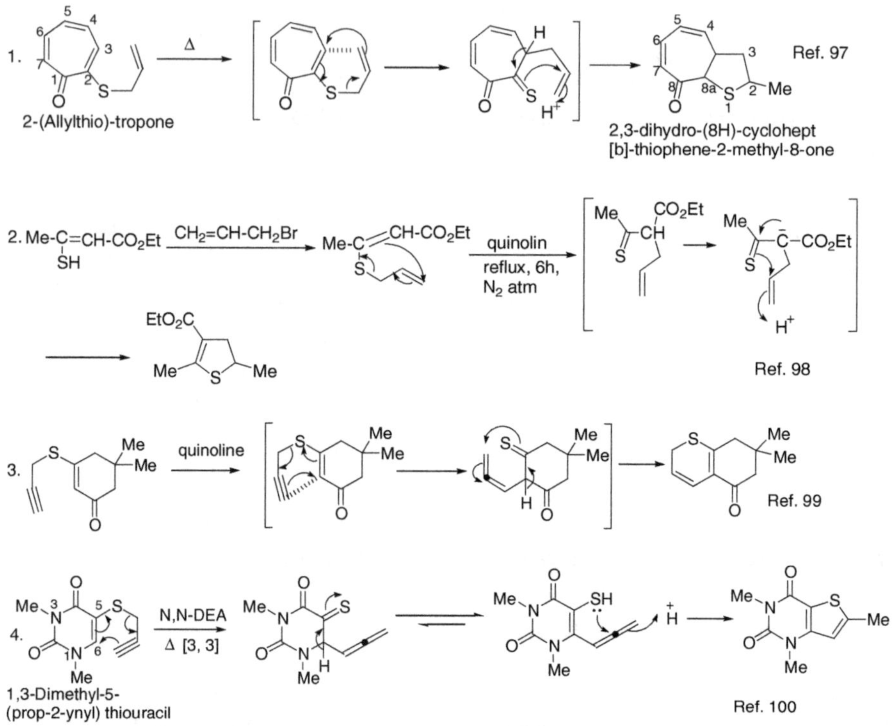

4.4.5.2 The Aza-Claisen Rearrangements

[3,3]-Sigmatropic rearrangements of N-allyl-N-vinylamines or N-allyl-N-arylamines are known as the *aza-Claisen* or *3-aza Cope rearrangements*. The rearrangements proceed through both chair and boat-like transition states, the chair TS being the preferred one. For example, **137** gives **138** [101].

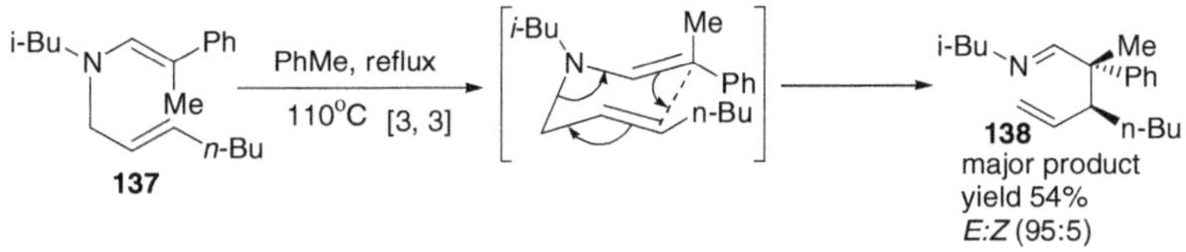

Ref. 101

The presence of a methyl substituent at C-4 of N-allyl-N-vinyl/aryl amines **139** and **140** improves the yield of the reaction by assuming the chair TS in preference [101, 102]. The presence of an electron-releasing substituent in the *p*-position of aromatic ring in amine **140a** increases the yield [102].

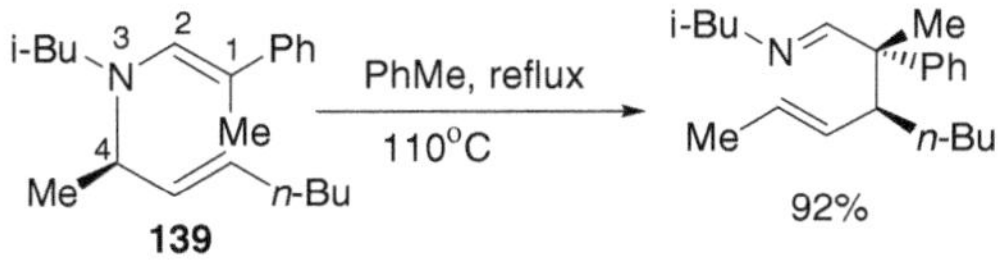

Ref. 101

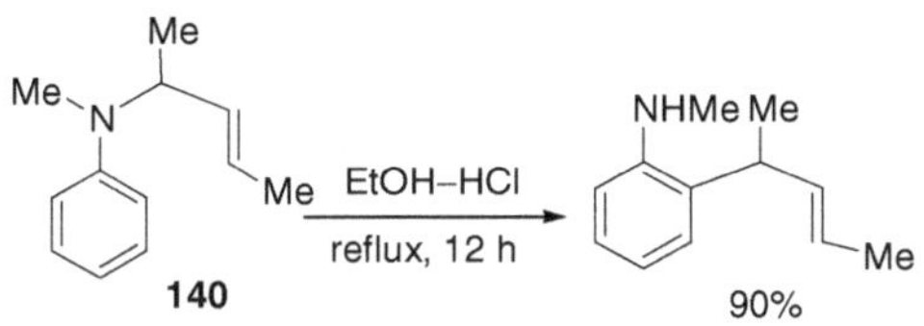

Ref. 102

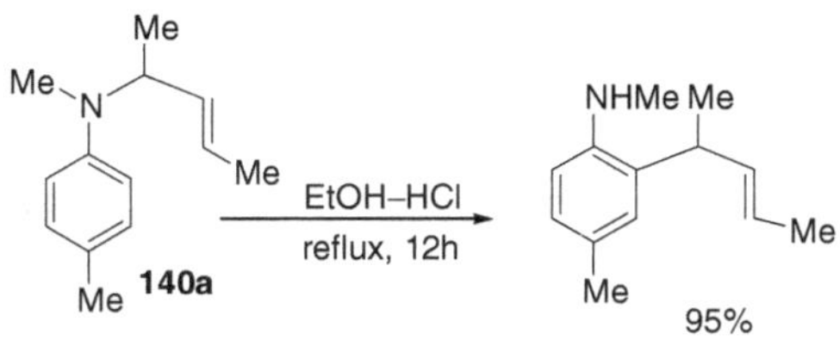

Ref. 102

The aza-Claisen rearrangement of chiral amine **141** gives diastereoselective product in moderate yield [103].

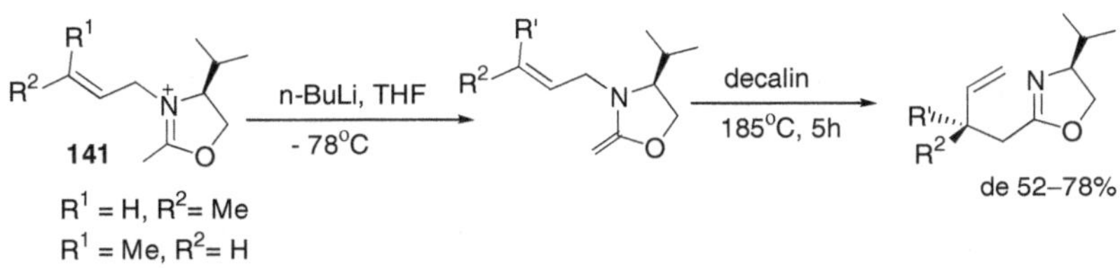

Ref. 103

The aza-Claisen rearrangement also occurs in N-allylic indoles **142** [104].

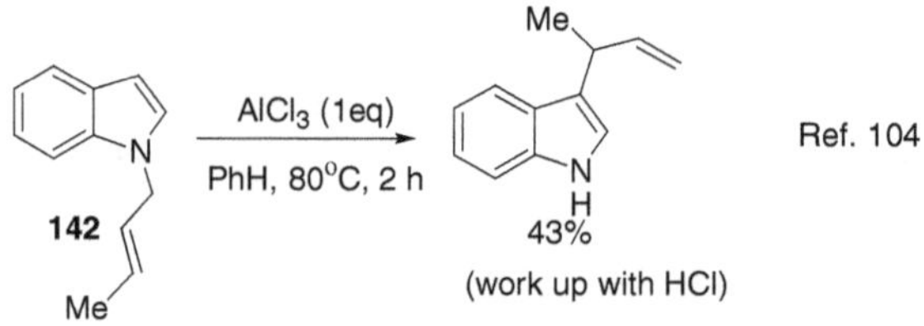

This aza-Claisen rearrangement is useful in the synthesis of medium-sized ring compounds. For example, **143** gives **144** in high yield [105].

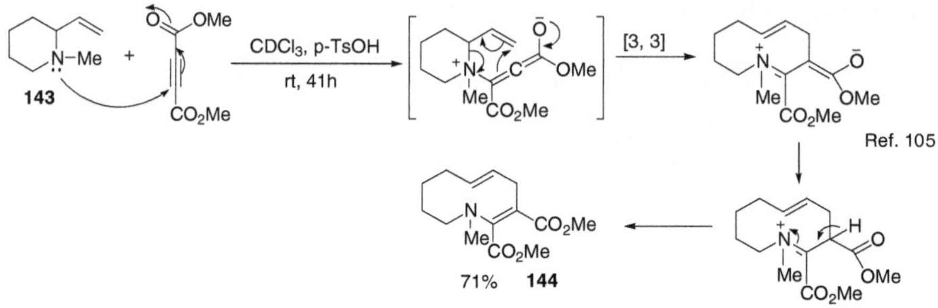

4.5 [2,3]-Sigmatropic Rearrangements

4.5.1 Overview of Different Types of [2,3]-Sigmatropic Rearrangements

Concerted [2,3]-sigmatropic rearrangements of allyl and benzyl ammonium ylides, allyl sulfonium ylides, allyl sulfoxides, selenoxides, amine oxides and anions of allyl ethers have been reported. These are illustrated by the following general examples:

a. Ammonium and sulfonium ylides

b. Allyl sulfoxides, selenoxides, and amine oxides

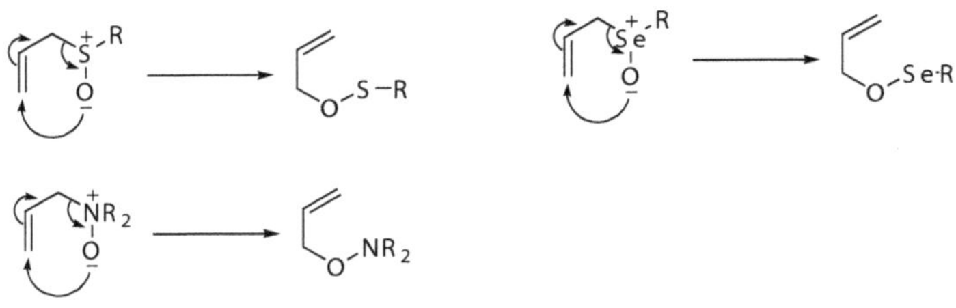

c. Anions of allyl ethers

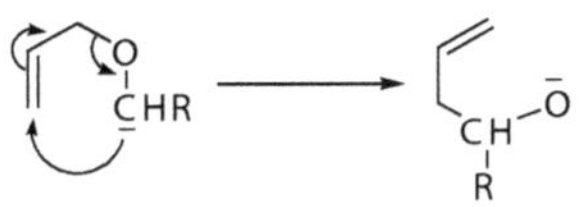

All of these rearrangements occur through a five-membered ring TS following Huckel-type topology of 6e process with an *endo*-orientation of the substituent of the migrating fragment. For example, in the rearrangement of an allyl sulfoxide, the preferred *endo*-TS **145** has the substituent on sulfur pointed toward allylic fragment. For a branched allyl sulfoxide, the bulkier group prefers a pseudoequatorial position in the TS **146** to give *E*-isomer as major product [106, 107].

preferred *endo* TS

145

less preferred *exo* TS

For example,

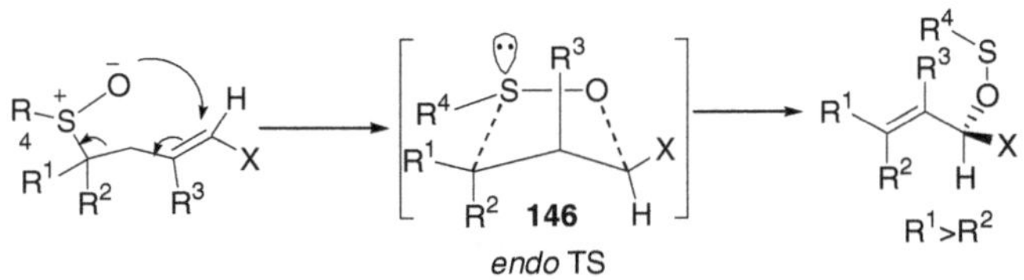

4.5.2 [2,3]-Sigmatropic Rearrangements of Allyl Ammonium Ylides

Allylic ammonium ylides generated in the presence of a strong base undergoes [2,3]-sigmatropic rearrangement. The following reactions are the illustrative examples [108–115]:

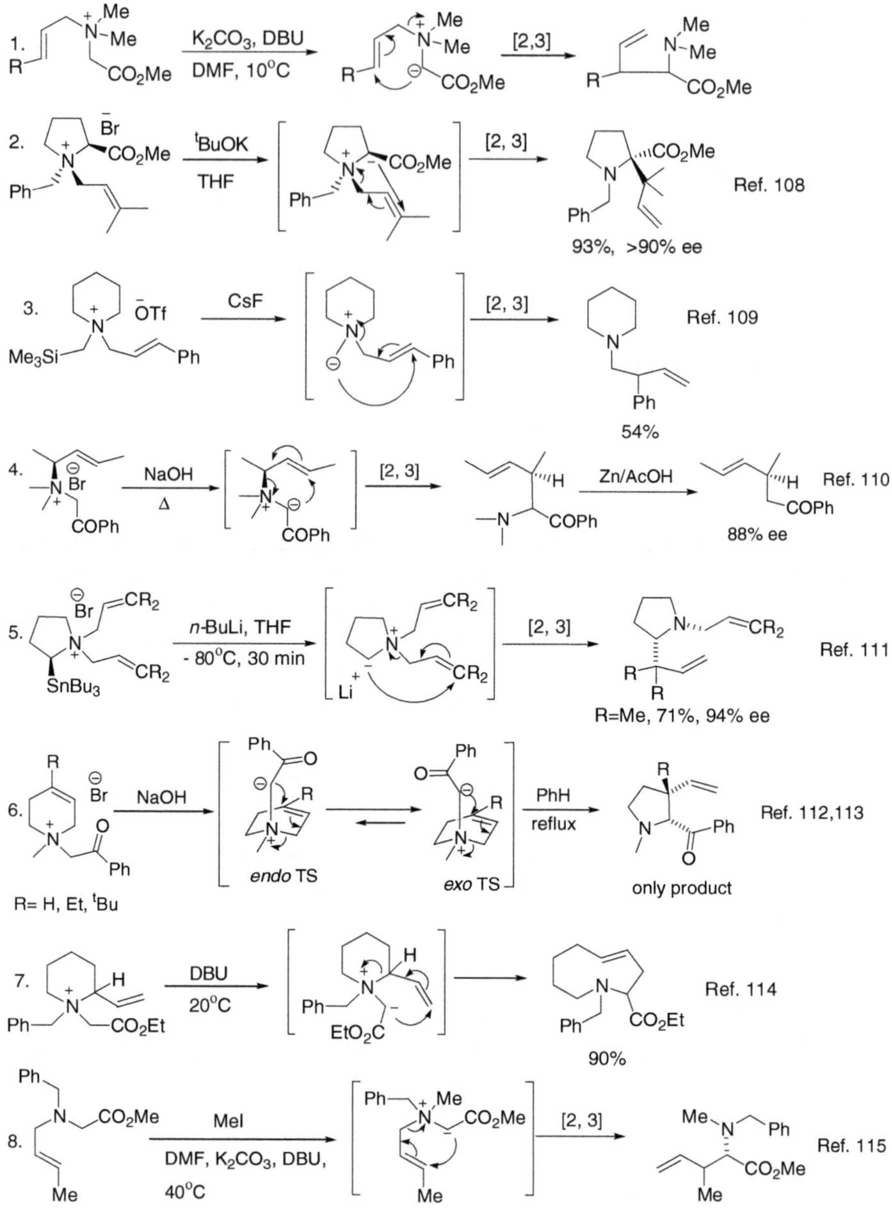
1.
K2CO3, DBU
DMF, 10°C
[2,3]
2.
tBuOK
THF
[2, 3]
Ref. 108
93%, >90% ee
3.
CsF
[2, 3]
Ref. 109
54%
4.
NaOH
Δ
[2, 3]
Zn/AcOH
Ref. 110
88% ee
5.
n-BuLi, THF
- 80°C, 30 min
[2, 3]
Ref. 111
R=Me, 71%, 94% ee
6.
NaOH
endo TS
exo TS
PhH
reflux
Ref. 112,113
only product
R= H, Et, tBu
7.
DBU
20°C
Ref. 114
90%
8.
MeI
DMF, K2CO3, DBU,
40°C
[2, 3]
Ref. 115
63%

4.5.3 *[2,3]-Sigmatropic Rearrangements of Benzyl Ammonium Ylides: The Sommelet–Hauser Rearrangement*

[2,3]-Sigmatropic rearrangements of benzyl quaternary ammonium salts are known as the *Sommelet–Hauser rearrangements*. The treatment of the benzyl quaternary ammonium salt with sodium amide or other alkali metal amide generates benzyl ammonium ylide. The following examples are illustrative [108, 116–119]:

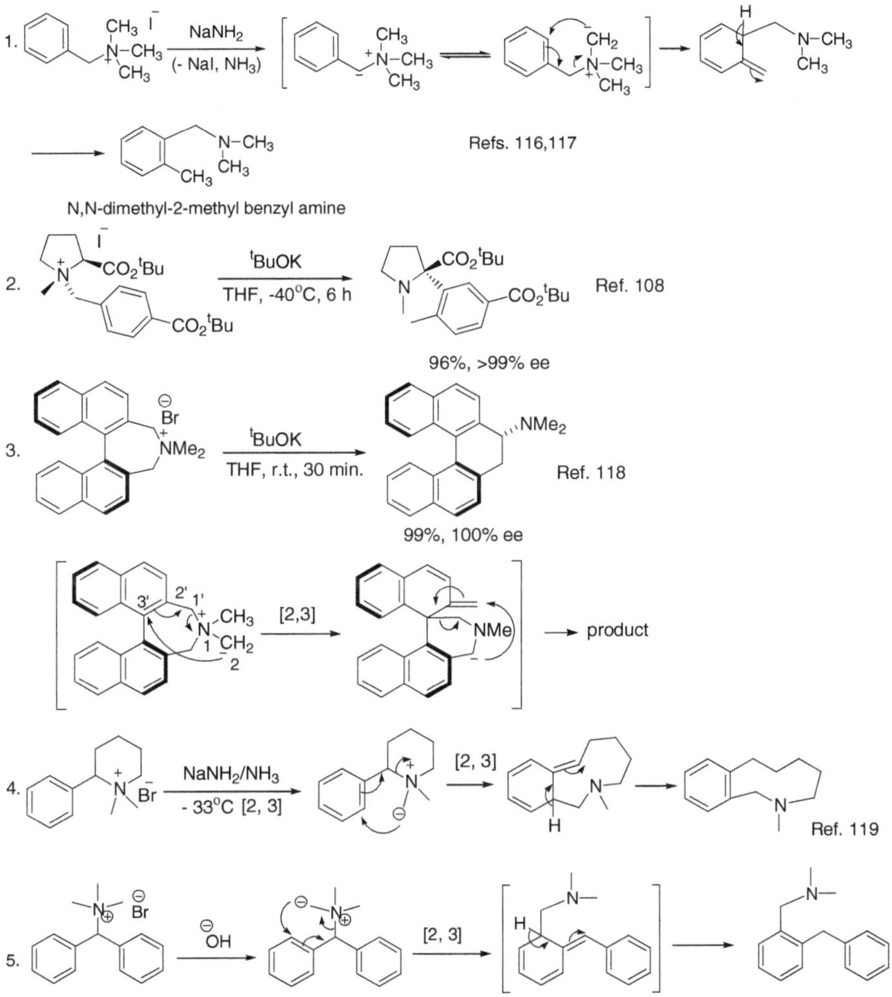

4.5.4 [2,3]-Sigmatropic Rearrangement of Allyl Sulfonium Ylides

Allylic sulfonium ylides undergo [2,3]-sigmatropic shifts. The presence of carbanion stabilizing substituent accelerates the reaction-rate by stabilization of the ylide. These reactions are utilized in the synthesis of medium-sized ring compounds. The following examples are illustrative [21, 120–123]:

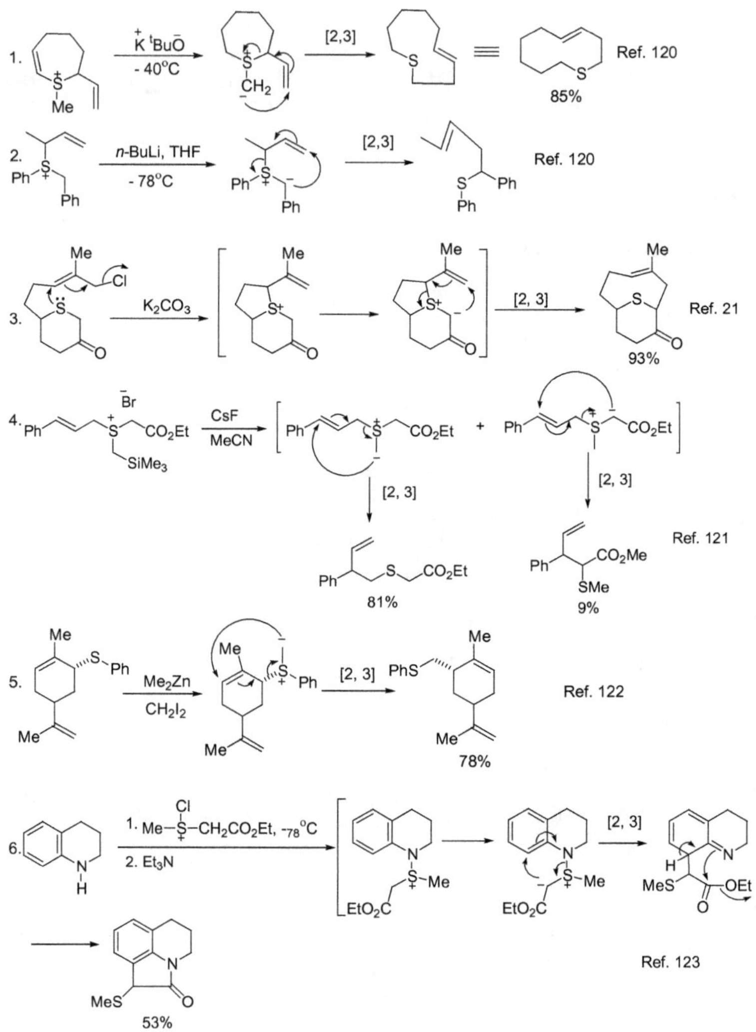

7. 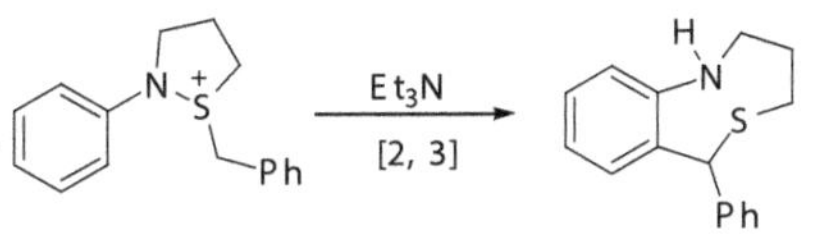

4.5.5 [2,3]-Sigmatropic Rearrangements of Allyl Sulfoxides: The Mislow–Evans Rearrangements

[2,3]-Sigmatropic rearrangements of allyl sulfoxides to allyl sulfenates are known as the *Mislow–Evans rearrangements*. Allylic sulfoxide with a bulky substituent at the allylic position was converted to the corresponding *E*-allylic alcohols with high stereoselectivity. Some illustrative examples are [124–127] as follows:

1. 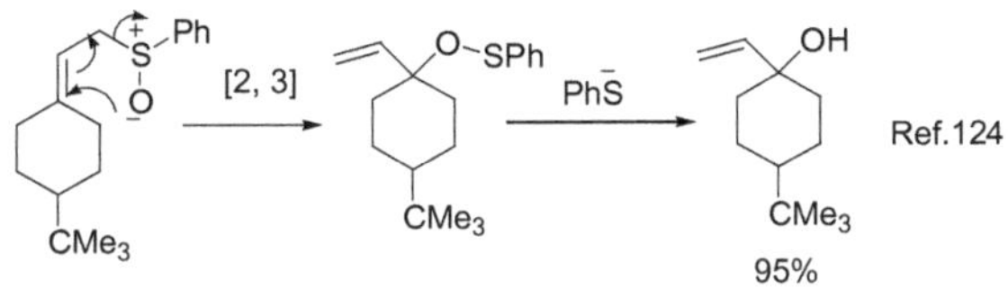

2. 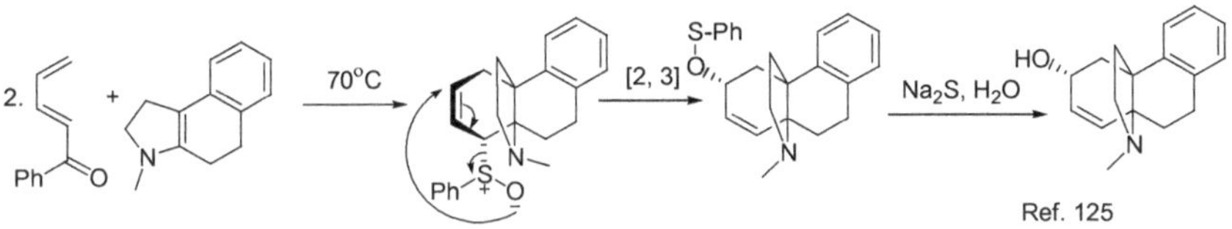

3. 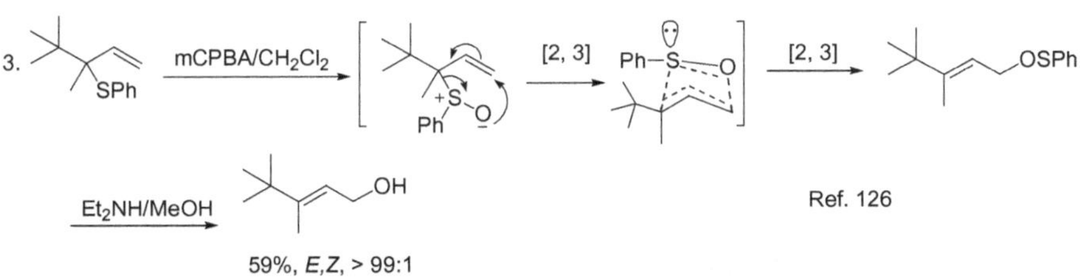

4. 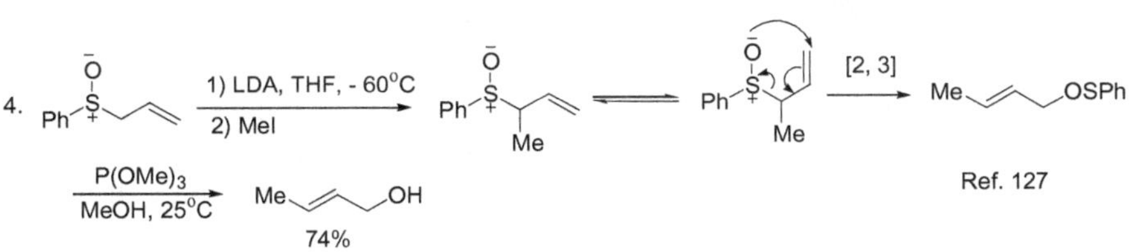

4.5.6 [2,3]-Sigmatropic Rearrangements of Allyl Selenoxides

Several thermal [2,3]-sigmatropic rearrangements of allyl selenoxides have been reported. The following examples are illustrative [128–130]:

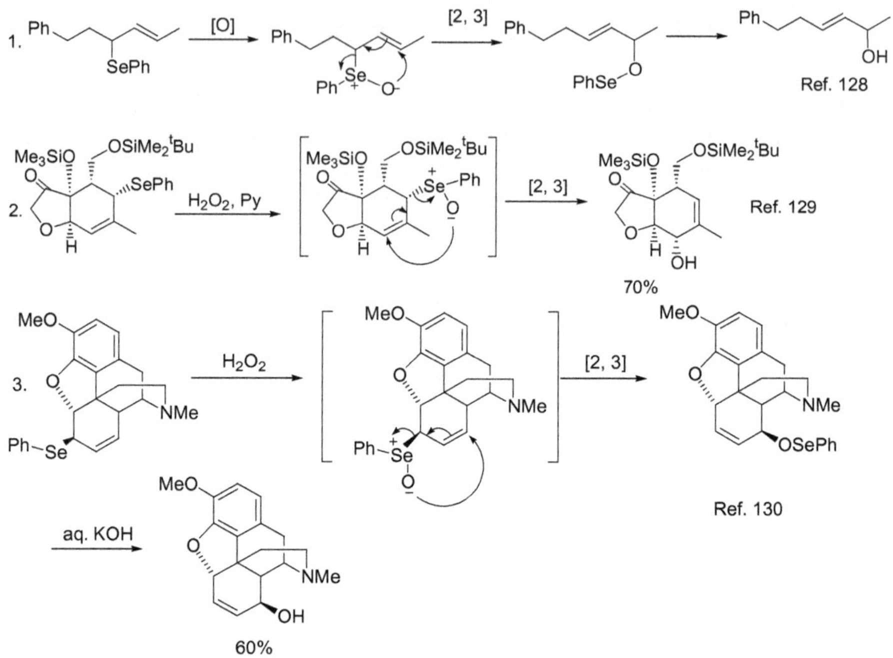

4.5.7 [2,3]-Sigmatropic Rearrangements of Anions of Allyl Ethers: The Wittig and Aza-Wittig Rearrangements

[2,3]-Sigmatropic rearrangements of the anions of allyl and benzyl ethers are known as the *Wittig rearrangements*. The anions are generated by treatment of allyl or benzyl ethers with a strong base. The following reactions are illustrative examples [21, 131–135]:

1. 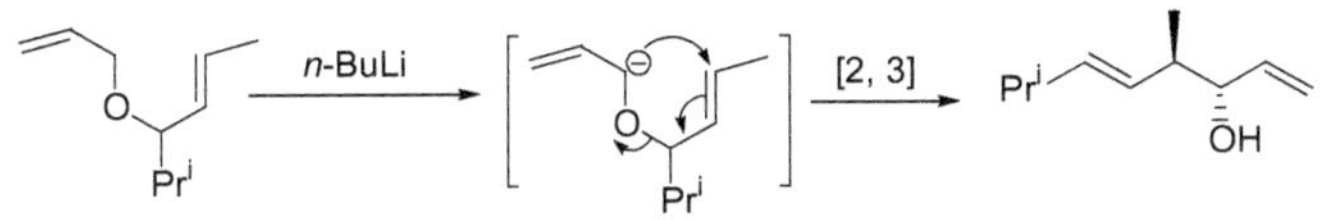

Ref. 21

2. 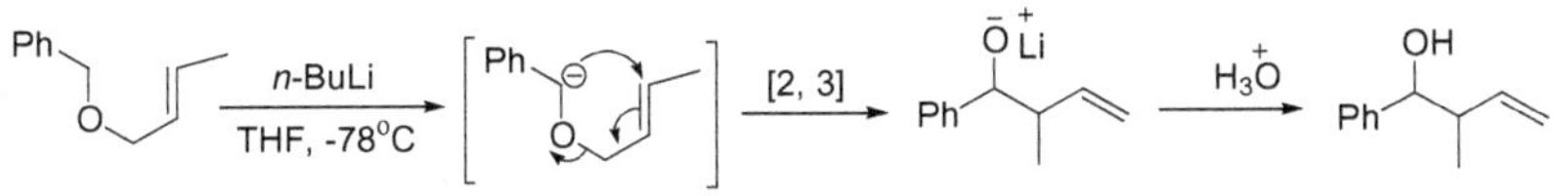

Ref.21

3. 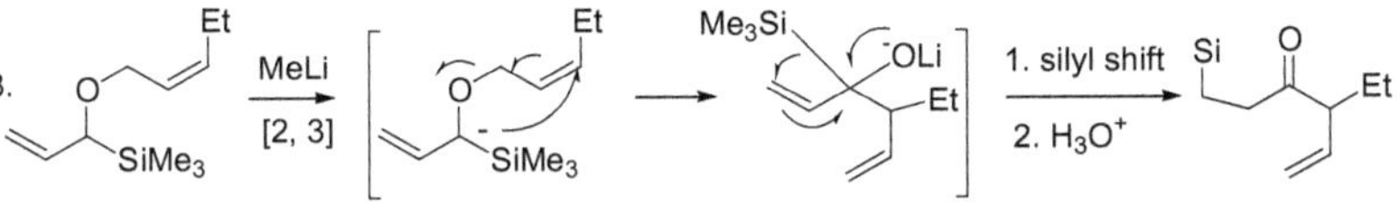

Ref. 131

4. 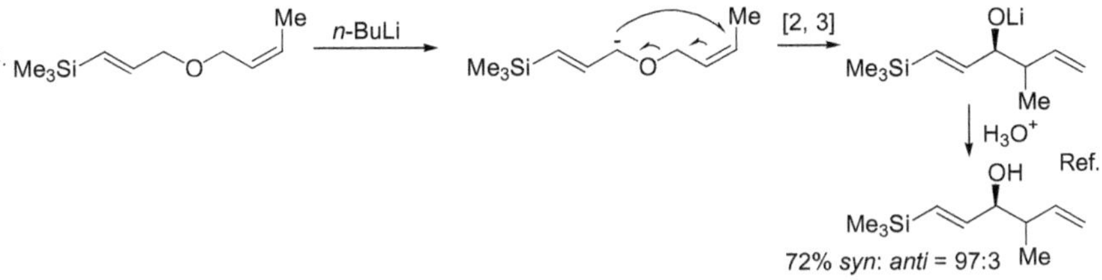

Ref. 132

5. 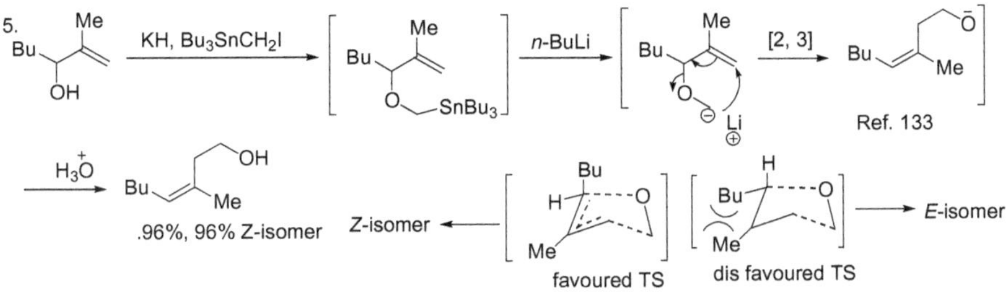

Ref. 133

6. 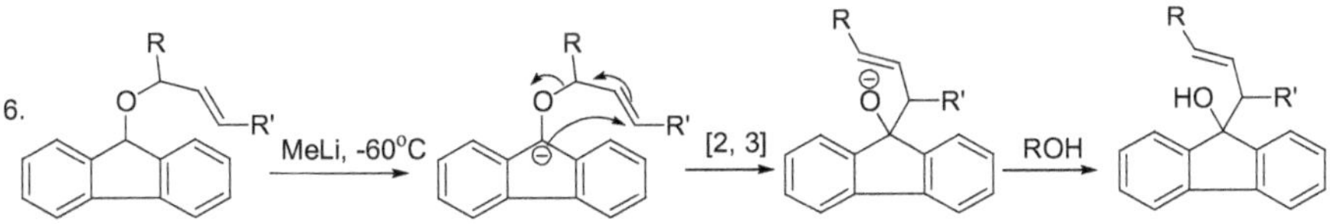

Ref. 134

7. 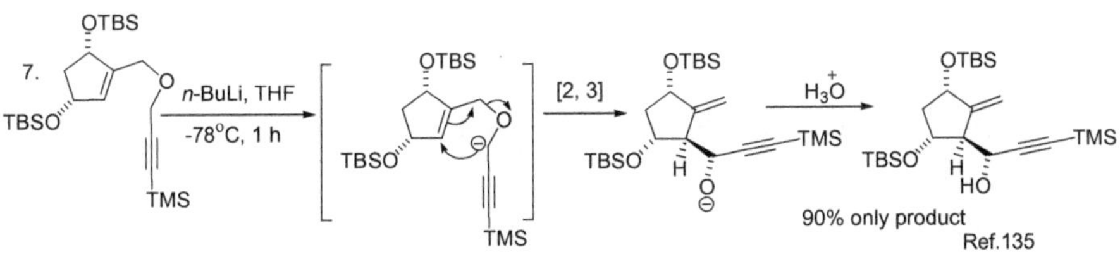

Ref.135

The [2,3]-sigmatropic rearrangements of the anions of vinyl aziridines are known as the *aza-Wittig rearrangements*. For example, **147** gives **148** [136].

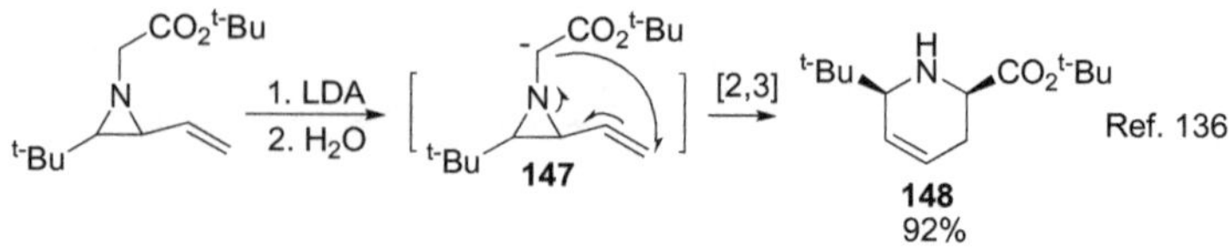

4.5.8 [2,3]-Sigmatropic Rearrangements of Allyl Amine Oxides: The Meisenheimer Rearrangement

[2,3]-Sigmatropic rearrangement of allyl amine oxides is known as the *Meisenheimer rearrangement*. This rearrangement provides O-allyl hydroxylamine derivatives. Some examples are [137, 138] as follows:

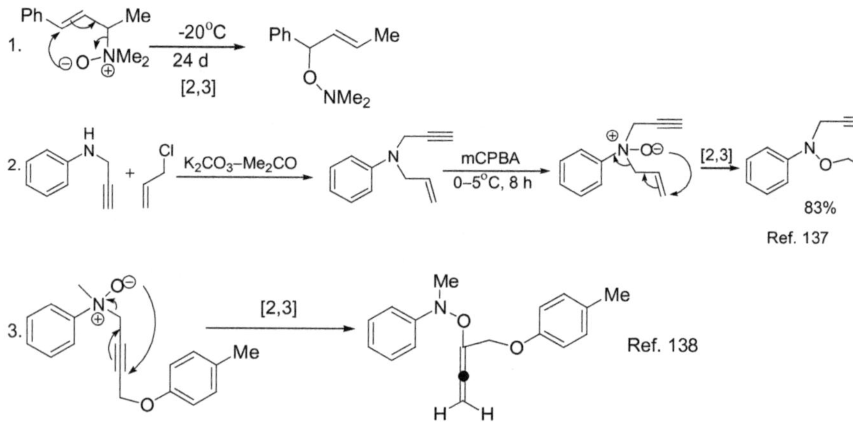

4.6 [3,5]-Sigmatropic Rearrangement

3,5-Sigmatropic shifts are not pericyclic because suprafacial overlapping of cyclic p-orbitals is not involved. In these shifts, the orbitals from the breaking bond and the lone pair overlap with the p-orbitals in the TS. These reactions have planar nonaromatic eight-centered cyclic transition states with orthogonal orbital overlapping, where p_z orbital of the nucleophilic site acts as HOMO and p-orbital of cyclohexadienyl group acts as LUMO. Either chair or flattened boat TS is involved with an antarafacial geometry. For example, acyloxycyclohexadienone **149** gives the major product **150** from [3,5]-rearrangement and minor product **151** from [3,3]-rearrangement [139].

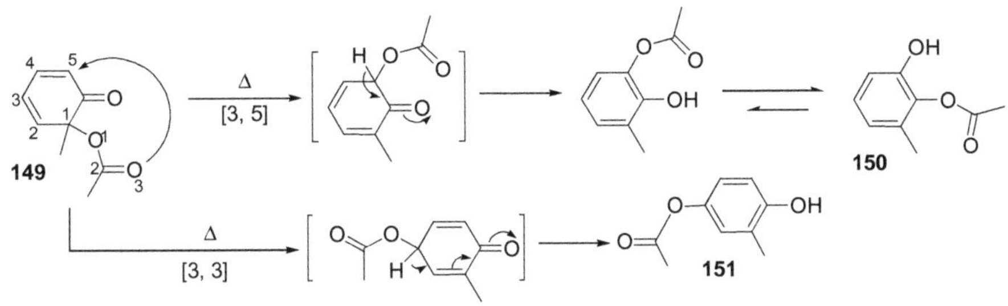

4.7 [4,5]-Sigmatropic Rearrangement

[4,5]-Sigmatropic shift is thermally allowed process as it involves 10 e (4n+2) process and TS is Huckel-type and aromatic. The ammonium salt **152a** on treatment with a base gives ammonium ylide **152**, which on [4,5] -shift affords **153**.

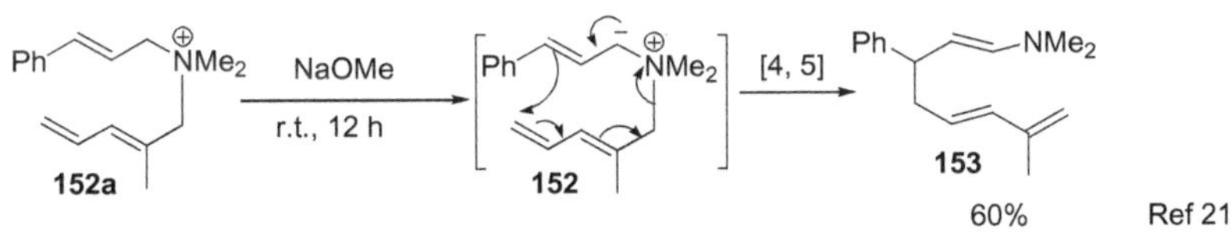

4.8 [5,5]-Sigmatropic Rearrangement

[5,5]-Sigmatropic shift is thermally allowed suprafacial process of 8πe and 2σe. The thermal rearrangements of aryl ethers **154 to 155**, 1,2-(1,3-butadienyl) cyclohexyl enolates **156 to 157** and hydrazobenzene **158** to p-benzidine **159** are illustrative

examples [140–142]. N,N′-Diarylhydrazide **160** with substituents at the *ortho*-position undergoes highly regioselective [5,5]-shift to give **161** in high yield [143].

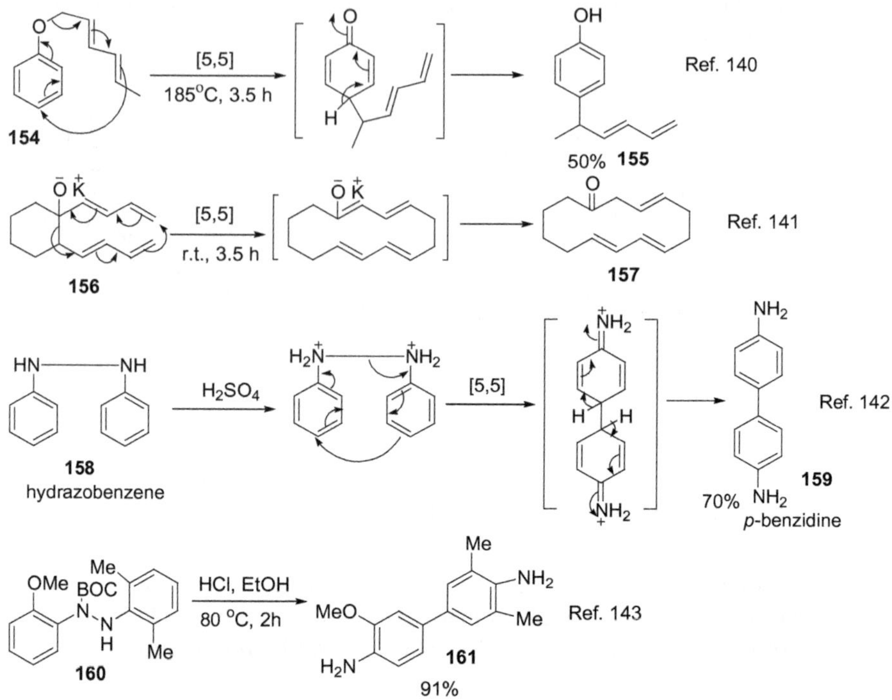

4.9 [9,9]-Sigmatropic Rearrangement

[9,9]-Sigmatropic shift is thermally allowed process and it involves a Huckel-type TS of 18e (16π and 2σe). For example, *bis*[4-(2-furyl)-phenyl] diazane **162** gives hydrochloride salt of 5,5′-*bis*(4-amino phenyl)-2,2′-bifuryl **163** in high yield in acidic solution [144].

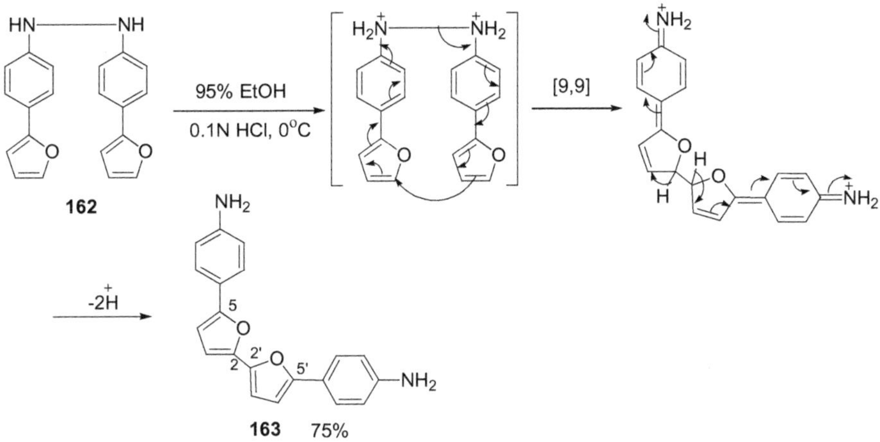

4.10 Problems

4.10.1. Indicate which of the following reactions are orbital symmetry allowed processes and which are forbidden. Also indicate the order of sigmatropic shift for each reaction. Each reaction may take place in one or more steps. Justify your answer.

(a)

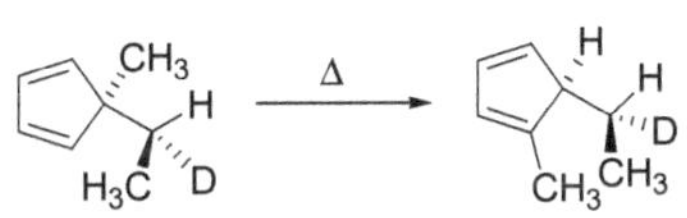

(b)

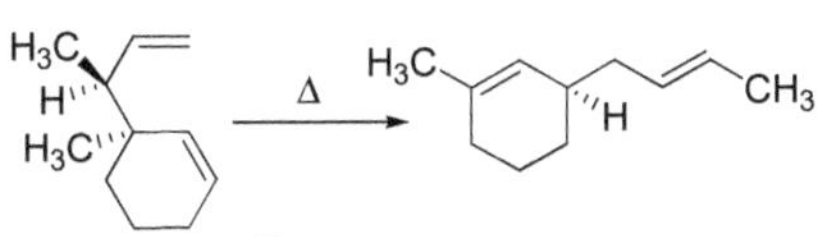

(c)

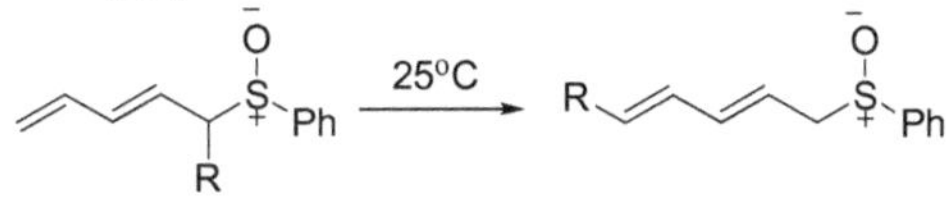

(d)

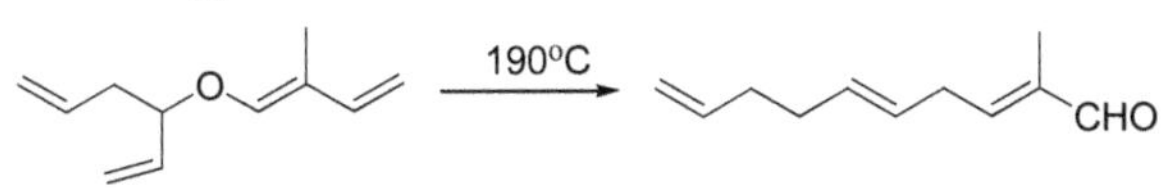

(e)

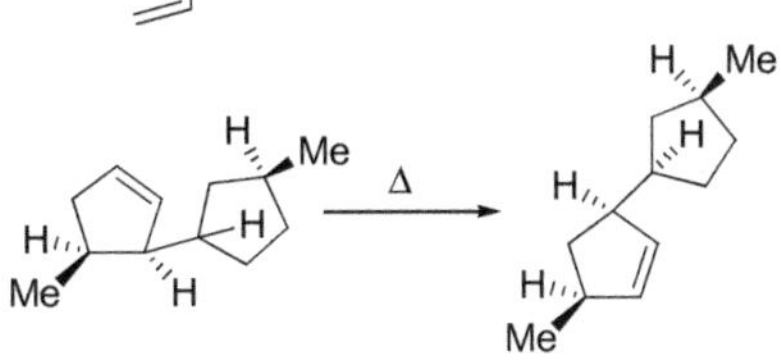

4.10.2. Suggest a mechanism for each of the following transformations. More than one step may be involved in each case. Predict the major and minor products when more than one product is expected.

(a) 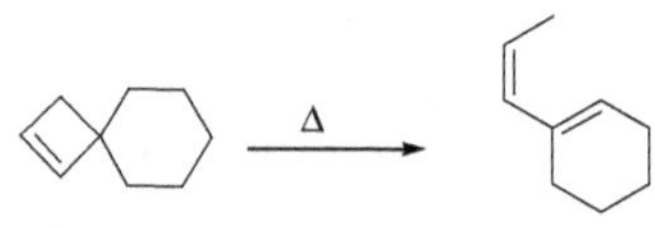

(b) 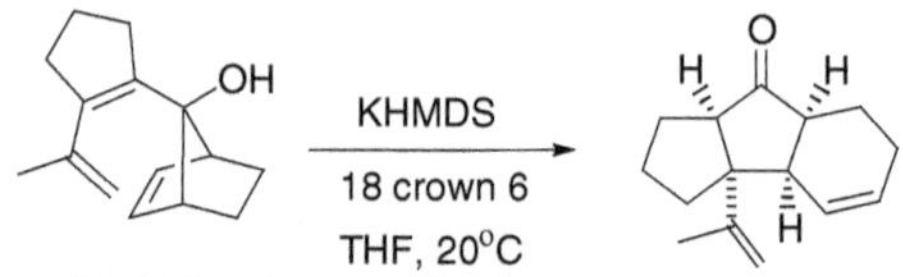

(Ref: Tetrahedron, **1988**, *44*, 3139)

(c) 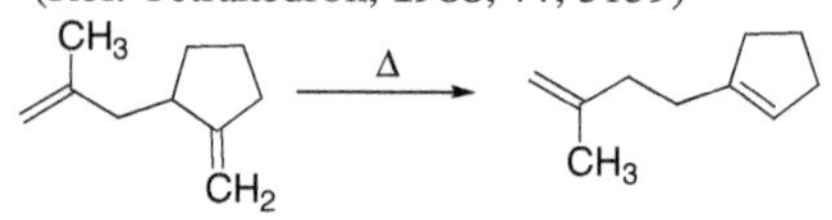

(d) 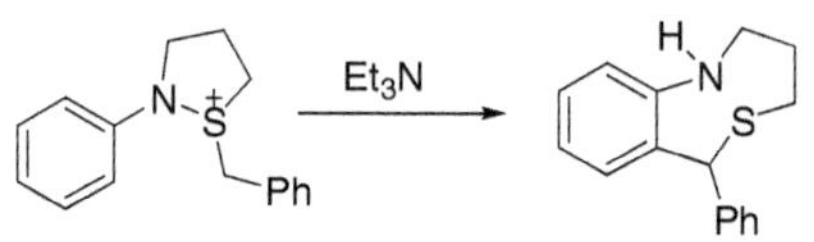

(e) 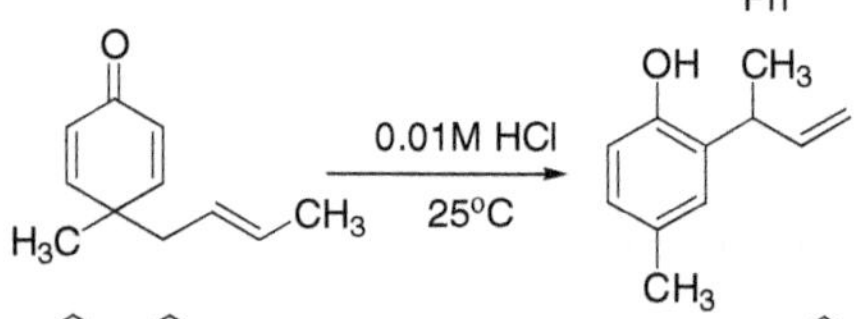

(f) 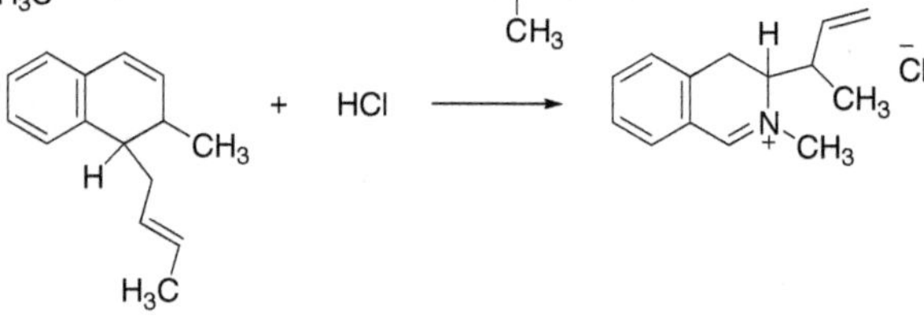

(g) 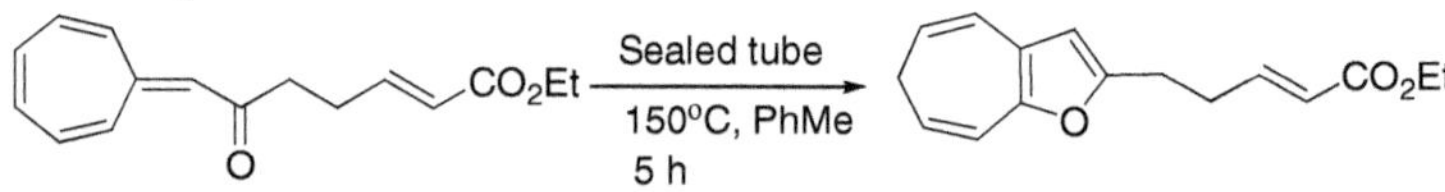

(Ref. Tetrahedron Lett., **1981**, *28*, 1367)

(h) 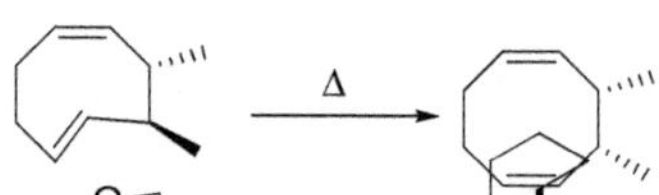

(i) 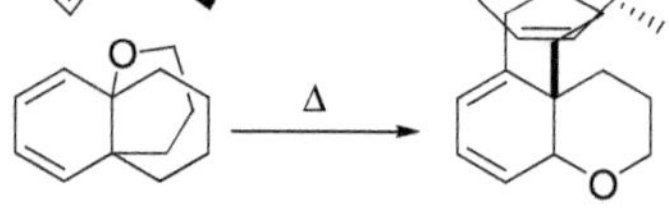

(j) 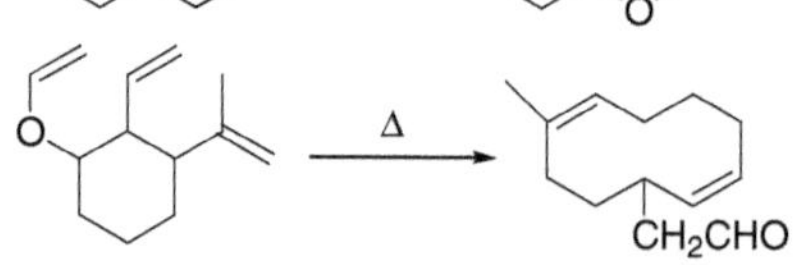

(k) 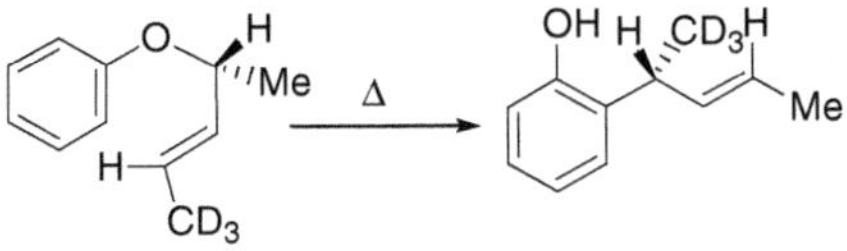

(l) 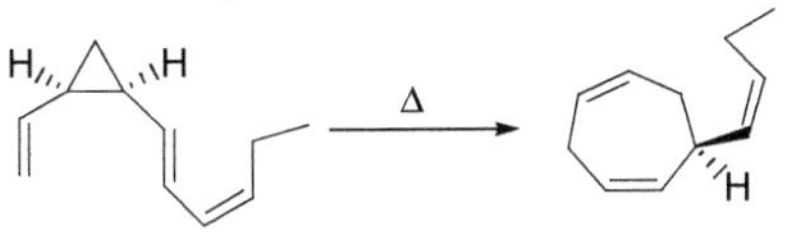

(m) 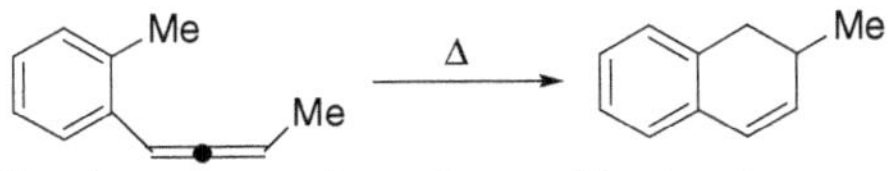

4.10.3. Predict the expected product with structure and stereochemistry for each of the following reactions:

(a) 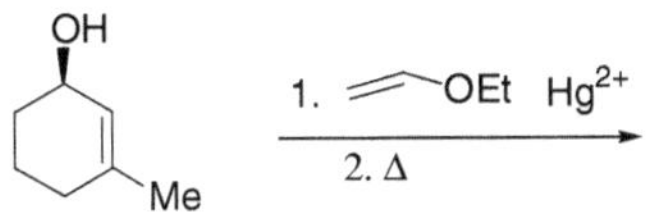

(b) 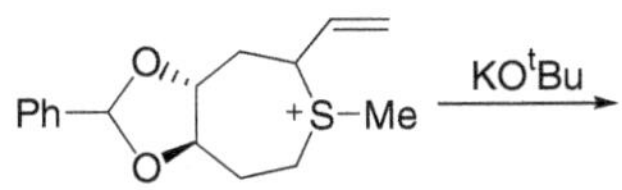

(c) 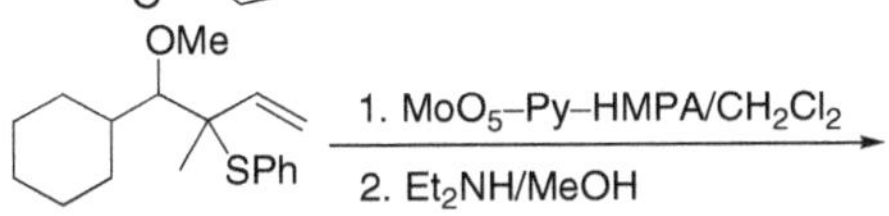

(d) 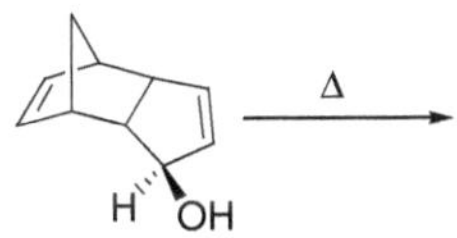

(e) 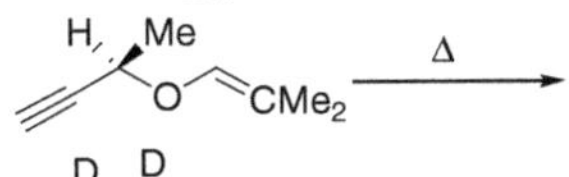

(f) 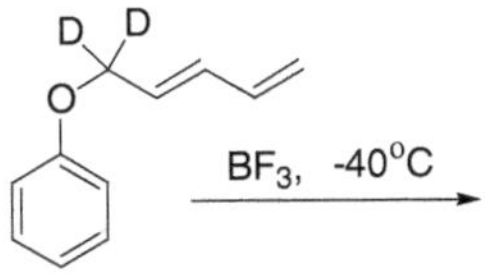

(g) 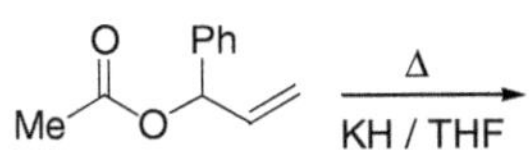

(h) 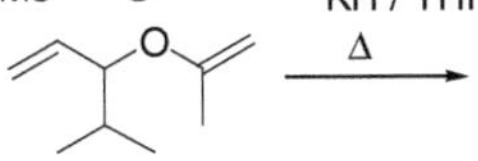

(i) 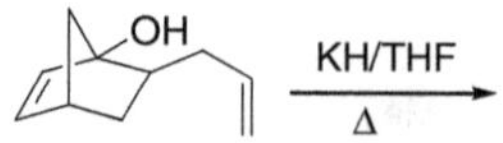

(j) 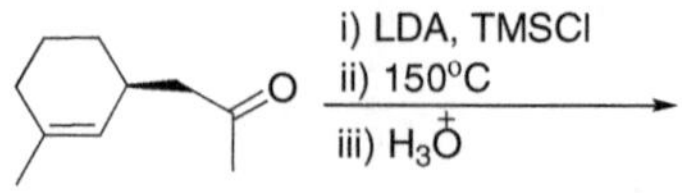

(k) 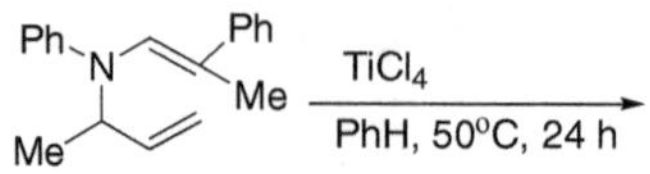

(l) 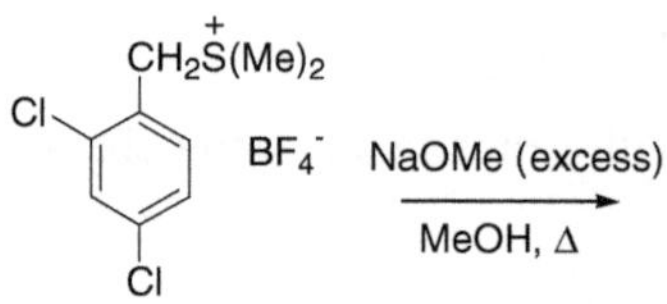

(m) 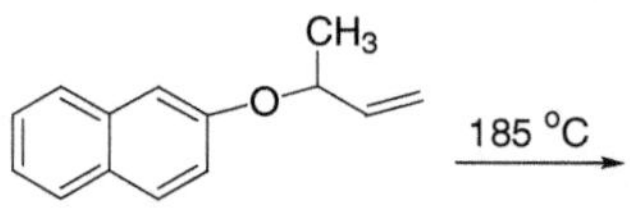

(n) 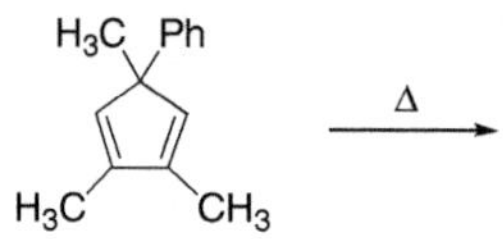

(o) 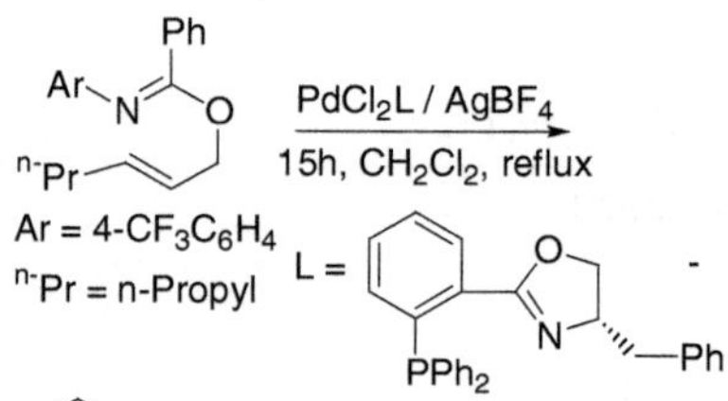

(p) 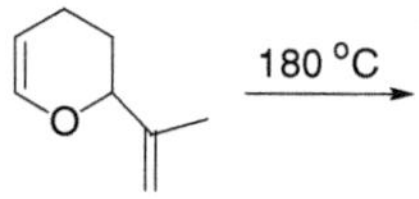

(q) 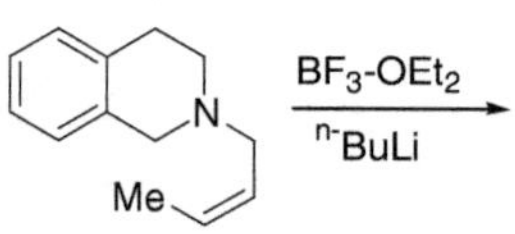

(r) 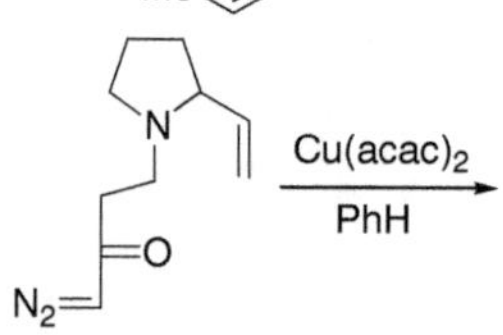

(s) 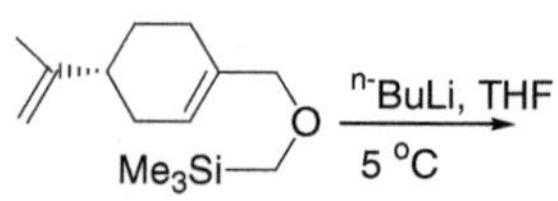

(t) 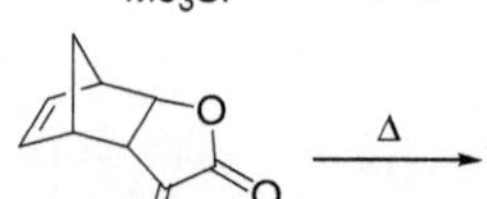

4.10.4. Suggest the mechanism for each of the following reactions, which may occur in two or more pericyclic steps not restricted to sigmatropic shifts

(a)

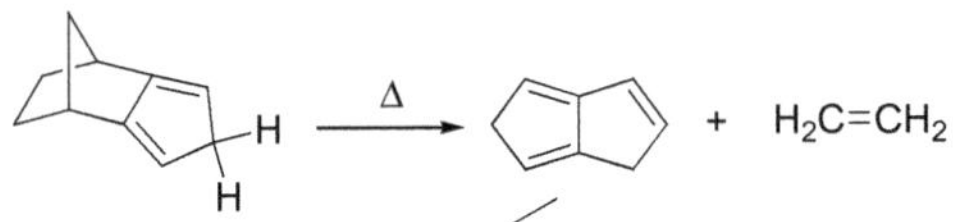

(b)

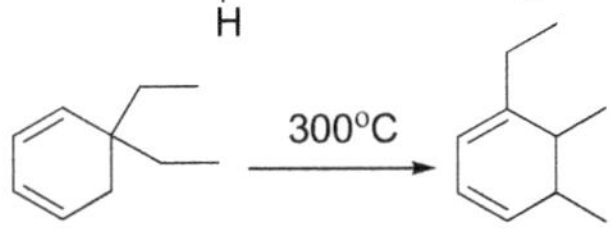

(c)

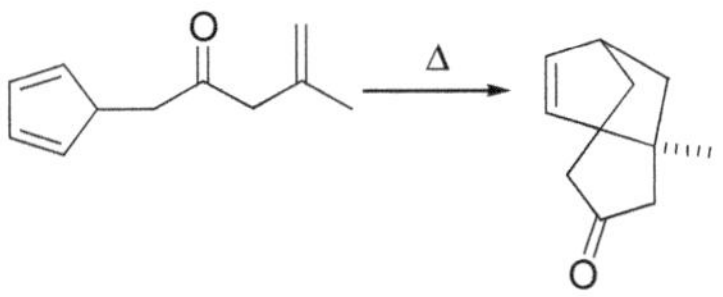

(d)

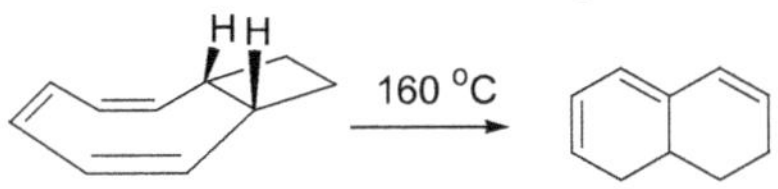

(e)

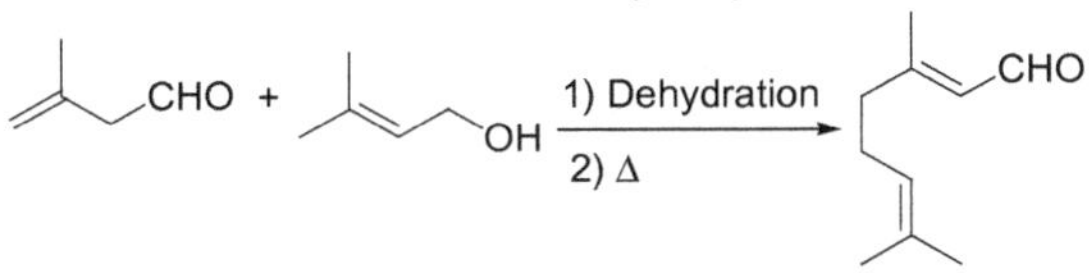

(f)

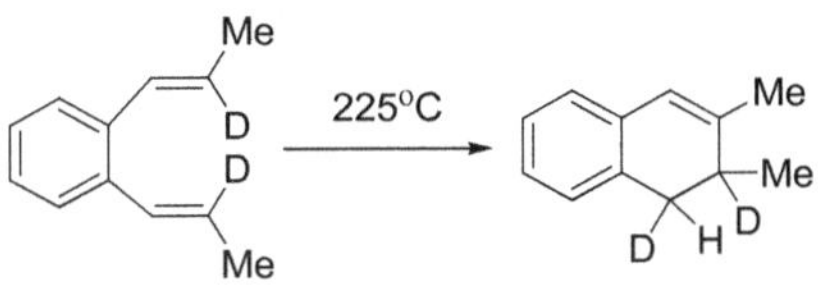

(g)

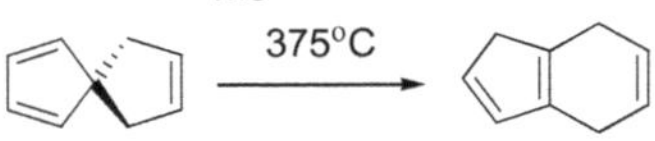

(h)

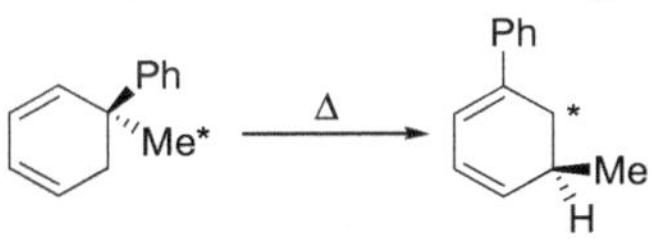

(i)

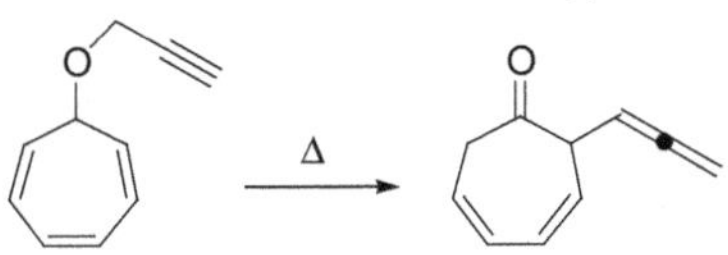

(j)

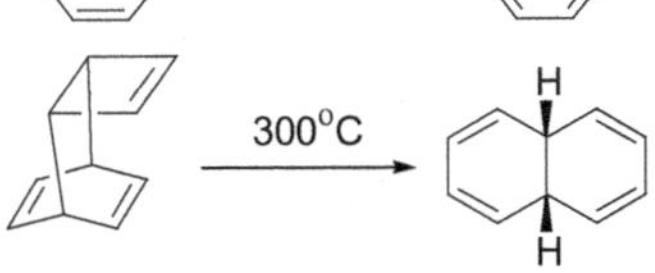

(k)

4.11 Further Reading

1. Hill RK (1991). In: Paquette LA (ed) comprehensive organic synthesis, vol 5. Pergamon, New York, p 827
2. Bruckner R (1991) [2,3]-Sigmatropic rearrangements. In: Winterfeldt W (ed) Comprehensive organic synthesis, vol 6. Pergamon, New York
3. Castro AMM (2004) Claisen rearrangement over the past nine decades. Chem Rev 104:2939
4. Nakai T, Mikami K (1986) [2,3]-Sigmatropic rearrangement of allyl ethers. Chem Rev 86:885

References

1. Woodward RB, Hoffmann R (1971) The conservation of orbital symmetry. Academic Press, New York, p 85
2. Woodward RB, Hoffmann R (1965) J Am Chem Soc 87:2511
3. Shea KJ, Stoddard GJ, England WP, Haffner CD (1992) J Am Chem Soc 114:2635
4. Berson JA (1968) Acc Chem Res 1:152; Berson JA, Nelson GL (1967) J Am Chem Soc 89:5503
5. Roth WR, Friedrich A (1969) Tetrahedron Lett 2607; Masamune S, Takada S, Nakatasuka N, Vukov R, Cain EN (1969) J Am Chem Soc 91:4322
6. Cookson RC, Kemp JE (1971) Chem Commun 385; Goldschmidt Z, Cramer B (1988) Chem Soc Rev 17:229
7. Cookson RC, Hudec J, Sharma M (1971) Chem Commun 107; Sharma M J Am Chem Soc 97:1153
8. Wolinsky J, Chollar B, Baird MD (1962) J Am Chem Soc 84:2775
9. Marvell EN, Caple G, Schatz B, Pippin W (1973) Tetrahedron 29:378; Skattebol L (1966) J Org Chem 31:2789
10. Roth WR (1964) Tetrahedron Lett 1009
11. Roth WR, Koeing J, Stein K (1970) Chem Ber 103:426
12. Heimgartner H, Zaindely J, Hansen HJ (1973) Helv Chim Acta 56:2924; Morrison RT, Boyd RN (2001) Organic chemistry, 6th edn. Prentice Hall, New Delhi
13. Mironov VA, Chizhov OS, Kimelfeld IM, Akhrem AA (1969) Tetrahedron Lett 499
14. Mironov VA, Sobolev EV, Elizarova AN (1963) Tetrahedron 19:1939
15. Boersma MAM, de Haan JW, Kloosterziel H, van de Ven LJM (1970) J Chem Soc, Chem Commun 1168
16. Fields DJ, Jones DW, Kneen G (1976) Chem. Commun 873; Miller LL, Greisinger R, Boyer RF (1969) J Am Chem Soc 91:1578; Manning C, Mc Clary MR, Mc Cullough JJ (1981) J Org Chem 46:919
17. Berson JA, Willcott MR (1966) J Am Chem Soc 88:2494
18. Barrack SA, Okamura WH (1986) J Org Chem 51:3201
19. Okamura WH (1983) Acc Chem Res 16:81
20. Kiefer EF, Tanna CH (1969) J Am Chem Soc 91:4478
21. Fleming I (2002) Pericyclic reactions. Oxford University Press, New York
22. Schlatmann JLMA, Pot J, Havinga E (1964) Rec Trav Chim 83:1173
23. Schweizer EE, Crouse DM, Dalrymple DI (1969) Chem Commun 354; Baldwin JE, Reddy VP (1987) J Am Chem Soc 109:8051

24. Hoeger CA, Johnson AD, Okamura WH (1987) J Am Chem Soc 109:4690; Okamura WH, Hoeger CF, Miller KJ, Reischl W (1988) J Am Chem Soc 110:973
25. Klarner FG (1972) Angew Chem Int Ed Engl 11:832
26. Jones LB, Jones VK (1968) J Am Chem Soc 90:1540
27. Hill RK, Gilman NW (1967) J Chem Soc Chem Commun 619
28. Dewar MJS, Wade LE (1973) J Am Chem Soc 95:290; (1977) Ibid 99: 4417; Wehrli R, Schmid H, Bellus DE, Hansen HJ (1977) Helv Chim Acta 60:1325
29. Dollinger M, Henning W, Kirmse W (1982) Chem Ber 115:2309
30. Cope AC, Hardy EM (1940) J Am Chem Soc 62:441
31. Lutz RP, Berg HAG (1980) J Org Chem 45:3915
32. Doering WVE, Roth WR (1962) Tetrahedron Lett 18:67; Goldstein MJ, Benzon MZ (1972) Ibid 94:7149
33. Wharton PS, Kretchmer RA (1968) J Org Chem 33:4258
34. Wender PA, Eissenstat MA, Filosa MP (1979) J Am Chem Soc 101:2196; Cupas C, Watts WE, Schleyer PVR (1964) Tetrahedron Lett 2503
35. Gajewski JJ, Benner CW, Hawkins CM (1987) J Org Chem 52:5198; Goldstein MJ, Benzon MS (1972) J Am Chem Soc 94:7147
36. Shea KJ, Phillips RB (1980) J Am Chem Soc 102:3156
37. Doering WVE, Roth WR (1963) Tetrahedron 19:715
38. Willcott MR, Cargle VH (1967) J Am Chem Soc 89:723
39. Vogel E (1958) Justus Liebigs Ann Chem 615:1; Hammond GS, De Boer CD (1964) J Am Chem Soc 86:899
40. Vogel E, Grimme W, Dinne E (1963) Angew Chem 75:1103
41. Schroder G, Oth JFM, Merenyi R (1965) Angew Chem Int Ed Engl 4:752; Gunther H, Pawliezek JB, Ulmen J, Grimme W (1972) Angew Chem Int Ed Engl 11:517; Doering WVE, Roth WR (1963) Tetrahedron 19:715
42. Schroder G, Oth JFM (1967) Angew Chem Int Ed Engl 6:414
43. Doering WVE, Ferrier BM, Fossel ET, Hartenstein JH, Jones M Jr, Klumpp G, Rubin RM, Saunders M (1967) Tetrahedron 23:3943; Gunther H, Runsink J, Schmickler H, Schmitt P (1985) J Org Chem 50:289
44. Cheng AK, Anet FAL, Mioduski J, Meinwald J (1974) J Am Chem Soc 96:2887; Moskau D, Aydin R, Leber W, Gunther H, Quast H, Martin HD, Hassenruck K, Miller lS, Grohmann K (1989) Chem Ber 122:925
45. Paquette LA (1990) Angew Chem Int Ed Engl 29:609; (1997) Tetrahedron 53:13971
46. Viola A, Iorio EJ, Chen KKN, Glover GM, Nayak U, Kocienski PJ (1967) J Am Chem Soc 89:3462
47. Evans DA, Golob AM (1975) J Am Chem Soc 97:4765; Evans DA, Baillargion DJ, Nelson JV (1978) J Am Chem Soc 100:2242
48. Evans DA, Golob AM, Mandel NS (1978) J Am Chem Soc 100:8170
49. Uychara T, Ueno M, Seki K, Tooya M, Sato T (1998) Terahedron Lett 39:8673
50. Evans DA, Ballallargeon DJ, Nelson JV (1978) J Am Chem Soc 100:2242
51. Gajewski JJ, Gee KR (1991) J Am Chem Soc 113:967
52. Hanna I, Gentric L, Huboux A, Zaghdoudi R (2003) Org Lett 5:3631
53. Hartley RC, Rutherford AP (2000) Tetrahedron Lett 41:737
54. Wei SY, Tomooka K, Nakai T (1993) Tetrahedron 49:1025
55. Tomooka K, Wei SY, Nakai T (1991) Chem Lett 43
56. Gentric L, Hanna I, Huboux A, Zaghdoudi R (2003) Org Lett 5:3631
57. Balakumar A, Janardhanam S, Rajagopalan K (1993) J Org Chem 58:5482
58. Paquette LA, Maynard GD (1991) Angew Chem Int Ed Engl 30:1368; Paquette LA, Maynard GD (1992) J Am Chem Soc 114:5018
59. Allin SM, Button MAC, Shuttleworth J (1997) Synlett 725
60. Allin SM, Button MAC, Baird RD (1998) Synlett 1117
61. Marshall JA, Babler JH (1969) J Org Chem 34:4186
62. Overman LE, Kakimoto M, Okazaki ME, Meier GP (1983) J Am Chem Soc 105:6622

63. Overman LE, Trenkle WC (1997) Isr J Chem 37:23
64. Overman LE, Kakimoto M, Okawara M (1979) Tetrahedron Lett 20:4041
65. Kakimoto M, Okawara M (1979) Chem Lett 1171
66. Chu M, Wu PL, Givre S (1986) Tetrahedron Lett 27:461
67. Zakarian A, Lu CD (2006) J Am Che Soc 128:5356
68. Claisen L (1912) Chem Ber 45:3157; Tarbell DS (1944) Org React 2:1
69. Ganem B (1996) Angew Chem Int Ed Engl 35:937; Severance DL, Jorgensen WL (1992) J Am Chem Soc 114:10966
70. Schuler FW, Murphy GW (1950) J Am Chem Soc 72:3155
71. Wilson SE (1975) Tetrahedron Lett 4651; Burgstahler AW, Nordin IC (1961) J Am Chem Soc 83:198
72. Hurd CD, Schmerling I (1937) J Am Chem Soc 59:107
73. Ryan JP, O'Connor PR (1952) J Am Chem Soc 74:5866
74. Marbet R, Saucy G (1967) Helv Chim Acta 50:2095; Burgstahler AW (1960) J Am Chem Soc 82:4681; Perrin CL, Faulkner DJ (1969) Tetrahedron Lett 2783
75. Miller B (1965) J Am Chem Soc 67:5115
76. Seki K, Tooya M, Sato T, Ueno M, Uyehara T (1998) Tetrahedron Lett 39:8673; Curtin DY, Johnson HW Jr (1956) J Am Chem Soc 78:2611
77. Wilson SR, Price MF (1984) J Org Chem 49:722
78. Wick AE, Felix D, Steen K (1964) Helv Chim Acta 47:2425; Sucrow W, Richter W (1971) Chem Ber 104:3679; Felix D, Gschwend-Steen K, Wick AE, Eschenmoser A (1969) Helv Chim Acta 52:1030
79. Seebach D, Amstutz R, Laube T, Schweizer WB, Dunitz JD (1985) J Am Chem Soc 107:5403; Ireland RE, Mueller RH, Williard AK (1976) J Am Chem Soc 98:2868
80. Koreeda M, Luengo JI (1985) J Am Chem Soc 107:5572
81. Johnson WS, Werthemann L, Bartlett WR, Brocksom TJ, Li T, Faulkner DJ, Petersen MR (1970) J Am Chem Soc 92:741
82. Henrick CA, Schaub R, Siddall JB (1972) J Am Chem Soc 94:5374
83. Gosteli J (1972) Helv Chim Acta 55:451
84. Abraham L, Korner M, Schwab P, Hiersemann M (2004) Adv Synth Catal 346:1281; Pollex A, Hiersemann M (2005) Org Lett 7:5705; Korner M, Hiersemann M (2007) Org Lett 9:4979; Uyeda C, Jacobsen EN (2008) J Am Chem Soc 130:9228; idem (2011) J Am Chem Soc 133:5062; Becker J, Butt L, von Kiedrowski V, Mischler E, Quentin F, Hiersemann M (2013) Org Lett 15:5982
85. Becker J, Butt L, von Kiedrowski V, Mischler E, Quentin F, Hiersemann M (2014) J Org Chem 79:3040
86. Malherbe R, Bellus D (1978) Helv Chim Acta 61:3096; Malherbe R, Rist G, Bellus D (1983) J Org Chem 48:860
87. Diederich M, Nubbemeyer U (1995) Angew Chem Int Ed Engl 34:1026; Deur CJ, Miller MW, Hegedus LS (1996) J Org Chem 61:2871; Kunng FA, Gu JM, Chao Y, Mariano PS (1983) J Org Chem 48:4262
88. Yoon TP, Dong VM, Mac Millan DWC (1999) J Am Chem Soc 121:9726
89. Panaras AA (1983) Tetrahedron Lett 24:3
90. Andrews PR, Smith GD, Young IG (1973) Biochemistry 12:3492
91. Ireland RE, Mueller RH, Willard AK (1976) J Am Chem Soc 98:2868
92. Chapleo CB, Hallett P, Lythgoe B, Waterhouse I, Wright PW (1977) J Chem Soc Perkin Trans 1(1211):1218
93. Overman LE, Knoll FM (1979) Tetrahedron Lett 321
94. Schlossarczyk H, Sieber W, Hesse M (1973) Helv Chim Acta 56:875
95. Subramanian RS, Balasubramanian KK (1988) Tetrahedron Lett 29:6797
96. Conroy H, Firestone RA (1956) J Am Chem Soc 78:2290

97. Oshima K, Takahashi H, Yamamoto H, Nozaki H (1973) J Am Chem Soc 95:2693
98. Takashita H, Uchida K, Mametsuka H (1983) Heterocycles 20:1709
99. Dalgaard L, Lawesson SO (1972) Tetrahedron 28:2051
100. Dalgaard L, Lawesson SO (1974) Acta Chem Scand B 28:1077
101. Majumdar KC, Jana NK (2000) Synth Commun 30:4183
102. Cook GR, Stille JR (1994) Tetrahedron 50:4105
103. Kurth MJ, Decker OHW (1985) J Org Chem 50:5769
104. Inada S, Nagai K, Takayanagi Y, Okazaki M (1976) Bull Chem Soc Jpn 49:833
105. Harrington PJ, Hegedus LS, McDaniel KF (1987) J Am Chem Soc 109:4335; Patterson JM, Wu A, Kook CS, Smith WT (1974) J Org Chem 39:486
106. Vedejs E, Gingras M (1994) J Am Chem Soc 116:579
107. Hoffmann RW, Maak N (1976) Tetrahedron Lett 2237; Reich HJ, Yelm KE, Hollowwitz S (1983) J Am Chem Soc 105:2503
108. Sato T, Otera J, Nozaki H (1989) J Org Chem 54:2779
109. Glaeske KW, West FG (1999) Org Lett 1:31
110. Vedejs E, Martinez GR (1979) J Am Chem Soc 101:6452
111. Hill PK (1984) In: Morrison JD (ed) Asymmetric synthesis. Academic Press, New York
112. Gawley RE, Zhang Q, Campagna S (1995) J Am Chem Soc 117:11817
113. Burn B, Coates B, Neeson SJ, Stevenson PJ (1990) Tetrahedron Lett 31:4351
114. Ollis WD, Sutherland IO, Thebtaranonth Y (1981) J Chem Soc Perkin Trans 1:1963
115. Vedejs E, Arco MJ, Powell DW, Renga JM, Singer SP (1978) J Org Chem 43:4831
116. Cokham I, Middleton ML, Taylor PL (1998) J Chem Soc Perkin Trans 1:2817
117. Sommelet M (1937) Compt Rend 205:56
118. Kantor SW, Hauser CR (1951) J Am Chem Soc 73:4122
119. Stara IG, Stary I, Tichy M, Zadava J, Hanus V (1994) J Am Chem Soc 116:5084
120. Morrison RT, Boyd RN Organic chemistry. Prentice Hall
121. Cere V, Paolucci C, Pollicino S, Sandri E, Fava A (1978) J Org Chem 43:4826
122. Vedejs E, Martinez GR (1979) J Am Chem Soc 101:6452
123. Bruckner R (1991) In: Trost BM, Fleming I (eds) Comprehensive organic synthesis, vol 6. Pergamon Press, Oxford, p 873
124. Gassman PG, Roos JJ, Lee SJ (1984) J Org Chem 49:717
125. Evans DA, Andrews GC, Sims CL (1971) J Am Chem Soc 93:4956
126. Evans DA, Bryan CA, Sims CL (1972) J Am Chem Soc 94:2891
127. Sato T, Otera J, Nozaki H (1989) J Org Chem 54:2779
128. Evans DA, Andrews GC (1974) Acc Chem Res 7:147
129. Reich HJ (1975) J Org Chem 17:2570; Clive DLJ, Chittatu G, Curtis NJ, Menchen SM (1978) Chem Commun 770
130. Crimmins MT, Hollis WG, Lever JG (1987) Tetrahedron Lett 28:3647
131. Kshirsagar TA, Moe ST, Portoghese PS (1998) J Org Chem 63:1704
132. Maleczka RE Jr, Geng F (1999) Org Lett 1:1115
133. Mikami K, Kishi N, Nakai T (1989) Chem Lett 1683
134. Still C, Mitra A (1978) J Am Chem Soc 100:1927
135. Nakazawa M, Sakamoto Y, Takahashi T, Tomooka K, Ishikawa K, Nakai T (1993) Tetrahedron Lett 34:5923
136. Ahman J, Somfai P (1994) J Am Chem Soc 116: 9781; (1995) Tetrahedron 51:9747
137. Majumdar KC, Jana GH (1998) Can J Chem 76:297
138. Majumdar KC, Roy B, Basu PK, Biswas P (2006) Indian J Chem 45B:719
139. Birney DM, Xu X, Ham S (1999) Angew Chem Int Ed 38:189
140. Frater G, Schmid H (1968) Helv Chim Acta 51:190; Alker D, Ollis WD, Shahriari-Zavareh H (1990) J Chem Soc Perkin Trans 1:1637

141. Wender PA, Ternansky RJ, Sieburth SM (1985) Tetrahedron Lett 26:4319; Jung ME, Davidov B (2001) Org Lett 3:3025
142. Shine HJ, Zmuda H, Park KH, Kwart H, Horgan AG, Brechbiel M (1982) J Am Chem Soc 104:2501
143. Kang HM, Lim YK, Shin IJ, Kim HY, Cho CG (2006) Org Lett 8:2047
144. Park KH, Kang JS (1997) J Org Chem 62:3794

Chapter 5
Group Transfer Reactions

5.1 Introduction

The transfer of one or more groups from one molecule to another in a concerted process is known as group transfer reaction. In most of the cases hydrogen is transferred. Only a few reactions of this class are known. Among them, the most common are ene reactions and diimide reduction.

5.2 The *Ene Reactions*

5.2.1 Overview of the Ene Reactions

The process of transfer of one hydrogen atom from an allylic alkene (an ene) to an electrophilic alkene (an enophile) followed by formation of a new σ-bond and migration of allylic double bond is known as *ene reaction* or *Alder ene reaction* [1]. The ene may be an alkene, alkyne, allene, cyclopropane ring or arene, whereas an enophile may be an alkene, carbonyl, thiocarbonyl, imino or diazo compound having an electron withdrawing substituent. Oxygen may also serve as an enophile. The reaction usually takes place from left to right, since a new σ bond is formed at the expense of the π bond of the enophile as depicted below.

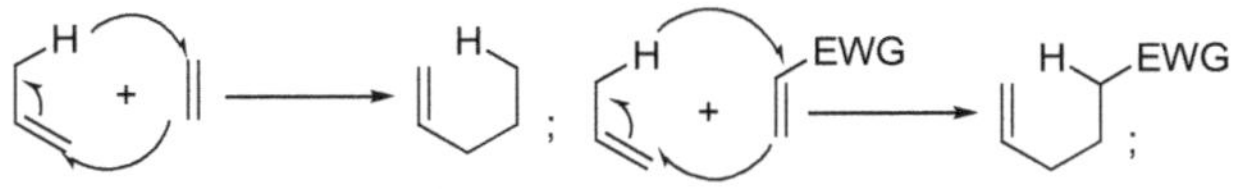

enophile: C=C, C=O, C=N

These reactions resemble Diels–Alder reactions, with one of the π bonds of the diene is replaced by a σ bond in the allylic alkene. These reactions are also considered as intermolecular 1,5-hydrogen shifts. The Lewis acid catalyzed ene

B. Dinda, *Essentials of Pericyclic and Photochemical Reactions*,
Lecture Notes in Chemistry 93, DOI 10.1007/978-3-319-45934-9_5

reactions occur at lower temperatures with high stereoselectivity compared to uncatalyzed ene reactions. The Lewis acid is attached to carbonyl function of the electron withdrawing group of the enophile and increases the electrophilic character of the enophile. The following examples are illustrative:

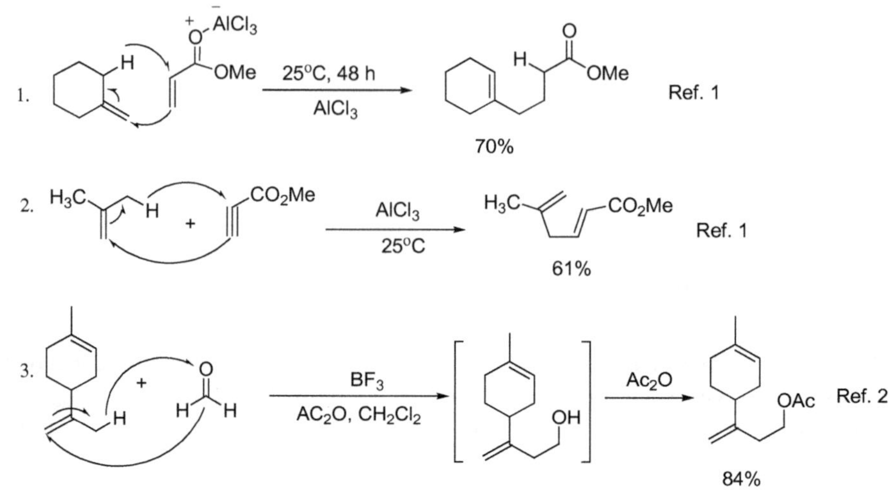

In entry 3, allylic function in the open chain takes part in ene reaction because reaction with endocyclic double bond will produce less stable TS.

When an unsaturated ketone in enol form undergoes ene reaction, the reaction is known as the *Conia ene reaction* [3]. The following examples [3–5] are illustrative.

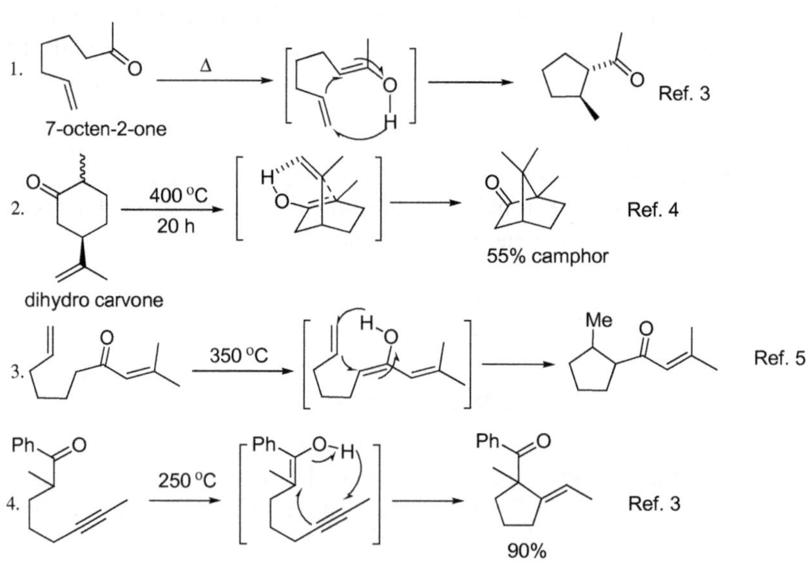

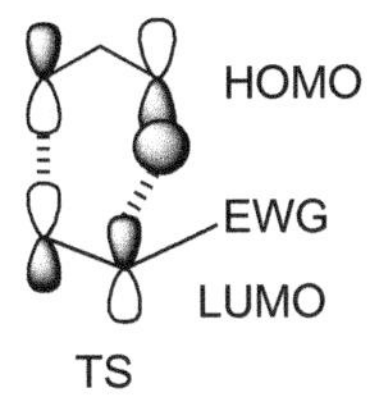

Fig. 5.1 Orbital interactions of ene and enophile in the TS of an ene reaction

5.2.2 *Stereochemistry and Regioselectivity*

Most of the ene reactions are concerted and orbital symmetry allowed processes involving all suprafacial transition states of 6e (4π and 2σ) with *endo* orientations of the electron withdrawing group as depicted in (Fig. 5.1). The addition of the ene to the enophile is stereospecific *syn*. The TS requires higher activation energy compared to that of Diels–Alder reaction because two σ-electrons of the allylic σ-bond are involved instead of a π-bond of a diene.

Regioselectivity of the ene reaction of an unsymmetrical enophile is governed by the orbital coefficient of carbons in allylic alkene and enophile. For example, in the reaction of propylene with methyl acrylate, the major and minor products are obtained as per orbital interactions of ene and enophile [4]:

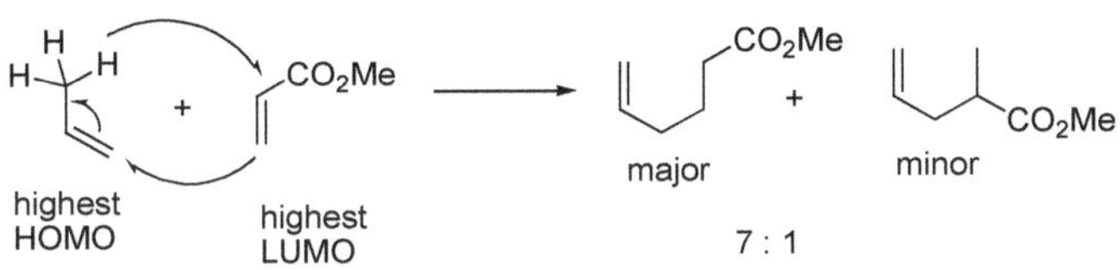

While in the ene reaction of 1-octene with methyl acrylate, steric interaction in the TS reduces the yield of major product.

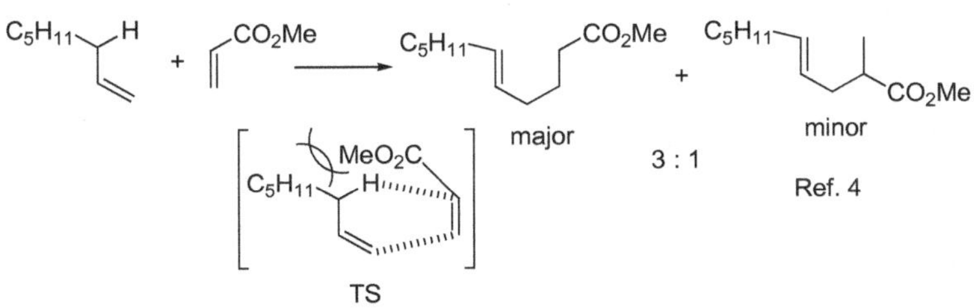

The presence of a germinal methyl group induces the methyl group at C-1 to deliver the hydrogen atom rather than from the vicinal methyl group at C-4.

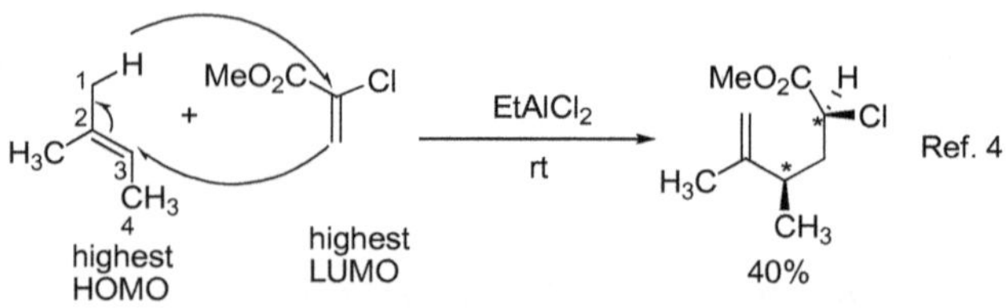

In some cases, the Woodward–Hoffmann rules are not applicable to ascertain the major product of ene reaction due to ring or steric strains. As for example, the *trans*-octa-1,6-diene **1** gives mainly the *cis* disubstituted cyclopentane (14:1, *cis:trans*) on the basis of favourable folded *endo* TS of lower energy, while *cis*-octa-1,6-diene **2** gives the major product through *exo* TS because *endo* TS is highly strained [6].

The *endo* TS of **1** is of lower activation energy as the hydrogens are on the same side of the folded bicyclic transition structure.

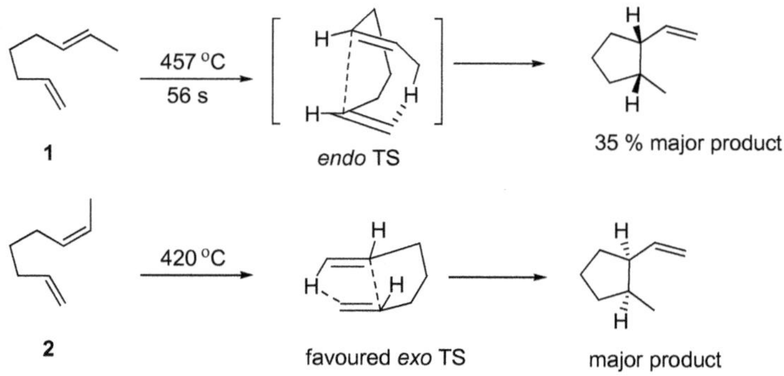

Similar folded TS is also observed in the gas phase intramolecular ene reaction of the carbonyl compound **3**.

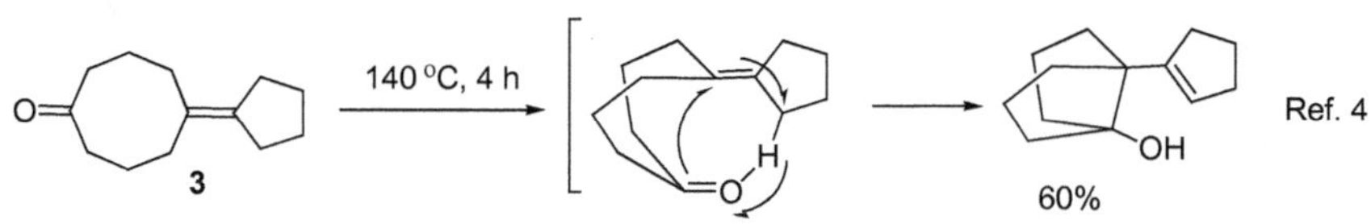

5.2.3 *Applications of Intermolecular-, Intramolecular- and Enantioselective-Ene Reactions*

Both intermolecular and intramolecular ene reactions have been utilized in intermediate steps for the synthesis of several bioactive organic compounds. Use of Lewis acid catalysts with chiral ligands provides diastereoselective products in high enantiomeric excess. Some of the ene reactions are illustrated.

(a) Intermolecular ene reactions

1.

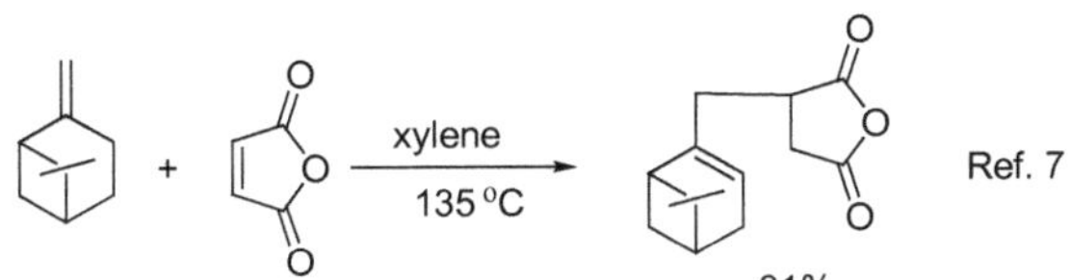

2.

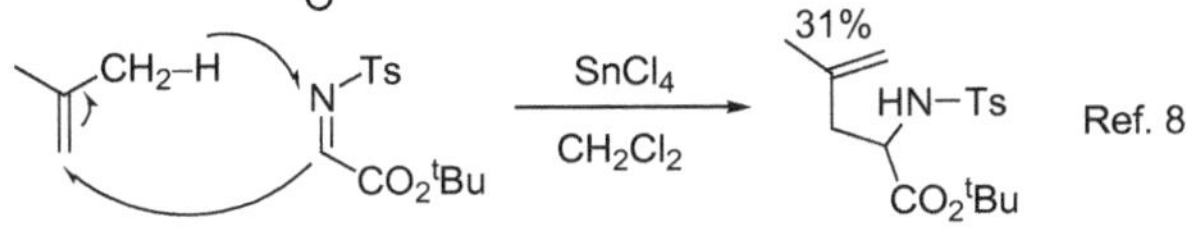

3.

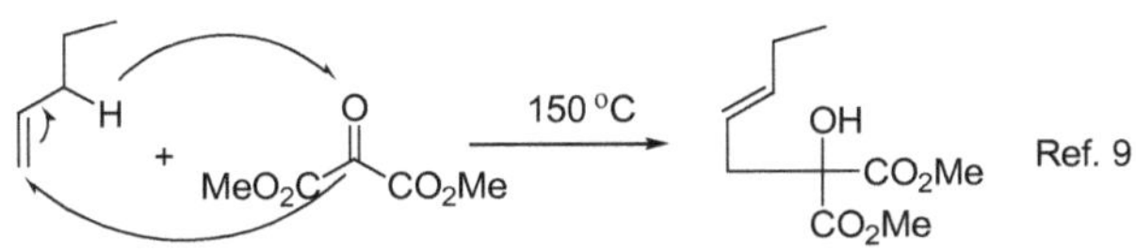

4.

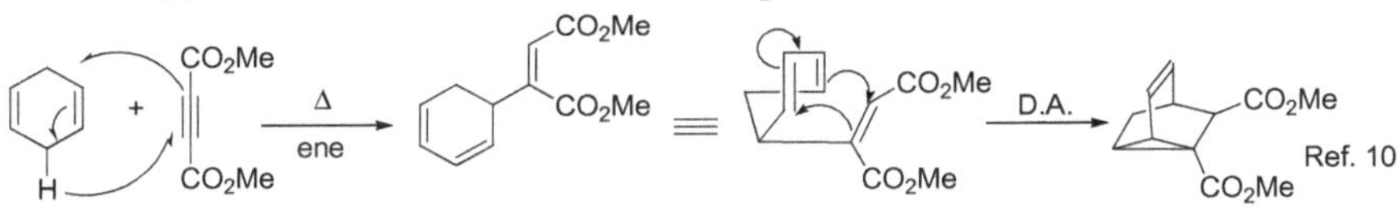

5.

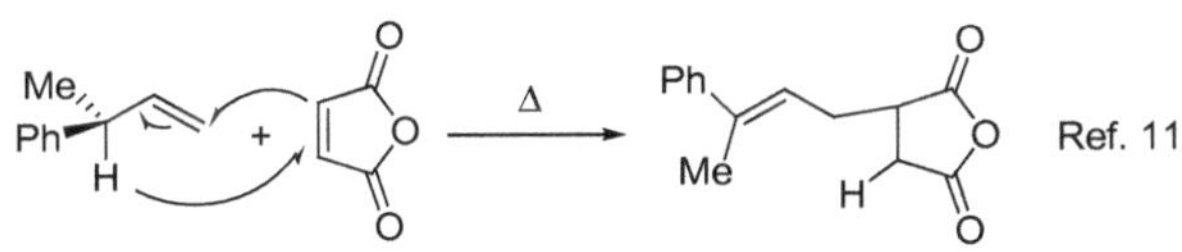

6.

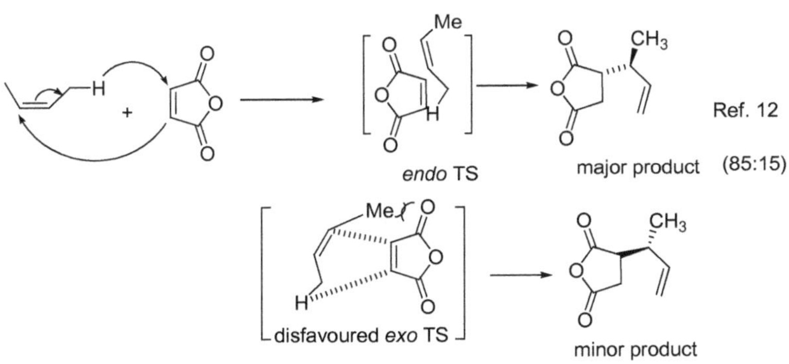

7. 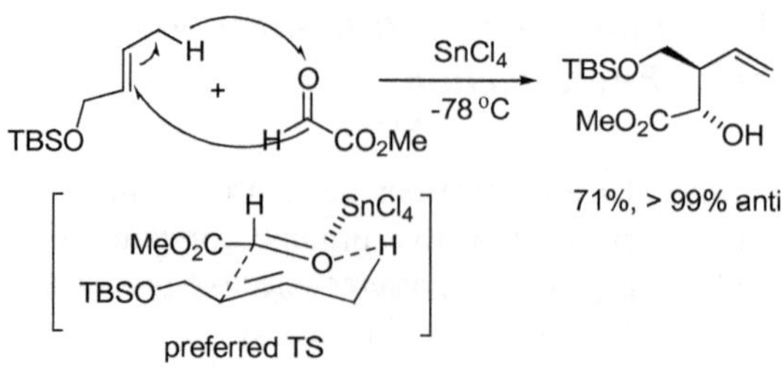

Ref. 12

8. 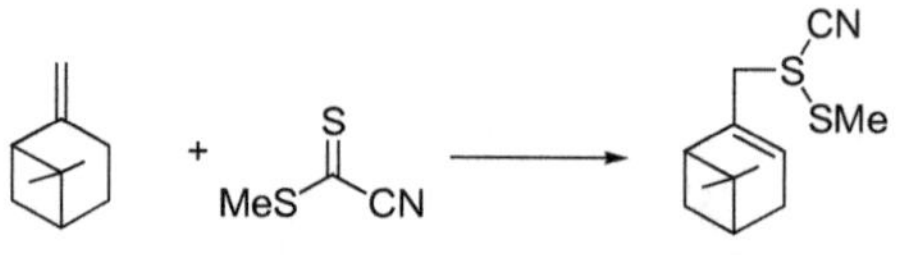

Ref. 13

9. 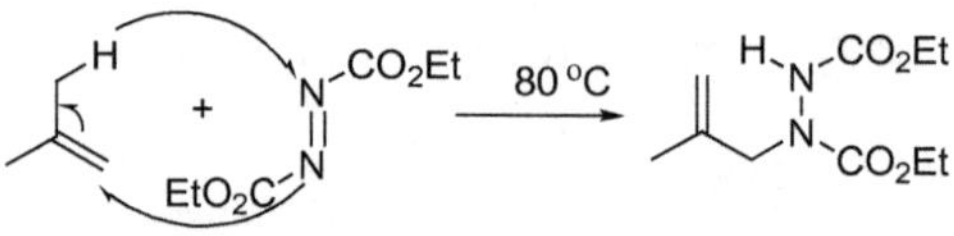

Ref.14

10. 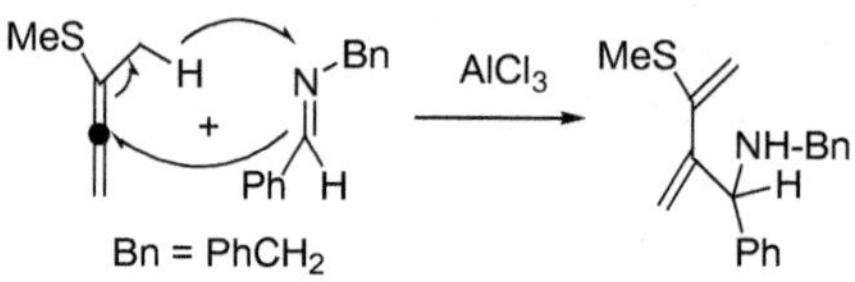

Ref. 15

11. 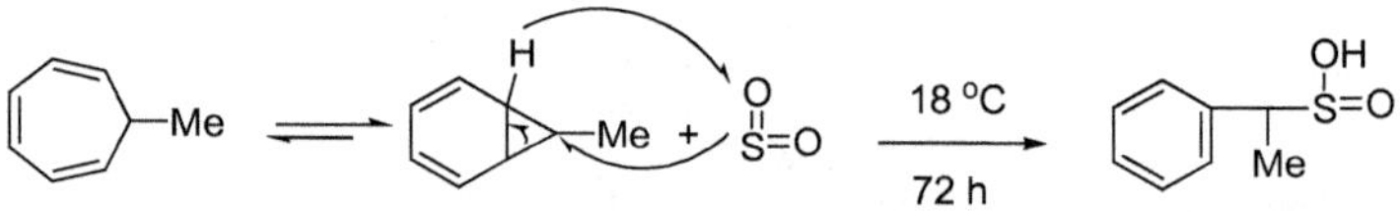

Ref. 16

(b) Intramolecular ene reactions

12. 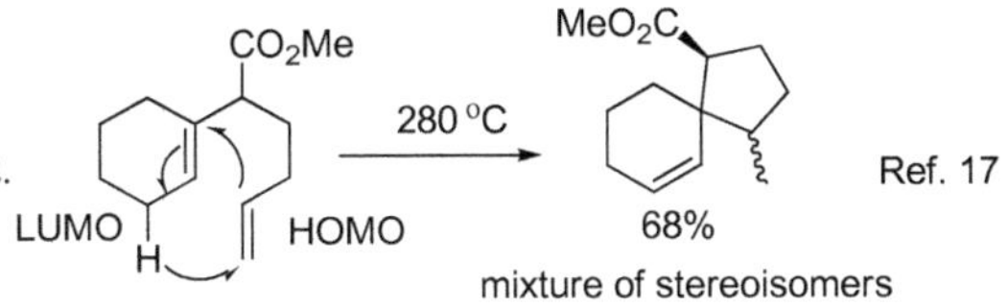

13. 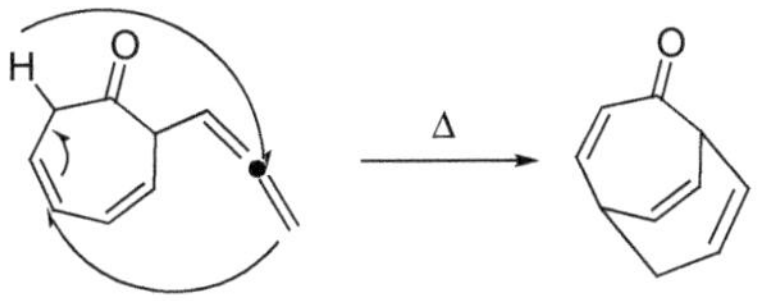

14. 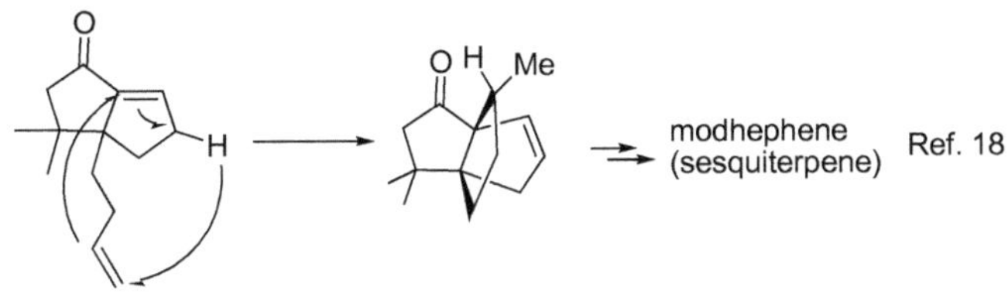

15. 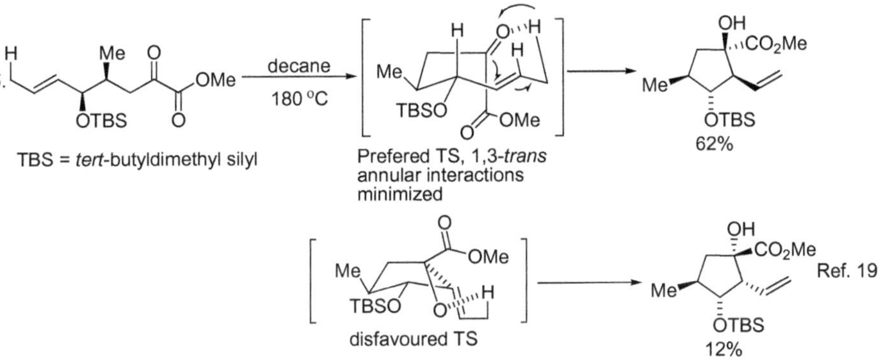

(c) Enantioselective ene reactions

16.

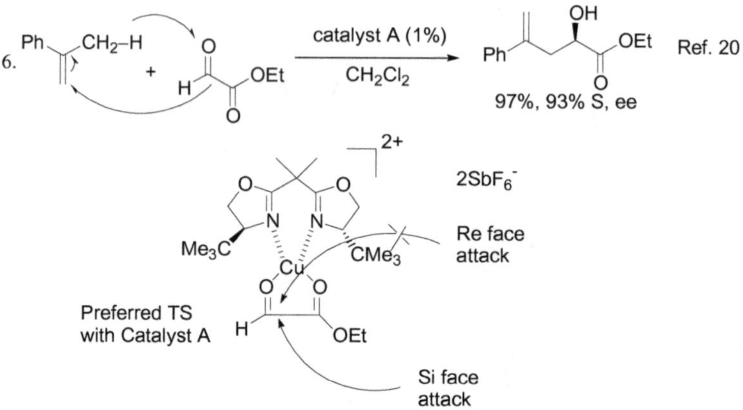

17.

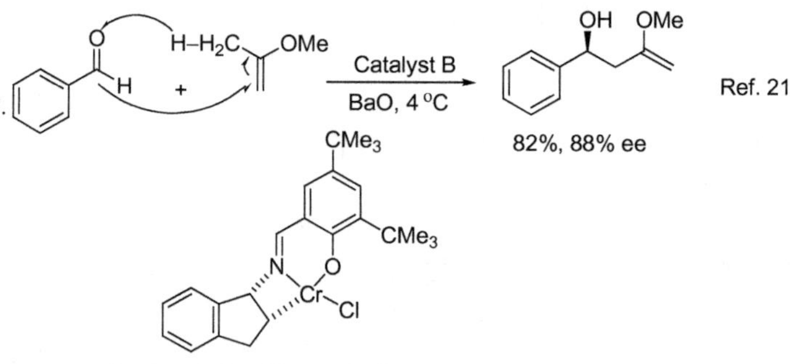

Catalyst B, Schiff base, Cr (III) complex

18.

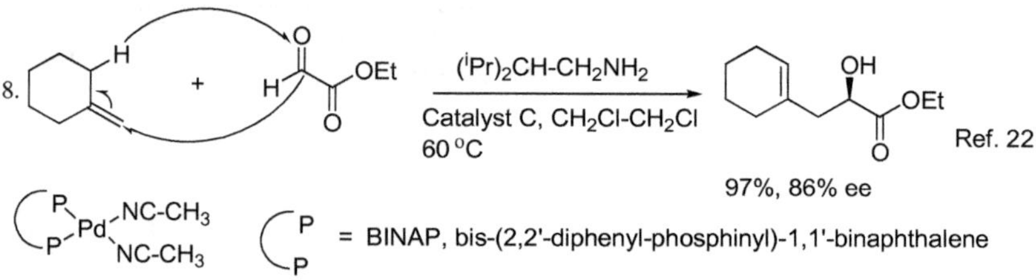

Catalyst C

19. 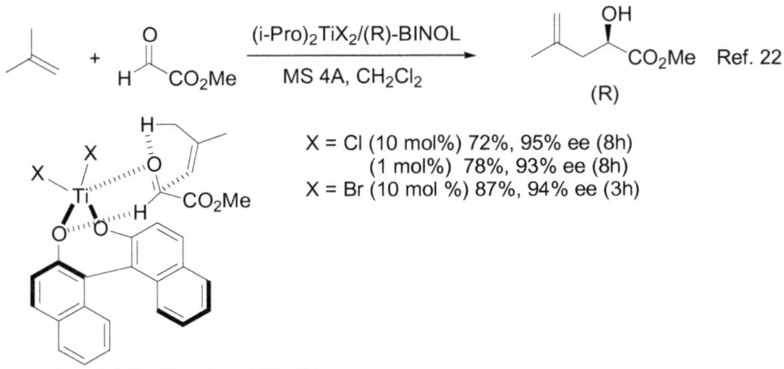

preferred TS with catalyst BINOL

5.3 The Metallo-Ene Reactions

The transfer of metal atoms such as lithium, magnesium, silicon, or palladium from an allylic organometallic compound to strong electrophilic olefin is known as the *metallo-ene reaction* [23].

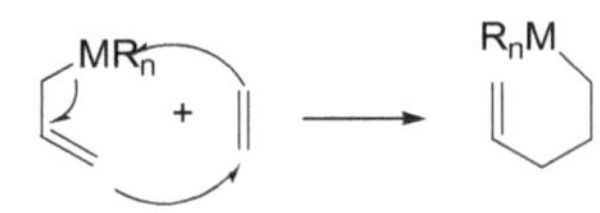

For example [24],

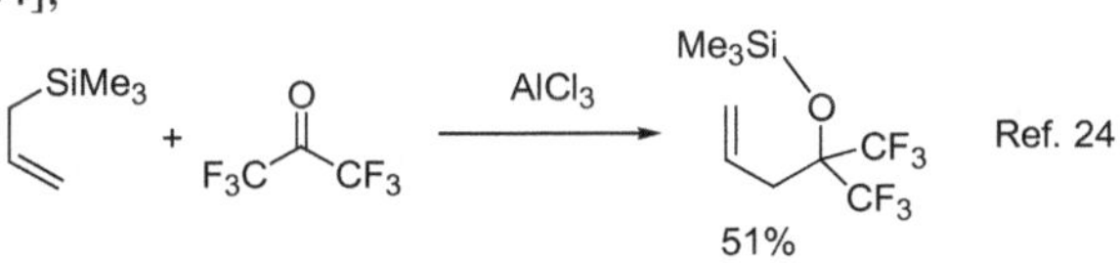

These reactions proceed through a concerted process of six-atom TS or an ionic process, where the intermediate is stabilized in polar solvent or by the catalyst. The intramolecular metallo-ene reactions are useful for synthesis of carbocyclic compounds. Some of the examples [25–27] of these reactions are illustrated:

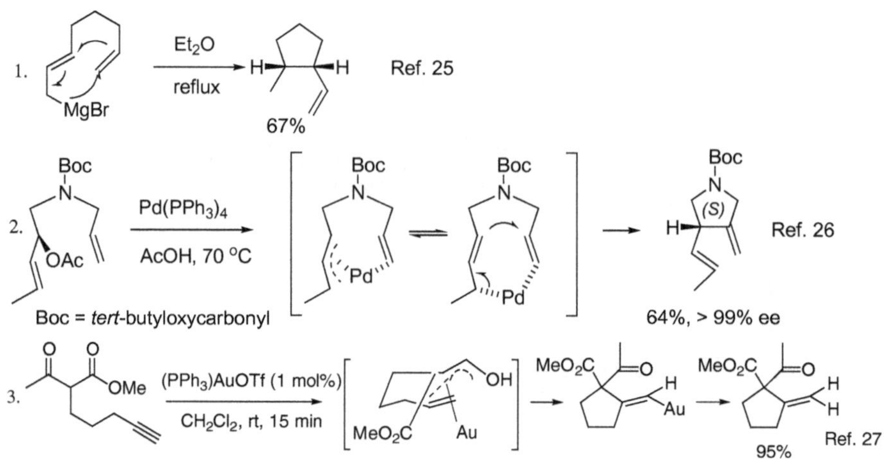

5.4 The Retro-Ene Reactions

The reversal of ene reactions are known as the *retro-ene reactions*. These reactions are favored at higher temperatures.

The principal synthetic value of these reactions is for synthesis of allenes, dienes, and other compounds, which would be difficult to obtain under normal conditions. Entries 1–4, illustrate the use of the retro-ene reaction to produce different classes of organic compounds [28–31].

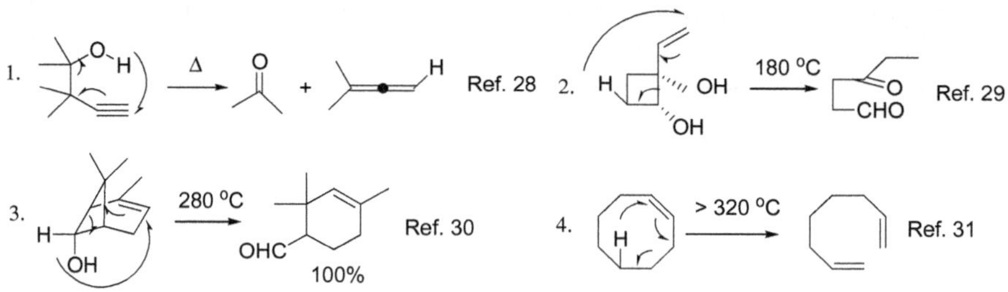

5.5 Diimide and Related Reductions

Diimide reduction of alkenes and alkynes are also group transfer reactions. Delivery of two hydrogen atoms to an alkyne or alkene takes place in a concerted process involving suprafacial delivery of two hydrogens in a TS. These reactions are pericyclic in nature.

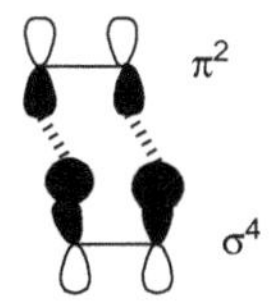

T.S.

Diimide is unstable species and is generated in situ by oxidation of hydrazine and its derivatives. Common methods for generation of diimide are [32]:

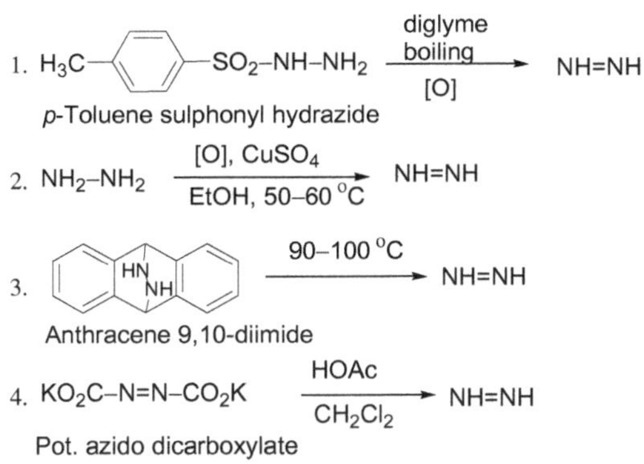

Diimide reduction is a metal-free catalytic hydrogenation process and does not cleave the sensitive O–O and N–O bonds in the substrate. Diimide reduces symmetrical double bonds, *e.g.*, C=C, N=N, O=O, etc.

Some of the important reactions of diimide are illustrated:

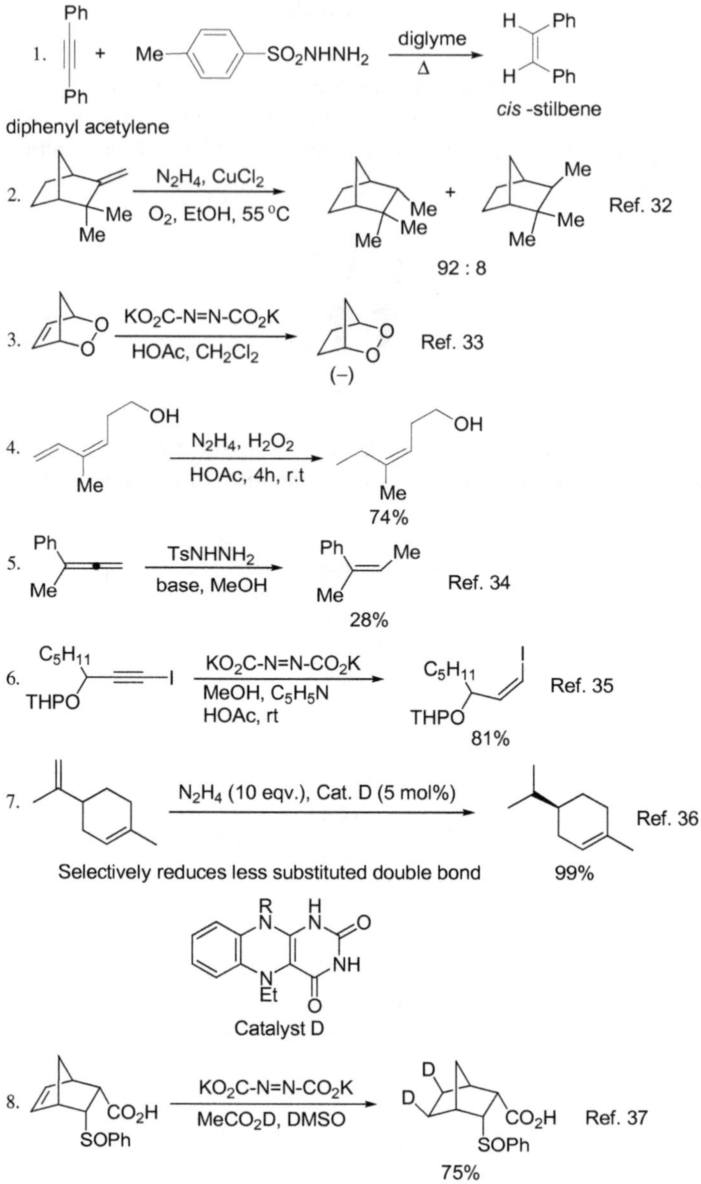

Among other group transfer reactions, the reduction of 1,2-dimethylcyclohexene with 9,10-dihydronaphthalene is used frequently [4].

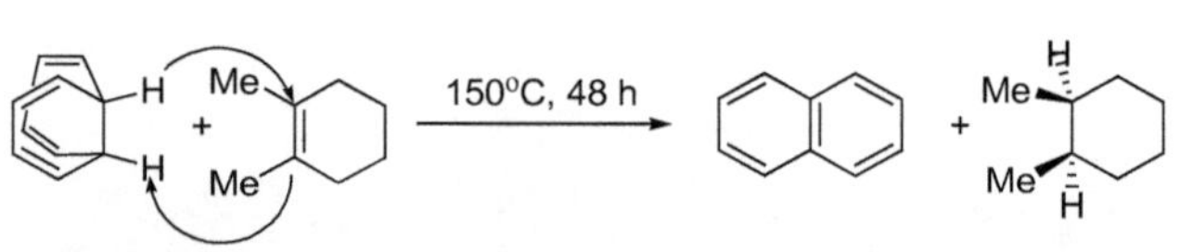

5.6 Thermal Elimination Reactions of Xanthates, N-Oxides, Sulfoxides and Selenoxides

Thermal β-elimination reactions of acetates, benzoates, xanthates, sulfoxides, selenoxides, and N-oxides are also group transfer reactions. All these elimination reactions are *syn*-stereospecific and proceed through a cyclic six membered—or five membered—ring transition state of 6e process by intramolecular transfer of hydrogen atom, where all the participating orbitals have suprafacial interactions. These reactions are fundamentally retro-group transfer reactions.

Pyrolysis of xanthate esters known as *Chugaev reaction* requires temperatures of 150–250 °C. For example, the pyrolysis of xanthates **4** and **5** gives more substituted olefins (Saytzeff products) as major products.

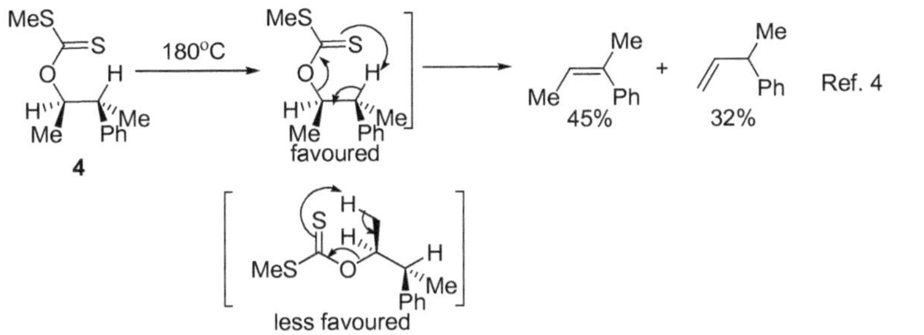

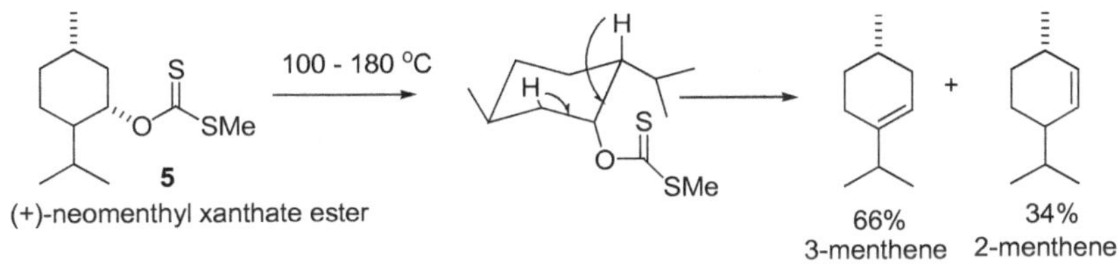

Pyrolysis of N-oxides known as *Cope elimination* takes place at lower temperatures (100–150 °C). The pyrolysis of sulphoxides and selenoxides takes place easily below 100 °C because of weaker C–S and C–Se bonds. As for example, *erythro*-N-oxide **6** and *erythro*-sulfoxide **7** on pyrolysis give *cis* and *trans* olefins as major product, respectively.

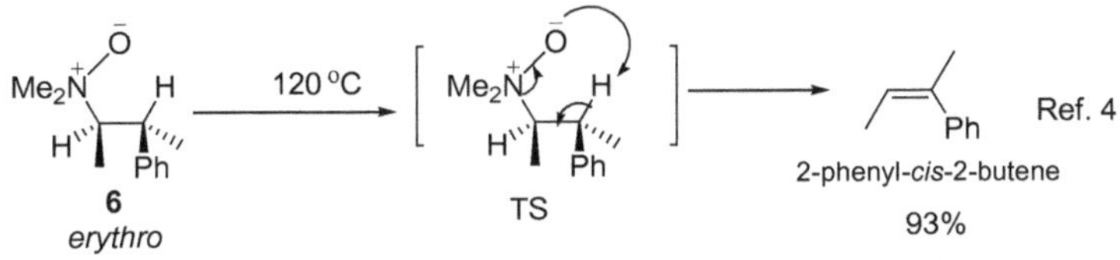

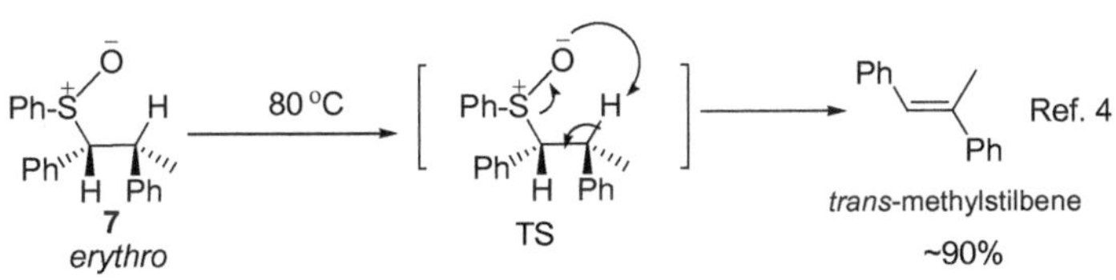

Similarly, *syn*-elimination of PhSeOH from selenoxide **8** and selenoxides derived from selenides **9–11** in presence of H_2O_2 occurs below rt to produce enones in high yields.

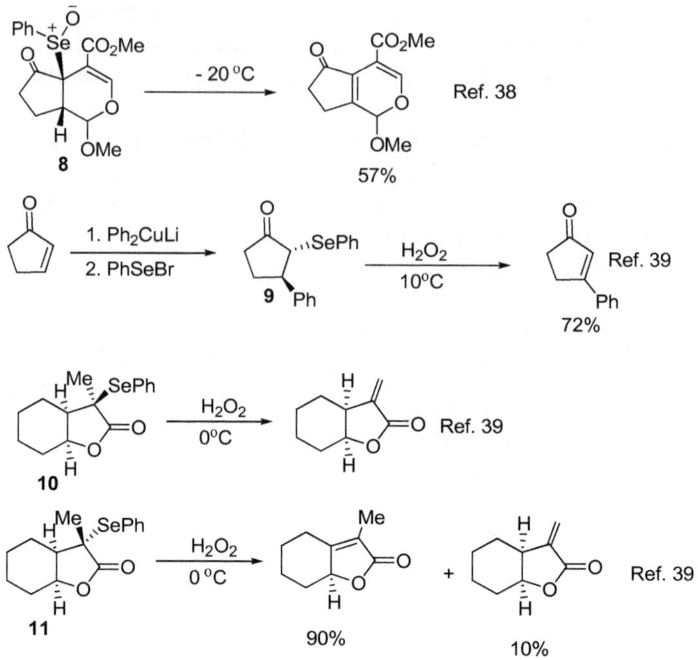

5.7 Problems

5.7.1. Suggest a mechanism of the following reactions: More than one step may be involved.

a. 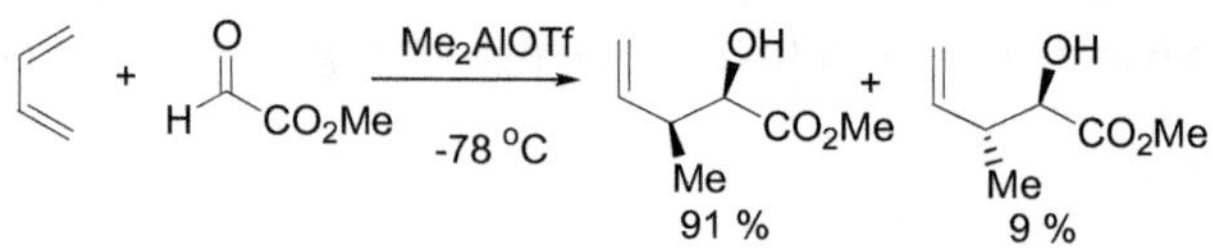

b. 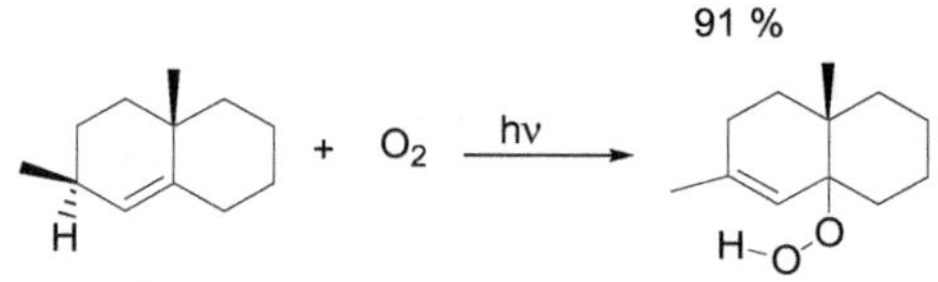

c. 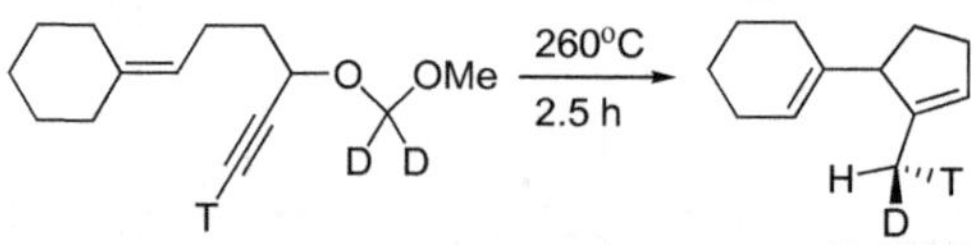

d. 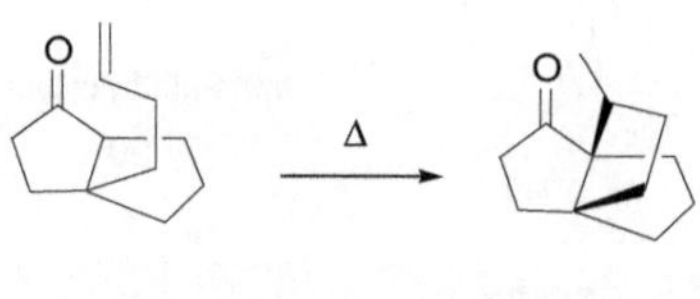

5.7.2. Predict the expected product with structure and stereochemistry.

a. 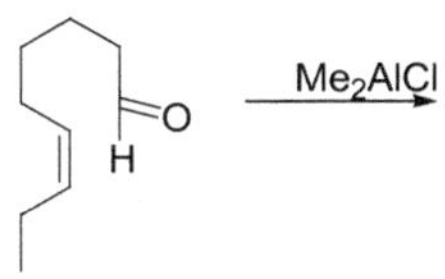

b. 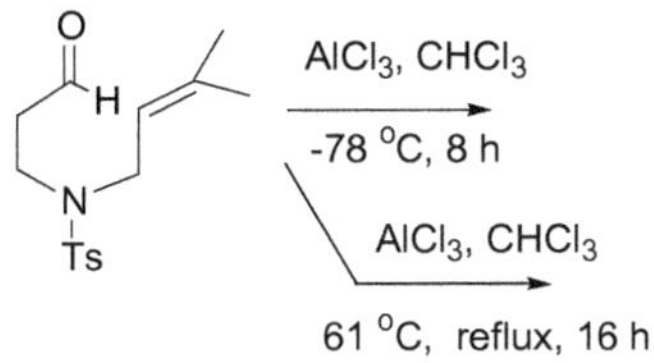

c. 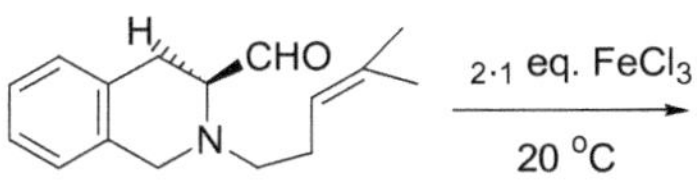

d. 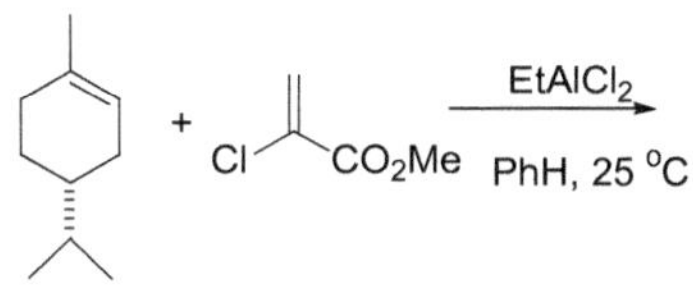

e. 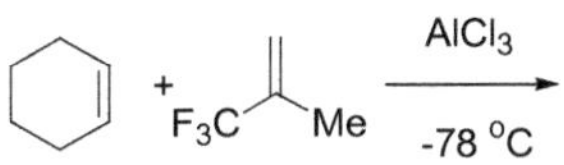

f. 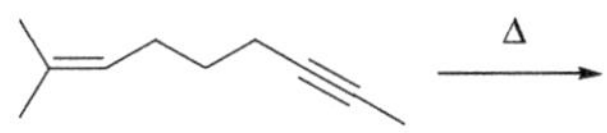

g. 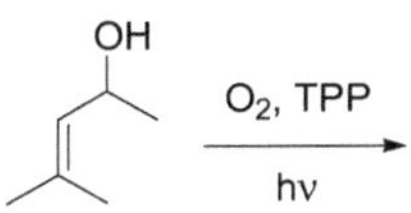

h. 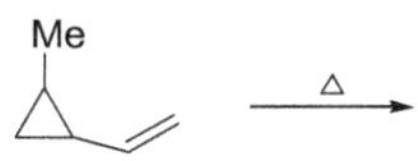

i. 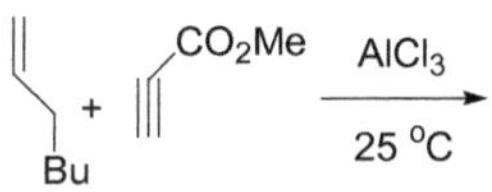

j. 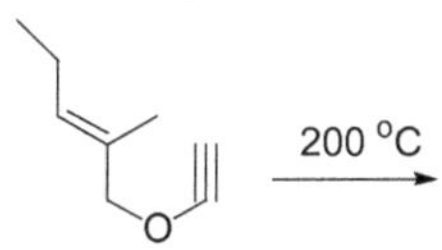

k. 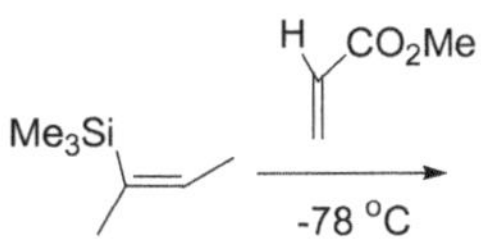

l. 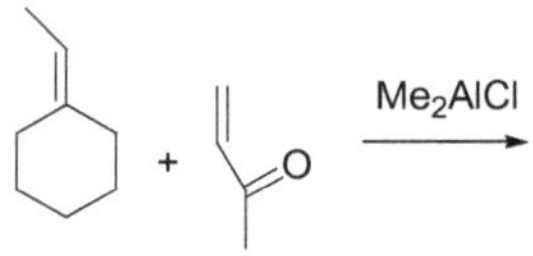

m. 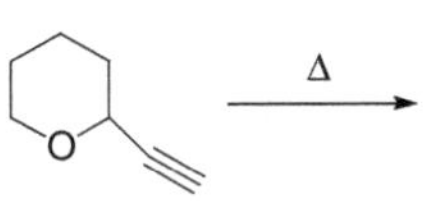

n. 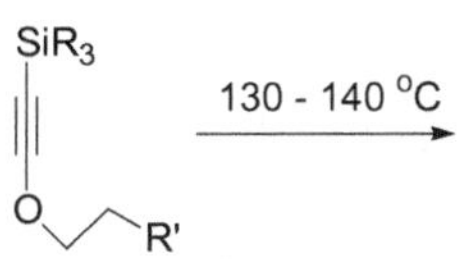

o. 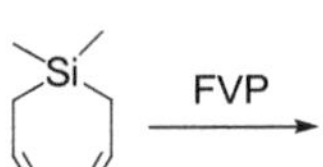

FVP = Flash Vacuum Pyrolysis

5.8 Further Reading

1. Hoffmann HMR (1969) The ene reaction. Angew Chem Int Ed Engl 8:556.
2. Oppolzer W, Snieckus V (1978) Intramolecular ene reactions in organic synthesis. Angew Chem Int Ed Engl 17:476

References

1. Alder K, Pascher F, Schmitz A, (1943) Chem Ber 76:27; Snider BB, Rodini DJ, Conn RSE, Sealfon S, (1979) J Am Chem Soc 101:5283
2. Blomquist AT, Himics RJ (1968) J Org Chem 33:156
3. Conia MJ, LePerchec P (1975) Synthesis 1
4. Fleming I (2002) Pericyclic reactions. Oxford University Press, New York, pp 84–87
5. Leyendecker F, Drouin J, Conia JM (1974) Tetrahedron Lett 2931
6. Huntsman WD, Solomon VC, Eros D (1958) J Am Chem Soc 80:5455
7. Arnold RT, Showell JS (1957) J Am Chem Soc 79:419
8. Achmatowicz O, Pietraszkiewicz P (1981) J Chem Soc Perkin Trans 2680
9. Salomon MF, Pardo SN, Salomon RG (1984) J Org Chem 49:2446
10. Giguere RJ, Namen AM, Lopez BO, Arepally A, Ramos DE, Majetich G, Defauw J (1987) Tetrahedron Lett 28:6553
11. Mislow K (1964) Introduction to stereochemistry. Benjamin, New Jersey (Chapter 3)
12. Mikami K, Shimizu M (1992) Chem Rev 92:1021
13. Bachrach SM, Jiang S (1997) J Org Chem 62:8319
14. Hoffman HMR (1969) Angew Chem Int Ed Engl 8:556
15. Hayashi Y, Shibata T, Narasaka K (1990) Chem Lett 1693
16. Lucchi O, Filipuzzi F, Lucchini V (1984) Tetrahedron Lett 25:1407
17. Oppolzer W, Mahalanabis KK, Ballig K (1977) Helv Chim Acta 60:2388
18. Oppolzer W (1981) Pure Appl Chem 53:1181
19. Schnabel C, Sterz K, Muller H (2011) J Org Chem 76:512
20. Evans DA, Tregay SW, Burgey CS, Paras NA, Vojkovsky T (2000) J Am Chem Soc 122:7936
21. Ruck RT, Jacobsen EN (2002) J Am Chem Soc 124:2882
22. Mikami K, Terada K, Takeshi N (1990) 112:3949; Corey EJ, Barnes-Seeman D, Lee TW, Goodman SN (1997) Tetrahedron Lett 37:6513
23. Oppolzer W, Pitteloud R, Strauss F (1982) J Am Chem Soc 104:6476
24. Abel EW, Rowley RJ (1975) J Organomet Chem 84:199
25. Felkin (1972) Tetrahedron Lett 22:2285
26. Oppolzer W (1990) Pure Appl Chem 62:1941
27. Kennedy-Smith JJ, Staben ST, Toste FD (2004) J Am Chem Soc 126:4526
28. Viola A, Collins J, Filipp N (1981) Tetrahedron 37:3772; Hoft H, Kirsch R (1985) Tetrahedron Lett 26:3327
29. Conia JM, Barnier JP (1971) Tetrahedron Lett 4981
30. Joulian D, Rouessac F(1972) J Chem Soc Chem Commun 314; Ohloff G (1970) Angew Chem Int Ed Engl 9:743
31. Hoffmann HMR (1969) Angew Chem Int Ed Engl 8:556
32. Norman ROC, Coxon JM (1992) Principles of organic synthesis, 3rd edn. ELBS with Chapman and Hall, Oxford; Corey EJ, Pasto J, Mock L (1961) J Am Chem Soc 83:2967
33. Adam W, Eggelte J (1977) J Org Chem 42:3987
34. Moro K, Ohki M, Sato A, Matsui M (1972) Tetrahedron 28:3739

35. Luethy C, Konstantin P, Untch JG (1978) J Am Chem Soc 100:6211
36. Smit C, Fraaije M, Minnard A (2008) J Org Chem 73:9482
37. Annunziata R, Fornasier R, Montanan F (1974) J Org Chem 39:3196
38. Callant P, Ongena R, Vandewalle M (1981) Tetrahedron 37:2085
39. Reich HJ, Wollowitz S (1993) Org Reactions 44:1

Part II
Photochemical Reactions

Chapter 6
Principles of Photochemical Reactions

6.1 Introduction

Photochemical reactions of organic compounds have attracted much interest in the recent times for its fascinating nature and wide applications in the synthesis of organic compounds. There are two key features of photochemical reactions which give them special importance over thermal reactions. First, the reactions take place in the excited state of the molecules having a large excess of energy compared to ground state, it is often possible to effect reactions which are thermodynamically unfavorable due to their ground-state reactants. Second, the reactions are usually carried out at low temperatures so that the products can be formed in cold. Hence, it is often possible to make highly strained ring systems by pumping out excess energy as light to overcome the activation energy barrier in their formations.

The photochemical reactions are usually carried out by irradiation with ultraviolet (λ 200–350 nm) and less frequently by visible lights. Therefore, these reactions are essentially limited to those in which at least one of the reactants is unsaturated or aromatic.

Another important criterion for these reactions is to carry out these reactions in pure fused quartz vessels, which can transmit UV radiations, whereas common borosilicate pyrex glass or sodium silicate glass vessels are unsuitable to use because they can transmit radiations only longer than 300 nm and absorb below this wavelength.

Now we have to understand the basic elements of photochemical reactions and the basic processes that are involved in light–matter (organic molecules) interactions.

Photochemical reactions of organic compounds are the chemical reactions that result from interactions between organic molecules and ultraviolet or visible light. In a photochemical reaction, the light of a particular wavelength is irradiated to excite the molecules. According to the quantum theory, both matter and light are quantised, and only certain specific energies of light are absorbed by a specific

B. Dinda, *Essentials of Pericyclic and Photochemical Reactions*,
Lecture Notes in Chemistry 93, DOI 10.1007/978-3-319-45934-9_6

organic molecule for its excitation. The absorption or emission of light occurs by the transfer of energy as photons. These photons have both wave and particle-like properties and the energy E of a photon is given by Planck's law,

$$E = h\nu$$

where h is Planck constant and is equal to 6.63×10^{-34} Js and υ is the frequency of oscillation of the photon in units of s^{-1} or Hertz (Hz).

$$\nu = c/\lambda$$

where c is the velocity of light and λ is the wavelength of oscillation of photon. Thus,

$$E = h\nu = hc/\lambda \quad (1)$$

Therefore, the energy of a photon is proportional to its frequency and inversely proportional to its wavelength. The energy of one mole of photons (6.02×10^{23} photons) is called an Einstein and is measured in units in kJ mol^{-1}. It is equal to Nhc/λ.

6.2 Light Sources Used in Photochemical Reactions

For ultraviolet region, mercury vapor lamps are suitable. These lamps mainly emit at 214, 254, 313 and 366 nm. Low-pressure mercury lamps strongly emit spectral line at 254 nm. For the visible region, a tungsten lamp or more powerful xenon arc is suitable. The composition of the radiation reaching the sample can be controlled by the filters in a spectrophotometer. For the reaction of aromatic compounds, the radiation at 254 nm is desired. At this wavelength, the energy of a photon is equal $2.86 \times 10^4/254 = 112.6$ kcal mol^{-1} is irradiated for excitation of a molecule. This energy is sufficient to rupture most of the single covalent bonds in an aromatic compound. Sometimes solid-state lasers (light amplification by stimulator emission of radiation) such as ruby laser and neodymium-doped yttrium aluminum garnet (Nd-YAG) laser, and gas lasers such as helium–neon laser and argon ion laser are used for generation of light of strong intensity.

6.3 Laws of Photochemistry

Two fundamental principles relating to light absorption by the organic molecules are the basis for understanding of their photochemical transformations.

1. **The Grotthuss–Draper law**: The law states that the only the fraction of light which is absorbed by a chemical entity can bring about the photochemical change.
2. **The Stark–Einstein law**: The law states that each molecule or atom absorbs one photon or one quantum of light for its excitation or activation, i.e., for a molecule, AB,

$$\underset{\text{groundstate}}{AB} + h\nu \rightarrow \underset{\text{excited state}}{AB^*}$$

This law is obeyed in the majority of cases but exception occurs when very intense light sources such as lasers are used for irradiation of a sample.

6.4 The *Beer–Lambert's Law of Light Absorption*

The extent of light absorbed by a substance depends on its molar absorption coefficient (ε). The fraction of light absorbed (I/I_0) by a substance is given by the Beer–Lambert law. The law states that the ratio of the intensity of the emergent light (I) and incident light (I_0) has an exponential relationship with the concentration (c) and path length (l) of the absorbing substance, i.e.,

$$I/I_0 = 10^{-\varepsilon cl}$$

Taking logarithm to the base 10 gives

$$\log(I/I_0) = -\varepsilon cl$$
$$\text{or} \quad \log(I_0/I) = \varepsilon cl$$

The left-hand quantity is the absorbance, A, and hence

$$A = \varepsilon cl$$

where c is the concentration of the substance in moles per liter, mol L^{-1}, and l is the path length in cm.

The higher the ε value, higher will be the intensity of the absorption. Usually the intensity of light absorption is high for allowed electronic transition process and low for forbidden transition process.

6.5 Physical Basis of Light Absorption by Molecules: The *Franck–Condon Principle*

Chromophores or chromophoric groups present in the molecules are responsible for the absorption of light. The absorption of UV or visible light by a molecule results in the promotion of an electron from its ground-state orbital to a higher excited state orbital. Normally the amount of energy necessary to make this transition depends mostly on the nature of the two orbitals involved and much less on the rest part of the molecule. When this electronic transition occurs, the absorbing chromophore undergoes an electronic dipole transition. This transition dipole moment lasts for the duration of the transition only and the intensity of the resulting absorption of light is proportional to the square of the transition dipole moment.

The total energy of a molecule is made up of its electronic energy and energy of its nuclear (vibrational and rotational) motion. It is expressed as

$$E_t = E_e + E_v + E_r$$

where the subscripts refer to the total energy, electronic energy, vibrational energy, and rotational energy, respectively. The energy gap between electronic states is much greater than that between vibrational states, which in turn is much greater than that between rotational states. Absorption of light by molecules causes transition of electrons from one electronic state to another much more rapidly than that of the nuclei because of their heavier mass compared to the mass of electrons. The electronic transition takes place so rapidly than that of the nuclei of the vibrating molecule and hence nuclei can be assumed to be fixed during this electronic transition period. This is called the *Franck–Condon principle* [1]. It states that absorption of light by a molecule causes an electronic transition within a stationary nuclear framework of the molecule.

Thus, the electronic transition by absorption of a photon is often referred to as a *vertical transition* or *Franck–Condon transition*. The electronic transition in a

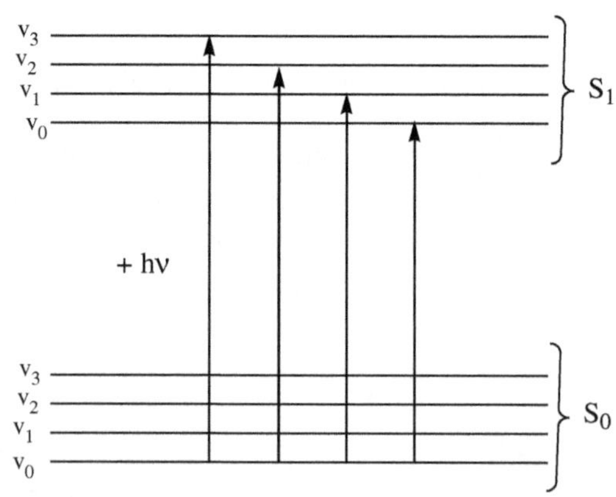

Fig. 6.1 Schematic diagram of the electronic ground state and the first excited electronic state of a diatomic molecule. The *vertical arrows* show vibronic transitions due to absorption of photons

molecule by absorption of light results in changes in both electronic and vibrational states of electron. Hence, such electronic transition is called vibronic transition of electron. Figure 6.1 shows the potential energy curve known as Morse potential energy curve of a diatomic molecule for its electronic ground state (S_0) and the first excited electronic state (S_1).

6.6 Electronic Transitions and Their Nomenclature

In principle, six types of electronic transitions, designated as $\sigma \rightarrow \sigma^*$, $\sigma \rightarrow \pi^*$, $\pi \rightarrow \pi^*$, $\pi \rightarrow \sigma^*$, $n \rightarrow \sigma^*$, and $n \rightarrow \pi^*$ are possible (Fig. 6.2). The $\sigma \rightarrow \sigma^*$ transition corresponds to absorption in the inaccessible far-UV (100–200 nm) and both $\sigma \rightarrow \pi^*$ and $\pi \rightarrow \sigma^*$ are obscured by much stronger $\pi \rightarrow \pi^*$ absorptions, and $n \rightarrow \sigma^*$ transition occurs by vacuum UV light (below 200 nm). So in photochemical reactions, only $\pi \rightarrow \pi^*$ and $n \rightarrow \pi^*$ transitions occur, which produce (π, π^*) and (n, π^*) electronically excited states, respectively.

Absorptions due to $\pi \rightarrow \pi^*$ and $n \rightarrow \pi^*$ transitions differ from one another in several important aspects as shown in Table 6.1.

In a molecule, the singlet electronic states are denoted by S_0, S_1, S_2, etc., of increasing energy and triplet electronic states as T_1, T_2, etc. S_0 indicates the ground state of singlet electron, whereas S_1, S_2, etc., and T_1, T_2, etc., are excited singlet and triplet states, respectively. The $n \rightarrow \pi^*$ transition is the lowest energy transition for most of saturated carbonyl compounds and this transition is known as $S_0 \rightarrow S_1$ and T_1. These transitions occur from the promotion of an electron from n molecular orbital to the π^* molecular orbital and these transitions are referred as ^{1}(n, π^*) and ^{3}(n, π^*) states, respectively. Similarly, S_2 and T_2 excited states arise from the

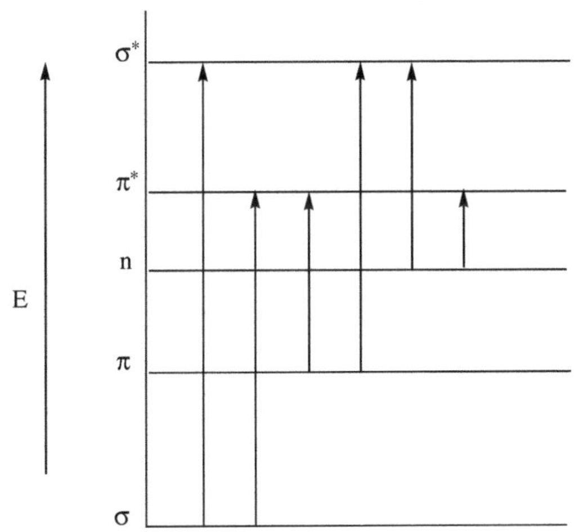

Fig. 6.2 Generalized ordering of molecular orbital energies of organic molecules and electronic transitions that occur by excitation with light

Table 6.1 Comparison of light absorptions due to $\pi \rightarrow \pi^*$ and $n \rightarrow \pi^*$ electronic transitions

	Absorptions due to $\pi \rightarrow \pi^*$ transitions	Absorptions due to $n \rightarrow \pi^*$ transitions
1	Occurs at shorter wavelengths (220–260 nm)	Occurs at longer wavelengths (270–350 nm)
2	Substitution moves the absorption to longer wavelength	Substitution moves the absorption to shorter wavelengths
3	Shows relatively strong absorption with ε_{max} of $\sim 10^3$–10^5 l mol^{-1} cm^{-1}	Shows relatively weak absorption with ε_{max} of ~ 1–10^2 l mol^{-1} cm^{-1}
4	Absorption band occurs at longer wavelength in a polar solvent than in non polar solvent (shows bathochromic shift)	Absorption band occurs at shorter wavelength in a polar solvent than in a non polar solvent (shows hypsochromic shift)

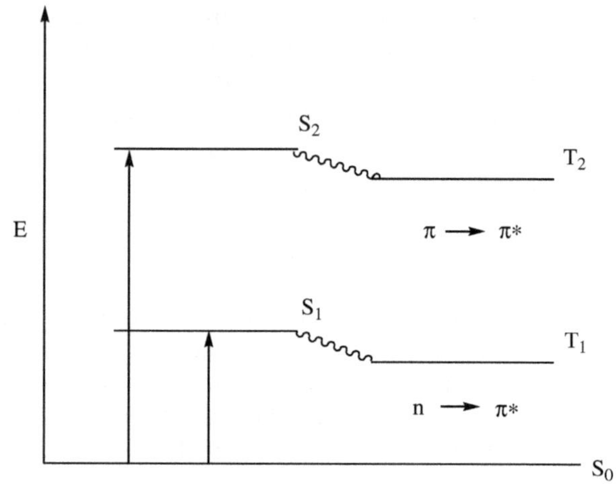

Fig. 6.3 Electronic states of molecular orbitals of an organic compound

promotion of an electron from its π MO to its π* MO and are referred to as $^1(\pi, \pi^*)$ and $^3(\pi, \pi^*)$ states, respectively.

These transitions are represented in Fig. 6.3.

6.7 Spin Multiplicity of Electronic States

The electronic state of a molecular orbital is specified by its spin multiplicity. Spin multiplicity is determined from the equation,

$$\text{Spin multiplicity} = 2S + 1$$

where S is the sum of the spin quantum numbers of the electrons present in an orbital.

A ground-state helium atom has a pair of electrons of opposite spin in the 1 s orbital ($1S^2$). Hence, total spin $S = ½ - ½ = 0$ and spin multiplicity = $(2S + 1) = 1$.

Thus, the **ground-state singlet** S_0 has spin multiplicity of 1.

In the lower excited state (S_1 and T_1) of helium atom, there are two possible spin configurations:

In the S_1 state, total spin $S = ½ - ½ = 0$.

So, its spin multiplicity is 1. In the T_1 state, total spin $S = ½ + ½ = 1$ and hence its spin multiplicity is 3.

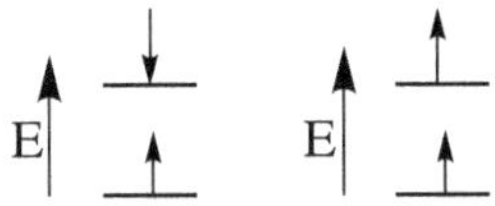

S_1- state T_1-state

6.8 The HOMO and LUMO Concept of Electronic Transitions

The excitation of a molecule promotes an electron from a filled molecular orbital to a vacant molecular orbital. The highest occupied molecular orbital (HOMO) is the filled nonbonding n molecular orbital of a saturated carbonyl compound. For an alkene, it is the bonding π molecular orbital, whereas for a diene, it is the bonding ψ_2 molecular orbital for thermal and ψ_3 molecular orbital for photochemical reaction. The lowest unoccupied molecular orbital (LUMO) is the antibonding π^* molecular orbital of a saturated carbonyl compound or alkene. For a diene, it is ψ_3 and ψ_4 molecular orbital, for thermal and photochemical reaction, respectively.

6.9 The Selection Rules for Electronic Transitions

Electronic transitions between energy levels in organic molecules are governed by some compulsions, known as **selection rules**.

1. ***Spin-selection rule***

   An electronic transition in which spin multiplicity of the election remains the same in both ground and excited states, i.e., $\Delta S = 0$, the transition is allowed. The change of spin multiplicity involves the change in angular momentum and such a change violates the law of conservation of angular momentum. Therefore, singlet–singlet and triplet–triplet transitions are allowed, whereas singlet–triplet and triplet–singlet transitions are forbidden. However, this rule is not obeyed in certain cases, most often when a heavy atom (such as iodine) is present in the molecule. In such cases, singlet–triplet promotions take place [2].

2. ***Orbital symmetry selection rule***
   An electronic transition proceeds more rapidly when the wave functions of the initial and final states closely resemble each other, i.e., the transition is allowed. In the $\pi \rightarrow \pi^*$ transition, π and π^* orbitals occupy the same regions of space and so the overlap between them is large. Thus, the $\pi \rightarrow \pi^*$ is allowed process. In $n \rightarrow \pi^*$ transitions, the n and π^* orbitals are perpendicular to each other and so they overlap to much smaller extent. Therefore, this transition is forbidden. In practice, this $n \rightarrow \pi^*$ transition is weakly allowed due to coupling interaction between vibrational and electronic motion in the molecule.

6.10 Physical Properties of Excited States: Jablonski Diagram

Electronically excited states of the molecules are short-lived because of excess energy content and try to deactivate through various photophysical and photochemical processes for return to their original ground states.

Photophysical relaxation processes may be classified as:

1. **Intramolecular processes**:
   a. **Radiative transitions**: These processes involve the emission of electromagnetic radiation from the excited state during return to ground state as fluorescence and phosphorescence, collectively known as luminescence process.
   b. **Radiationless transitions**: These are the internal conversions without emission of electromagnetic radiations such as from S_2 to S_1, intersystem crossing (S_1 to T_1), and vibrational cascade ($v_4 \rightarrow v_3$, $v_3 \rightarrow v_2$, $v_2 \rightarrow v_1$, etc.) processes.
2. **Intermolecular processes**:
   a. **Vibrational relaxation**: The excited molecules undergo rapid collision with one another and with solvent molecules to produce molecules in the lowest vibrational levels of a particular electronic energy level (say, S_2, S_1, T_2, T_1, etc.)
   b. **Energy transfer**: The excited molecule acts as donor or photosensitizer and transfers excess energy to another molecule, known as acceptor or quencher.
   c. **Electron transfer**: The photoexcited molecule donates its electron to a ground-state acceptor molecule to form an ion pair. This ion pair results in quenching of energy from the excited donor molecule.

Jablonski diagram

Jablonski diagram illustrates the electronic states, properties and relaxation processes of an excited organic molecule (Fig. 6.4) [3]. The electronic states are arranged vertically by energy and grouped horizontally by spin multiplicity.

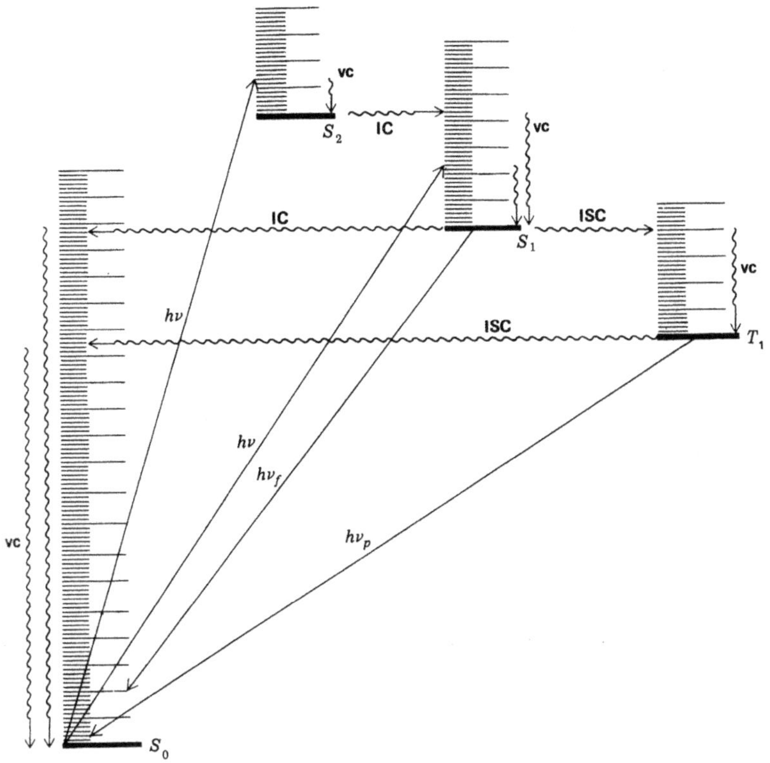

Fig. 6.4 Modified Jablonski diagram for an organic molecule showing ground and excited states and intramolecular photophysical processes from excited states. Radiative processes—fluorescence ($h\nu_f$) and phosphorescence ($h\nu_p$) are shown in *straight lines*, radiationless processes—internal conversion (IC), inter system crossing (ISC), and vibrational cascade (vc) are shown in *wavy lines*. Adapted with permission from (Smith MB, March J 2006 March's Advanced Organic Chemistry: Reactions, Mechanisms and Structures, 6th Ed., John Wiley, New York). Copyright (2007) John Wiley & Sons

Nonradiative transitions are indicated by squiggly arrows and radiative transitions by straight arrows.

The Jablonski diagram shows the following:

1. The electronic states of the molecule and their relative energies. Singlet electronic states are S_0, S_1, S_2, etc., and triplet electronic states are T_1, T_2 etc.
2. A molecule may be excited to either S_2 or S_1 state. In liquids and solids, the higher S_2 state rapidly drops to S_1 state by internal conversion ($\sim 10^{-13}$–10^{-11} s). S_1 state may undergo intersystem crossing to T_1 state or return to S_0 state by fluorescence. In addition, it may take part in chemical reaction or photofragmentation.

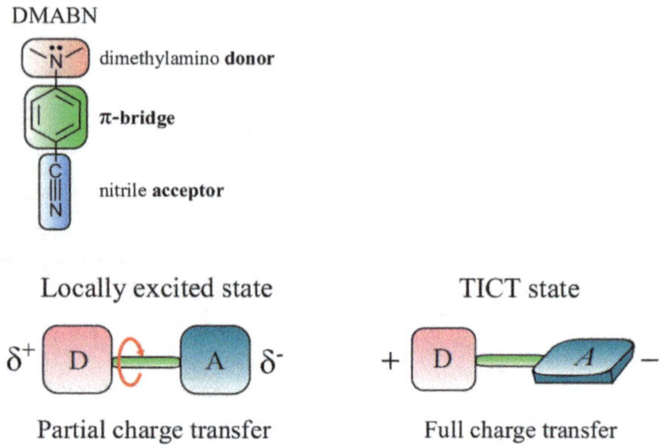

Fig. 6.5 Intramolecular energy transfer of dimethylaminobenzonitrile by TICT process

3. T_1 state may take part in chemical reaction or may return to S_0 state by phosphorescence.
4. In each S_2, S_1 and T_1 states, it undergoes vibrational relaxation by vibrational cascade, e.g., S_2 ($v = 3$) $\rightarrow$ S_2 ($v = 0$)

In addition to these photophysical processes, twisted intramolecular charge transfer (TICT) process (non-radiative process) also takes place to return to ground state. Molecules of compounds such as dimethylaminobenzonitrile (DMABN) are flexible. These are planar in the ground state, but are twisted in excited state. The twisted conformation is able to transfer its full electronic charge from one part to another in polar solvents and thus deactivated to ground state (Fig. 6.5) [4].

6.11 Lifetimes of Electronic Excited States

Lifetime of radiative S_1 state is given by:

$$\tau_0 = 10^{-4}/\varepsilon_{\max},$$

where τ_0 has units of s and $\varepsilon_{\max}$ has units of l mol^{-1} cm^{-1}. Thus, for [1] (π, π*) transitions, [1] τ_0 is 10^{-9}–10^{-6} s, whereas for [1] (n, π*) transitions, [1] τ_0 is 10^{-6}–10^{-3} s.

Lifetime of radiative T_1 state is relatively long. In general, [3] (π, π*) states have long lifetimes (1–10^{2} s), whereas [3] (n, π*) states have short lifetimes (10^{-4}–10^{-2} s).

6.12 Efficiency of Photochemical Processes: Quantum Yield of Photochemical Reaction

After photon absorption, the excited molecules undergo several competing processes including photophysical processes, and hence only a fraction of the excited molecules undergo photochemical reaction. Therefore, the fraction of the molecules that chemically react relative to those that are excited is called the quantum yield of the photochemical reaction. It is denoted by φ.

φ = Number of molecules reacted in a given time/Number of photon absorbed by the molecules in a given time.

According to the Stark–Einstein law, φ should be equal to 1. In practice, in most cases, it is less than 1. For instance, if the quantum yield is 0.01, then only one hundredth of the molecules that are excited undergo photochemical reaction. In chain reactions, secondary processes occur and hence their φ is greater than 1. For example, in the photo dissociation of acetone, quantum yield of the reaction may be 1 or 2 depending on the number of bonds broken.

$$CH_3COCH_3 \xrightarrow[\lambda > 266\ nm]{h\nu} (CH_3)_2CO^* \longrightarrow {}^*CH_3 + CH_3C^*O$$

$$\phi = 1$$

$$CH_3COCH_3 \xrightarrow[\lambda < 193\ nm]{h\nu} (CH_3)_2CO^* \longrightarrow 2\ {}^*CH_3 + CO$$

$$\phi = 2$$

The quantum yield for some photophysical events such as fluorescence can also be defined from an excited fluorophore [5].

φ = No of photons emitted by a fluorophore/No of photons absorbed by the fluorophore.

6.13 Intramolecular Process of Excited States: Fluorescence and Phosphorescence

6.13.1 Fluorescence and Its Measurement

Fluorescence is the radiative emission of light energy from an excited molecule for its return to ground state of same spin multiplicity, i.e., from S_1 to S_0 state. Fluorescence is a spin-allowed transition process and occurs strongly in a relatively short time in the order of picoseconds to microseconds. The fluorescence emission spectra are almost the mirror image of the absorption spectra. Only one peak is

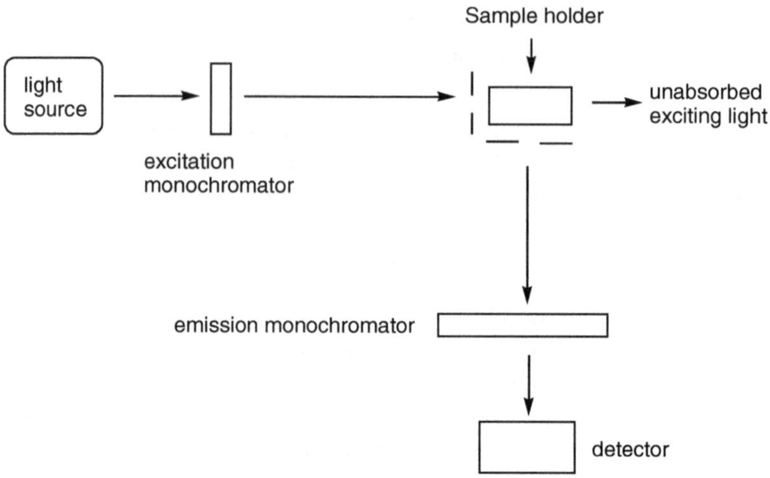

Fig. 6.6 Basic components of a spectrofluorometer

common among these spectra that occurs from transitions between the lower vibrational levels of the two states, i.e., S_1 (v_0) to S_0 (v_0) and vice versa and this peak is called the 0–0 band. In solution, this 0–0 band may appear at slightly different wavelengths due to solvation effect in the two states. Because of the possibility of fluorescence, any photochemical reaction in S_1 state occurs very fast before the occurrence of fluorescence.

The intensity of fluorescence is measured for quantitative analysis of fluorescent compounds present in different clinical and industrial samples.

The intensity of fluorescence is directly proportional to the concentration of the fluorescent compound. If the target compound is not fluorescent, then it is converted into a fluorescent derivative by reaction with a suitable (nonfluorescent) reagent. The fluorescence emitted by the fluorescent compound is measured using a spectrofluorometer [6]. Most of the modern spectrofluorometers employ diffraction grating monochromators to select the appropriate wavelengths for maximum excitation and emission. The basic components of a fluorometer are: a light source, an excitation monochromator, a sample holder, an emission monochromator, and a fluorescence detector as shown in Fig. 6.6.

Light source

The most commonly employed lamps are medium- and high-pressure mercury lamp or xenon arc lamp, having an output covering the whole UV–Visible spectrum range. Xenon arc lamp operated stroboscopically is preferred for its continuous output. The lamp is operated in a current of air to disperse the toxic ozone formed from oxygen on exposure to UV radiation.

Excitation monochromator
The slit width of the monochromator is adjusted to select the wavelength for maximum absorption by the sample and allow its transmission for excitation of the sample.

Sample holder
The majority of the fluorescence assays are carried out in solution contained in a circular or square cuvette, made from quartz material. The cuvette is placed normal to the incident beam. The resulting fluorescence is collected from the front surface of the cuvette, at right angle to the incident beam.

Emission monochromator
The slit width of the monochromator is adjusted to get maximum emission of fluorescence.

Detector
Photomultiplier tubes (two) are used to cover the complete UV–Visible range of emission spectra. The output from the detector is amplified and displayed on a readout digital device. A continuous sensitivity adjustment is useful in the measurement of sample of widely differing concentrations.

Concentration range of sample
The concentration of the sample solution is adjusted to provide a solution of absorbance <0.1 A to minimize reabsorption effects.

6.13.2 Kasha's Rule for Fluorescence

Kasha rule states that the fluorescence from organic compounds usually originates from the lowest vibrational levels ($v = 0$) of the lowest excited singlet state (S_1). The exception to this rule is the hydrocarbon azulene, where fluorescence originates from S_2 to S_0 [7]. This is due to large $S_2 - S_1$ energy gap, which results in the slowing down of S_2 to S_1 internal conversion.

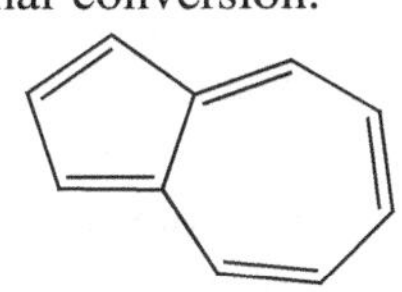

Azulene

6.13.3 Vavilov's Rule for Fluorescence

Vavilov rule states that the fluorescence quantum yield is independent of the wavelength of the exciting light. If a molecule in S_1 ($v = 0$) undergoes fluorescence emission and has fluorescence, intersystem crossing and internal conversion having rate quantum yields φ_{f}, φ_{isc} and φ_{ic}, respectively, then

$$\varphi_{\mathrm{f}} + \varphi_{\mathrm{isc}} + \varphi_{\mathrm{ic}} = 1$$

Since φ_{ic} is much smaller than φ_{f} and φ_{isc},

$$\varphi_{\mathrm{f}} + \varphi_{\mathrm{isc}} \approx 1$$

This mathematical expression is known as **Ermolev rule.**

6.13.4 Phosphorescence and Its Measurement

Phosphorescence is a radiative emission of light energy from an excited state of a molecule during its return to ground state of different spin multiplicity, i.e., T_1 to S_0 state. This is a spin-forbidden transition process and hence occurs at a much smaller rate than that of fluorescence and is less intense. Most of the organic molecules from their S_1 state can undergo an *intersystem crossing* (ISC) to their lowest triplet state [8]. For example, almost 100 % of the molecules of excited benzophenone from their S_1 state cross over to T_1 state [9]. For this reason, benzophenone is preferred as effective photosensitizer. Because of the slow rate of phosphorescence, the lifetime of T_1 state is longer and hence T_1 state is susceptible to between the molecules of oxygen or other impurity molecule. To observe phosphorescence, it is necessary to prevent the diffusion process of quenching.

T_1 lies at lower energy level than S_1 and hence phosphorescence spectrum is always found at longer wavelengths than the fluorescence spectrum. The phosphorescence of a compound is usually measured in solid solution rather than fluid solution to avoid quenching process, which may arise by diffusion of either two T_1 molecules or the T_1 molecule and a dissolved oxygen molecule or some impurity molecule. The phosphorescence is determined using a rotating can phosphoroscope (Fig. 6.7) [10]. It consists of a hollow cylinder having one or more slits which are equally spaced in the circumference. The sample solution in a tube is inserted in the cylinder and is allowed to freeze in liquid nitrogen (77 K) so that a clear glassy solid is formed. The rotating can is then rotated by a variable-speed motor. During the rotation of the rotating can, the sample is first illuminated by the light source and then darkened. Whenever, there is darkness, the phosphorescence radiation passes to the monochromator and is measured. The rotation of the rotating can is set so that the path of the detector is blocked when the exciting light reaches the sample and open when the exciting light is blocked and the decayed phosphorescence reaches the detector.

Ethanol is an excellent solvent for polar molecules. Small quantities of acid or base are added to produce a clear transparent solid. For nonpolar or less polar

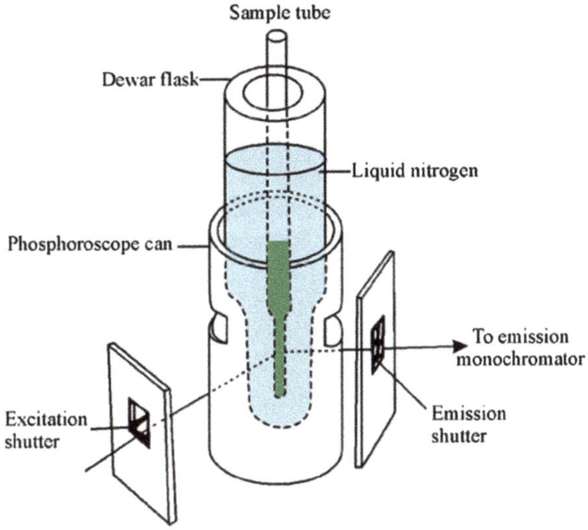

Fig. 6.7 Schematic diagram of a rotating can phosphoroscope with shutter system

compounds, a mixture of diethyl ether, isopentane and ethanol in the ratio of 5:5:2, commonly called EPA, is used as an excellent solvent.

Both fluorescence and phosphorescence are emitted by the compound. The two forms of luminescence are separated by exploiting the fact that T_1 states are much longer lived than S_1 states and so phosphorescence persists long after the fluorescence has decayed. Moreover, phosphorescence occurs at longer wavelengths. For the study of phosphorescence of a very few compounds in room temperatures, the compound is taken in a transparent polymer matrix such as perspex. A large number of organic compounds with conjugated ring systems have the property of phosphorescence emission and thus the phosphorimetry provides an excellent method for their analysis in trace amounts in clinical and industrial samples.

6.14 Intermolecular Physical Processes of Excited States: Photosensitization Processes

6.14.1 Photosensitization/Quenching and Excimer/Exciplex Formation

A molecule in an excited state (S_1 or T_1) may transfer its excess energy all at once to another molecule of same or different type (in ground state) in the environment, in a process of deactivation, called a quenching process and the acceptor molecule which receives the energy from the excited molecule is called a quencher and excited molecule is called a photosensitizer [11]. If the excited molecule on transfer of energy excited the quencher molecule, the process is called a photosensitization process. The intermolecular energy transfer from an excited molecule to another acceptor molecule in a photosensitization process takes place generally in two ways. These are: a triplet excited state generates another triplet state, and singlet excited state generates another singlet state. Both these processes obey the Wigner spin conservation rule, which states that the total electron spin does not change after energy transfer process. These processes are represented as:

$$\begin{array}{lll} {}^{3}\mathrm{D}^{*} + {}^{1}\mathrm{A} & \rightarrow & {}^{3}\mathrm{A}^{*} + {}^{1}\mathrm{D} \quad \text{(triplet}-\text{triplet energy transfer)} \\ {}^{1}\mathrm{D}^{*} + {}^{1}\mathrm{A} & \rightarrow & {}^{1}\mathrm{A}^{*} + {}^{1}\mathrm{D} \quad \text{(singlet}-\text{singlet energy transfer)} \end{array}$$

where D* is the excited donor molecule and A is the acceptor molecule in ground S_0 state

The triplet–triplet energy transfer normally requires a collision between the molecules, whereas, singlet–singlet energy transfer takes place over a relatively long distance (about 40 A). Triplet–triplet energy transfer is a very important process of energy transfer because triplet states are usually much difficult to prepare by direct irradiation than singlet states and lifetimes of triplet states are much longer than singlet states.

6.14.2 The Stern–Volmer Equation for Determination of Quenching Rate

The efficiency of a quencher can be evaluated by means of Stern–Volmer equation by measuring the fluorescence intensity in the absence and presence of the quencher.

Let us consider a quenching process of S_1-excited state in presence of a quencher Q.

$S_1 \rightarrow S_0 + h\nu$: fluorescence

$S_1 \rightsquigarrow T_1$: intersystem crossing

$S_1 + Q \rightsquigarrow S_0 + Q$: quenching

In absence of quencher, fluorescence takes place unimolecularly.

So, rate of fluorescence emission,

$$\begin{aligned} {}^{1}J_{\text{total}} &= (K_{\text{f}} + K_{\text{isc}} + K_{\text{ic}})[S_1] \\ &= {}^{1}K_{\text{total}}[S_1], \end{aligned}$$

where K_f, K_{isc}, and K_{ic} are the rate constants of fluorescence, intersystem crossing, and internal conversion, respectively.

In presence of quencher, fluorescence takes place in bimolecular process.

So,

$$Q_J = K_q[S_1]\,[Q],$$

where Q_J is the rate of deactivation by quenching and K_q is the rate constant.

The overall rate of deactivation is given by the sum of the rates of unimolecular and bimolecular processes

$$^{Q}J_{total} = {}^{1}J_{total} + Q_J = {}^{1}K_{total}[S_1] + K_q[S_1]\,[Q]$$

If $^{Q}\varphi_f$ and φ_f are the fluorescence quantum yields in presence and absence of a quencher, then

$$\begin{aligned} ^{Q}\varphi_f = J_f/{}^{Q}J_{total} &= K_f[S_1]/({}^{1}K_{total} + K_q[S_1]\,[Q] \\ &= K_f/\left({}^{1}K_{total} + K_q[Q]\right), \end{aligned}$$

where J_f is the rate of fluorescence emission and is equal to K_f $[S_1]$and $\varphi_f = K_f/{}^{1}K_{total}$

Therefore, $\varphi_f/{}^{Q}\varphi_f = ({}^{1}K_{total} + K_q[Q]/{}^{1}K_{total} = 1 + \left(K_q[Q]/{}^{1}K_{total}\right).$

The lifetime of $S_1 = {}^{1}\tau = 1/{}^{1}K_{total}$.

So,

$$\begin{aligned} \varphi_f/{}^{Q}\varphi_f &= 1 + K_q^{1}\tau[Q] \\ &= 1 + K_q[Q] \end{aligned}$$

where K_q is the Stern–Volmer quenching constant.

If the fluorescence quantum yields φ_f in the absence of quencher and $^{Q}\varphi_f$ at different concentration of Q are measured, the Stern–Volmer plot of $\varphi_f/{}^{Q}\varphi_f$ against $[Q]$ will give a straight line of slope K_q and intercept 1 (Fig. 6.8)

The Stern–Volmer method may also be used for monitoring of phosphorescence quenching.

6.14.3 Deviation from Stern–Volmer Kinetics

Stern–Volmer method for the study of the rate of decrease of fluorescence and phosphorescence in the presence of a quencher does not provide satisfactory results when the interactions between donor and acceptor are maximum at a certain concentration of the donor or acceptor. Further increase of donor or acceptor concentration shows a non-linear interaction between them. Possibly static quenching has dominant role in the luminescence emission and binding of quencher to donor is

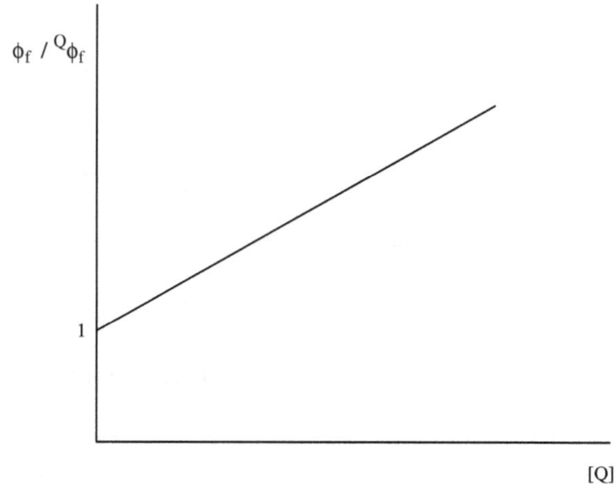

Fig. 6.8 Stern–Volmer plot of fluorescence quenching

the major factor for this deviation. For example, the binding of dissolved organic matter (DOM, humic acids) to phenanthrene shows a non-linear fluorescence spectrum at higher concentration of phenanthrene or DOM [12]. The concentration limit depends on the nature of donor and quencher.

6.14.4 The Excimers and Exciplexes

Excimer

The excited state dimer of a compound is called excimer. For example, a high concentrated solution of pyrene in toluene is irradiated with UV light, an excimer is formed.

$$\begin{array}{lcll} ^1P^* + P & \rightarrow & ^1[PP]^* & \text{(excimer)} \\ \downarrow & & \downarrow & \\ P + h\nu & & P + P + h\nu' & \end{array}$$

P denotes ground-state pyrene.

Exciplex

Exciplex is an excited complex formed by the reaction of an excited molecule of a compound with a quencher molecule. For example, when a solution of anthracene (*A*) in presence of diethylamine is irradiated with visible light, an exciplex is formed. Ground-state N,N-diethyl aniline acts as a quencher (*Q*).

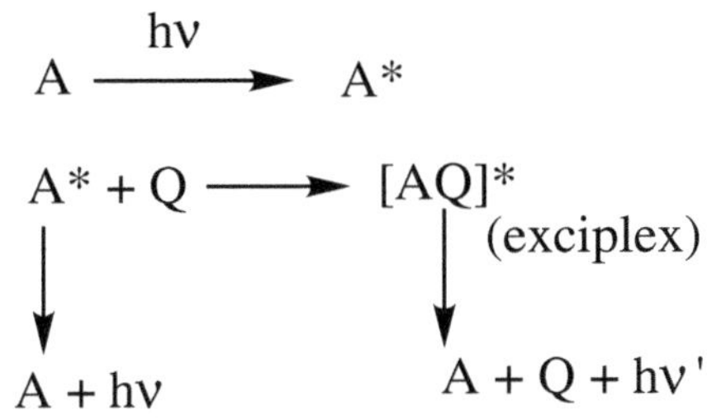

6.14.5 *Long-Range Energy Transfer Process: The FRET Process*

Forster proposed a theory for the transfer of energy between two fluorescent chromophores, known as Forster resonance energy transfer (FRET) process [13]. In this process, the energy transfer takes place by the dipole–dipole (Coulombic) interactions between the transition dipoles created between the electrons of the donor and acceptor molecules on absorption of light. This mechanism can only occur where spin multiplicity is conserved in energy transfer process. Singlet–singlet energy transfer occurs by this mechanism as the donor molecule (excited singlet to singlet) and acceptor molecule (singlet to excited singlet) undergo no change of spin multiplicity, resulting in the creation of large transition dipoles. This process is equivalent to the energy transfer process in a transmitter-antenna system. The relaxation of excited donor molecule to its ground state creates a transition dipole, which simultaneously induces a transition dipole in electronic excitation of singlet ground-state acceptor molecule into its excited singlet state. Thus, the coupling of donor and acceptor transition dipoles requires an equal energy for this long-range non-radiative energy transfer. This energy transfer process is sometimes called resonance energy transfer because the energies of the coupled transitions are identical, or in a state of resonance. The electronic movements in this energy transfer process are shown in Fig. 6.9.

$$^{1}\mathrm{D}^{*} + {}^{1}\mathrm{A} \rightarrow {}^{1}\mathrm{D} + \mathrm{A}^{*}$$

According to the Forster theory, the probability of energy transfer falls off inversely with the sixth power of the distance between the donor and the acceptor

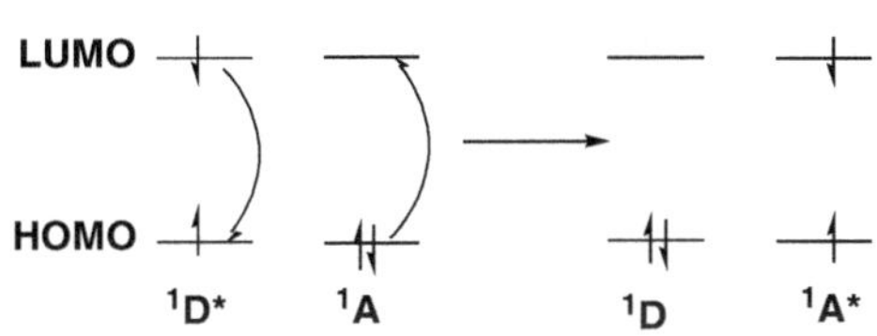

Fig. 6.9 Electronic movements occurring in the long-range singlet–singlet energy transfer process

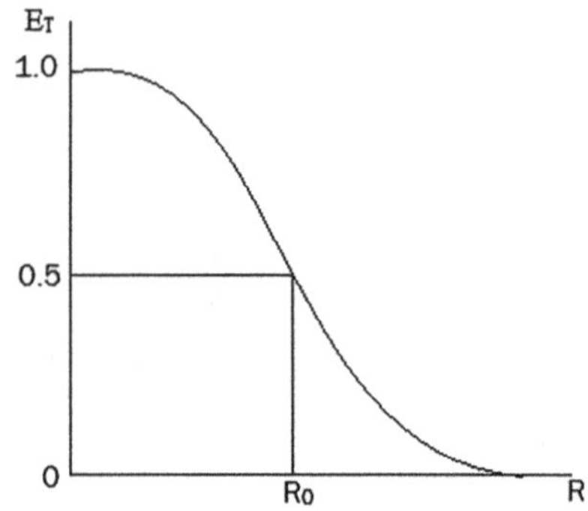

Fig. 6.10 The dependence of efficiency of energy transfer E_T on donor–acceptor distance R, as per Forster theory in a FRET process

molecules. The efficiency of resonance energy transfer, E_T increases with decreasing distance R, according to equation,

$$E_T = R_0^6/\left(R_0^6 + R^6\right)$$

where R_0 is the critical transfer distance, characteristic for a particular donor–acceptor pair and R is the distance between D* and A. At R_0 the efficiency of energy transfer is 50 % (Fig. 6.10).

This type of energy transfer frequently occurs in biological macromolecules such as proteins. The study of absorption and fluorescence spectra of green fluorescent protein (GFP) and blue fluorescent protein (BFP) isolated from living cells of jellyfish is useful to determine the donor–acceptor distances for FRET to occur. The BFP absorbs light at wavelength 380 nm and acts as donor, while GFP absorbs light and emits fluorescence at wavelength 510 nm and acts as acceptor. When these donor and acceptor fluorophores are labeled and their dynamic protein interactions in the living cells are visualized, it is found that in their normal conformation, the distance between them is 12 nm and no FRET occurs (Fig. 6.11). Under certain circumstances of cellular functioning, they are brought closer

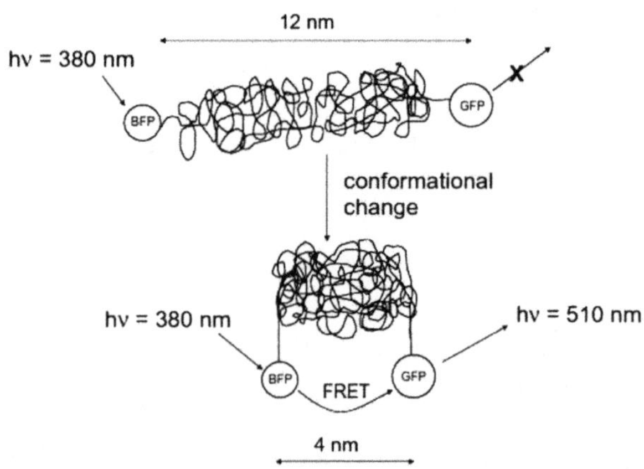

Fig. 6.11 Conformational change occurs in green fluorescent protein (GFP) of jellyfish during fluorescence emission. Adapted with permission from (Wardle B 2009 Principles and applications of photochemistry, Wiley, p. 102). Copyright (2009) John Wiley & Sons

together within a distance of 4 nm, the excitation of donor (BFP) at 380 nm gives emission fluorescence from acceptor at 510 nm.

6.14.5.1 *Efficiency of Energy Transfer in FRET* Process

The efficiency of energy transfer in FRET process depends on the following factors:

a. The relative fluorescence intensity of the donor in the absence and presence of the acceptor. The higher the fluorescence intensity (F_D) in the absence of acceptor, relative to that (F_{DA}) in the presence of acceptor, higher will be the efficiency of energy transfer.

$$E_T = 1-(F_{DA}-F_D)$$

where E_T denotes the efficiency of energy transfer and F_{DA} is the fluorescence intensity in the presence of the acceptor.

b. Similarly, E_T depends on relative fluorescence quantum yield of the donor in the absence (φ_D) and the presence (φ_{DA}) of the acceptor.

$$E_T = 1-(\varphi_{DA} - \varphi_D)$$

c. E_T depends on relative fluorescence lifetime of the donor in the absence (τ_D) and in the presence (τ_{DA}) of the acceptor.

$$E_T = 1-(\tau_{DA} - \tau_D)$$

6.14.6 Short-Range Energy Transfer Process: The Dexter Theory of Energy Transfer

David L. Dexter proposed a theory of energy transfer between donor and acceptor molecules from their close approach (within 10 Å), so that their electron orbitals can overlap to exchange the electrons between them. This theory is sometimes called short-range electron exchange or collisional energy transfer theory [14]. The distance that makes the energy transfer to occur between molecules D and A is almost comparable to their collisional diameter. For this reason, this theory is referred to collisional energy transfer theory.

The energy transfer by this exchange process occurs when the molecules have spin conservation, that is, the total electron spin does not change after the energy transfer.

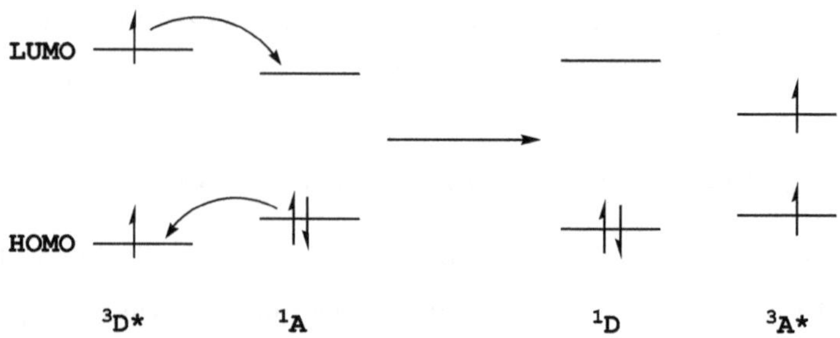

Fig. 6.12 Electron movements in Dexter short-range (triplet–triplet) energy transfer process

$$^{1}\mathrm{D}^{*} + {}^{1}\mathrm{A} \rightarrow {}^{1}\mathrm{D} + {}^{1}\mathrm{A}^{*}$$
$$^{3}\mathrm{D}^{*} + {}^{1}\mathrm{A} \rightarrow {}^{1}\mathrm{D} + {}^{3}\mathrm{A}^{*}$$

It means that an excited singlet molecule will produce another excited singlet molecule and an excited triplet molecule will produce another excited triplet molecule after their energy transfer. The singlet–singlet energy transfer can occur when the long-range Coulombic interaction takes place between the donor and acceptor molecules. Thus, the Dexter theory of energy transfer is applicable to triplet–triplet energy transfer process because this energy transfer process requires orbital overlap for exchange of their electron. The electron movement in this exchange process is shown in Fig. 6.12.

6.14.6.1 Rate of Dexter Energy Transfer

The rate constant for the Dexter exchange mechanism is given by:

$$K_{\mathrm{ET}}(\mathrm{exchange}) = 4\pi^2/h(H^{\mathrm{en}})^2 J_{\mathrm{D}}$$

where H^{en} is the electronic coupling between donor and acceptor, exponentially dependent on distance.

$$H^{\mathrm{en}} = H^{\mathrm{en}}(0)\ \exp[-\beta^{\mathrm{en}}(r_{\mathrm{DA}} - r_0)/2]$$

where β^{en} is the attenuation factor exchange energy transfer and r_{DA} is the distance between D and A.

J_{D} is the overlap factor and h is Planck constant.

This equation is simplified as

$$K_{\mathrm{ET}}(\mathrm{exchange}) = K\ \exp(-2r_{\mathrm{DA}})$$

Thus, the rate is dependent on the distance. It is observed that this mechanism operates when the r_{DA} is 5–10 A.

6.14.6.2 *The Triplet–Triplet Energy Transfer in Photosensitization Process*

The excitation of benzophenone in solid solution at 77 K with light of wavelength 366 nm emits phosphorescence. When naphthalene is added to this solid solution, the benzophenone phosphorescene is replaced by naphthalene phosphorescence even naphthalene does not absorb photons from light of wavelength 366 nm. This quenching process takes place by formation of exciplex between excited benzophenone and ground-state naphthalene followed by triplet–triplet energy transfer and emission of phosphorescence from the triplet naphthalene [15].

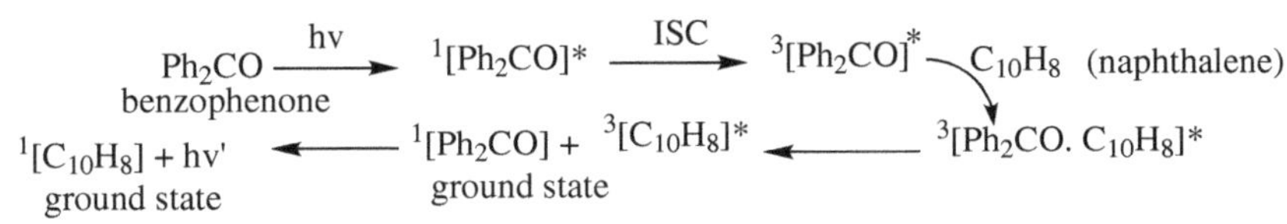

6.14.6.3 Applications of Triplet–Triplet Energy Transfer Process

The principle of Dexter energy transfer is frequently applied for commercial manufacture of white organic light-emitting diodes with fluorescent tubes and energy up conversion systems such as blue light emission [16] and while light emission [17], by triplet–triplet annihilation.

6.14.6.4 Triplet–Triplet Annihilation

Triplet–triplet annihilation (TTA) is an important process of exchange energy transfer. Both donor and acceptor molecules in their triplet states exchange their energy to produce their singlet states (Fig. 6.13).

$$^3D^* + {}^3A^* \rightarrow {}^1D + {}^1A^*$$

The fluorescence observed in triplet–triplet annihilation is known as P (pyrene)-typed delayed fluorescence because it was first observed in pyrene. This mechanism is utilized to produce high-energy light device simply using medium-energy light. The accepted mechanism of this exchange process is:

Absorption: $S_0 + hv \rightarrow S_1$

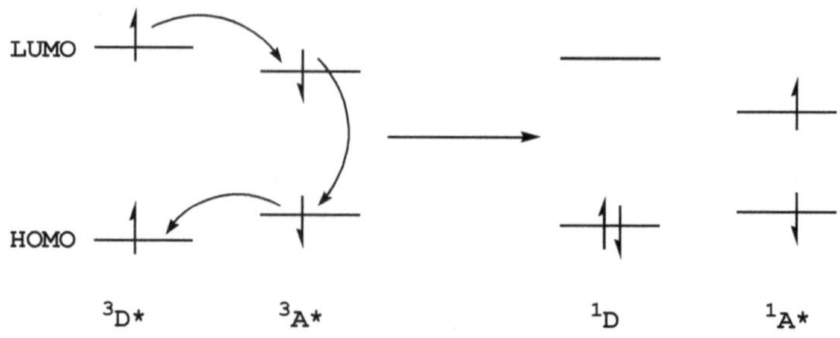

Fig. 6.13 Electron movement in a triplet–triplet annihilation process

Intersystem crossing: $S_1 \xrightarrow{\text{ISC}} T_1$

Triplet–triplet annihilation: $T_1 + T_1 \rightarrow E + S_1 + S_0$

Delayed fluorescence: $S_1 \rightarrow S_0 + hv$

The higher energy S_1 state is responsible for emission of high-energy light as fluorescence [18].

6.14.7 Photodynamic Tumor Therapy Using Singlet Oxygen

The ground state of oxygen molecule is triplet ($^3O_2$) and its lowest excited state is singlet state ($^1O_2$), which is difficult to generate by direct irradiation of the triplet ground state. In such case, energy transfer occurs in the reverse direction involving triplet excited state to singlet excited state in a spin-forbidden process. In photodynamic therapy, a photosensitizer(S) such as s, chlorophylls and dyes is injected in the blood, when it spreads out in different tissues including tumor cells. The tumor cells are exposed to laser light at a longer wavelength (~700–850 nm) corresponding to the absorption maximum of the sensitizer. It causes excitation of the sensitizer to its excited singlet state. The excited singlet state of the sensitizer is converted to excited triplet state by intersystem crossing. The excited triplet state of the sensitizer undergoes energy transfer to triplet oxygen producing singlet oxygen. The resulting singlet oxygen is toxic and oxidizes substances within the tumor cells, destroying the tumor in the process. The triplet sensitizer also undergoes photochemical hydrogen abstraction with organic molecules within tumor cells producing a number of radical species for the destruction of tumor [19].

The most common porphyrins used in this tumor therapy are protofrin and verteporfin having the generalized porphyrin structure (Fig. 6.14). It is extensively used in the treatment of oesophageal and lung cancer.

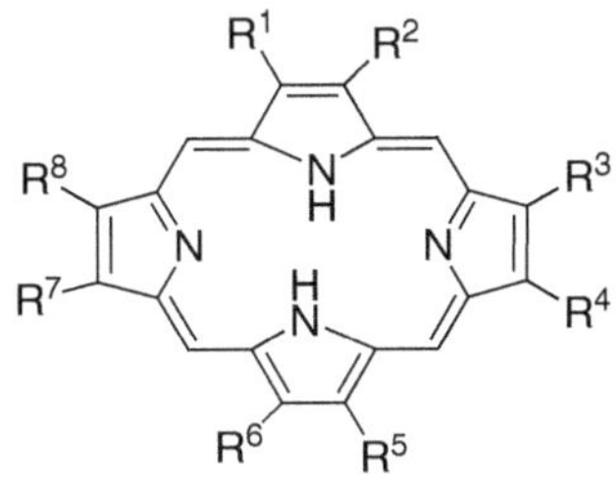

Fig. 6.14 Generalized structure of porphyrin. The *R* groups represent different side groups attached to the porphyrin ring

The energy transfer process takes place in the following steps:

$$^{1}\mathrm{P} + h\nu \rightarrow {}^{1}\mathrm{P}^{*} \xrightarrow{\mathrm{ISC}} {}^{3}\mathrm{P}^{*}$$

$$^{3}\mathrm{P}^{*} + {}^{3}\mathrm{O}_2 \rightarrow {}^{1}\mathrm{P} + {}^{1}\mathrm{O}_2^{*}$$

^{1}P represents porphyrin sensitizer in the ground state.

6.14.8 *Photo-induced Electron Transfer (PET) Process*

Photo-induced electron transfer (PET) process occurs in nature in the photosynthesis of bacteria and higher plants. The absorption of light by a molecule promotes the molecule to its higher electronic state, which makes the molecule a better electron donor or better electron acceptor state than its ground state. The transfer of electron from donor excited molecule to another ground-state molecule or from ground-state donor molecule to excited acceptor molecule is called the photo-induced electron transfer (PET) process. The PET process changes the redox properties of the molecules and involves a weak orbital overlapping and charge separation among them. Thus, the excited molecule (A) becomes a better reducing agent or oxidizing agent and interacts with a ground-state molecule B. It is a primary photochemical process in supramolecules.

$\mathrm{A}^{*} + \mathrm{B} \rightarrow \mathrm{A}^{\cdot +} + \mathrm{B}^{\cdot -}$ (reductant nature of A* and the process is oxidative electron transfer)
$\mathrm{A}^{*} + \mathrm{B} \rightarrow \mathrm{A}^{\cdot -} + \mathrm{B}^{\cdot +}$ (oxidant nature of A* and the process is reductive electron transfer)

Both these oxidative and reductive electron transfer processes are represented in molecular orbital interactions (Fig. 6.15)

6.14.8.1 Application of PET Process in Molecular Fluorescence Switch

This photo-induced electron transfer process among the molecules may be utilized as fluorescence switching. A fluorophore having a macrocyclic unit, on irradiation

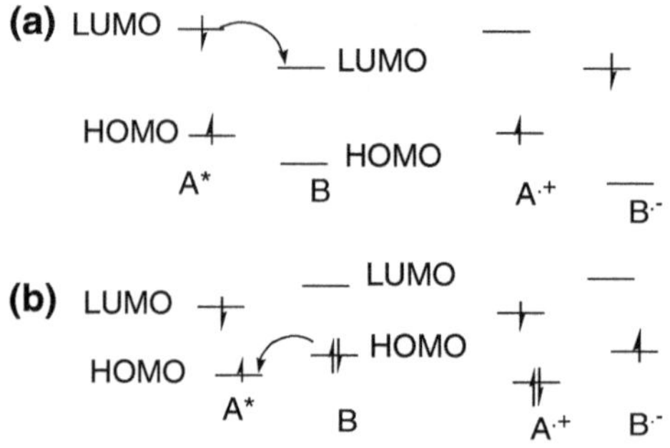

Fig. 6.15 Molecular orbital representation of electron transfer in a PET process. **a** Oxidative electron transfer, where B is electron poor acceptor molecule, and **b** reductive electron transfer, where B is electron-rich donor molecule

with light causes no emission of fluorescence. But on insertion of potassium cation in the macrocyclic unit of the fluorophore, it results in emission of fluorescence. Thus, potassium cation sensor acts as a molecular switch for the fluorescence [20]. The Fig. 6.16 illustrates the principle of the process using anthracene fluorophore.

The principle of PET process with potassium cation (K^+) sensor can be explained as follows. Excitation of the fluorophore causes the promotion of one electron from the HOMO to the LUMO. It enables the flow of one electron from the HOMO of donor macrocyclic unit to fluorophore, resulting in quenching of fluorescence. On binding with K^+ cation, the HOMO of the macrocyclic unit is of lower energy than that of fluorophore anthracene. Hence, PET is no longer possible and fluorescence quenching is stopped (Fig. 6.17).

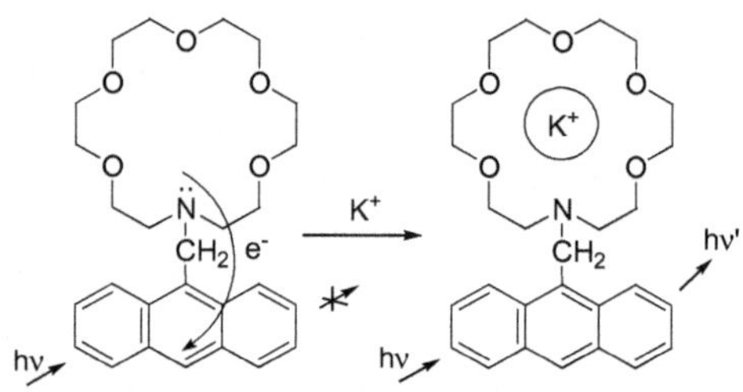

Fig. 6.16 Potassium cation sensor as a molecular fluorescence switch in a PET process of anthracene fluorophore having a macrocyclic donor unit

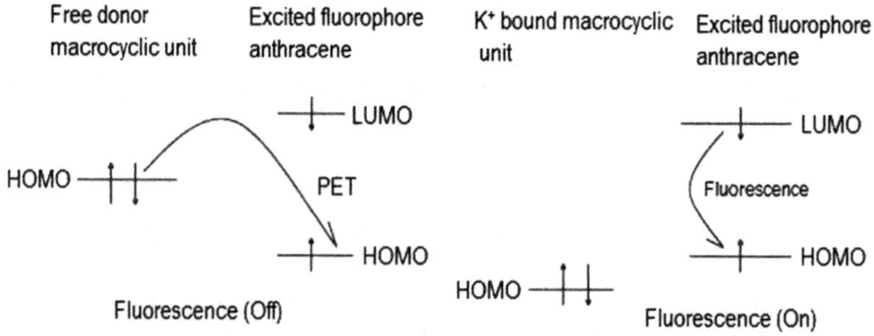

Fig. 6.17 Principle of PET process in K^+ bound sensor

The fluorescence switching devices are useful in the manufacture of optoelectronic materials and in the study of biological dynamics, living cells imaging and biomaterial sensors [21].

6.14.9 *The Marcus Theory of Electron Transfer*

The basic assumption of the Marcus theory of electron transfer process is that the reactants needed a weak interaction between them for this process to operate. The Marcus theory considers the reaction rate theory, potential energy surfaces and reorganization of the system to explain the electron transfer process [22]. The potential energy curves of an electron transfer reaction for the initial (i) and final (f) states of the system are represented by parabolic curves (Fig. 6.18). These curves quantitatively relate the rate of electron transfer to the reorganizational energy (λ) and the free energy changes for the electron transfer process (ΔG^0) and activation ($\Delta G^{\#}$).

In a polar solvent, the solvent dipoles are arranged around the molecules taking part in PET. The solvent reorganization is required to accommodate and stabilize the changed species (Fig. 6.19) and this reorganization process requires some energy from the system.

The free energy change, ΔG^0, of an electron transfer process is the driving force of the process. The free energy of activation, $\Delta G^{\#}$, is needed to reach the transition state, #. It is related to the reorganizational energy, λ of the system. From the geometry of the parabolas:

$$\Delta G^{\#} = \left(\Delta G^0 \pm \lambda\right)^2 / 4\lambda$$

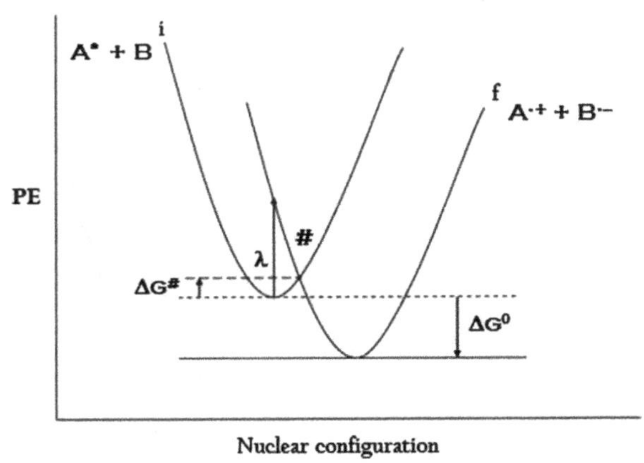

Fig. 6.18 Potential energy (PE) description of an electron transfer reaction. The parabolic curves intersect at the transition state (#)

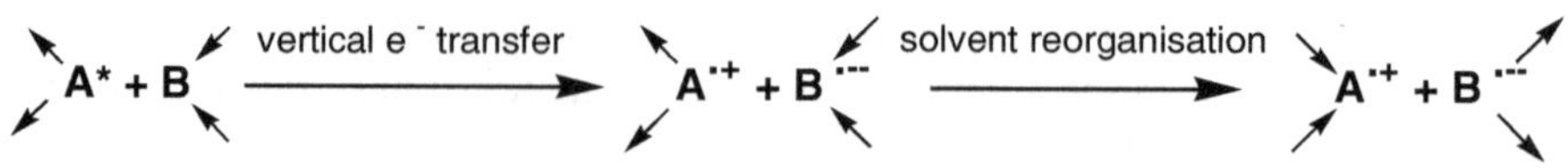

Fig. 6.19 Reorganization of polar solvent dipoles during PET process

According to the collision theory, the rate constant K_{ET} is given by

$$\begin{aligned} K_{\mathrm{ET}} &= A_{\exp}\left(-\Delta G^{\#}/\mathrm{RT}\right) \\ &= A_{\exp}\left[-\left(\Delta G^{0} \pm \lambda\right)^{2}/4\lambda\right]/\mathrm{RT} \end{aligned}$$

Or,

$$\ln K_{\mathrm{ET}} = \ln A + \left[-\left(\Delta G^{0} \pm \lambda\right)^{2}/4\lambda\right]/\mathrm{RT}$$

This equation based on Marcus model gives the relation between the kinetics (K_{ET}) and thermodynamic driving force (ΔG^0) of PET process. Analysis of this equation gives three distinct kinetic regions, as shown in Fig. 6.20, depending on ΔG^0.

a. **Normal region**: The PET process rate increases with increase of ΔG^0.
b. **Activationless region**: The change of ΔG^0 has negligible effect on the rate process.
c. **Inverted region**: The rate of PET process decreases with increase of ΔG^0

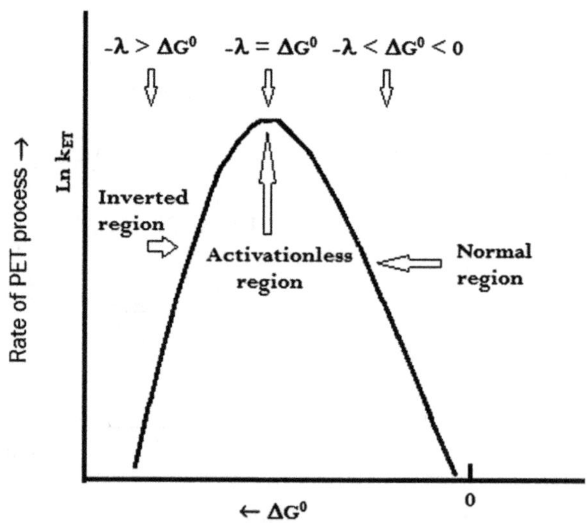

Fig. 6.20 Free energy change, ΔG^0 dependence of electron transfer rate, K_{ET} according to Marcus theory of electron transfer process

The equation for K_{ET} can be generalized to

$$\log_{10} K_{ET} = 13 - 0.6(R - 3.6) - 3.1\left(\Delta G^0 \pm \lambda\right)^2/\lambda$$

where R is the distance between the redox centers, i.e., center to center reactant distance.

The maximum rate of electron transfer occurs when $\lambda = -\Delta G^0$, and the equation is

$$\log_{10} K_{ET(Max)} = 13 - 0.6(R - 3.6).$$

6.14.9.1 Calculation of K_{ET} for an Electron Transfer System

Let us consider two groups A and B in a protein molecule having E^0/mV of −100 and −90, respectively. Their internuclear distance is 11.8 Å. The protein on irradiation with light undergoes PET process from A to B.

Therefore, ΔG^0 of the system = −100 − (−90) = −10 mV.

λ for intramolecular electron transfer is 750 mV.

Using the expression for K_{ET},

$$\begin{aligned} \log K_{ET(A\to B)} &= 13 - 0.6(11.8 - 3.6) - 3.1(0.01 + 0.75)^2/0.75 \\ &= 5.82 \\ K_{ET(A\to B)} &= 6.60 \times 10^5\ \mathrm{s}^{-1} \end{aligned}$$

6.14.9.2 Evidence of Inverted Region in a Dyad

A dyad is a supramolecular structure consisting of a donor and an acceptor component. In a fullerene–porphyrin-based dyad, C_{60} is the acceptor component and porphyrin is the donor component. On photoirradiation of this dyad, PET process takes place in the normal region and back-electron transfer (BET) from $C_{60}-$ to Zn Por+ in the inverted region (Fig. 6.21) [23].

Fig. 6.21 Normal and inverted regions of Marcus equation for electron transfer process in a Zinc porphyrin—C_{60} dyad

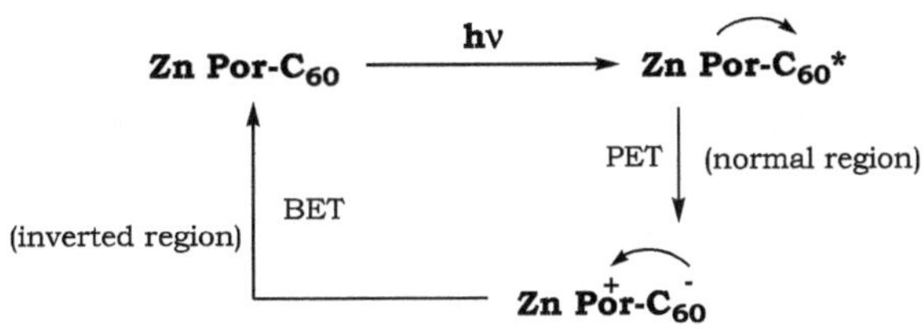

6.15 Photochemical Reactions and Their Kinetics

Organic molecules in their singlet and triplet excited states can undergo photochemical reactions. Singlet excited states have very short lifetimes and triplet states have relatively longer lifetimes and hence most of the photochemical reactions occur through triplet excited states [24]. Excited molecules undergo unimolecular or bimolecular reactions in a single step (concerted process) or in two or multistep processes involving one or more intermediates. Most of the photochemical reactions proceed through photolytic cleavage into radicals followed by radicals coupling, isomerisation, dimerization, hydrogen abstraction, elimination and rearrangements [25].

Absorption of a photon by an organic molecule, *R*, leads to the formation of an electronically excited state, R^* of the molecule.

$$R + hv \rightarrow R^*$$

The excited state R^* may react in any one of the two ways:

In a concerted process (i.e., in a single step) gives the product *P*:

$$R^* \rightarrow P$$

These concerted processes include a series of pericyclic reactions from S_1 (π, π*) via cyclic transition states, where σ or π bonds are cleaved and formed simultaneously.

In two or multistep process, one or more intermediates I are formed:

$$R^* \rightarrow I \rightarrow P$$

or

$$R^* \rightarrow I_1 \rightarrow I_2 \rightarrow I_3, \text{ etc} \rightarrow P$$

These include free radical reactions involving diradical intermediate or intermediates from either S_1 (π, π*; n, π*) or T_1 (π → π*, n, π*). The most common photochemical reactions are the reactions of carbonyl compounds, alkenes, and aromatic compounds as well as chain reactions of hydrocarbons.

All these photochemical reactions involve funnel-like conical intersections (CIs) in their electronic excited and ground-state potential energy surfaces, where the cones touch each other to give reactive intermediates and products (Fig. 6.21). These conical intersections are analogous to transition state in thermal reactions [26] (Fig 6.22).

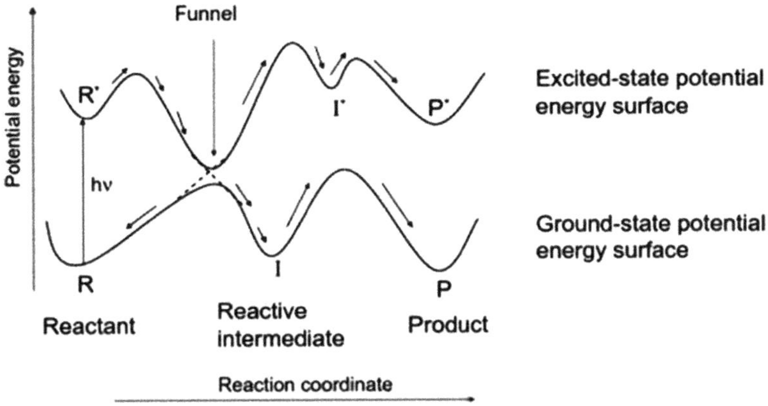

Fig. 6.22 Change of potential energy surfaces for excited-state and ground-state molecules. Adapted with permission from (Turro NJ 1991 Modern Molecular Photochemistry, University Science Books). Copyright (1991) University Science Books

6.15.1 Determination of the Excited State Configuration

Luminescence spectra (fluorescence and phosphorescence) measurements provide information about the configuration of excited states (S_1 or T_1) involved in the reactions. Moreover, the lifetimes of the excited states provide us the information about their origin—($\pi \rightarrow \pi^*$) or ($n \rightarrow \pi^*$). The typical radiative lifetimes of singlet $^1(n \rightarrow \pi^*)$ states are $10^{-6} - 10^{-3}$ s and of $^1(\pi \rightarrow \pi^*)$ states are $10^{-9} - 10^{-6}$ s, whereas of typical triplet $^3(n \rightarrow \pi^*)$ states are $10^{-4} - 10^{-2}$ and $^3(\pi \rightarrow \pi^*)$ states are $1–10^2$ s.

6.15.2 Determination of the Yield of Products

The ratio of the products formed in a photochemical reaction in the absence and in the presence of a quencher can be determined from the ratio of φ/φ_q of the Stern–Volmer equation related to triplet state quenching:

$$\varphi/\varphi_q = 1 + K_q{}^3\tau[Q]$$

where φ and φ_q are the quantum yields of product formation without and with the quencher, respectively; K_q is the rate constant for quenching; $[Q]$ is the concentration of the quencher and $^3\tau$ is the triplet lifetime in the absence of the quencher. The Stern–Volmer equation has a linear form and the quantity $K_q{}^3\tau$ is measured as the slope of the plot of φ/φ_q against $[Q]$ for different quencher concentrations.

The value of K_q in the range of 10^9–10^{10} mol^{-1} dm^3 s^{-1} is found in most cases and hence a value of $^3\tau$ may be obtained.

6.15.3 Determination of the Lifetime of Intermediates

The kinetics of a photochemical reaction is determined by the laser flash photolysis technique developed by Norrish and Porter in 1949 [27]. They won 1967 Nobel Prize in Chemistry for this invention. The basis of this flash photolysis is to irradiate the sample solution with a very short, intense pulse of light in the nanosecond timescale using Q-switching and then to monitor the changes in the sample system with time by some spectroscopic methods (absorption or emission spectroscopic study). The monitoring device is fast enough to observe the transient states before it decay. The mechanism of a photochemical reaction relating to a number of free radical species and their electronic excited states can be explored. The development of mode-locked titanium sapphire laser, sub-picosecond (ps) pulses can be produced and is usually pumped by continuous wave (CW) argon or Nd-YAG (neodymium–yttrium–aluminum garnet) lasers. The pulses coming from this laser up to the order of a few femtoseconds (fs, 10^{-15} s) will be useful to detect the singlet state free radicals of usual lifetimes 5–100 fs.

Whereas Q-switched Nd-YAG laser pulses are useful to monitor the triplet state diradicals of usual life times of 10–800 ns (10^{-9} s), the pump-pulse or probe pulse technique is useful to study the photochemical change that occurs in the picoseconds time scale. It includes *cis–trans*-isomerisation, internal conversion [S_1 ($v = 4$) to S_1 ($v = 0$)], energy transfer and electron transfer processes, etc. This method also enables to determine the order of a photochemical reaction.

6.15.4 Low-Temperature Matrix Studies

Many photochemical reactions are carried out at low temperatures as low as 4 K to slow down the reaction rate for the study of the lifetimes of the reactive intermediates. The most useful matrix materials are solid argon, solid neon and solid nitrogen. The initial photoproduct is trapped within a rigid matrix that inhibits the decay of the reactive species in diffusion process. For example, δ-hydroxy-α, β, γ, δ-unsaturated valerolactone **1** on photochemical decomposition gives cyclobutadiene **2** and carbon dioxide. The intermediate and the products of this reaction are characterized in low-temperature matrix isolation process.

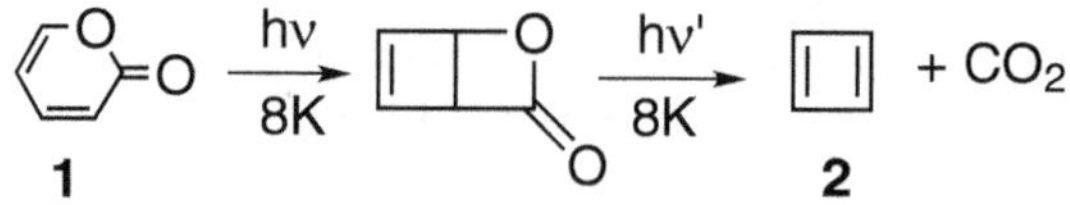

1 **2**

6.16 Further Reading

1. Wardle B (2009) Principles and applications of photochemistry. Wiley, New York.
2. Wayne CE, Wayne RP (2005) Photochemistry. Oxford University Press, Oxford.
3. Turro NJ, Ramamurthy V, Scaiano JC (2010) Modern molecular photochemistry of organic molecules. University Science Books, Sausalito.
4. Balzani V, Credi A, Silvi S, Vanturi M (2006) Artificial nanomachines based on interlocked molecular species: recent advances. Chem Soc Rev 35:1135.

References

1. Franck J (1926) Trans Farad Soc 21: 536; Condon E (1926) Physical Rev 27: 640; ibid (1928) 32: 858
2. Kozier JC, Cowan DO (1978) Acc Chem Res 11:334
3. Jablonski A (1933) Nature 131:839
4. Lippert E, Luder W, Moll F, Nagele W, Boos H, Prigge H, Blankenstein IS (1961) Angew Chem 73:695
5. Williams ATR, Winfield SA, Miller JN (1983) Analyst 108:1067
6. Guilbault GG (1973) Practical fluorescence: theory, methods and techniques. Marcel Dekkar, New York; Wehry EL (1976) Modern fluorescence spectroscopy, vol 1. Plenum Press, New York
7. Turro NJ (1991) Modern molecular photochemistry. University Science Books, New York
8. Sharf B, Silbey R (1970) Chem Phys Lett 5:314
9. Moore WM, Hammond GS, Foss RP (1961) J Am Chem Soc 83:2789
10. Schulman SG (1977) Fluorescence and phosphorescence spectroscopy: physiochemical principles and practice. Pergamon, Elmsford, New York
11. Turro NJ, Dalton JC, Weiss DS (1969) Org Photochem 2:1
12. Pan B, Xing B, Liu W, Xing G, Tao S (2007) Chemosphere 69:1555
13. Forster T (1948) Annalen der Physik 437:55; Barigelletti F, Flamigi L (2000) Chem Soc Rev 29:1
14. Dexter DL (1951) J Chem Phys 21:836
15. Bayrakceken F (2008) Spectrochemica Acta Part A 71:603
16. Reineke S, Lindner F, Schwartz G, Seidler N, Walzer K, Lussem B, Leo K (2009) Nature 459:234
17. Monguzzi A, Mezyk J, Schtognella F, Tubino R, Meinardi F (2008) Physical Rev B 78:195112
18. Singh Rachford TN, Castellano FN (2010) Coodination. Chem Rev 254:2560
19. Josefsen LB, Boyle RW (2008) Metal—Based Drugs 276109; Richter A, Waterfield E, Jain AK, Sternberg E, DolphinD, Levy JG (1990) Photochem Photobiol 144:221
20. Prasanna de Silva A, Nimal Gunaratne HQ, Gunnlaugsson T, Huxley AJM, Mc Coy CP, Rademacher JT, Rice TE (1997) Chem Rev 97:1515
21. Bolzani V, Credi A, Raymo FM, Stoddart JF (2000) Angew Chem Int Ed 39:3348; Bolzani V, Credi A, Mattersteig G, Matthews OA et al (2000) J Org Chem 65:1924
22. Marcus RA (1964) Ann Rev Phys Chem 15:155
23. Kuciauskas D, Lin S, Seely GR, Moore AL, Moore TA, Gust D, Drovetskaya T, Reed C, Boyd PDW (1996) J Phys Chem 100:15926

24. Wagner PJ, Hammond GS (1968) Adv Photochem 5:21
25. Calvert JG, Pitts JN (1966) Photochemistry. Wiley, New York
26. Haas Y, Zilberg S (2001) J Photochem Photobiol 144:221
27. Norrish RGW, Porter C (1949) Nature 164:658

Chapter 7
Photochemistry of Alkenes, Dienes, and Polyenes

7.1 Introduction

Alkenes, dienes, and polyenes on absorption of light are activated to their electronic high-energy singlet and triplet states. These electronic excited states before return to ground states undergo various chemical reactions. These reactions have a conspicuous role in the areas of material engineering, nanotechnology, and supramolecular chemistry. These reactions may occur on direct irradiation or in the presence of a sensitizer. These reactions are broadly classified into four types, *cis–trans*-isomerizations, electrocyclic reactions, cycloaddition reactions, and rearrangement reactions. These reactions may occur in a concerted process or in a non-concerted stepwise process.

7.2 *Cis–Trans*-Isomerizations

7.2.1 Cis–Trans-*Isomerizations of Alkenes*

The *cis–trans*-isomerization is the classic photochemical reaction of olefins. The photochemical reaction of an alkene or diene depends on the extinction coefficient, ε, of that compound at any wavelength of light used for the reaction. A *trans*-isomer is thermodynamically more stable than its *cis*-isomer and hence has higher extinction coefficient in longer wavelength of light and low extinction coefficient in shorter wavelength of light. Thus, in a photochemical reaction of an alkene at higher wave length, *trans*-isomer is isomerized to its *cis*-form and maintains a photochemical equilibrium when both *cis*- and *trans*-isomers are in equilibrium and no further conversion occurs. Similarly, when the reaction is carried out at lower wavelength, *cis*-isomer is converted into its *trans*-isomer to attain a photochemical equilibrium. For a particular wavelength of a monochromatic light, the composition

B. Dinda, *Essentials of Pericyclic and Photochemical Reactions*,
Lecture Notes in Chemistry 93, DOI 10.1007/978-3-319-45934-9_7

of the photochemical equilibrium or photostationary state for a *cis–trans*-isomerization is given by

$$\frac{[trans]}{[cis]} = \left(\frac{\varepsilon_c}{\varepsilon_t}\right)\left(\frac{\phi_{c\to t}}{\phi_{t\to c}}\right)$$

The *cis–trans*-isomerization of alkenes is believed to take place via an excited state in which the two sp^3 carbons are twisted by 90° with respect to each other. This twisted geometry is referred to the perpendicular (p) state. This *p*-state geometry is believed to be the minimum energy geometry for both the singlet (s-) and triplet (t) excited states. A molecule in the p-state may undergo bond rotation in either direction to return to either *cis*- or *trans*-ground states (Fig. 7.1). The return to ground state from singlet excited state requires repairing of electrons by a nonradiative process, whereas return to ground state from the triplet excited state requires an intersystem crossing. Unconjugated alkenes absorb light in far-UV region (175–200 nm), while conjugated alkenes absorb in UV region (220–330 nm).

For example, irradiation of *trans*-stilbene **1** at 310–320 nm gives a photostationary state having more than 90 % of *cis*-isomer [1].

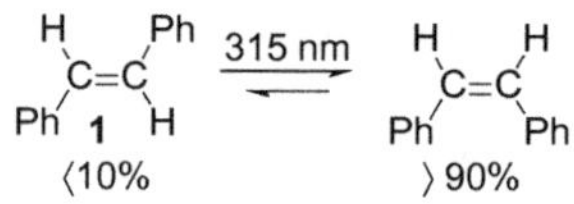

Detailed study on the composition of the photostationary state indicated that it is controlled by the wavelength of irradiation. Direct irradiation leads to isomerization via singlet state intermediate and irradiation in the presence of a sensitizer involves triplet state intermediate. Direct irradiation involves HOMO–LUMO interaction of the ethylenic double bond of stilbene. The study on the composition of photostationary state of *cis*- and *trans*-stilbenes using various photosensitizers revealed that the use of sensitizers (e.g., benzil, fluorenone, and 1-naphthyl phenyl ketone) having triplet excitation energies of 52–58 kcal/mol selectively excites *E*-stilbene resulting higher *Z:E* ratios, while use of sensitizers (e.g., benzophenone and acetophenone) having triplet excitation energies above 65 kcal/mol brings the *Z:E* ratios slightly higher than 1 [2].

Z- and *E*-2-Butenes in liquid neopentane undergo *cis–trans*-isomerization and it is competitive with their photochemical [2+2]-cycloadditions [3] . When the alkene solution is sufficiently diluted with an inert solvent, the isomerization becomes the dominant reaction. The triplet energy of sensitizer PhH is higher (~84 kcal/mol) than that for 2-butene (~80 kcal/mol) and hence triplet energy transfer is

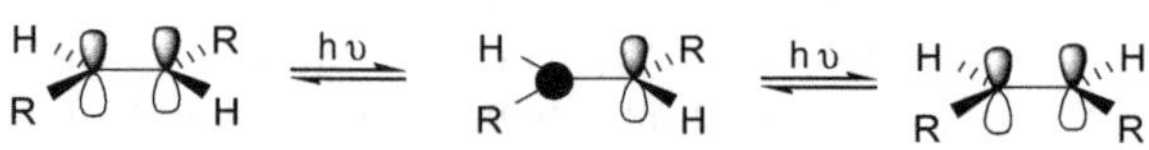

Fig. 7.1 Mechanism of photochemical *cis–trans*-isomerization of alkenes

favorable, while in case of ketone triplet having energy less than 70 kcal/mol, the reaction proceeds through a 1,4-diradical intermediate.

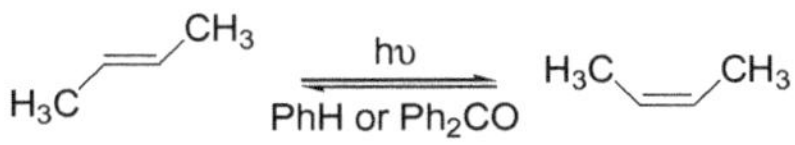

Among cycloalkenes, cyclohexenes, cycloheptenes, and cyclooctenes and their 1-methyl derivatives undergo *cis–trans*-isomerization on irradiation. *Trans*-isomers of cyclohexenes and cycloheptenes are very unstable because of high ring strain and hence are isolated by trapping in hydroxylic solvents like methanol [4]. For example, the photo-induced isomerization of *cis*-1-methylcyclohexene **2** to *trans*-isomer is trapped in methanol as methanol adduct [4].

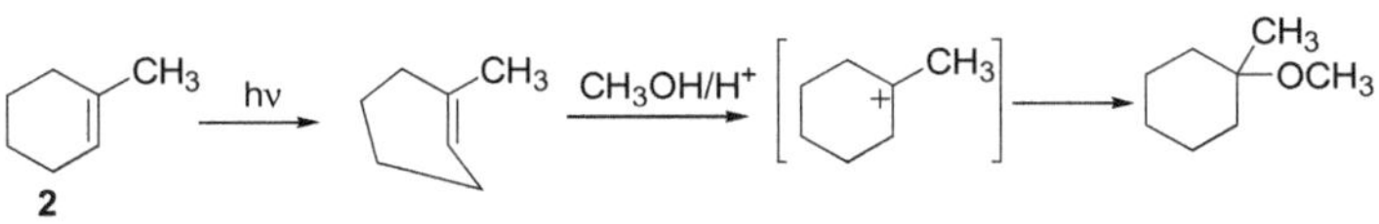

Irradiation of *cis*- and *trans*-cyclooctenes yields approximately equal amounts of both isomers in a photostationary state because of their almost equal extinction coefficients. Both these isomers can be isolated [5].

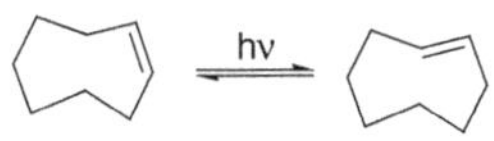

Cycloheptenes and cyclooctenes on irradiation, in addition to *cis–trans*-isomerization, undergo ring contraction and carbene insertion reactions [6, 7].

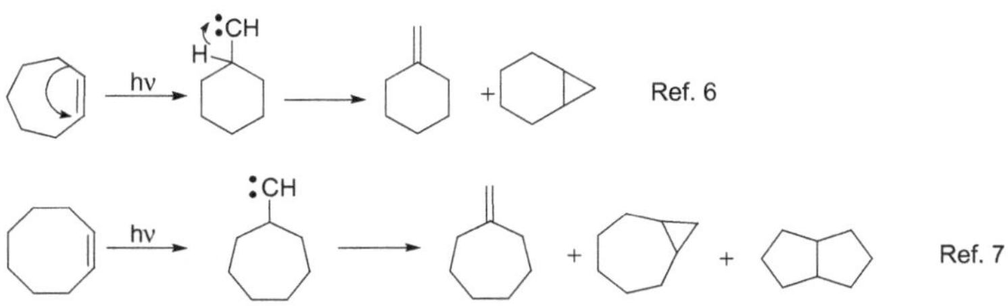

Norbornene **3** on irradiation results very unstable *trans*-isomer, but in methanol gives stable products by hydrogen abstraction and other radical coupling processes [8].

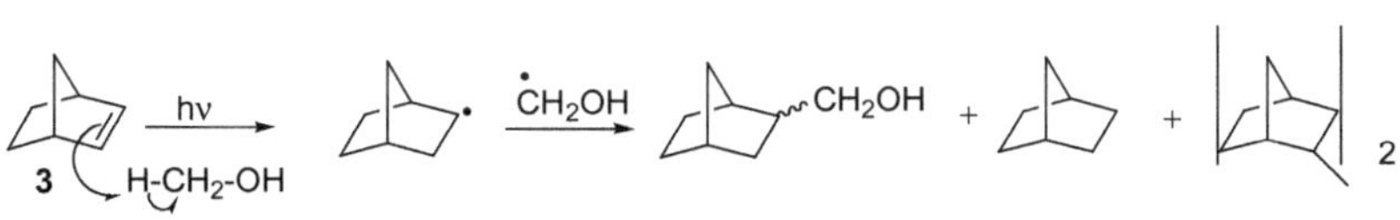

The photolysis of cyclopentene and norbornene **3** in inert solvent pentane results in several rearranged products from hydrogen abstraction and radical coupling [9].

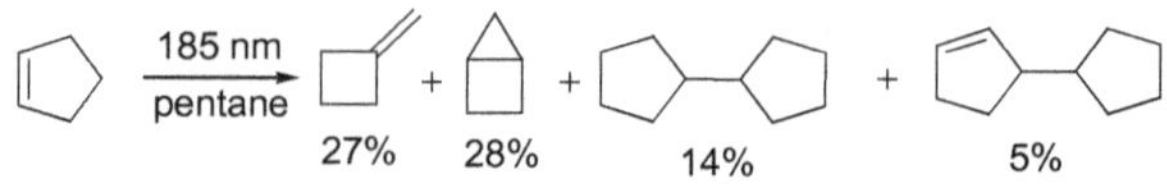

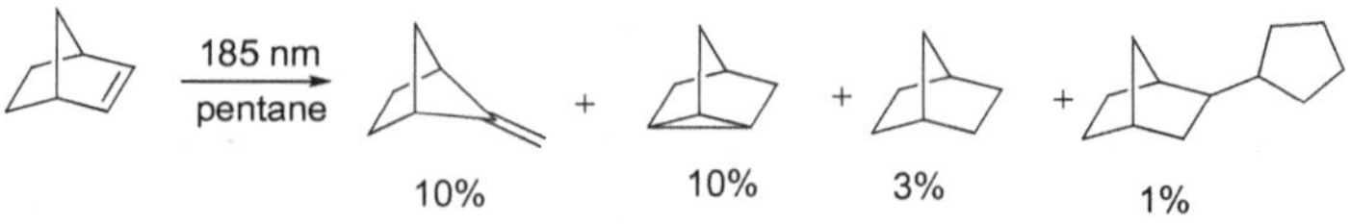

7.2.2 Cis–Trans-*Isomerization of Dienes*

Photosensitized *cis–trans*-isomerization of 1,3-pentadiene **4** has been studied in detail. It is observed that sensitizers whose triplet excitation energies are above 60 kcal/mol give photostationary mixtures containing about 55 % of the *trans*-isomer, while sensitizers having triplet excitation energies below 60 kcal/mol give a variety of photostationary mixtures containing 65–80 % *trans*-isomer, and are less efficient as sensitizers. Possibly the sensitizers having E_t greater than 60 kcal/mol transfer its triplet energy to either *cis*- or *trans*-1,3-pentadiene triplet excited states in a diffusion-controlled process, so that the composition of photostationary mixtures depends on the decay processes of the pentadiene triplets, while sensitizers of E_t less than 60 kcal/mol transfer their energy inefficiently to 1,3-pentadiene triplets, resulting the photostationary mixtures of high content of *trans*-isomer because the system prefers to produce the *cis*-diene triplet. Benzophenone-induced photo excitation of 1,3-pentadiene **4** results in both *cis–trans*-isomerization and dimerization [10].

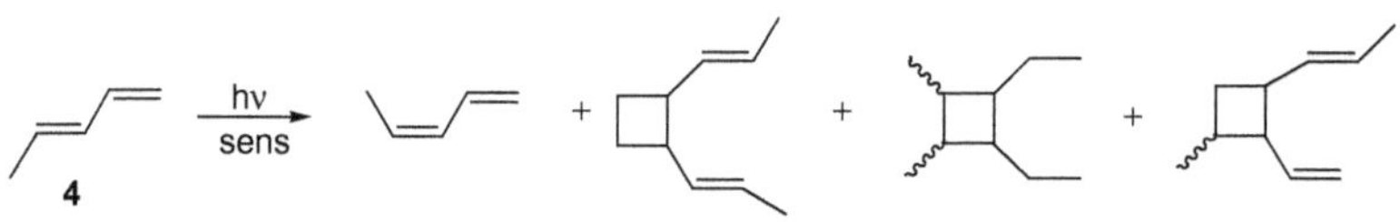

While direct irradiation of *E,E*-2,4-hexadiene **5** gives only *E,Z*-2,4-hexadiene from singlet excited state, triplet-sensitized reaction gives both *E,Z*- and *Z,Z*-2,4-hexadienes. The singlet state reaction proceeds with just one terminal double bond rotation involving allylic methylene or cyclopropane methylene diradical with just one double bond rotation, whereas the triplet excited state reaction proceeds with "double" double bond rotation [11].

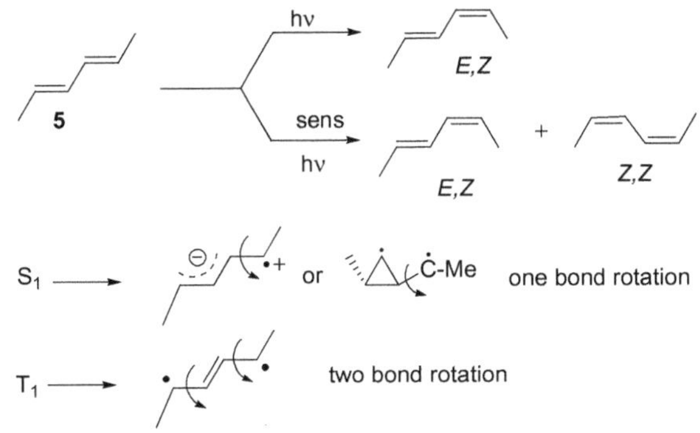

7.3 Photochemical Electrocyclic and Addition Reactions

Photochemical-concerted electrocyclic reactions of alkenes and dienes are symmetry allowed processes. For alkene system, the HOMO (SOMO) is the excited alkene π^* orbital and the LUMO is the π^* of ground-state alkene. The interactions of HOMO and LUMO produce two new σ bonds in the product, cyclobutane.

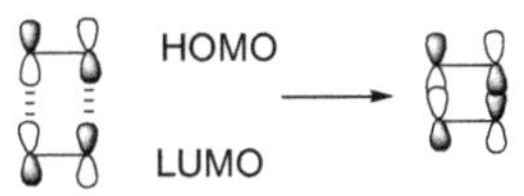

Similarly, photochemical reaction of 1,3-butadiene into cyclobutene and its reverse process are symmetry allowed processes in disrotatory mode of cyclization and opening of ring.

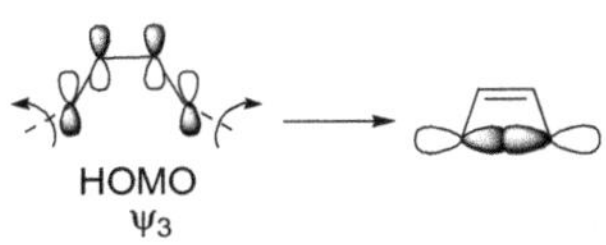

Experimental results on the photochemical reaction of cyclobutene to butadiene ring opening indicated that the reaction is not straightforward concerted. The ring opening is accompanied by fragmentation products, acetylene and ethylene and rearrangement to methylene cyclopropane [12]. Similarly, *trans*- and *cis*-3,4-dimethylcyclobutenes **6** on direct irradiation give mixtures of three isomeric 2,4-hexadienes in different ratios [13]. The photochemical cyclization of butadiene also gives a mixture of products [14].

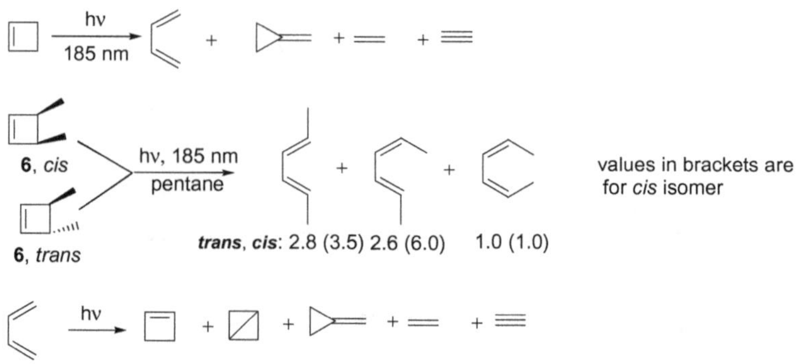

These fragmentation and rearrangement products from the photolysis of butadiene and cyclobutene can be interpreted by considering the different excited states and their return to the ground states through conical intersections (CIs). In addition to singlet and triplet excited states, a Rydberg excited state is also observed from irradiation of alkenes and dienes. A Rydberg state is an excited state in which one electron in a nonbonding orbital is far from immediate neighborhood bond being excited. The Rydberg states are designated by symbol "R" and occur at energies similar to and sometimes lower than those required for $\pi \rightarrow \pi^*$ transitions.

The Rydberg state of alkene is designated as π, R(3s), to specify that one electron from the π bond remains in the π orbital and the other has been shifted to the Rydberg orbital, which resembles the 3s orbital of a helium atom. The Rydberg state of an alkene is considered to have a partially ionic structure such that a net positive charge remains on a carbon atom. The Rydberg state of ethylene is shown below.

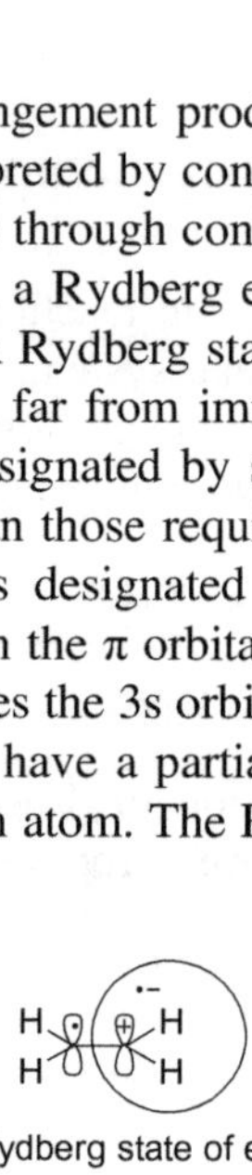

a Rydberg state of ethylene

The CI is analogous to transition state in thermal reactions. In the photolysis of butadiene, the CI is considered as a tetraradicaloid, which on different modes of electron repairing produces cyclobutene, bicyclo [1.1.0]-butane, and a carbene. The carbene intermediate on rearrangement and fragmentation gives cyclopropylmethylene, acetylene, and ethylene.

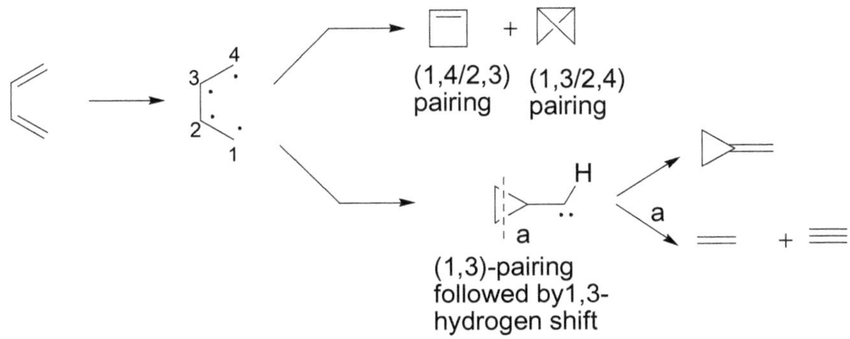

The transformations of 3,4-dimethylcyclobutenes **6** are believed to occur through a twisted CI in which all the π electrons are unpaired similar to that formed from excited singlet butadiene. Passage through this CI results in a mixture of stereoisomers [15].

Cyclic dienes such as cyclooctadiene **7**, where *s-cis*-conformation is dominant, undergo electrocyclization to give the dominant products. Similarly, acyclic conjugated diene such as 2,3-dimethyl-1,3-butadiene **8** undergoes electrocyclization on direct irradiation from the *s-cis*-diene conformation preferentially [16].

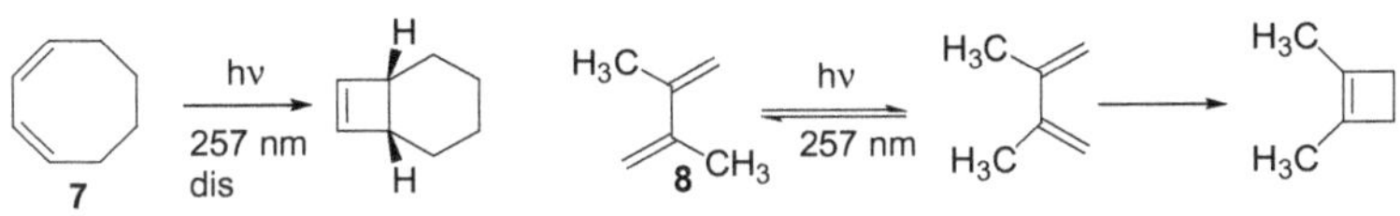

1,3-Pentadiene **4** on direct irradiation undergoes *cis–trans*-isomerization as well as cyclization to 1,3-dimethylcyclopropene and orbital symmetry allowed electrocyclization to 3-methylcyclobutene [17].

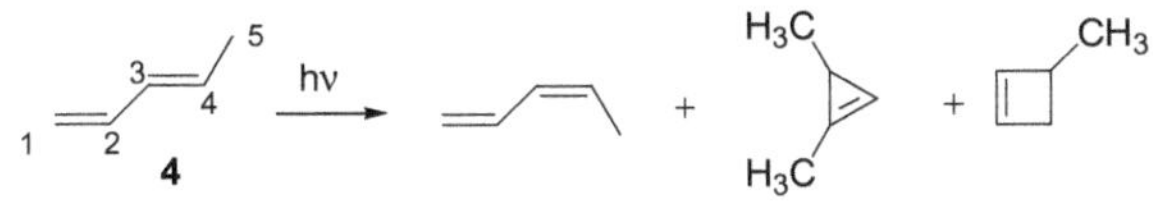

The cyclopropene arises from (2,4)-pairing followed by hydrogen migration.

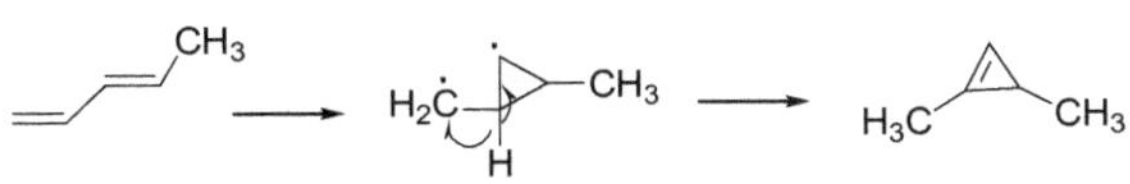

2,4-Hexadienes **9** on direct irradiation undergo both *cis–trans*-isomerization and stereospecific electrocyclic reaction [18].

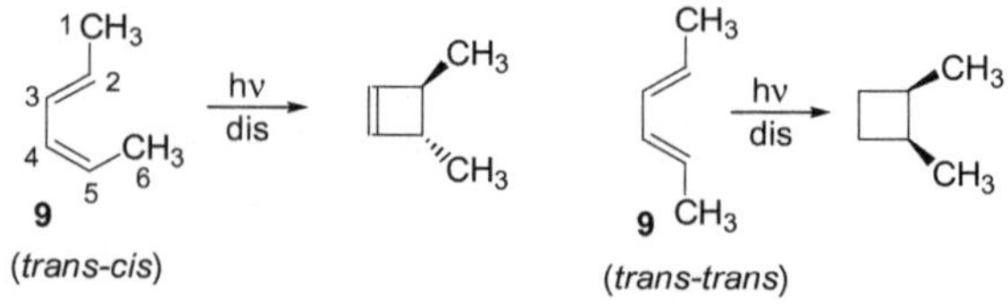

Cyclic heteroannular diene **10** on irradiation also produces cyclobutene by either electrocyclization or by isomerization in one ring to *E*-isomer, followed by [2+2]-cycloaddition reaction [19]. The isomerization of one ring into *E*-isomer was verified by trapping in MeOH.

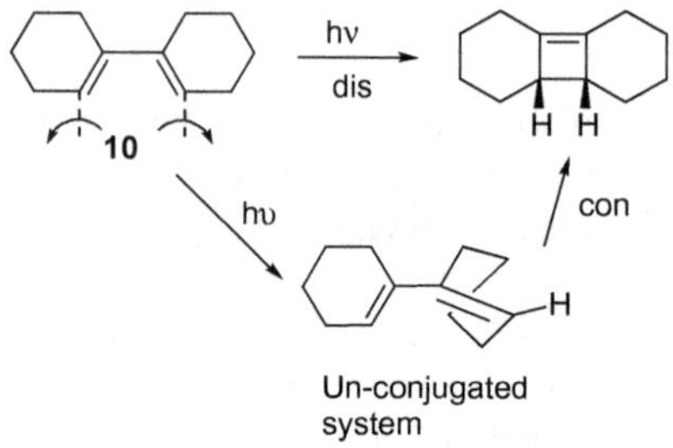

Electron-withdrawing or electron-donor substituent in 2 and/or 3 position of acyclic dienes directs the reactions to proceed through both concerted and non-concerted processes. For example, 2-cyano-1,3-butadiene **11** gives **12** and **13** [20].

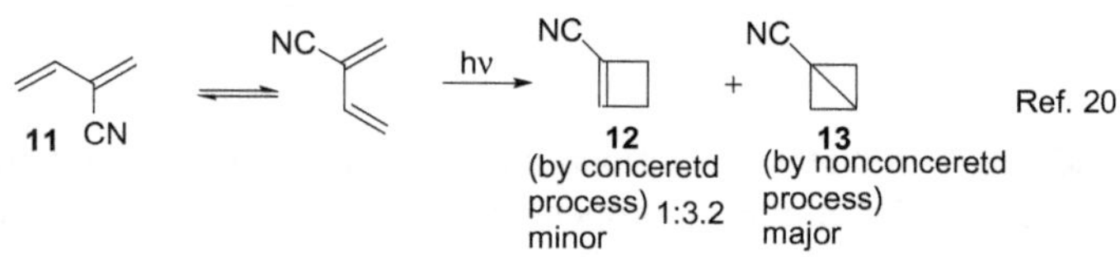

The major bicyclic compound, 1-cyano-bicyclo [1.1.0]-butane **13** is formed through a diradical intermediate.

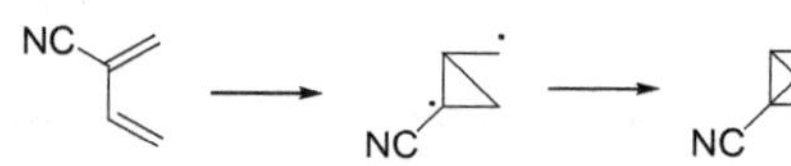

2,3-Di-*tert*-butyl-1,3-butadiene **14** gives only **15** in a non-concerted process due to steric factor [21].

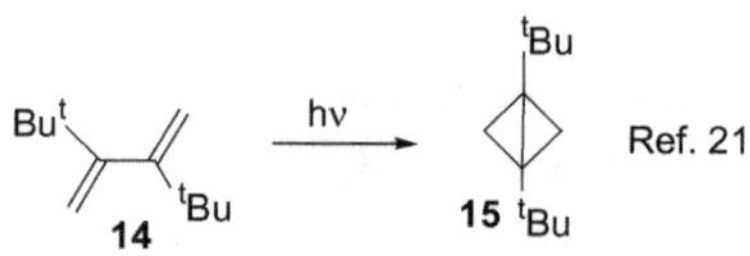

Photosensitized reaction of 1,3-butadiene gives a mixture of *cis*- and *trans*-1,2-divinyl cyclobutanes and 4-vinyl cyclohexene [22].

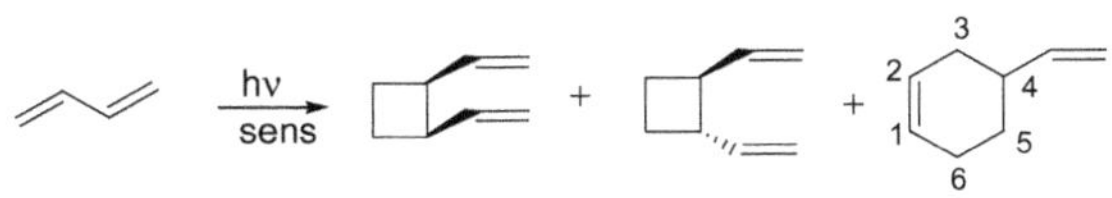

Ref. 22

The ratio of the products, divinyl cyclobutanes to cyclohexene depends on the triplet energy (E_t) of the sensitizer used. With sensitizer $E_t > 60$ kcal/mol, both *s-trans* and *s-cis* are excited having a preference to *s-trans*. With sensitizer E_t ~ 55 kcal/mol, the *s-cis* is preferentially excited. Excited *s-trans* gives only cyclobutanes, while the excited *s-cis* conformer gives both cyclobutane and cyclohexene. For instance, sensitizer benzophenone of E_t 68.5 kcal/mol gives divinylcyclobutanes and cyclohexene in a ratio of 92:8, whereas sensitizer benzil of E_t 53.7 kcal/mol, gives divinylcyclobutanes and cyclohexene in a ratio of 55:45 [22].

1-Aryl cyclobutene **16** on direct irradiation undergoes cycloreversion through zwitterionic S_1 excited state to give arylacetylene. When the reaction is carried out in MeOH (hydroxylic solvent) only addition product is found [23].

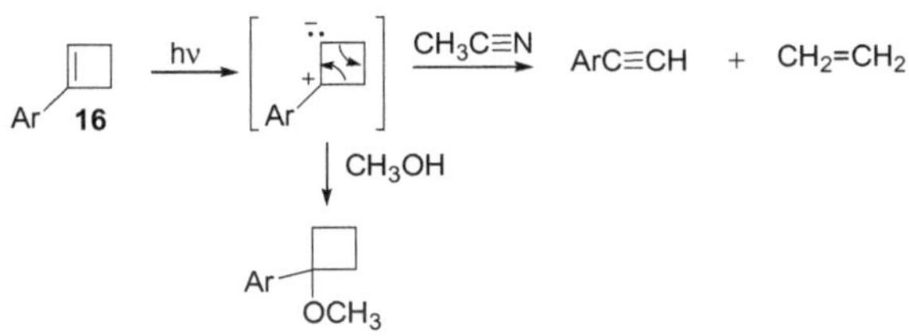

Cyclohexenes and cycloheptenes containing an acyloxy group at the allylic or homoallylic positions give regiospecific addition products in polar solvents due to formation of intermediate oxonium ions of different stabilities. For example, methyl-3-cyclohexene-1-carboxylate **17** in acetic acid on benzene-sensitized irradiation gives methyl *trans*-4-acetoxy cyclohexane carboxylate **18** as major product along with small amount of methyl *trans*-3-acetoxy cyclohexane carboxylate **19** [24].

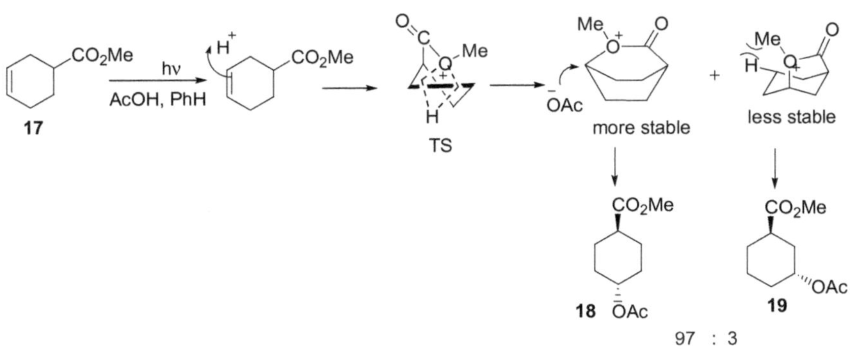

Such regioselectivity is not observed in *cis*-**20** and *trans*-4-cyclohexene-1,2-dicarboxylates because the participation of oxygen functions in the TS is not significant in the stability of the TS.

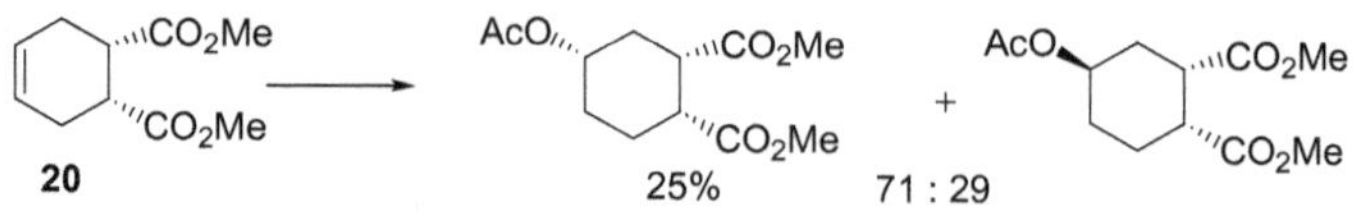

The photochemistry of hexatriene system is interesting. Unsubstituted 1,3,5-hexatriene **21** undergoes *cis–trans*-isomerization, whereas substituted hexatrienes undergo cyclization by both concerted and non-concerted processes [25]. A triene **22** having conformation that is *s-cis* at both C(2)–C(3) and C(4)–C(5) as well as *Z* at C(3) and C(4) is favorable to electrocyclization [26]. An extended conformation of 1,3,5-hexatriene is not favorable for electrocyclic process. For example, trienes **23–25** on irradiation give products in non-concerted processes [27].

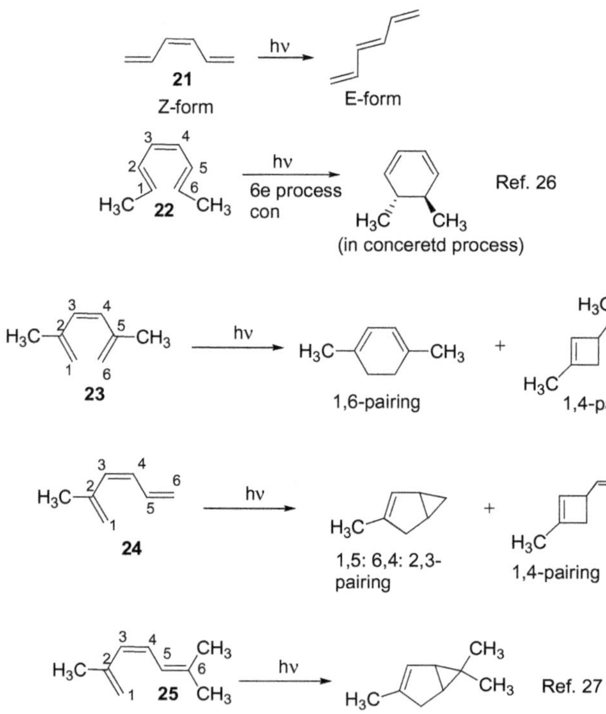

The products of the trienes (**23–25**) are formed in non-concerted processes from a tetraradicaloid CI via a diradical intermediate [28]. The formation of products from a triene depends on its ground-state conformation.

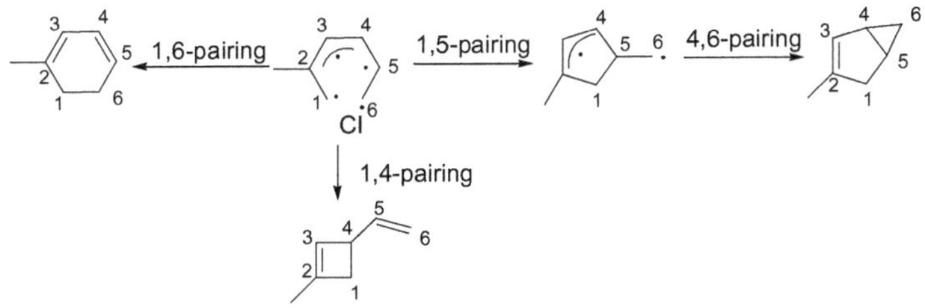

7.4 Photochemical [2+2]-Cycloaddition and Dimerization Reactions

Open-chain alkenes undergo dimerizations in stereospecific $[\pi_s^2 + \pi_s^2]$ paths in low yields on irradiation. For example, photodimerization of *Z*-2-butene **26** gives two products in which *cis*-geometry of methyl groups is retained. Similarly, the cycloadducts from irradiation of *E*-2-butene **27** retains the *trans*-geometry of methyl groups [29]. Irradiation of *Z*-2-butene also gives cycloadduct with *E*-stilbene in a similar manner [29].

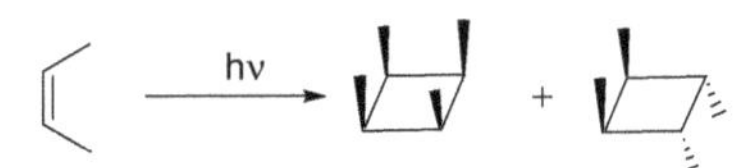

26

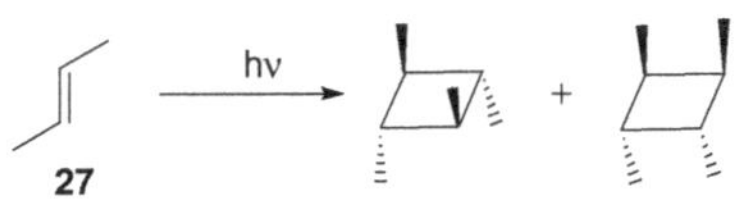

27

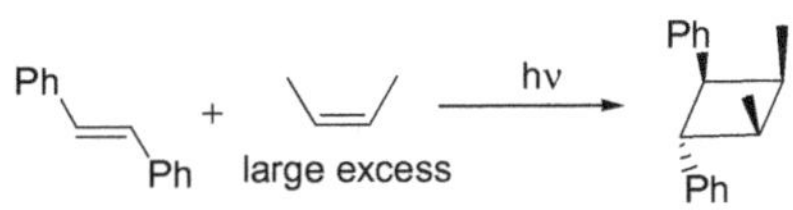

Cyclohexene **27a** undergoes photo-induced dimerization in the presence of xylene sensitizer to give a mixture of stereoisomers in non-stereospecific manner [30].

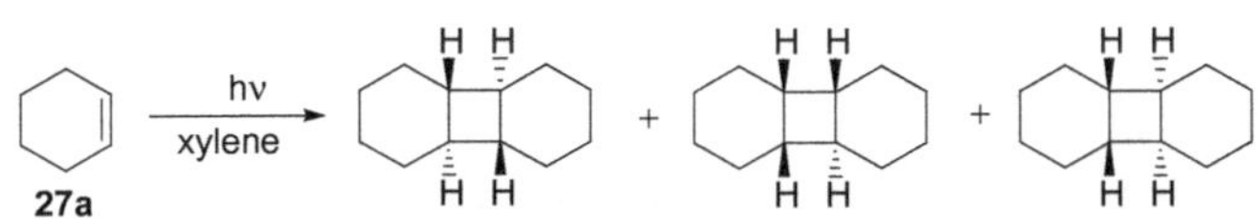

27a 2% 2.6% 1%

1-Phenylcyclohexene **28** also undergoes dimerization on irradiation directly or in the presence of sensitizer to give tail-to-tail products. This indicates that both singlet and triplet excited states diradical intermediates are involved in the formation of the products [31].

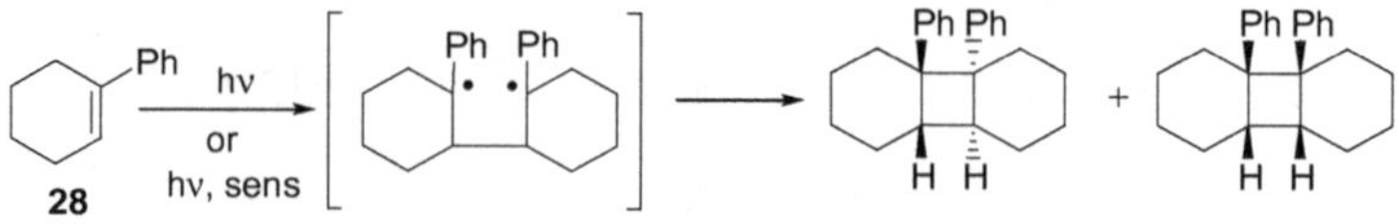

Stereocontrolled photoaddition of allene to cyclopent-1-ene-1-carboxaldehyde **29** has been utilized in the introduction of exocyclic methylene group in steviol **30** [32].

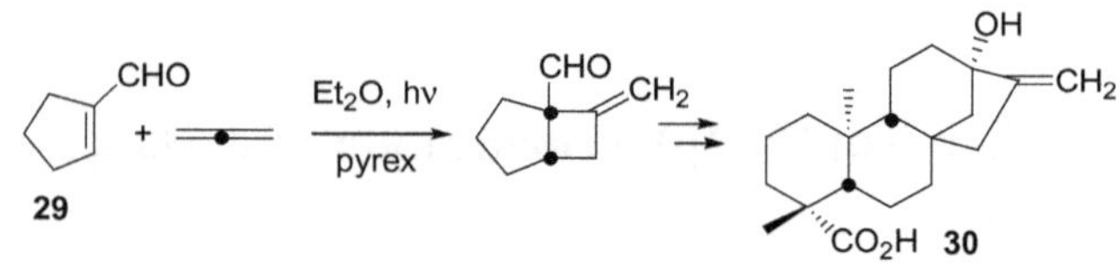

Intramolecular [π^2+π^2]-cycloadditions are also observed in cyclic 1,4-dienes **31** [33] and **32** [34] and indene **33** [35].

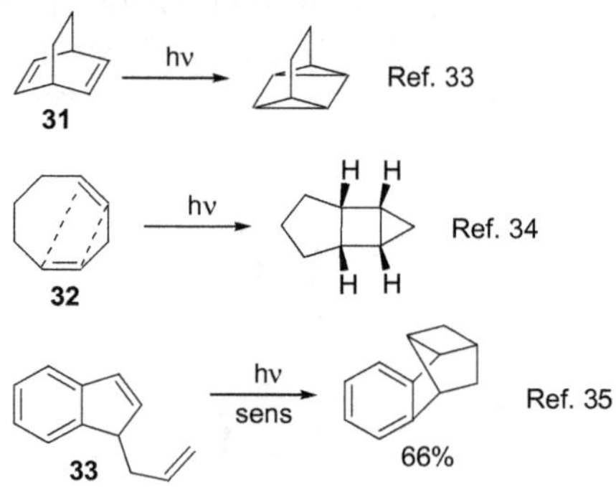

Conjugated dienes such as cyclopentadiene undergoes photodimerization through triplet sensitization in the presence of benzophenone to give both [4+2] and [2+2]-cycloaddition products in equal proportion [36].

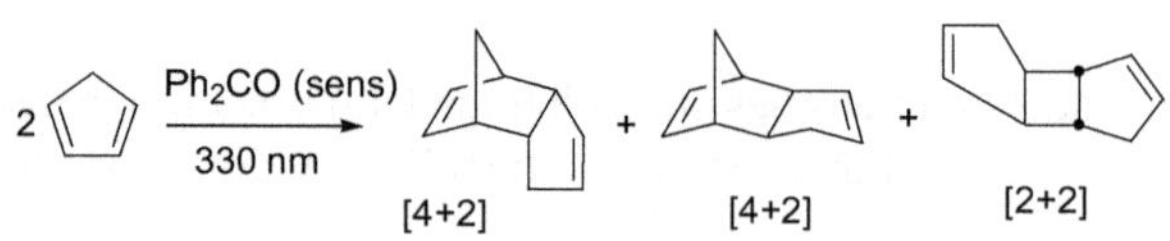

7.5 Photochemical Rearrangements

Several alkenes and dienes undergo rearrangement reactions on irradiation. For example, methylene cyclopropane **34** in acetonitrile on irradiation gives **35** as major product along with minor products **36** and **37** (about 3 %) via the formation of 2,2-diphenylcyclobutylidene intermediate [37].

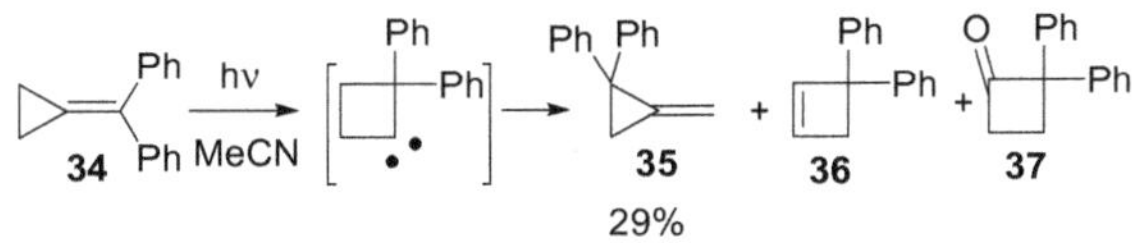

5,5-Diphenyl-cyclohexa-1,3-diene **38** on direct irradiation undergoes electrocyclic ring opening to give **39** as major product, and on photosensitization gives **40** as major rearrangement product [38]. The rearrangement proceeds by 1,2-sigmatropic shift of one of the phenyl rings in the triplet excited state followed by the formation of a diradical intermediate and consequent formation of a cyclopropane ring. This rearrangement is known as the *di-π-methane rearrangement.*

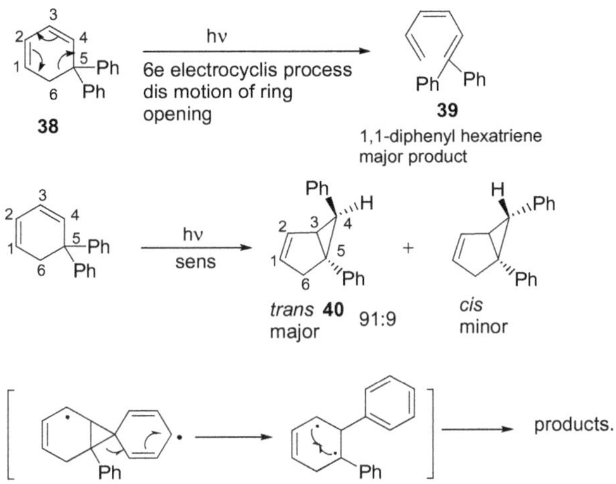

7.5.1 The di-π-Methane Rearrangements

Direct or sensitized irradiation of 1,4-dienes and other related molecules in which two π systems are separated by an sp^3 carbon atom gives vinyl or π-substituted cyclopropanes as major products. These reactions are known as the *di-π-methane rearrangements or Zimmerman rearrangements* because Howard Zimmerman group of the University of Wisconsin, USA had made major works on these reactions [39]. The rearrangements are believed to undergo apparent 1,2-vinyl migrations, followed by formation of a new σ bond between the lateral carbon atoms, giving rise to vinyl cyclopropanes. The π system may be a vinyl, aromatic ring, acetylenic, or allenyl moiety. For examples, 1,4-dienes **41** and **43** give **42** and **44**, respectively [39, 40].

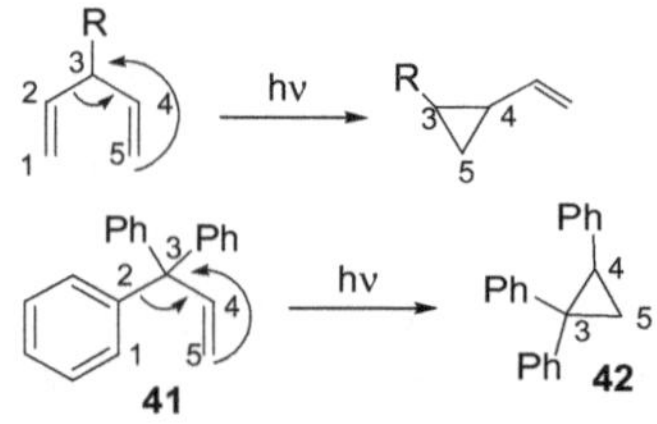

Ref. 39

Phenyl substituted vinyl group has higher preference of cyclization than methyl substituted vinyl group because of higher stabilization of radical character.

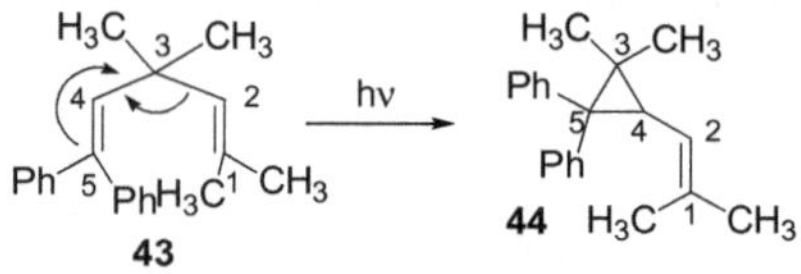

Ref. 40

7.5.1.1 Mechanism and Stereochemistry

Di-π-methane rearrangement may occur from either a singlet or a triplet excited state [41]. The reaction involving singlet excited state occurs in a concerted process and this mechanism is followed in acyclic dienes and cyclic systems in which concerted process is sterically feasible by free rotation of the π systems, whereas the reaction from the triplet excited state occurs in various bicyclic systems where both the π systems are in rigid structural environments and prohibited for free rotation. The singlet reaction is analogous to sigmatropic shifts of π system. The reaction proceeds by a 1,2-sigmatropic shift of one π system followed by reorientation of π electrons to form a new σ bond between the lateral carbon atoms in a concerted process. As the new σ bond is formed between C(3) and C(5) using back lobe of the C(2)–C(3) σ bond by disrotatory motion, *anti* to the migrating π system results in the inversion of configuration at C(3). The process is photochemically allowed because the orbital array corresponds to Mobius topology of 6e process (two σ bonds and one π bond) with one phase change between the C(1) and C(2) positions as depicted in Fig. 7.2. The uncyclized π system retains its *E*- or *Z*-configuration present in the starting material. In Fig. 7.2, the double bond between C(1) and C(2) retains its *E*- or *Z*-geometry in the product.

Fig. 7.2 The orbital array of di-π-methane rearrangement through singlet excited state

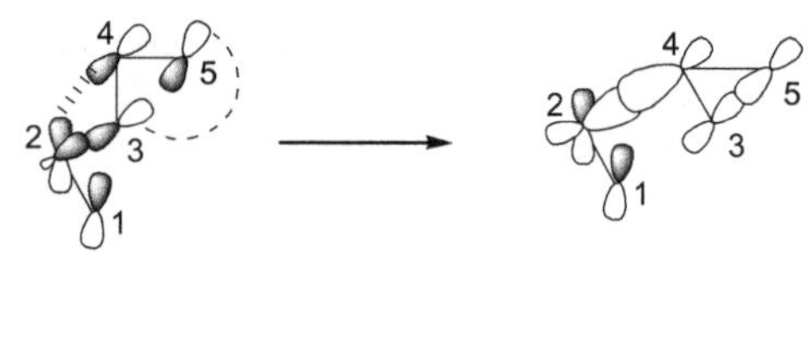

The following examples confirmed the predicted stereochemical course of the rearrangement [42, 43].

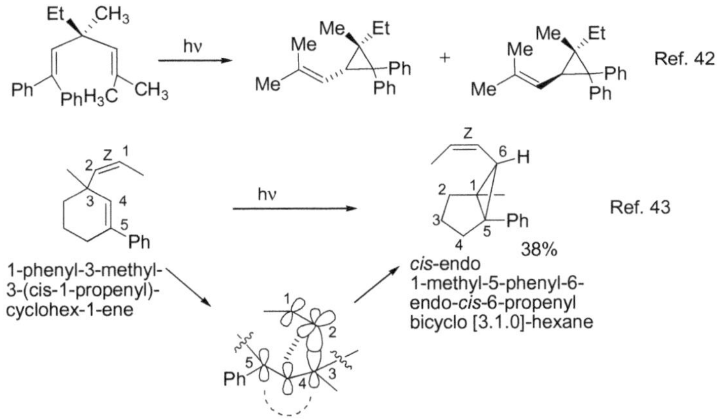

The triplet reaction proceeds through the formation of a cyclopropyl diradical intermediate. The less stabilized diradical center utilizes its odd electron density to open the cyclopropane ring and forms a more stabilized 1,3-diradical intermediate. The new 1,3-diradical intermediate gives the cyclopropane derivative as major product of the reaction. Thus, in a 1,4-diene system, a terminus substituted with aryl groups will cyclize in preference to an unsubstituted or alkyl-substituted terminus because of the greater stability of the diradical intermediate by delocalization with aryl groups. The following example of **43** demonstrated the mechanism of the triplet reaction [40].

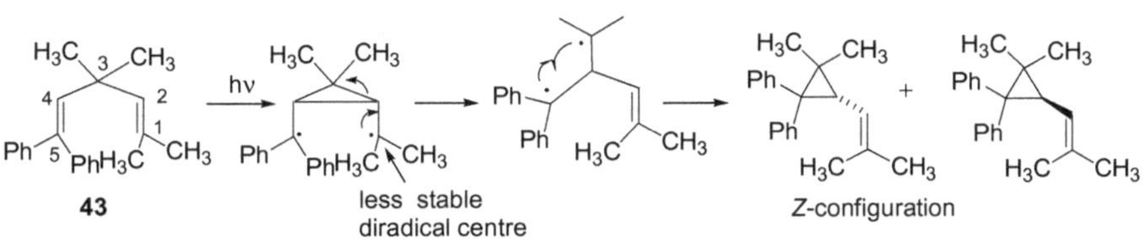

7.5.1.2 Regioselectivity

The rearrangement proceeds rapidly in 1,4-dienes when phenyl or other groups are present as substituents in the vinylic parts. In allylic aromatic analogs, aromatic ring migrates in preference to a vinyl group. When two different substituents are present in two positions of the vinylic systems, the regioselectivity of the reaction depends on the relative stability of the diradical species. For example, compound **44** gives compound **45** through a diradical intermediate, where phenyl group stabilizes the diradical much more compared to carbomethoxy group [44].

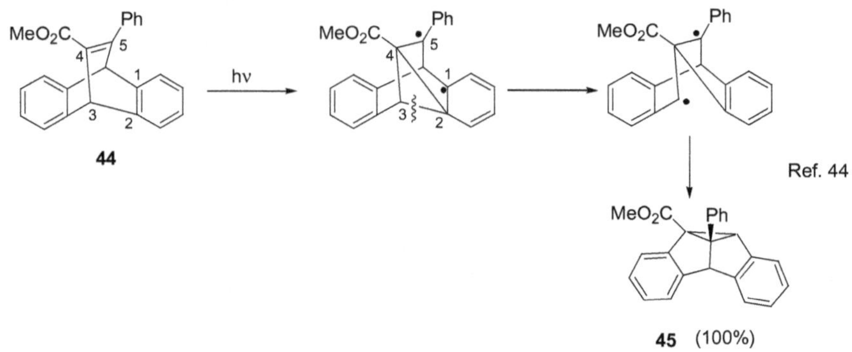

In dihydrobenzobarrelene **46**, field effect of the substituent controls regioselectivity of the rearrangement. The exclusive formation of product **47** from *syn*-7-hydroxy-7,8-dihydrobenzo barrelene **46** can be interpreted by greater stabilization of the diradical **48** due to field effect of hydroxyl group over the competitive diradical **49** [45].

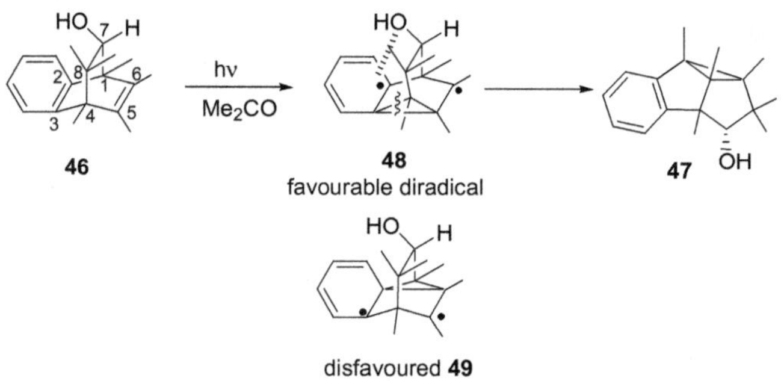

7.5.1.3 Substituent Effect on Central Sp^3-Carbon

When the central sp^3-carbon of 1,4-diene is unsubstituted, the di-π-methane rearrangement is less favorable. The reaction of 1,1,5,5-tetraphenyl-2,4-dideuterio-1,4-pentadiene **50** illustrates the fact [46].

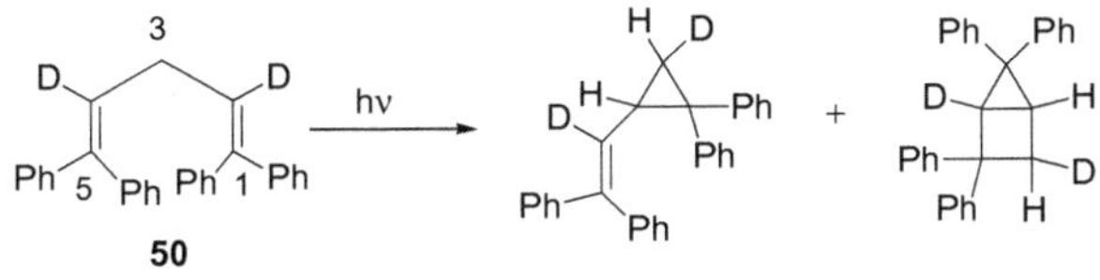

In this case, the stable diradical is formed by hydrogen atom migration from C(3) because it produces a more stable allylic radical. The formation of products can be explained by the following mechanistic path.

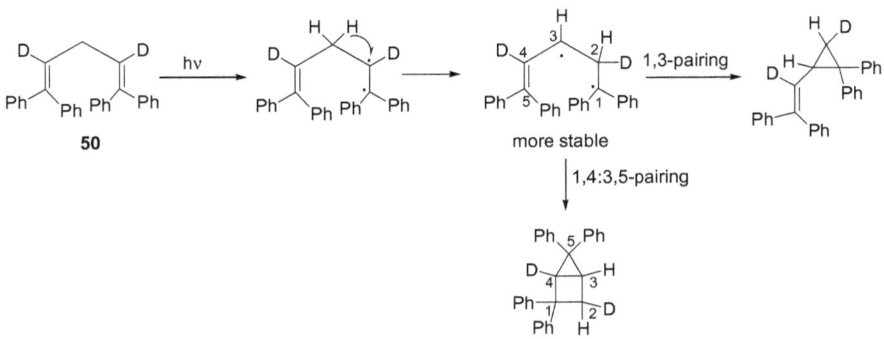

Similarly, 1,5-diphenyl-1,4-pentadiene **51** on irradiation undergoes intramolecular [2+2]-cycloaddition in preference to di-π-methane rearrangement [47].

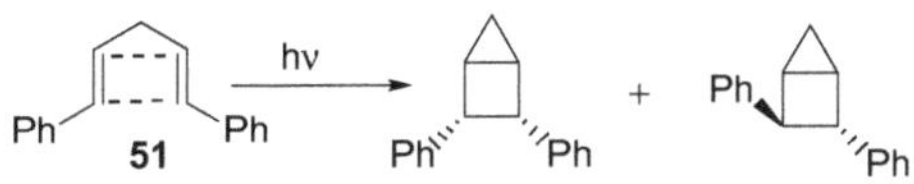

Delocalized aryl substituents on the central sp^3-carbon accelerates the reaction rate by stabilization of 1,3-diradical. For example, the diene **52** gives **53** in high yield [48].

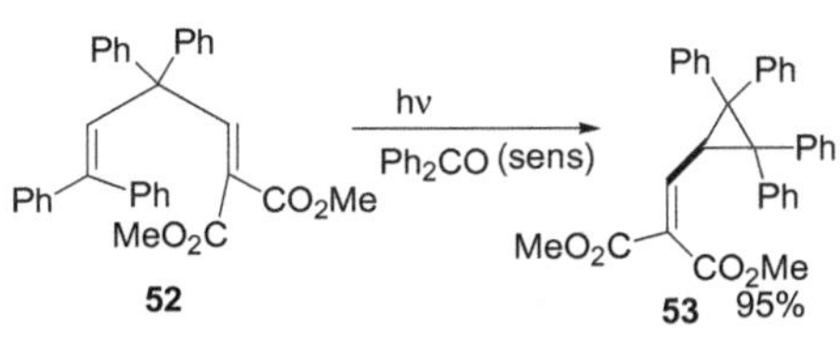

The reaction of **52** in direct irradiation gives a different product **54** [48]. Possibly in the singlet state, better zwitterionic nature of carbomethoxy-substituted vinyl group makes it to take part in cyclization in preference to diphenyl-substituted vinyl group. In this case, the formation of cyclopropane ring is controlled by the dipolar nature of the diradical.

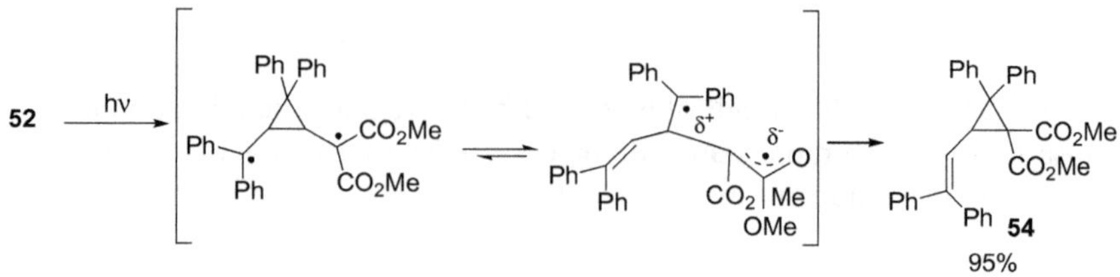

Some other typical examples of the di-π-methane rearrangements [49–61] are:

1. 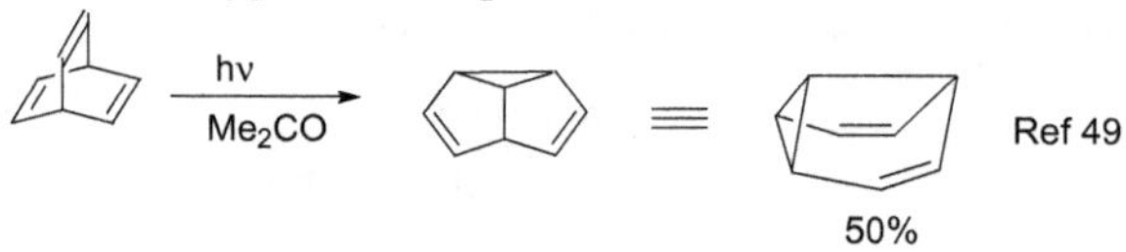

2. 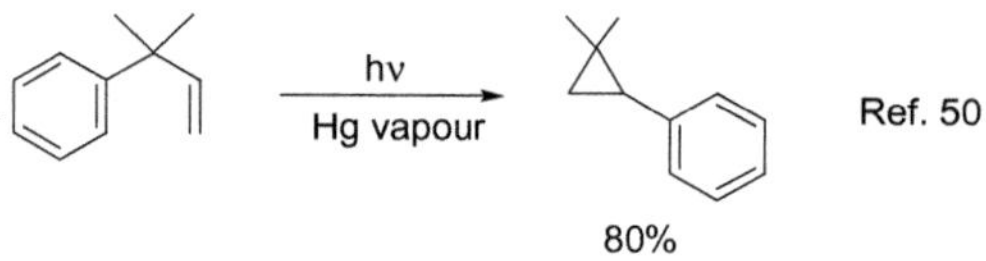

3. 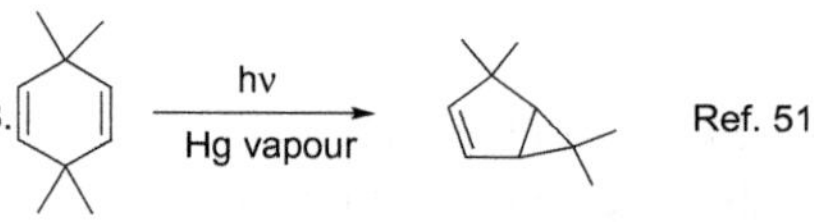

4. 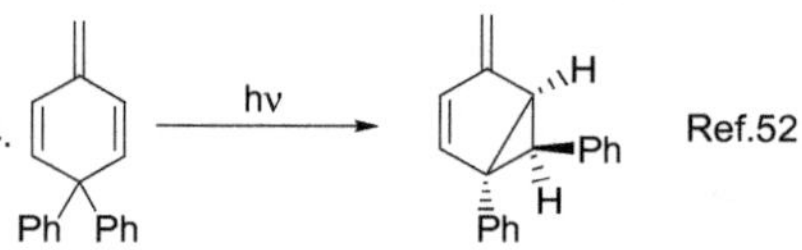

5. 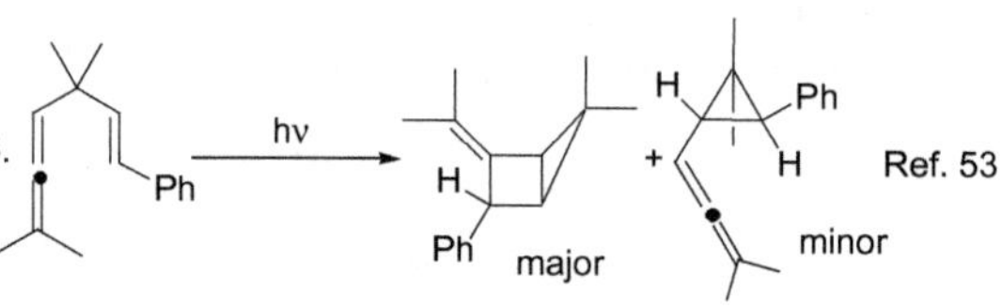

6. 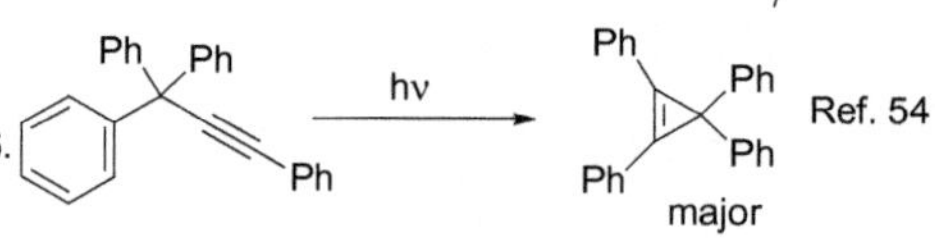

7. 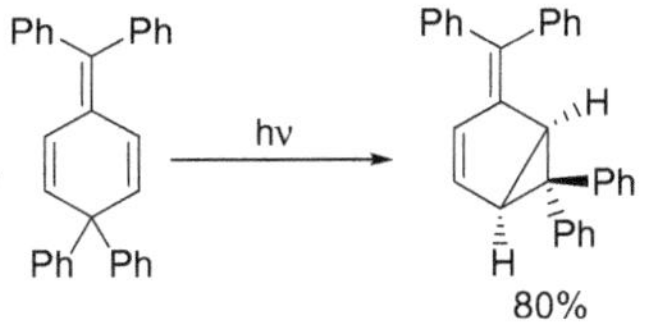 Ref. 55

8. 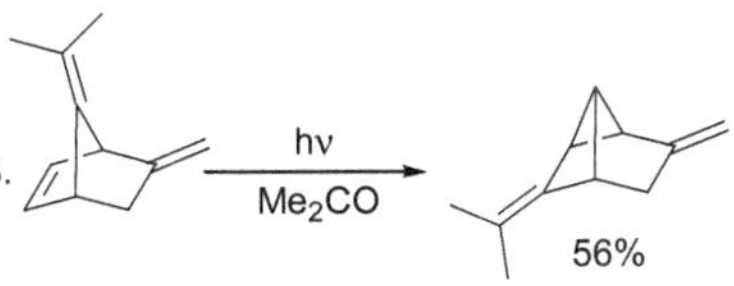 Ref. 56

9. 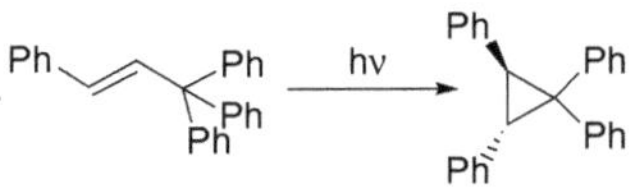 Ref. 53

10. 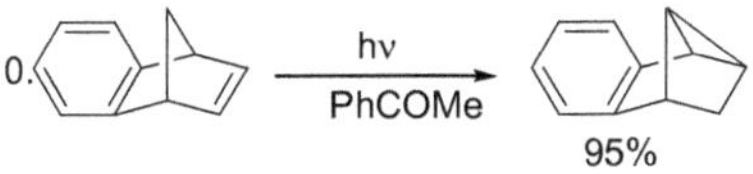 Ref. 57

11. 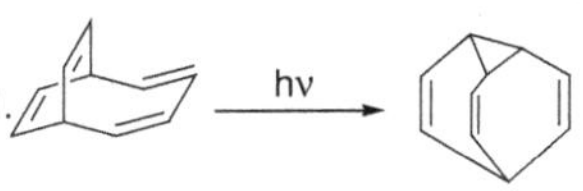 Ref. 58

12. 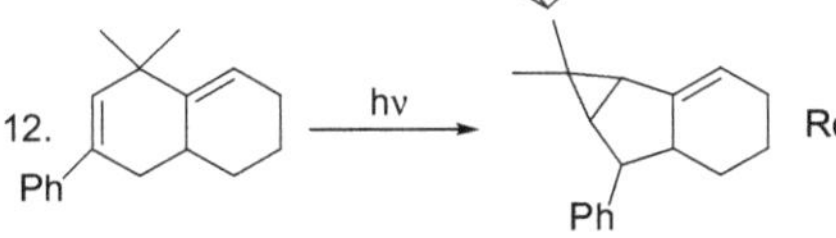 Ref. 59

13. 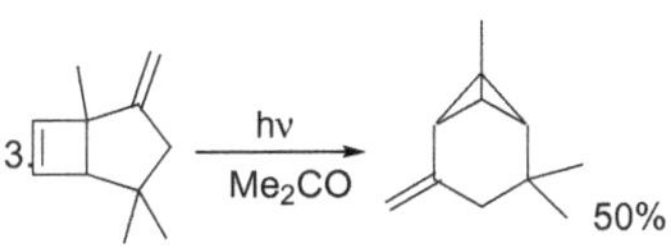 Ref. 60

14. 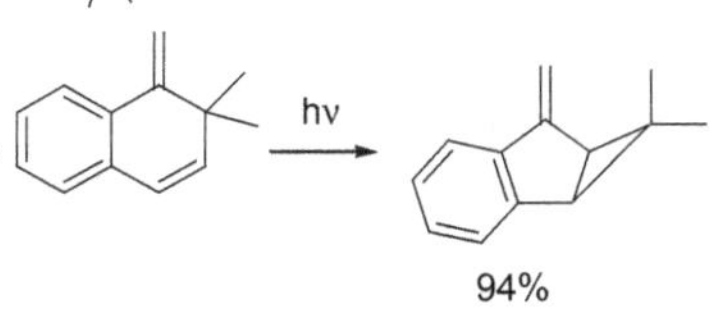 Ref. 61

7.5.2 The *aza-di-π-Methane* Rearrangements

Photochemical rearrangement reactions of 1-aza-1,4-dienes and 2-aza-1,4-dienes from their triplet states to form corresponding cyclopropylimines are known as the *aza-di-π-methane rearrangements* [62].

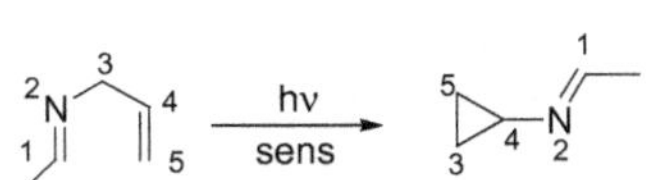

cyclopropylimine

The rearrangement is analogous to the di-π-methane rearrangement and is considered as 1,2-shift of imino group from C(3) to C(4), followed by a σ bond formation between C(3) and C(5). For example, the photorearrangement of β,γ-unsaturated imine **55** gives cyclopropyl imine **56**, which on hydrolysis gives cyclopropane aldehyde **57** [62].

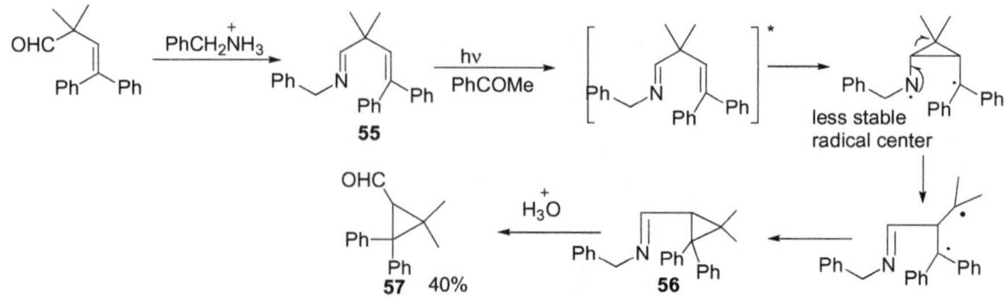

This reaction is useful for synthesis of cyclopropane carboxylic acids. Similarly, the β,γ-unsaturated oxime acetate **58** gives **59** in high yield [63].

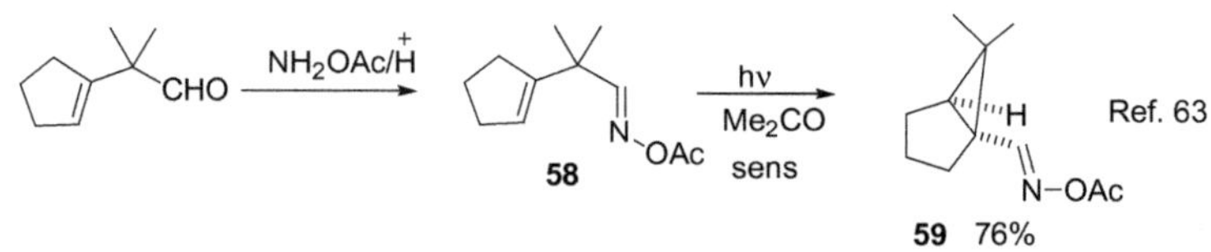

The presence of an electron-withdrawing group in the oxime ester improves the yield of the reaction. For example, oxime **60** gives **61** in 90 % yield [64].

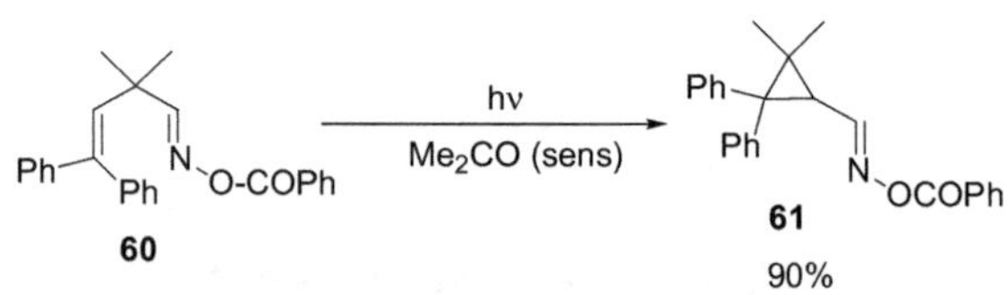

7.5.3 The tri-π-Methane Rearrangements

The light-induced rearrangement reactions of trivinyl methanes to give divinyl cyclopropane and vinyl cyclopentene derivatives are known as the *tri-π-methane rearrangements*. The reaction involves both singlet and triplet excited states. For sterically congested molecular systems, the triplet excited states are involved and proceed through the formation of cyclopropyl dicarbinyl diradical intermediates, which rearrange to more stable vinyl-allyl-carbinyl diradicals by cyclopropane ring opening and give the products. The ratio of cyclopentene to cyclopropane derivatives in a reaction depends on the involvement of the excited state. The following examples of *tris*-diphenylvinyl methanes (**62**, **65** and **68**) are illustrative [65–67]:

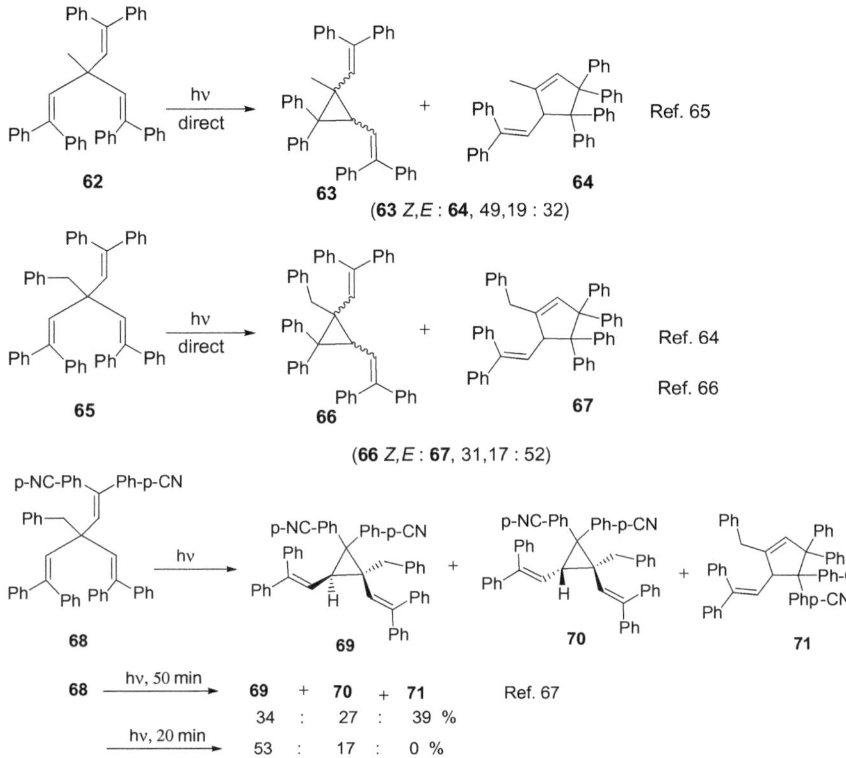

In direct irradiation, the products are formed through a single intermediate, whereas for sensitized reactions, the products are formed through two-stage intermediates following the di-π-methane path. In two-stage process, in the second stage, the *transoid* intermediate is dominant due to its greater stability and provides the major product of the reaction. The formation of the products can be rationalized by the following mechanism:

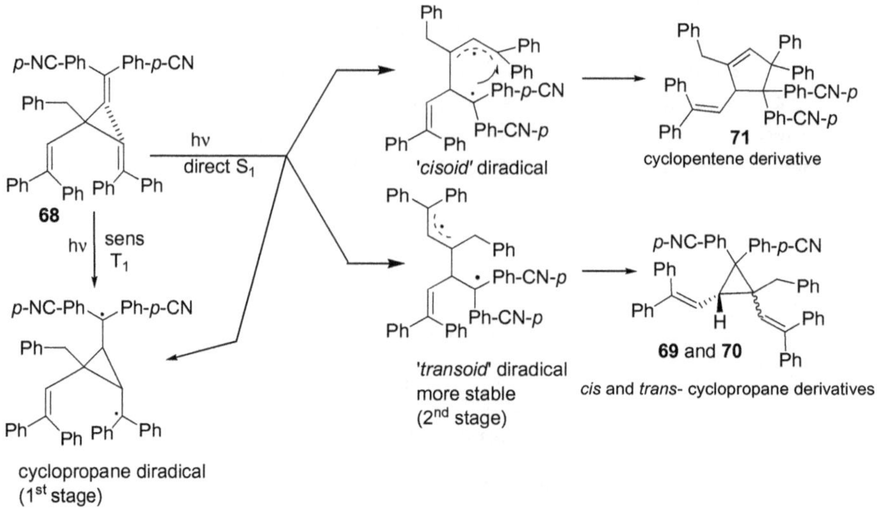

7.6 Problems

7.6.1. Suggest a mechanistic rationalization for each of the following reactions and mention the major product when more than one product is formed

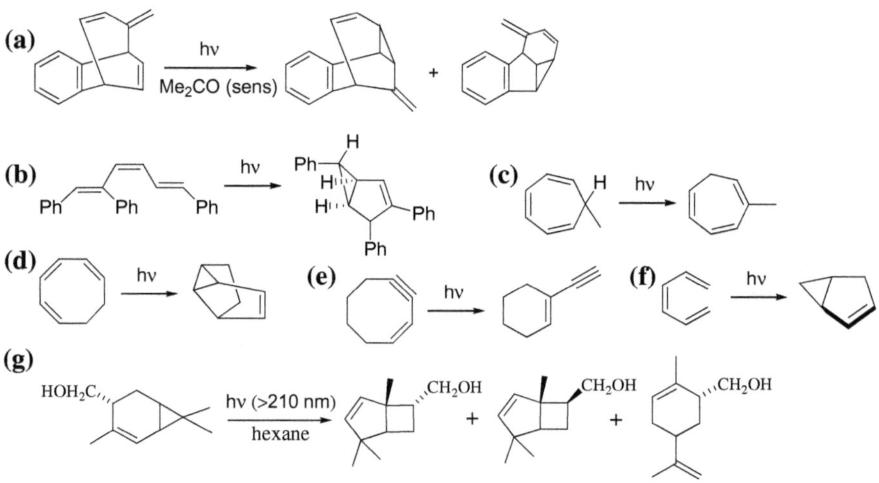

7.6.2. Predict the structure(s) of the principal product(s) formed in the following reactions:

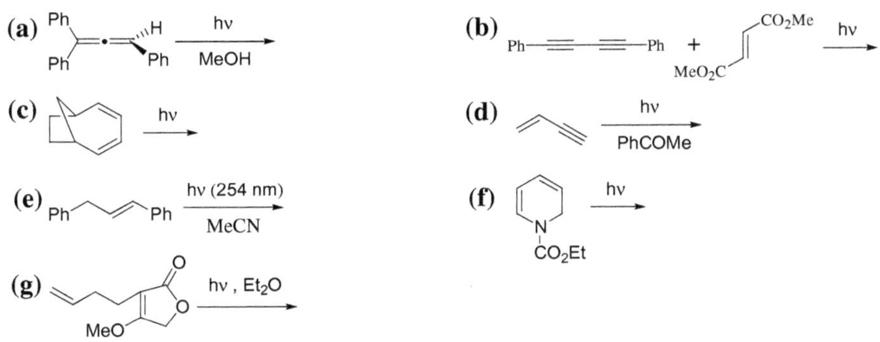

7.7 Further Reading

1. Coyle JD (1986) Introduction to organic photochemistry. Wiley, New York
2. Kagan J (1993) Organic photochemistry, principles and applications. Academic Press, New York
3. Coxon JM, Halton B (2011) Organic photochemistry, 2nd edn. Cambridge University, London
4. Gilbert A, Baggot J (1991) Essentials of molecular photochemistry. Blackwell Scientific Publications, Oxford
5. Mayo PD (ed) (1980) Rearrangements in ground and excited states, vol 3. Academic, New York
6. Mattay J, Griesbeck AG (1994) Photochemical key steps in organic synthesis. Wiley- VCH

References

1. Saltiel J, D'Agostino JT, Megarity ED, Metts L, Neuberger KR, Wrighton M, Zafiriow OC (1973) Org Photochem 3:1; Mazzucato U, Aloisi GA, Bartocci G, Elisei F, Galiazzo G, Spalletti A (1995) Med Biol Environ 23: 69
2. Hammond GS, Saltiel J, Lamola AA, Turro NJ, Bradshaw JS, Cowan DO, Counsell RC, Vogt V, Dalton C (1964) J Am Chem Soc 86:3197; Yamauchi S, Azumi T (1973) J Am Chem Soc 95:2709
3. Yamazaki H, Cventanovic RJ (1969) J Am Chem Soc 91:520; Yamazaki H, Cventanovic RJ, Irwin RS (1976) J Am Chem Soc 98:2198

4. Kropp PJ, Reardon EJ Jr, Gaibel ZLF, Williard KF, Hattaway JH Jr (1973) J Am Chem Soc 95:7058; Marshall JA (1969) Acc Chem Res 2:33
5. Srinivasan R, Brown KH (1978) J Am Chem Soc 100:2589; Squillacote A, Bergman A , DeFelippis J (1989) Tetrahedron Lett 30: 6805
6. Inoue Y, Takamura S, Sakurai H (1975) J Chem Soc Chem Commun 577
7. Kropp PJ, Mason JD, Smith GFH (1985) Can J Chem 63:1845
8. Kropp PJ (1969) J Am Chem Soc 91:5783
9. Srinivasan R, Brown KH (1978) J Am Chem Soc 100:4602; Inoue Y, Mukai T, Hakushi T (1982) Chem Lett 1045
10. Saltiel J, Townsend DE, Sykes A (1973) J Am Chem Soc 95:5968
11. Synder JJ, Tise FP, Davis RD, Kropp PJ (1981) J Org Chem 46:3609; Saltiel J, Rousseau AD, Sykes A (1972) J Am Chem Soc 94:5903; Saltiel J, Metts L, Wrighton M J Am Chem Soc 92:3227
12. Leigh WJ, Zheng K, Nguyen N, Werstiuk NH, Ma J (1993) J Am Chem Soc 113:4993
13. Clark KB, Leigh WJ (1987) J Am Chem Soc 109:6086
14. Fuss W, Schmid WE, Trushin SA (2001) Chem Phys Lett 342:91
15. Bernardi F, Olivucci M, Robb MA (1990) Acc Chem Res 23:405; Bernardi F, Olivucci M, Ragazos IN, Robb MA (1992) J Am Chem Soc 114: 2752
16. Leigh WJ, Zheng K, Clark KB (1991) J Org Chem 56:1574; Laarhoven WH (1987) Org Photochem 9:129; Squillacote ME, Semple TC (1990) J Am Chem Soc 112:5546
17. Boue S, Srinivasan R (1970) J Am Chem Soc 92:3226
18. Srinivasan R (1968) J Am Chem Soc 90:4498
19. Saltiel J, Marchand GR, Bonneau R (1987) J Photochem 28:367; Daino Y, Hagiwara S, Hakushi Y (1989) J Chem Soc Perkin Trans 2:275
20. Gale DM (1970) J Org Chem 35:970; Olivucci M, Bernardi F, Ottani S, Robb MA (1994) J Am Chem Soc 116:2034
21. Garavelli M, Frabboni B, Fato M, Celani P, Bernardi F, Robb MA, Olivucci M (1999) J Am Chem Soc 121:1537
22. Liu RSH, Turro NJ Jr, Hammond GS (1965) J Am Chem Soc 87:3406; Dilling WL, Kroening RD, Little JC (1970) J Am Chem Soc 92:928
23. Leigh WJ, Postigo JA (1995) Can J Chem 73:191
24. Leong TY, Imagawa T, Kimoto K, Kawanishi M (1973) Bull Chem Soc Jpn 46:596
25. Vroegop PJ, Lugtenburg J, Havinga E (1973) Tetrahedron 29:1393; Havinga E (1973) Experientia 29:1181
26. Fonken GJ (1962) Tetrahedron Lett 549
27. Courtot P, Rumin R, Salaun JY (1977) Pure Appl Chem 49:317; Courtot P, Rumin R (1976) Tetrahedron 32:441
28. Garavelli M, Celani R, Fato M, Bearpark MJ, Smith BR, Olivucci M, Robb MA (1997) J Phys Chem A 101:2023
29. Yamazaki H, Cvetanovic RJ (1969) J Am Chem Soc 91:520
30. Kropp PJ, Snyder JJ, Rawlings PC, Fravel HG Jr (1980) J Org Chem 45:4471
31. Unett DJ, Caldwell RA, Hrncir DC (1996) J Am Chem Soc 118:1682
32. Ziegler FE, Kloek JA (1977) Tetrahedron 33:373
33. Dauben WG, Cargill RL (1961) Tetrahedron 15:197
34. Moon S, Ganz CR (1968) Tetrahedron Lett 6275
35. Padwa A, Goldstein S, Pulwer M (1982) J Org Chem 47:3893
36. Takahashi Y, Mori Y, Nakamura A, Tomioka H (2005) Tetrahedron Lett 46:8415; Norman ROC, Coxon JM (1993) Principles of organic synthesis, ELBS with Chapman & Hall, Oxford, p 504
37. Zimmerman HE, Epling GA (1972) J Am Chem Soc 94: 8749; Swenton JS, Hyatt JA, Walker TJ, Crumrine AL (1971) J Am Chem Soc 93:4808; Su MD (1995) J Org Chem 60:6621
38. Zimmerman HE, Grunewald GL (1966) J Am Chem Soc 88:183; Zimmerman HE, Binkley RW, Givens RS, Grunewald GL, Sherwin MA (1969) J Am Chem Soc 91:3316

39. Griffin GW, Marcantonio AF, Kristinsson H (1965) Tetrahedron Lett 2951; Hixon S (1972) Tetrahedron Lett 1155
40. Zimmerman HE, Pratt AC (1970) J Am Chem Soc 92:1409
41. Zimmerman HE, Mariano PS (1969) J Am Chem Soc 91:1718; Mariano PS, Steitle RB, Watson DG, Peters MJ, Bay E (1976) J Am Chem Soc 98:5897
42. Zimmerman HE, Robbins JD, McKelvey RD, Samuel CJ, Sousa LR (1974) J Am Chem Soc 96:4630
43. Mariano PS, Ko JK (1972) J Am Chem Soc 94:1766
44. Rattray G, Yang J, Gudmundsdottier AD, Scheffer JR (1993) Tetrahedron Lett 34:35
45. Hart H, Murray RK (1969) J Am Chem Soc 91:2183
46. Zimmerman HE, Pincock JO (1973) J Am Chem Soc 95:2957
47. Block E, Orf HW (1972) J Am Chem Soc 94:8438
48. Zimmerman HE, Factor RE (1981) Tetrahedron 37 (Suppl-1):125.
49. Zimmerman HE, Grunewald GL (1966) J Am Chem Soc 88:183; Demuth M, Raghavan PR, Carter C, Nakano K, Schafner K (1980) 63:2434
50. Zimmerman HE, Swafford RL (1984) J Org Chem 48:3069
51. Reusch W, Frey DW (1967) Tetrahedron Lett 5193
52. Zimmerman HE, Samuelson GE (1969) J Am Chem Soc 91:5307; (1967) J Am Chem Soc 89:5971
53. Griffin GW, Coveli J, Petterson RC, Dodson RM, Close C (1965) J Am Chem Soc 87:1410; Griffin GW, Marcantonio AF, Kristinsson H (1965) Tetrahedron Lett 2951
54. Halton B, Kulig M, Perreten J, Gibson DM, Griffin GW (1971) J Am Chem Soc 93:2327
55. Zimmerman HE, Diehl DR (1979) J Am Chem Soc 101:1841
56. Ipaktschi J (1972) Chem Ber 105:1840
57. Edman JR (1969) J Am Chem Soc 91:7103
58. Jones MJ Jr, Reich SD, Scott LT (1970) J Am Chem Soc 92 :3118
59. Zimmermann HE, Tolbert LM (1975) J Am Chem Soc 97:5497
60. Ipaktschi J (1972) Chem Ber 105:1996
61. Zimmermann HE, Factor RE (1980) J Am Chem Soc 102:3538
62. Armesto D, Martin JF, Preg- Ossorio R, Horspool WM (1982) Tetrahedron Lett 23:2149; Armesto D, Horspool WM, Martin JF, Prez- Ossorio R (1986) J Chem Res S:46; M:0631
63. Armesto D, Ramos A (1993) Tetrahedron 49:7159
64. Armesto D, Horspool WM, Manchano MJ, Ortiz MJ (1990) J Chem Soc Perkin Trans 1 2348; (1992) idem 2325
65. Zimmerman HE, Cirkva V (2001) J Org Chem 66:1839
66. Zimmerman HE, Cirkva V (2000) Org Lett 2:2365
67. Zimmerman HE, Novork T (2003) J Org Chem 68:5056

Chapter 8
Photochemistry of Carbonyl Compounds

8.1 Introduction

Carbonyl compounds undergo various photochemical reactions in both gas and liquid phases. The reactive excited states of saturated ketones are the n → π* states, whereas that of conjugated ketones are π → π* states. Both these n → π* and π → π* transitions of carbonyl compounds may occur by singlet or triplet excited states. Both singlet and triplet excited states of a carbonyl compound react in different rates to give same type of products in different ratios. For saturated ketones, the activation energies for singlet and triplet excited states are about 80–85 and 75–80 kcal/mol, respectively, and hence require UV light of wavelengths of about 270–280 nm (near-UV region), whereas for conjugated ketones, the activation energy for singlet and triplet excited states is below 80 kcal/mol, in the range 45–78 kcal/mol, and require light in the far-UV region (310–330 nm). The excited carbonyl groups have radical characters at both carbon and oxygen and the dipole moments of the excited states are reduced compared to that of ground states due to transfer of electron density from an orbital localized on oxygen to the orbital of carbon atom [1]. The excited states of saturated and conjugated ketones can be represented as hybrid structures of diradical and dipolar forms.

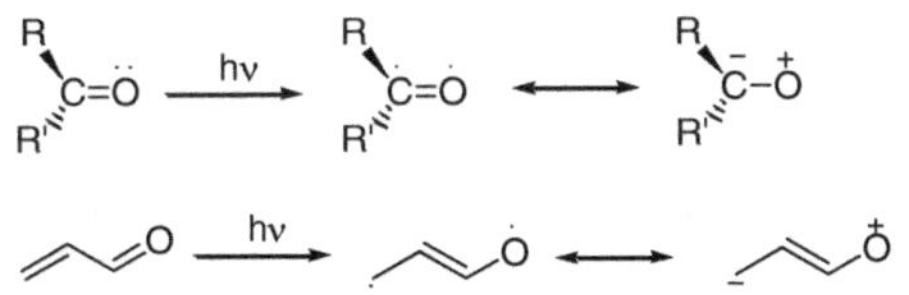

The high reactivity of the excited states of carbonyl compounds is due to half-filled orbital of oxygen.

electronic structure of a carbonyl group in its n,π* excited state

B. Dinda, *Essentials of Pericyclic and Photochemical Reactions*,
Lecture Notes in Chemistry 93, DOI 10.1007/978-3-319-45934-9_8

Important photo-induced reactions of carbonyl compounds are the reduction of carbonyl compounds by hydrogen abstraction, fragmentation, cycloaddition to alkenes and rearrangement.

8.2 Hydrogen Abstraction and Fragmentation Reactions

One of the common reactions of photo-excited carbonyl compounds is hydrogen abstraction from solvent or some other hydrogen donor substrate. The hydrogen donor substrate may be other compound or the same compound. The hydrogen abstraction generates free radical intermediates. These free radicals may react among themselves or with ground-state molecules to give the products. Many aromatic ketones on irradiation undergo abstraction of hydrogen atom from alcohols to give pinacol type diols in a photoreduction process. For example, benzophenone is reduced to pinacol-like diol **1** on photoirradiation in 2-propanol [2]. The quantum yield of the reaction is 2.0. It indicates that two molecules of benzophenone are reduced by photoexcitation of its one molecule. The reaction is believed to take place through triplet state of the carbonyl compound in the following steps:

$$Ph_2C{=}O \xrightarrow{h\nu} \underset{\text{triplet}}{[Ph_2C{=}O]^*}$$

$$[Ph_2C{=}O]^* + (CH_3)_2CHOH \longrightarrow Ph_2\dot{C}OH + (CH_3)_2\dot{C}OH$$

$$(CH_3)_2\dot{C}OH + Ph_2C{=}O \longrightarrow Ph_2\dot{C}OH + (CH_3)_2C{=}O$$

$$2Ph_2\dot{C}OH \longrightarrow \underset{\mathbf{1}}{Ph_2C(OH){-}C(OH)Ph_2}$$

The photoreduction efficiency of *ortho*-alkyl benzophenone derivatives is greatly reduced by intramolecular enolization process, known as *photoenolization reaction*. For example, *ortho*-ethyl benzophenone **2** on photoirradiation in deuterated hydroxylic solvents gives deuterated ethyl benzophenone **3** by photoenolization without reduction [3].

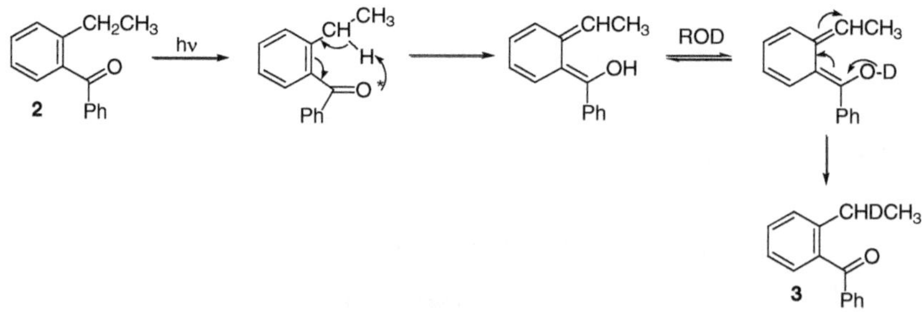

For some aromatic ketones, the reactive dienols undergo electrocyclization to cyclobutenols [4].

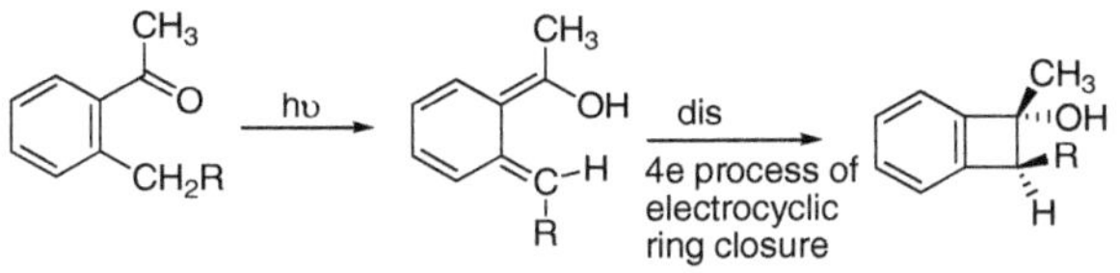

The reactive enols **4** may be trapped as Diels–Alder adducts **5**, for example, with dimethyl acetylenedicarboxylate [4]:

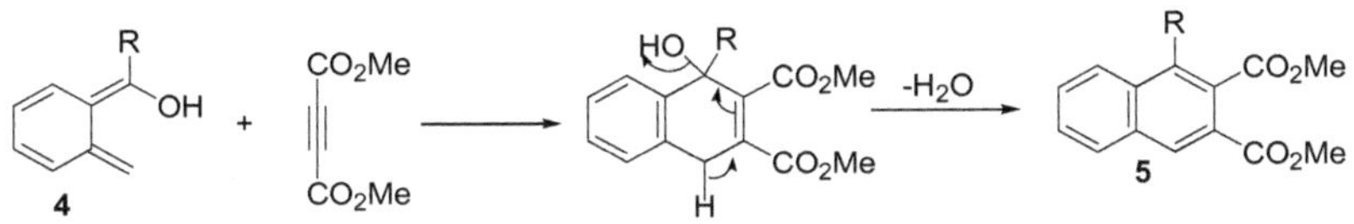

Another important photochemical reaction of both aliphatic and aromatic ketones is the fragmentation reaction. Unconjugated ketones on photoexcitation undergo α-cleavage followed by decarbonylation and subsequent reactions of alkyl radicals to give product(s). All these processes are collectively known as *Norrish type-1 cleavage* reactions [5]. These reactions take place both in gaseous and liquid phases.

$$R{-}\overset{O}{\overset{\|}{C}}{-}R' \xrightarrow{h\nu} R{-}\dot{C}{=}O + \dot{R'} \longrightarrow \dot{R} + CO + \dot{R'} \longrightarrow R{-}R'$$

The energies of α-cleavage and subsequent coupling, disproportionation and hydrogen abstraction processes depend on the structure of the ketones and stability of the radical fragments that are ejected. For examples, dibenzylketone **6** and *t*-butyl ketone **7** undergo photolytic α-cleavage readily in solution to give a mixture of products [6, 7]. In unsymmetrical ketones, α-cleavage preferably takes place at the site of alkyl group that can form relatively more stable free radical.

$$\underset{\mathbf{6}}{PhCH_2{-}\overset{O}{\overset{\|}{C}}{-}CHPh_2} \xrightarrow{h\nu} \underset{25\%}{PhCH_2CH_2Ph} + \underset{50\%}{PhCH_2CHPh_2} + \underset{25\%}{Ph_2CHCHPh_2} \quad \text{Ref. 6}$$

$$PhCH_2{-}\overset{O}{\overset{\|}{C}}{-}CHPh_2 \xrightarrow{h\nu} PhCH_2{-}\overset{O}{\overset{\|}{C}}{\cdot} + \underset{\text{more stable radical}}{\cdot CHPh_2}$$

$$PhCH_2{-}\overset{O}{\overset{\|}{C}}{\cdot} \longrightarrow Ph\dot{C}H_2 + CO$$

$$3\,Ph\dot{C}H_2 + 3\,\dot{C}HPh_2 \longrightarrow PhCH_2CH_2Ph + PhCH_2CHPh_2 + Ph_2CHCHPh_2$$

$$\underset{\mathbf{7}}{(H_3C)_3C{-}\overset{O}{\overset{\|}{C}}{-}CH_3} \xrightarrow{h\nu} H_3C{-}\overset{O}{\overset{\|}{C}}{\cdot} + \cdot C(CH_3)_3 \longrightarrow (CH_3)_3CH + (CH_3)_2C{=}CH_2 + CH_3CH{=}O \quad \text{Ref. 7}$$

The formation of the products can be explained as follows:

$$(CH_3)_3\dot{C} + H{-}CH_2{-}\dot{C}(CH_3)_2 \longrightarrow (CH_3)_3CH + (CH_3)_2C{=}CH_2$$

$$H_3C{-}\overset{O}{\overset{\|}{C}}\cdot + H{-}CH_2{-}\dot{C}(CH_3)_2 \longrightarrow (CH_3)_2C{=}CH_2 + CH_3CH{=}O$$

Cyclic ketones **8–10** also undergo similar α-cleavage, decarbonylation and hydrogen abstraction reactions to give products [6, 8].

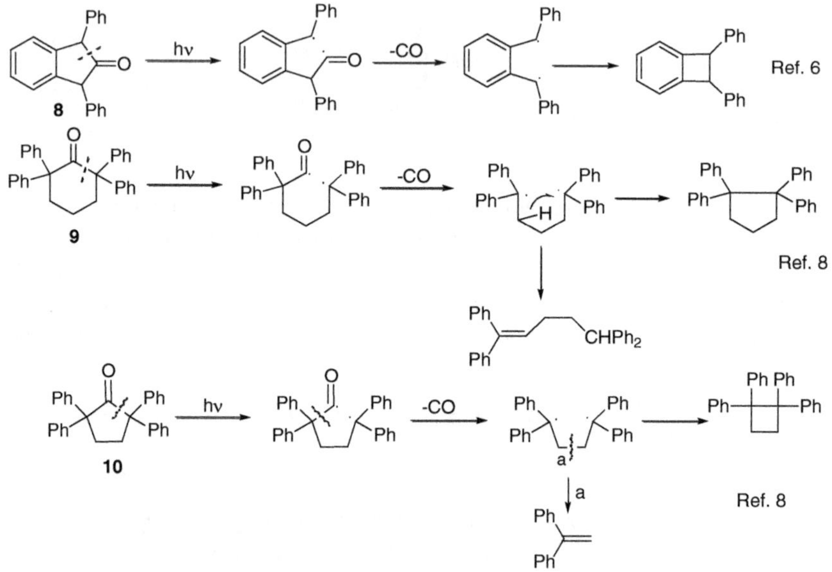

Some cyclic ketones undergo α-cleavage followed by intramolecular hydrogen abstraction to form unsaturated aldehydes. Usually abstraction of hydrogen takes place from δ-carbon. In addition to aldehyde, ketene is also formed involving another path in some cyclic ketones. The following examples [9, 10] are illustrative:

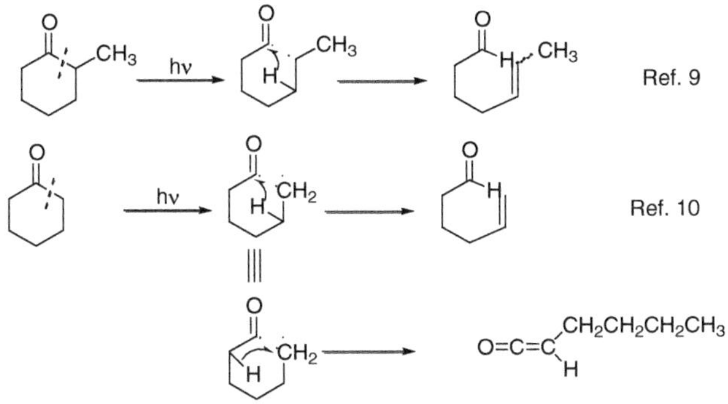

Cyclic ketones **11** and **13** on photoirradiation in hydroxylic solvents give acetals **12** and **14**, respectively, by α-cleavage followed by formation of carbene and subsequent reaction with solvent [11].

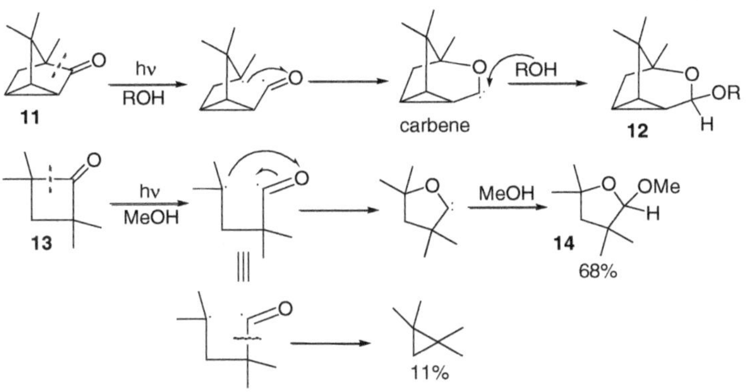

Aliphatic and aromatic ketones having propyl or longer alkyl group as a carbonyl substituent on photoirradiation undergo intramolecular hydrogen abstraction preferably from γ-carbon to give diradicals, which on β-cleavage or ring closure give ketones and cyclobutanol derivatives. The β-cleavage results in the formation of an alkene and a new carbonyl compound.

All these reaction processes are collectively known as *Norrish type-II cleavage reactions* [12]. The hydrogen abstraction occurs through a chair-like conformation to generate a diradical intermediate. Usually the β-cleavage is the dominant reaction for both aryl and allyl ketones, but high yields of cyclobutanols may be found when favorable gauche conformation of the diradical intermediate is feasible [13]. For example, aromatic ketones **15** and **16** give major products from the β-cleavage [13].

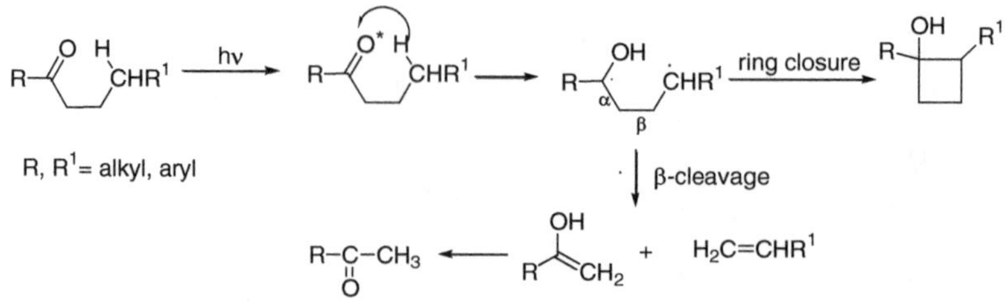

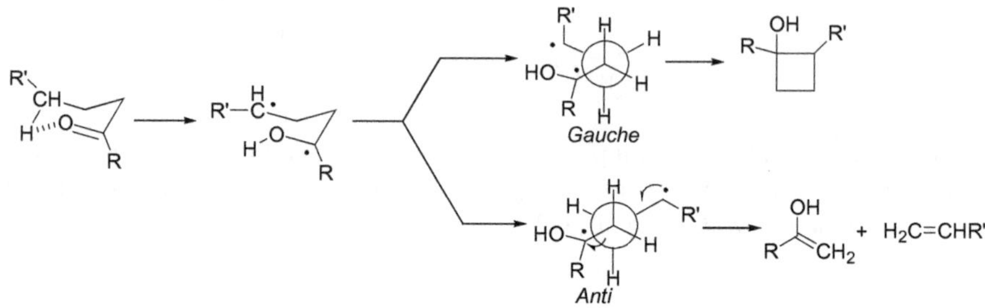

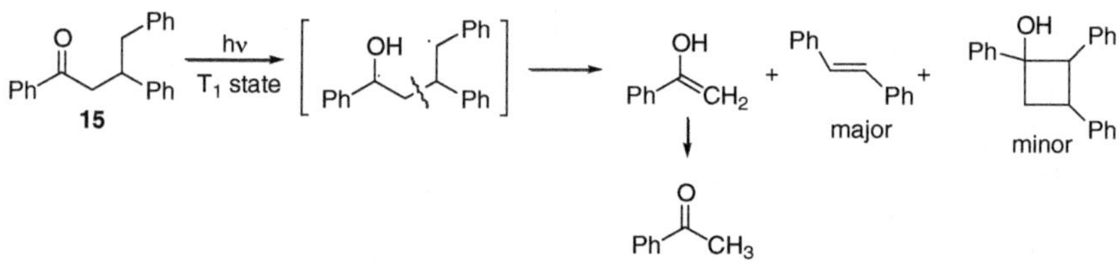

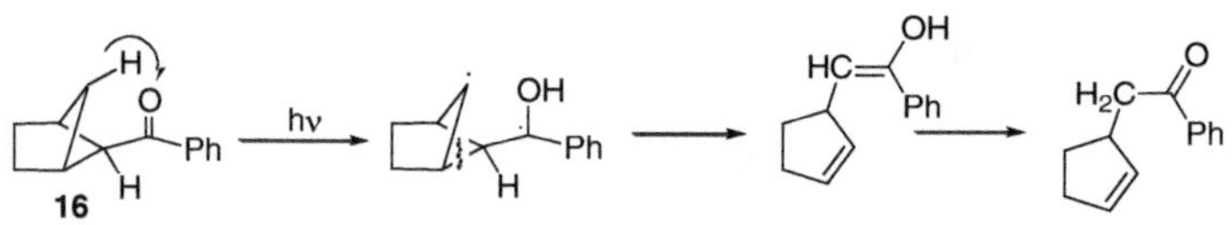

Ref. 13

Whereas aromatic ketones **17** and **19** with an α-substituent give cyclobutanol **18** and cyclopropyl ketone **20** as major products, respectively [4, 13].

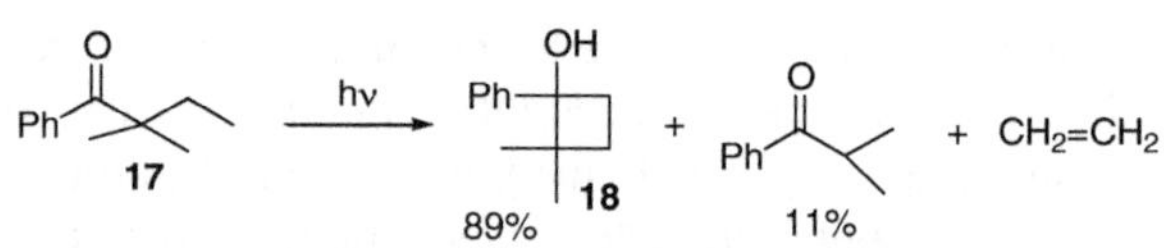

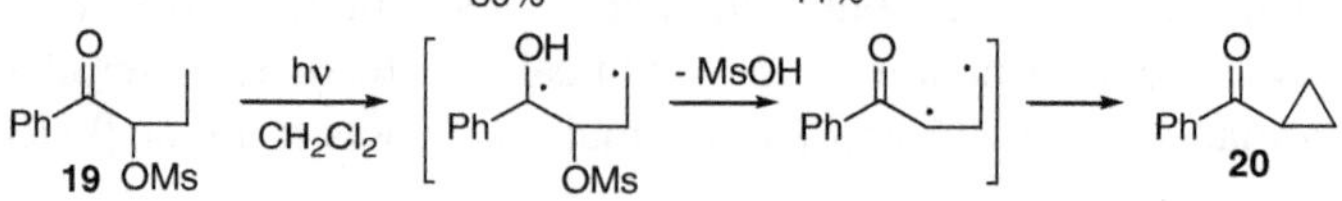

In aliphatic and allylic ketones, the reaction proceeds through both singlet and triplet excited states, but in aryl ketones, intersystem crossing is very fast and triplet state is more effective. For examples, aromatic ketones give major β-cleavage products from their triplet excited states. Aliphatic 2-hexanone **21** gives acetone, propylene and 1,2-dimethylcyclobutanol by both β-cleavage and ring closure processes [14]. Unsaturated aliphatic isomesityl oxide **22** gives major product cyclobutane **23** and minor products by α-cleavage [14].

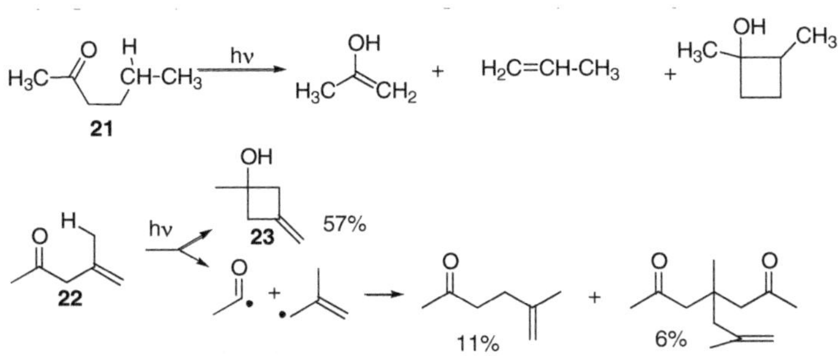

Diastereomeric aliphatic ketones **24** undergo *Norrish type-II reactions* to give stereospecific *cis*-and *trans*-alkenes **25** and **26** from their singlet excited states by exclusive β-cleavage process [15].

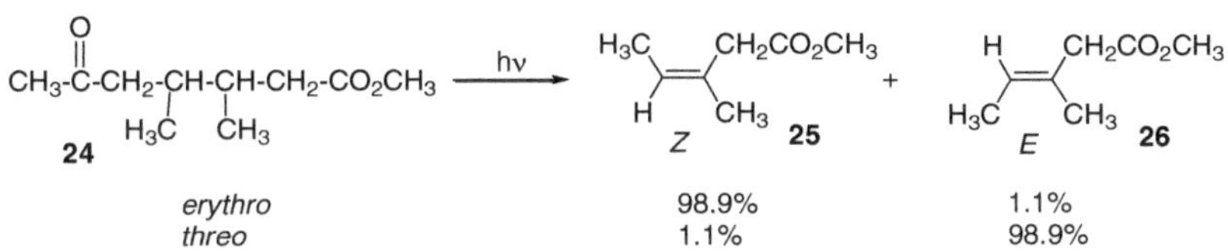

High yields of cyclobutanols are obtained in the reactions of cyclic ketones **27** and **28** [16, 17].

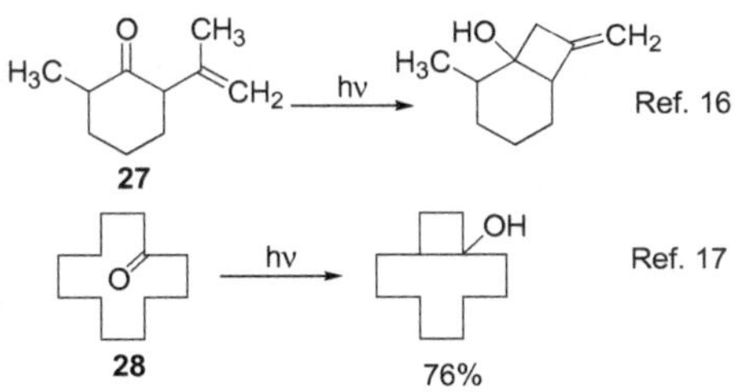

In ketone **29**, γ-hydrogen is not available and hence δ-hydrogen abstraction takes place to give cyclopentanol **30** as major product.

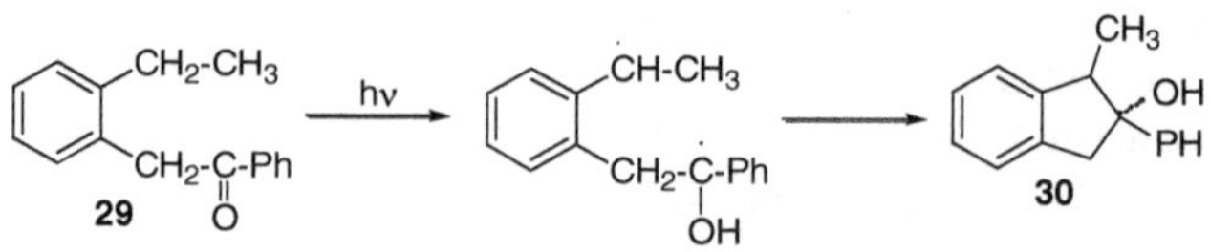

Ref. 18

α-Diketone **31** and β,γ-unsaturated ketones undergo γ-hydrogen abstraction to give only cyclization products **32** and **33** because in each case, one radical site of the diradical intermediate is allylic type such that its π system is orthogonal to the C (2)–C(3) bond and prevents β-cleavage process [19].

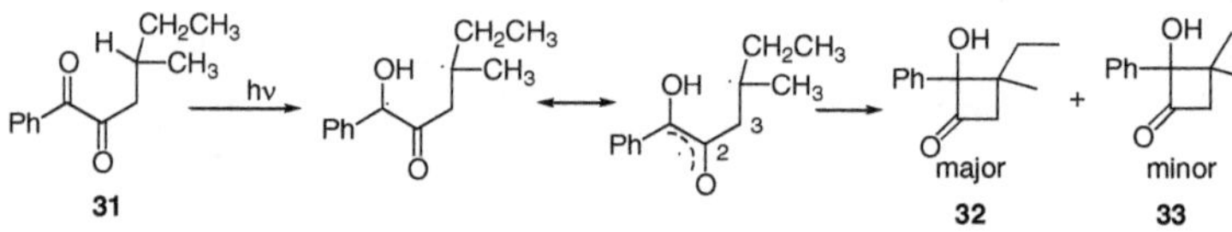

Ref. 19

Acyclic α,β-unsaturated ketone **34** on *Norrish type II reactions* gives both normal product **35** and a by-product **36** due to resonance hybrid of the diradical intermediate [20].

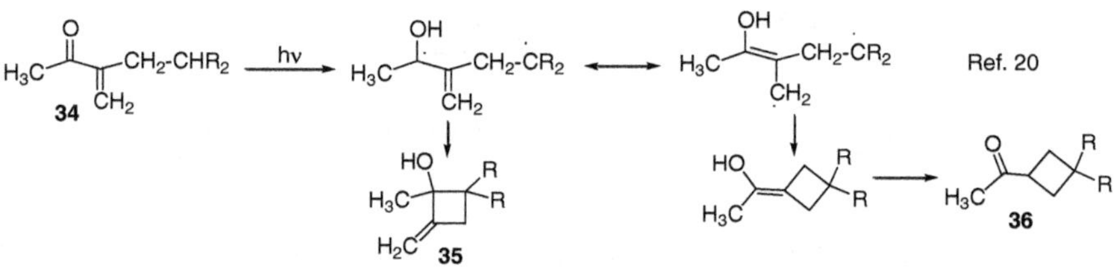

Ref. 20

Alicyclic α,β-unsaturated ketone **37** on irradiation yields major product **38** from *Norrish type II reactions* and minor products **39** and **40** from other paths [21].

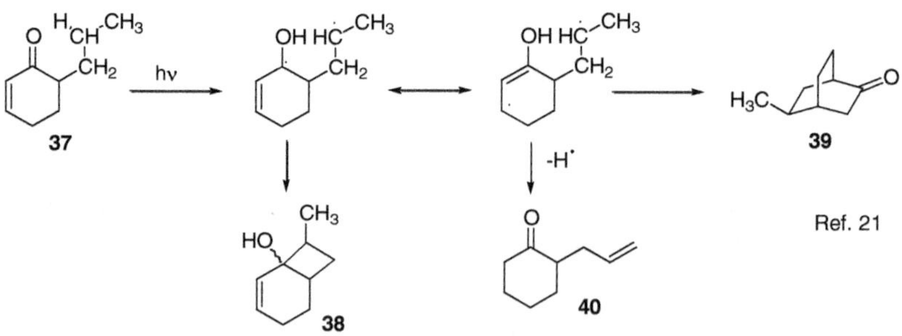

Ref. 21

In ketones **41** and **42**, hydrogen abstraction takes place from β- or ε-carbon [22, 23].

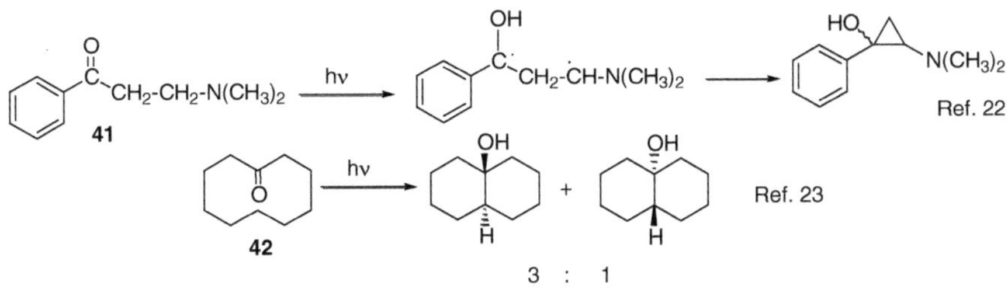

Norrish Type II reactions of 2-benzylcyclohexanones **43–45** are interesting. In 2-benzylcyclohexanone **43**, the exclusive β-cleavage gives the product, whereas 2-benzoyl-2-methylcyclohexanone **44** gives the major product from cyclobutanol ring opening. 2-(2-Methylbenzoyl)-cyclohexanone and 2-(2-methylbenzoyl)-2-methylcyclohexanone **45** give hexahydroanthracenone derivative **48** as the only product. The direction of hydrogen abstraction and stability of the intermediate diradicals control the reaction course [24–26].

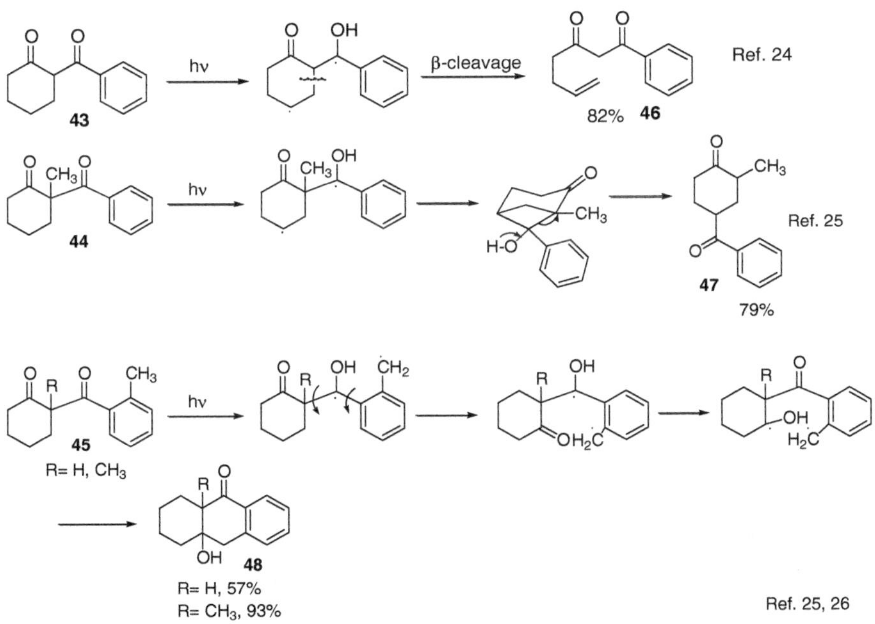

α,β-Unsaturated ketone **49** in the presence of catalytic amount of acid or base undergoes hydrogen abstraction from γ-carbon followed by isomerization to give β,γ-unsaturated ketone **50** [27].

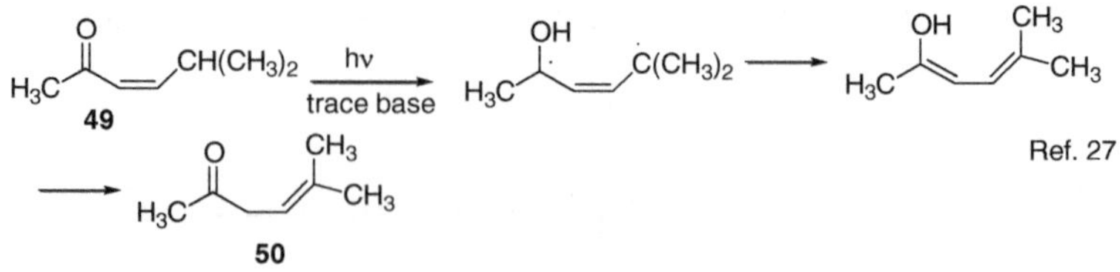

β,γ-Unsaturated cyclic ketones **51** and **52** on irradiation undergo α-cleavage followed by isomerization of the diradical intermediates to yield the major products **53** and **54**, respectively [28, 29].

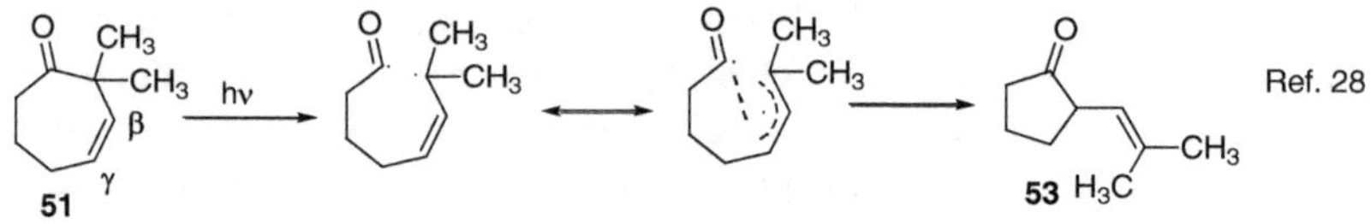

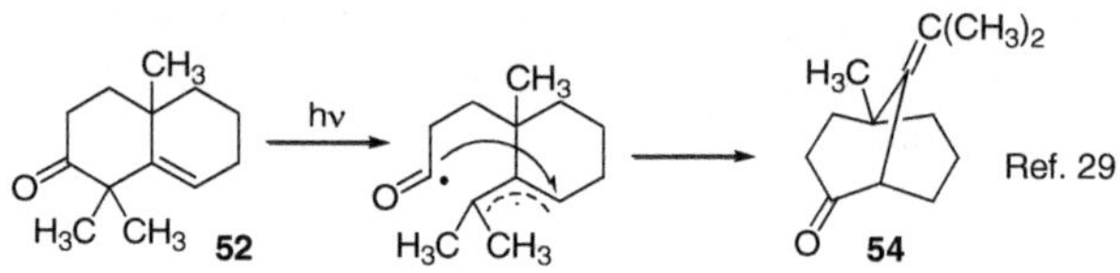

Irradiation of cyclopentenone in hydrocarbon solvents results in intermolecular hydrogen abstraction, followed by recombination of the resulting radicals to give the addition products [30].

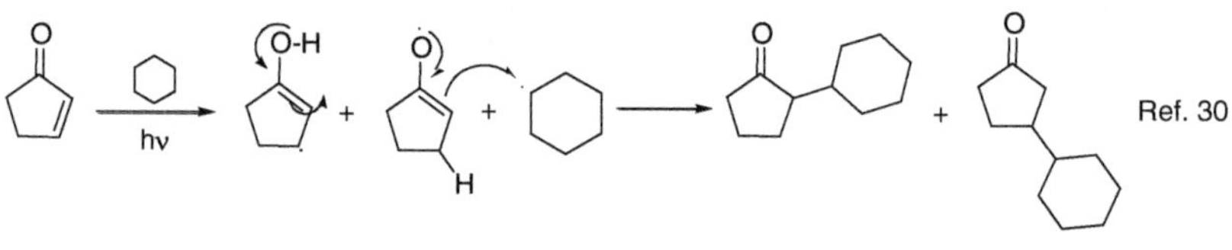

Cyclohexenone **55** and cycloheptenones undergo photo-induced acid-catalyzed addition reactions with hydroxylic solvents via *cis–trans*-isomerization [31].

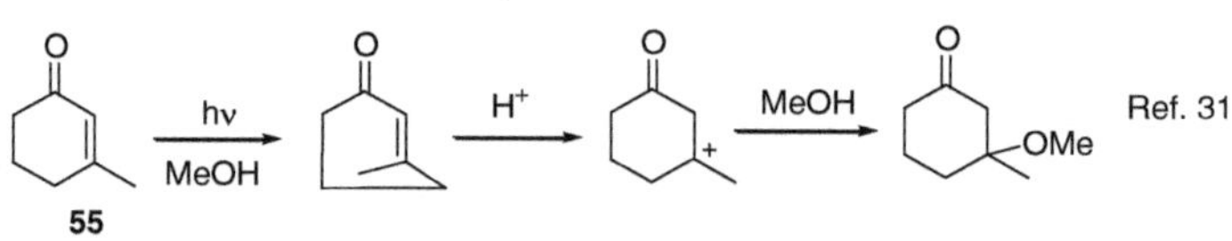

8.3 Cycloaddition and Rearrangement Reactions of Unsaturated Carbonyl Compounds

Cyclopentenone **56** having a propyl-like substituent undergoes photosensitized intramolecular hydrogen abstraction to give rearrangement product **57** and cycloadduct **58** [30].

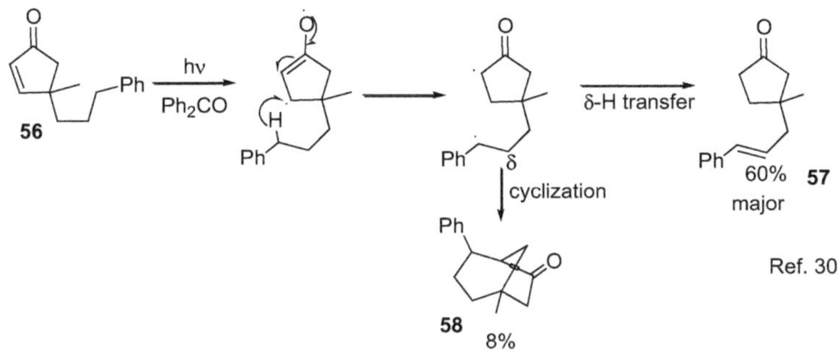

α,β-Unsaturated cyclic ketone **59** undergoes intermolecular cycloaddition reactions with alkenes and dimerization on irradiation. The reactions take place in a π–π* triplet excited state via the formation of 1,4-diradical intermediates. The stability and relative efficiency of the diradicals determine the regioselectivity of the addition products [32, 33].

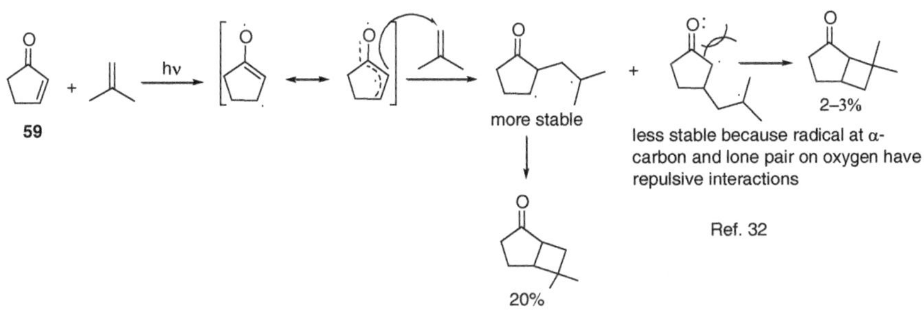

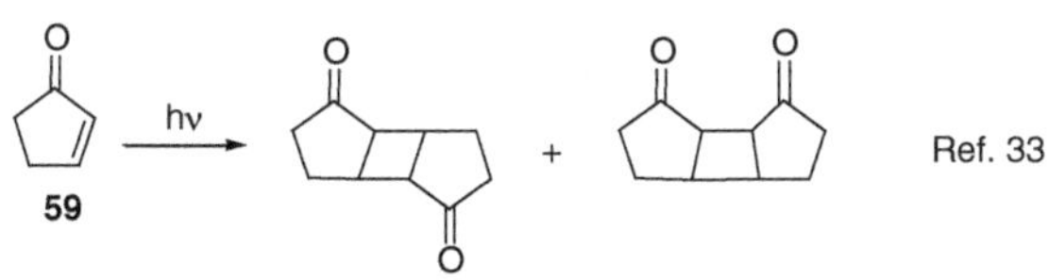

The dimerization occurs via the formation of a 1,4-diradical intermediate followed by addition with a ground-state molecule.

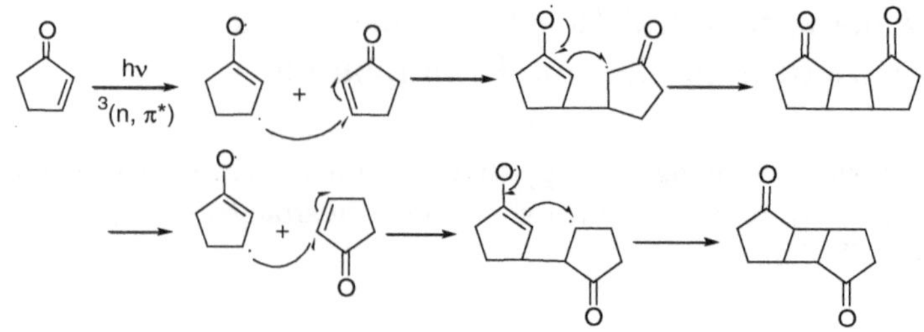

The stereochemistry of ring junction was studied in the cycloaddition of cyclohexenone with isobutene. The *trans*-ring closure is the preferred cyclization process because of minimum steric repulsion [34].

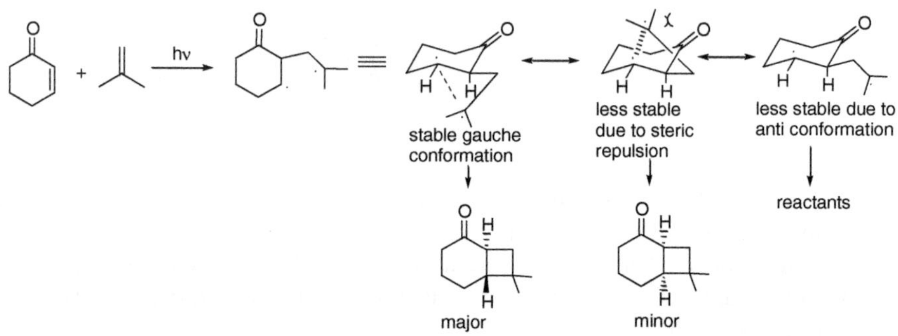

4,4-Dialkylcyclohexenone **60** undergoes photo-induced *lumiketone rearrangement* to form cyclopropane derivative **61** [35].

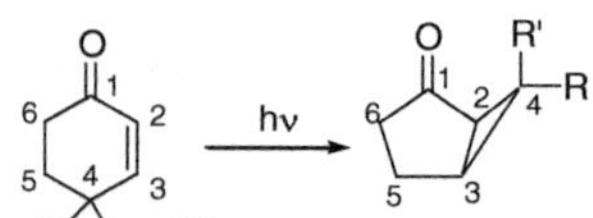

60 **61**

The rearrangement is stereospecific with inversion of configuration at C(4). It is a photochemically allowed $[\pi^2_a + \sigma^2_a]$ cycloaddition process. The mechanism is very similar to the di-π-methane rearrangement. It involves 1,2-shift of C(5)–C(4) σ bond to C(3) followed by formation of a new σ bond between C(2) and C(4) using back lobe of C(4)-p-orbital.

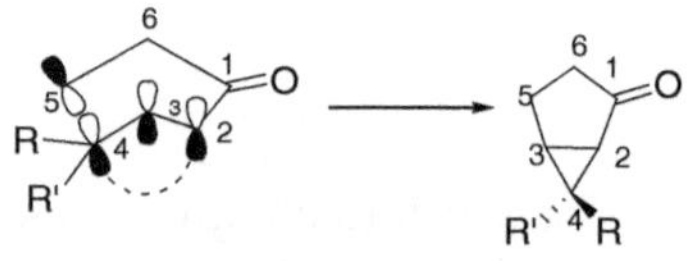

An alternative mechanism via triplet excited state was also proposed by Zimmerman with chiral diradical intermediate.

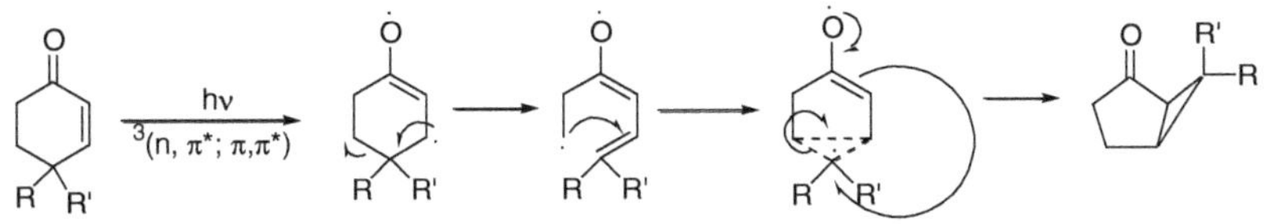

For example, cyclohexanone **62** gives stereoisomeric products **63** and **64** with inversion of configuration at C(4) [36].

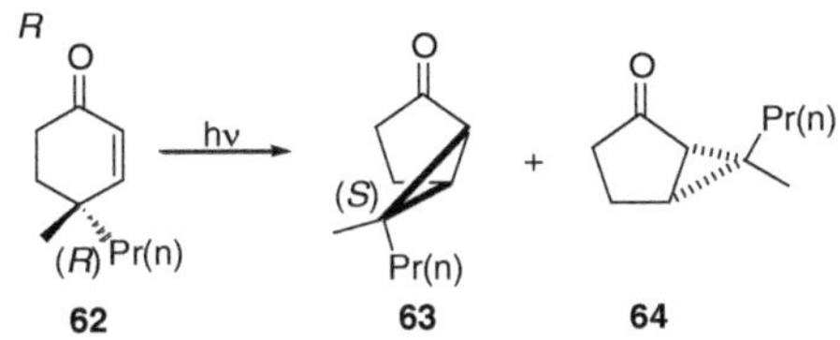
Ref. 36

This lumiketone rearrangement also occurs in steroid **65** and 4-alkyl-4-aryl cyclohexenone **66** to produce **67** and **68**, respectively [37, 38].

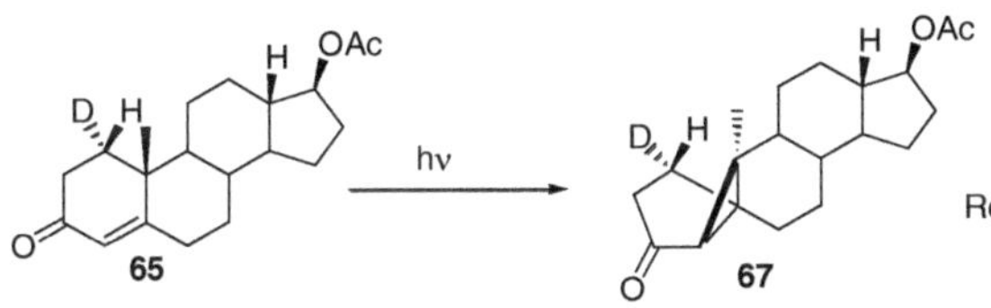
Ref. 37

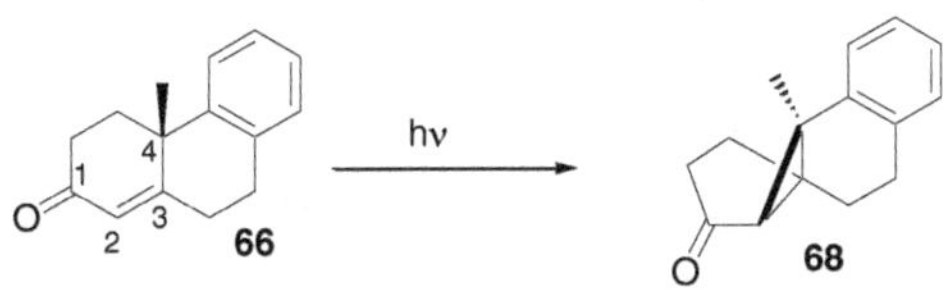
Ref. 38

4,4-Diphenylcyclohexenone **69** on irradiation gives products **70** and **71** via di-π-methane rearrangement from n → π^* transition of carbonyl group [39].

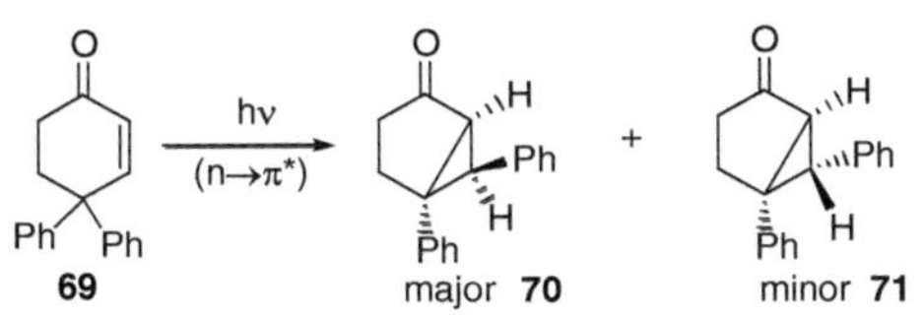
Ref. 39

In 4,4-diarylcyclohexenone **72**, the aryl group having electron-withdrawing substituent migrates in preference to phenyl group to give major product **73** [40].

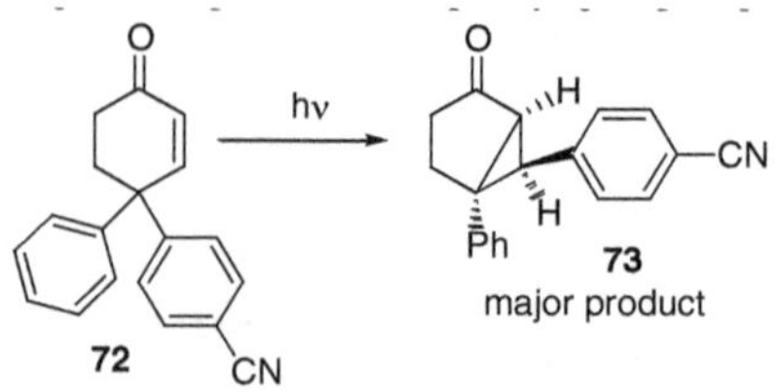

Ref. 40

These compounds undergo photo-induced rearrangement from a triplet excited state via a TS similar to that of di-π-methane rearrangement. The reaction is a stereospecific concerted process. In the TS, the migration of one aryl group from C (4) to C(3) and bridging between C(2) and C(4) occur in a concerted mechanism to yield kinetically controlled *endo-product* as major product [39].

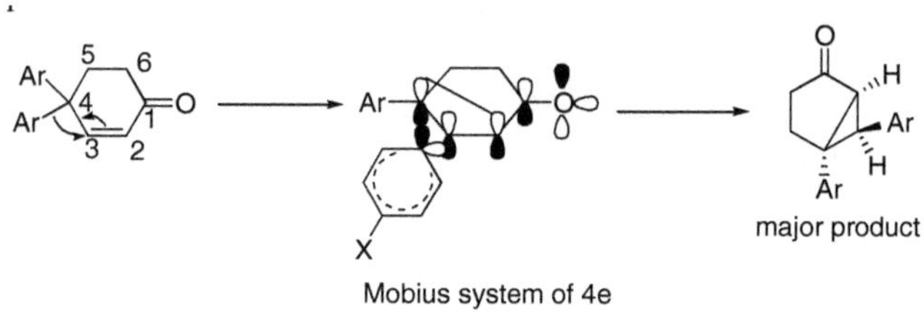

Alternative mechanism involving diradical intermediate was also proposed [39].

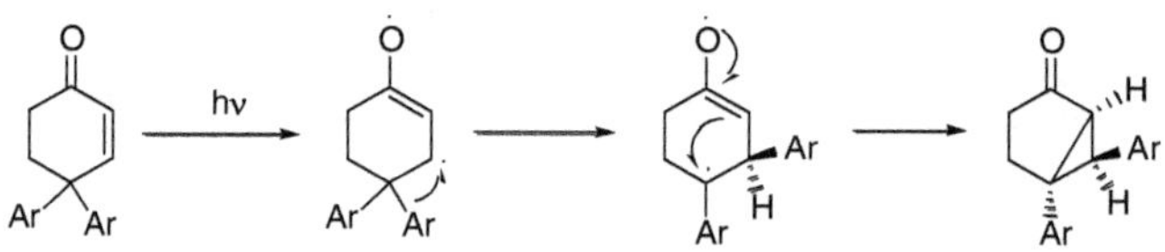

Cyclohexenone **74** with alkyl and vinyl substituents in C(4) gives major product **75** by migration of vinyl substituents. Similarly with alkyl and aryl substituents in C (4) position, aryl substituent will migrate in preference to alkyl substituent to give the major product of the reaction. In both cases, the migrating vinyl or aryl group stabilizes the radical character of the TS.

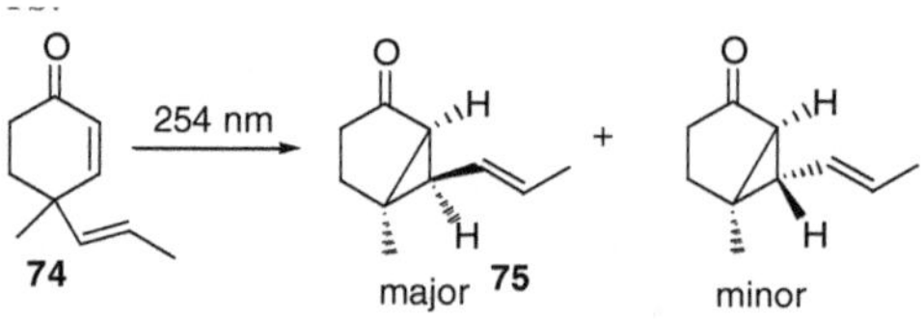

Ref. 39

β,γ-Unsaturated ketones and aldehydes undergo photo-induced rearrangement reactions to give cyclopropyl-ketones and aldehydes, respectively. These reactions are known as the *oxa-di-π-methane* (ODPM) *rearrangements* [41]. These reactions occur in both acyclic and cyclic systems. The following examples are illustrative [41–44]:

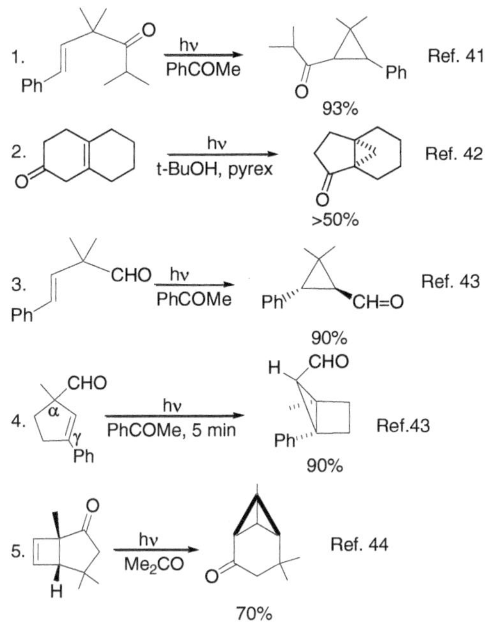

These reactions proceed through the $\pi \rightarrow \pi^*$ excited singlet and triplet states of the carbonyl compounds. The efficiency of the reactions depends on aryl substitution at the γ-carbon and disubstitution at the α-carbon. Possibly, the substituents at these carbons stabilize the intermediate triplet diradical. The efficiency of the intersystem crossing of singlet S_1 state to triplet T_1 state also determines the outcome of the reaction [45].

In singlet state, the reaction is believed to take place by 1,2-sigmatropic shift of acyl group followed by sigma bonding between C(4) and C(2).

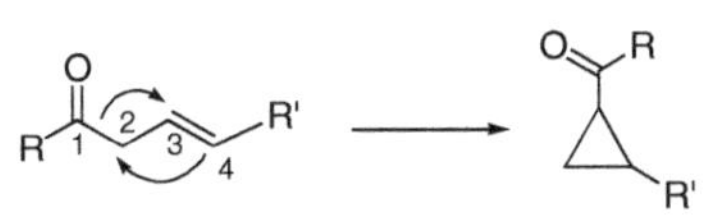

The reaction of triplet excited state of the carbonyl compound is believed to proceed through a σ bond formation between C(1) and C(3) followed by cleavage of the cyclopropane ring in the resulting diradical to form a relatively more stable diradical and subsequent cyclization of the diradical.

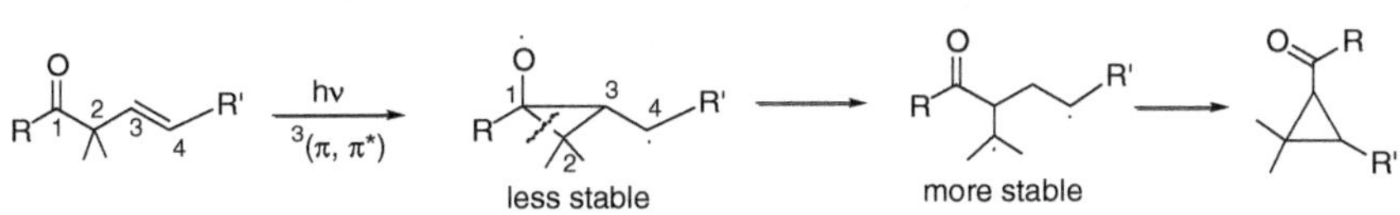

Some β,γ-unsaturated ketones undergo both 1,2- and 1,3-acyl migrations in photoexcitations. For example, 1,2-dimethylcyclopent-2-enyl methyl ketone **76** gives **77** and **78**.

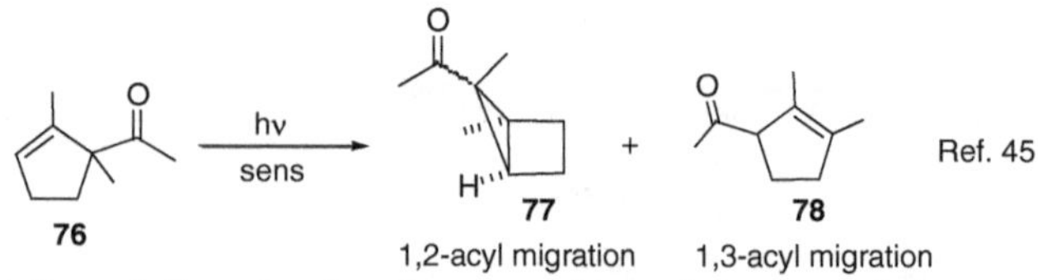

The product from 1,3-acyl migration was formed from a triplet excited state through the formation of new σ bond between C(1) and C(4), followed by cleavage of the cyclobutane ring of the diradical.

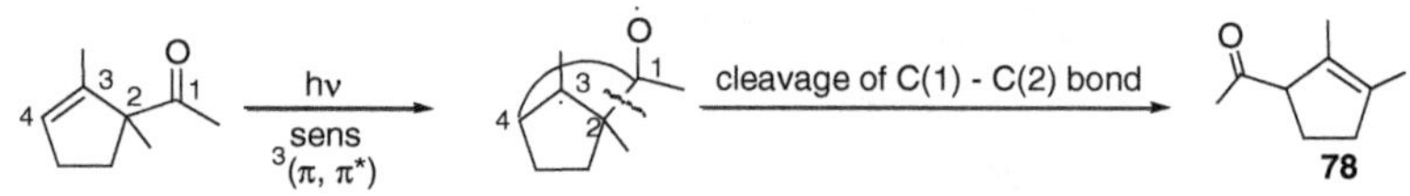

In β,γ-unsaturated ketones, where both di-π-methane and oxa-di-π-methane possibilities are available, only one of these paths is favored due to steric interaction and stabilization of 1,3-diradical. Thus, enone **79** gives **80** solely by oxa-di-π-methane path [46].

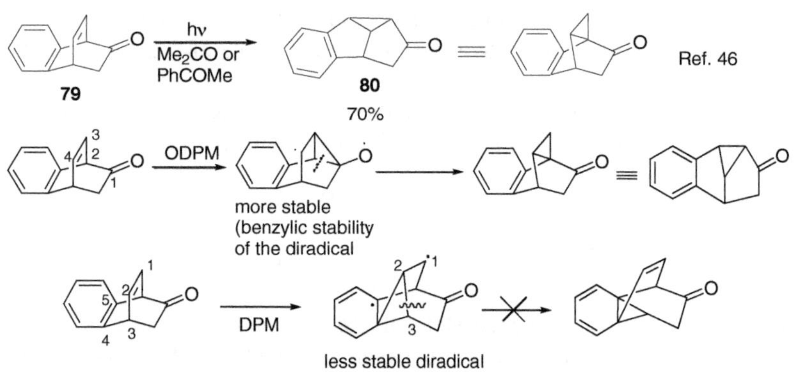

The dienone **81**, where both DPM and ODPM processes may be feasible, the DPM process is preferred due to involvement of weaker ethylenic π system and charge transfer stabilization of its 1,3-diradical [47].

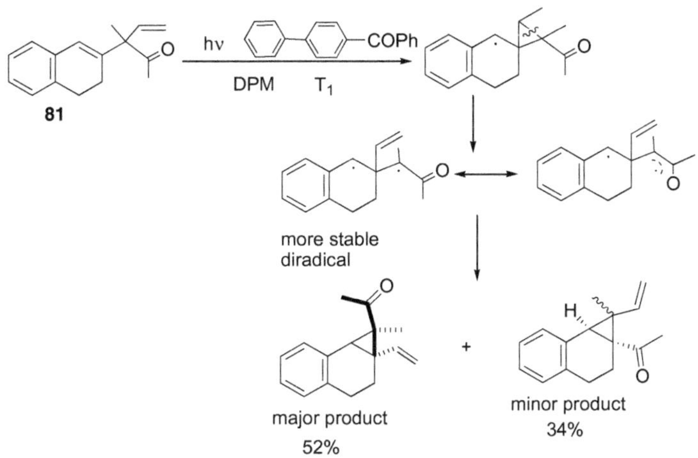

Spirocyclic β,γ,δ,ε-unsaturated ketone **82** undergoes ODPM rearrangement to give two photoproducts **83** and **84** from its triplet excited state on irradiation at λ ≥ 340 nm. The major product comes from larger spin orbit coupling interaction and better orbital overlaps due to less atomic motion, whereas **82** on irradiation at 254 nm undergoes electrocyclic ring opening of cyclohexadiene ring to give trienone **85** and on irradiation at 300 nm gives aromatic aldehyde **86** by α-cleavage and β-H abstraction [48].

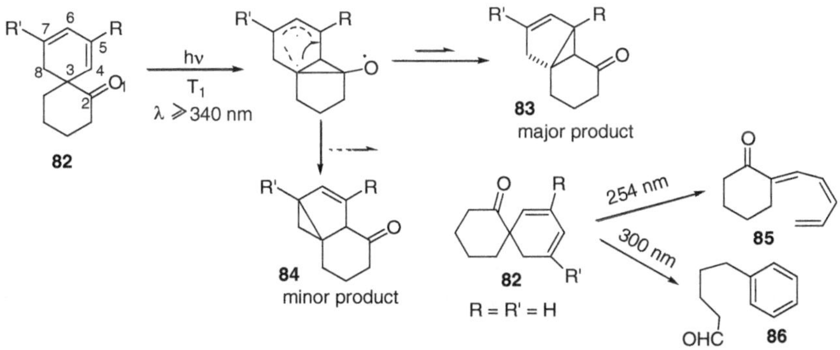

The presence of a carbomethoxy or hydroxymethyl group at C(2) position of the 1,3-diene system of the spiroketone **87** gives only one photoproduct **88** with inversion of configuration at the chiral C(6) [49].

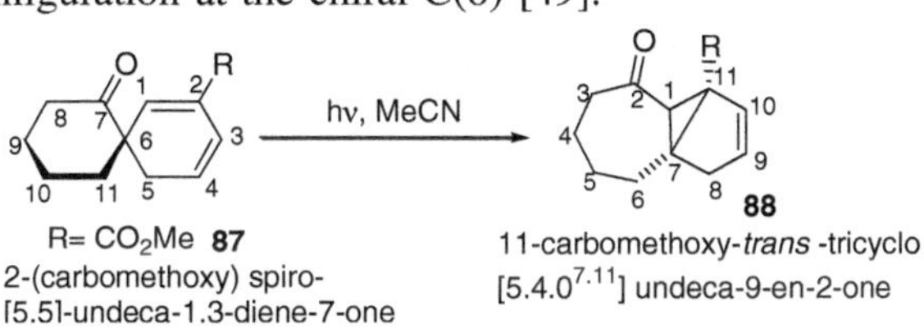

In *bis*-β,γ-unsaturated ketones, the products are obtained from two successive oxa-di-π-methane rearrangements. For example, **89** gives **90** [50].

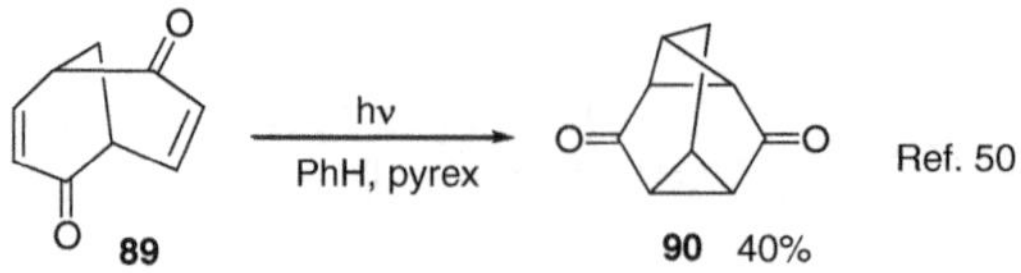

Some other typical examples [51–55] of ODPM rearrangement are:

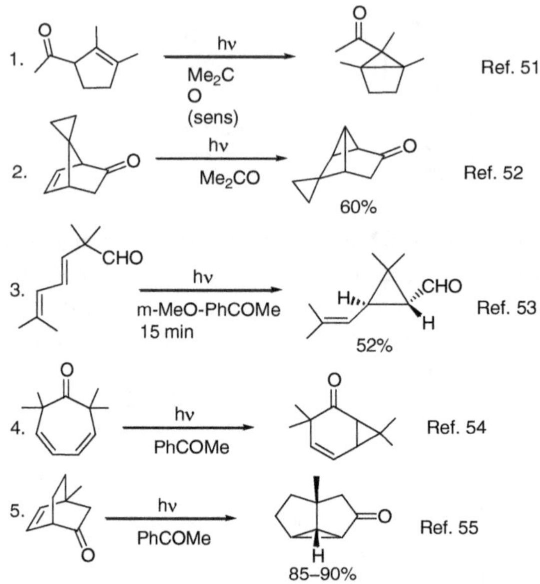

Cross-conjugated cyclohexadienones on photoirradiation undergo rearrangements to give cyclopentenones fused with a cyclopropyl ring, called lumiketones. These rearrangements are known as *Lumiketone rearrangements*. Detailed study on the mechanism of the reactions indicated that the reaction proceeds from a triplet excited state via the formation of a ground-state zwitterion intermediate. The zwitterion undergoes 1,4-charge transfer followed by migration of a sigma bond C(3)–C(4) from C(3) to cation site C(6) to give the product. For example, 4,4-diphenylcyclohexadienone **91** gives lumiketone **92** as major product on photolysis [56].

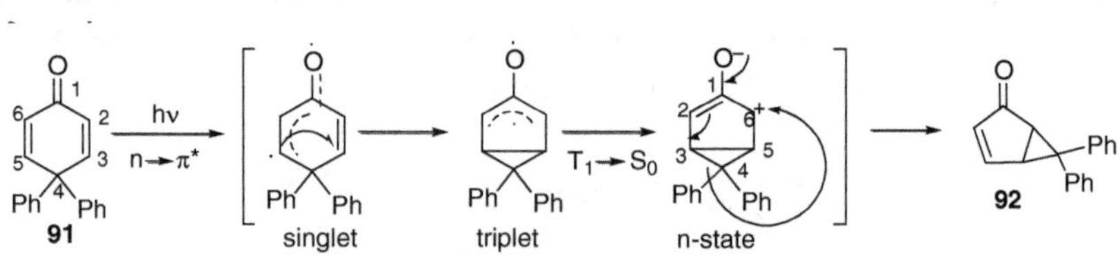

An alternative mechanism was also proposed.

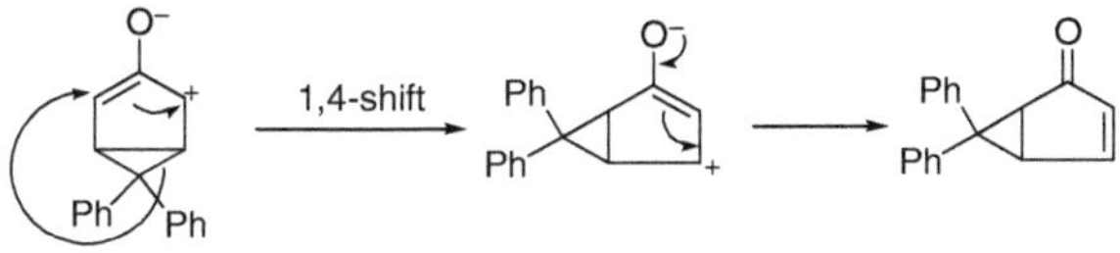

The Woodward–Hoffmann rules predict an inversion of configuration at C(4) for such [1, 4]-sigmatropic shift.

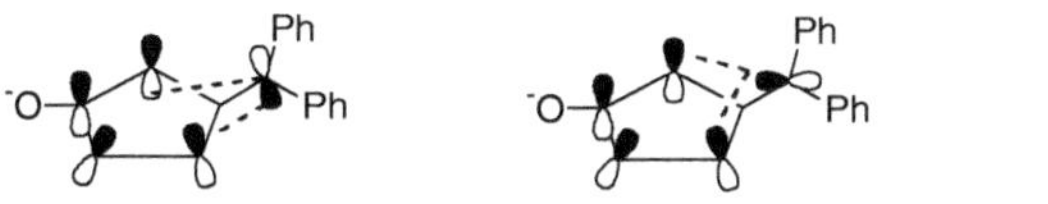

4e Mobius system (one node) 1,4-shift with inversion is allowed

4e Huckel system (0 node) 1,4-shift with retention is forbidden

The rearrangement of α-bromoketone **93** in the presence of a base gives **94**, with inversion of configuration at C(4) [57].

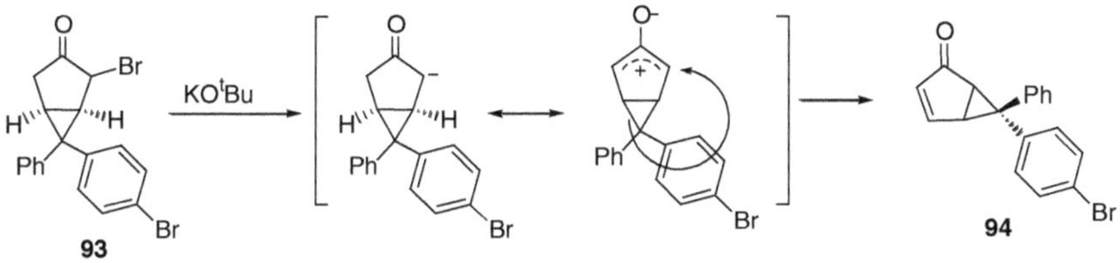

Direct irradiation of 4,5,5-triphenylcyclohex-2-en-1-one **95** gives four products **96–99**, whereas irradiation of 4-methyl-5,5-diphenylcyclohex-2-en-1-one **100** gives only one product **101** on photorearrangement [58].

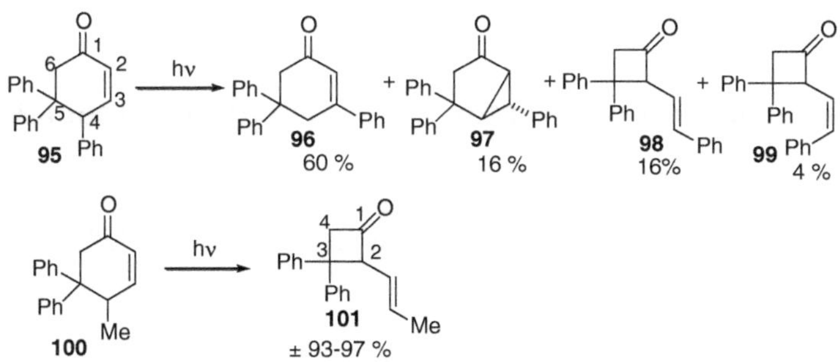

Both these cyclohexenones react with their triplet excited states through diradical intermediates. In both cases, the cyclobutanone products were formed by the homolytic cleavage of C(4)–C(5), followed by attack of diphenyl methyl radical

center on C-2. In the former cyclohexenone **95**, bicyclic ketone and rearranged cyclohexenones were formed by a phenyl migration mechanism.

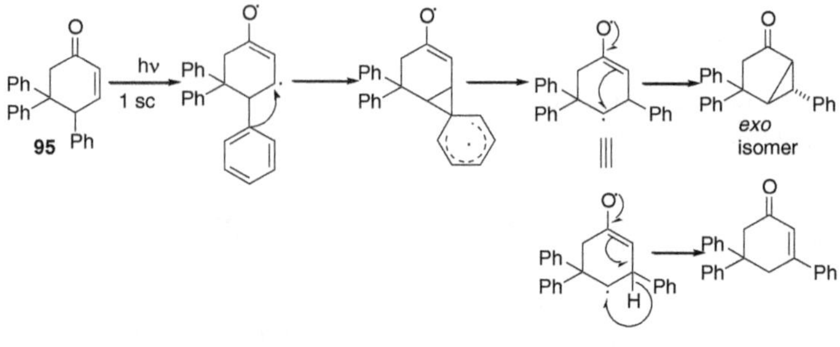

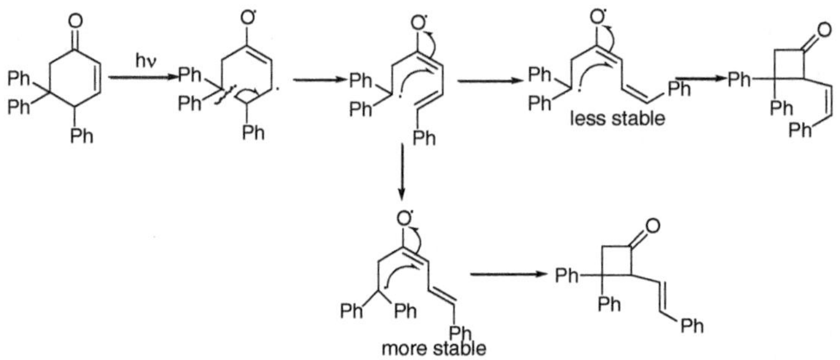

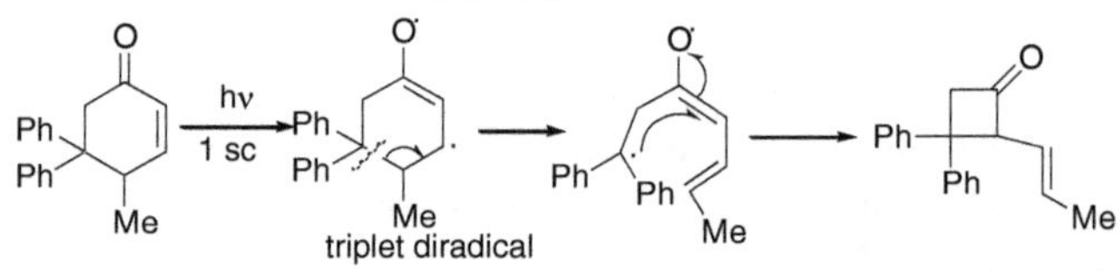

8.4 Isomerization of Unsaturated Carbonyl Compounds

6,6-Dimethyl-2,4-cyclohexadienones **102** undergo photoisomerization to **103** by cleavage of the 1,6-bond from their n,π* singlet excited states [59].

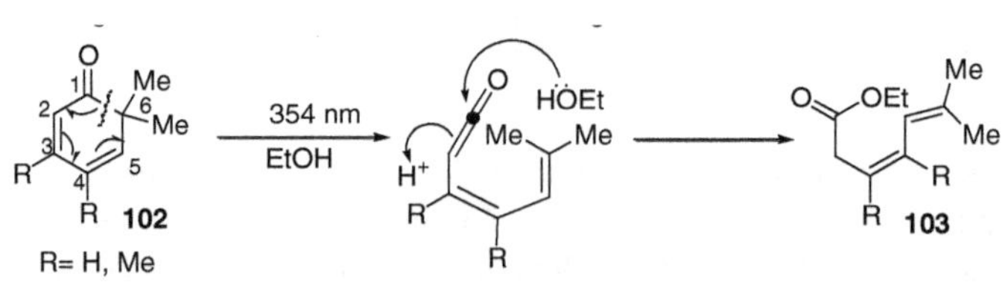

α-Tropolone methyl ethers **104** on photoisomerization and hydrolysis give cyclopentenone derivatives **105** [60].

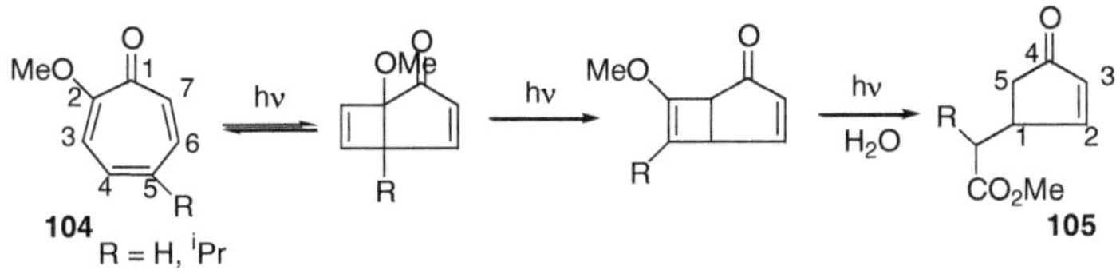

104 R = H, ^{i}Pr **105**

The isomerization is believed to take place by formation of a new σ bond between C(2) and C(5), followed by formation of a zwitterion and its rearrangement. The resultant isomer undergoes nucleophilic addition of H_2O and cleavage of cyclobutene ring to give **105**.

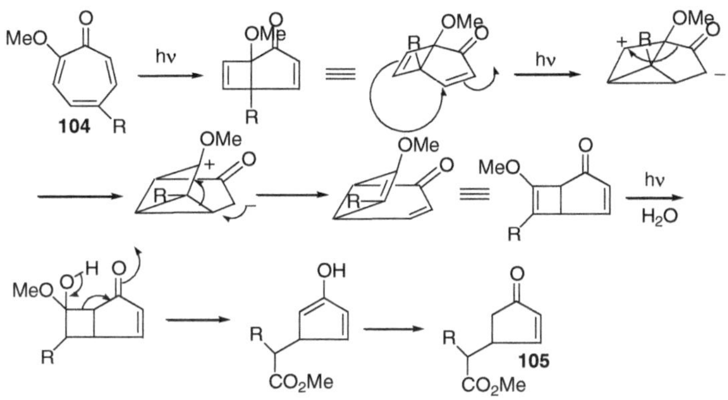

8.5 Cycloaddition Reactions of Carbonyl Compounds with Alkenes

The photochemical [2 + 2]-cycloaddition reactions of carbonyl compounds with alkenes are known as the *Paterno–Buchi reactions* This reaction was first reported by the group of E. Paterno and G. Buchi on the reaction of benzaldehyde with 2-

methyl-2-butene to give 3,4,4-trimethyl-2-phenyloxetane **106** [61]. Benzophenone reacts with isobutene and *E*/*Z*-2-butene to give isomeric oxetanes **107–110** in a regioselective manner [62, 63].

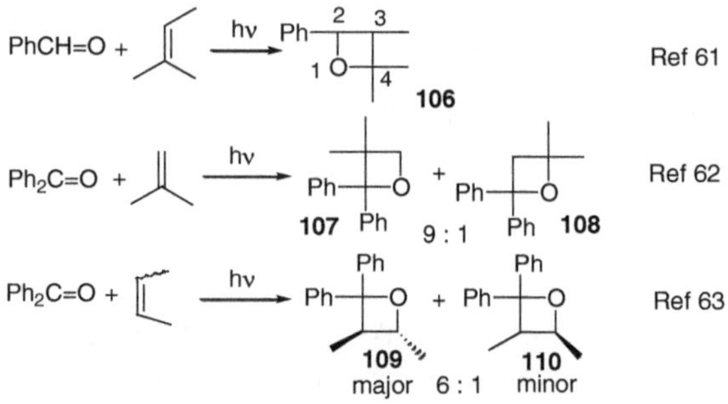

The study of the mechanism of the reactions indicated that for aromatic carbonyl compounds, the reaction occurs through a triplet excited state of the carbonyl compound, whereas for aliphatic carbonyl compounds through both singlet and triplet excited states of the carbonyl compound. The reaction is stereospecific for aliphatic carbonyl compounds and gives *syn* adduct. For cyclic alkenes, kinetically controlled *endo*-isomer is the major product. The regioselectivity of this cycloaddition reaction depends on the stability and steric interactions of the intermediate diradical. In the reaction of benzophenone with isobutene, the major product is derived from the stable diradical.

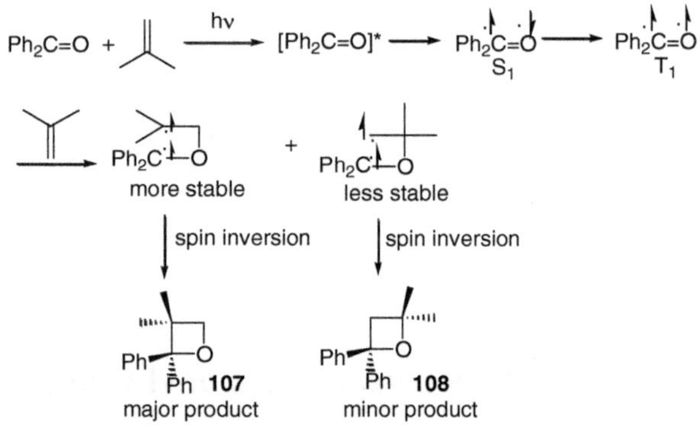

The lifetime of S_1 is too short and so oxetane formation is much faster than C–C bond rotation. The formation of oxetane can be explained from FMO approach. The frontier orbital interactions between half occupied n orbital of carbonyl oxygen atom (LUMO) with the π orbital of electron rich alkene (HOMO) take place to form a C,C-diradical.

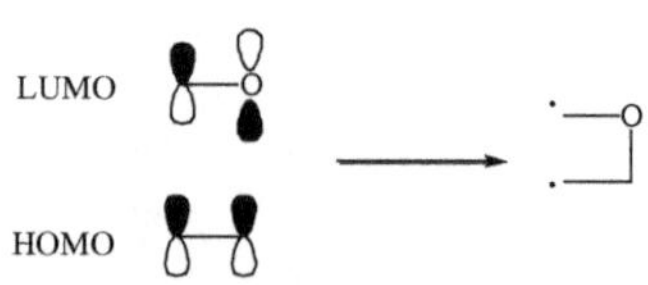

The reaction of benzophenone with dihydrofuran gives regioselective and almost single product **111** [64].

The overall yield of the reaction is 98 % as a >98:2 regioisomeric mixture. The major isomer is 88:12 *endo*/*exo*-mixture. It can be explained from consideration of the stability of plausible triplet diradicals and the ISC of the conformers in the formation of the product. The *endo*-conformer shows favorable spin–orbit interactions for formation of a sigma bond in the product, whereas *exo*-conformer faces steric restriction for this orbital interaction.

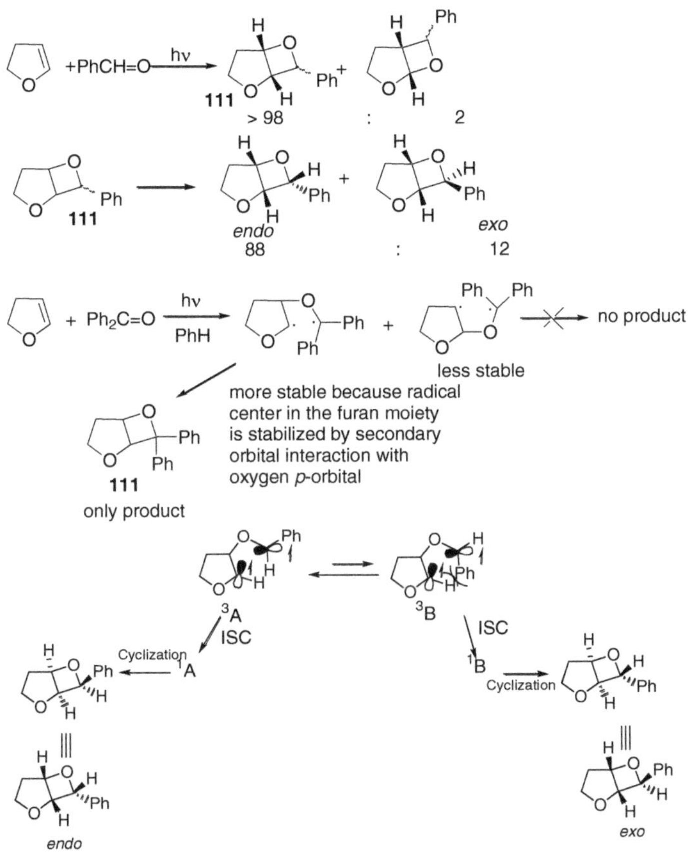

The reaction of furan with benzaldehyde gives unusual *exo*-product **112** in high yield by reversal of regioselectivity and stereoselectivity [65]. The triplet diradical **113** is more stable than **114** by 16.5 kcal/mol due to allylic stability. The *exo*-conformer of triplet diradical **113** undergoes favorable ISC to give singlet diradical of enlarged lifetime due to secondary orbital interactions and gives product **112** of high diastereoselectivity, *exo/endo*, 98:2.

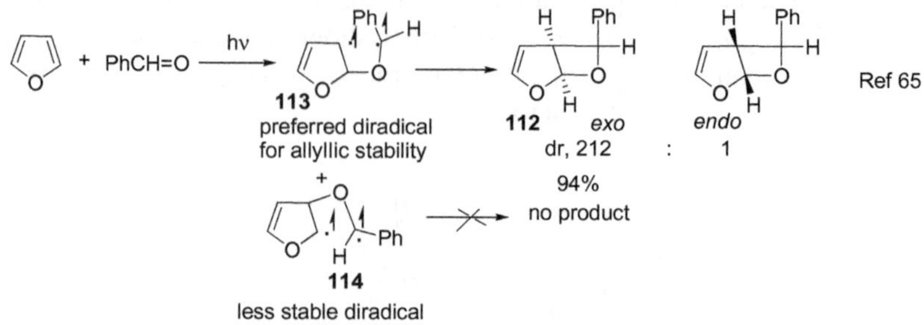

The *exo*-product **112** results from the diradical **113** via intersystem crossing (ISC) from triplet to singlet conformer (^{1}C) and ring closure as follows:

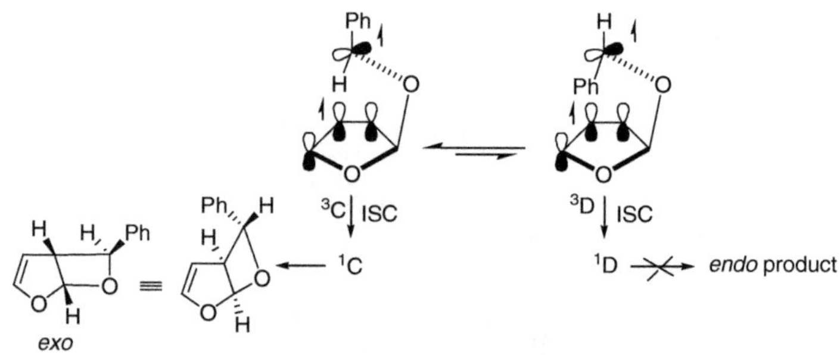

The reaction of dihydrofuran with *β*-naphthaldehyde gives high *exo*-selective product. Possibly the singlet excited state of the carbonyl compound was responsible for such *exo*-selectivity [66].

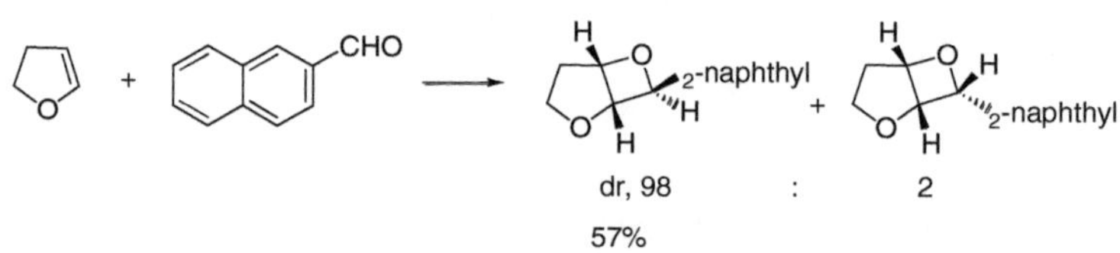

Irradiation of enol ether of dihydropyran **115** in the presence of benzaldehyde gives major diastereoselective *exo*-product **116** due to steric and stereoelectronic factors [67] and kinetically controlled *endo*-product **117** as minor product.

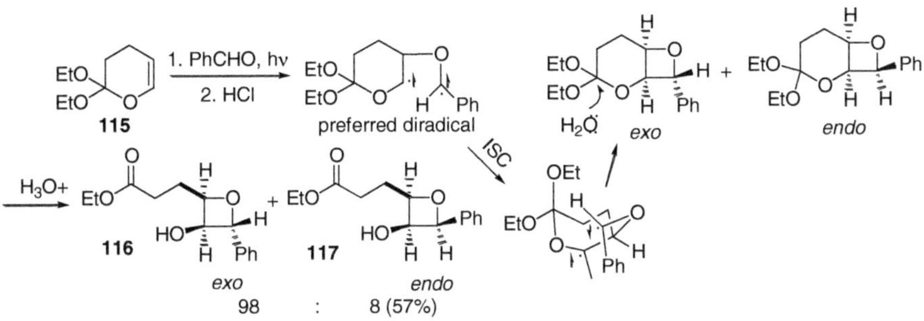

Homobenzvalene **118** on irradiation in the presence of ethyl phenylglyoxalate gives regio- and stereoselective *endo*-product **119** due to absence of steric interaction.

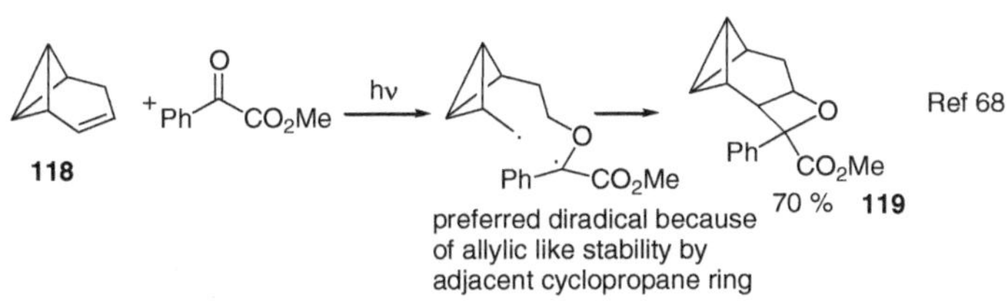

High *cis*-diastereoselectivity was observed in the photoaddition reaction of chiral allyl alcohol **120** with propionaldehyde **121** [69]. Possibly the hydrogen bonding between singlet and triplet excited states of propionaldehyde and the substrate in the exciplex favored the formation of the *threo*-isomer **122**, whereas 1,3-allylic strain prevents the formation of the *erythro*-isomer **123**. The *cis*-diastereoselectivity can be explained by consideration of its optimal conformation where steric interactions are minimized. The effect of the hydrogen bonding in the stereoselectivity of the product was rationalized on the fact that when the reaction was carried out in MeOH, such diastereoselectivity was dropped. The preferred conformation for *cis*-*threo*-diastereoselectivity is controlled by hydrogen bonding-induced gauche steric interactions with chiral methyl and strong spin–orbit coupling in the triplet diradical intermediate for facile ISC to singlet state [69].

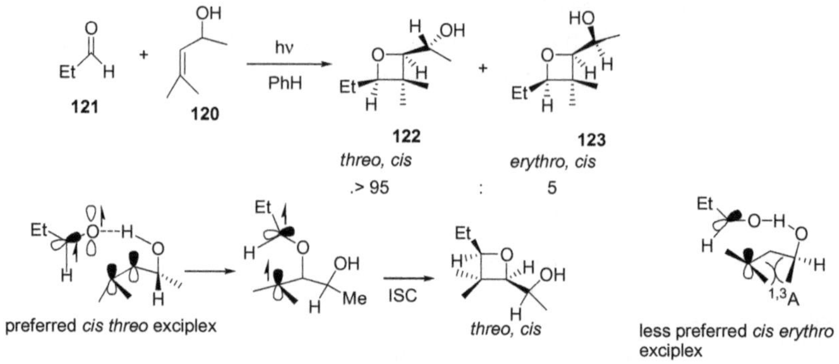

Cis–trans-selectivity of *threo*-isomer is controlled by spin–orbit coupling interaction of the diradical.

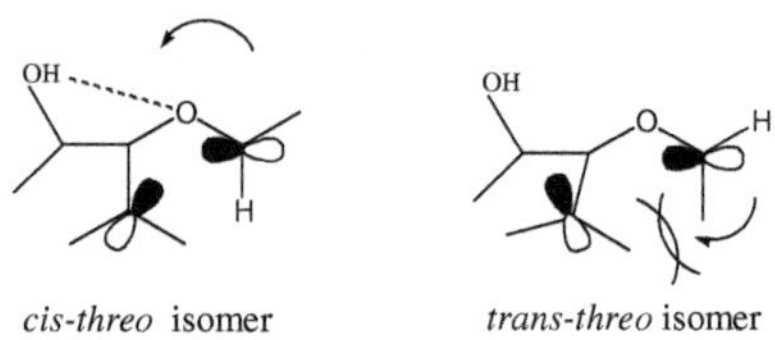

Similar to furan, photocycloaddition of 1,3-cyclohexadiene **124** to propanal gives exclusively *exo*-oxetane **125** from the reaction of singlet excited propanal to a ground-state 1,3-cyclohexadiene [70].

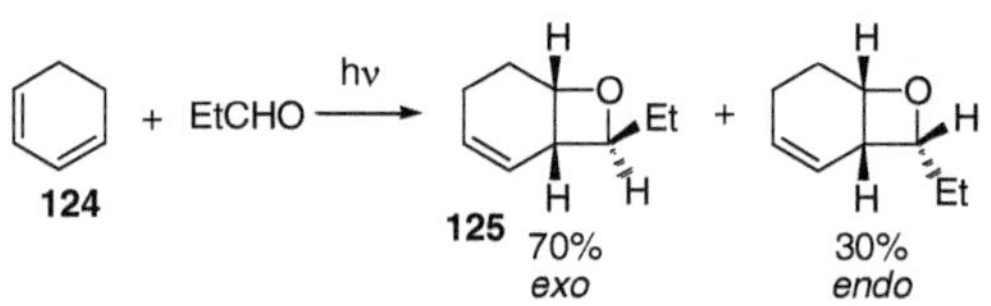

In contrast to furan, the exo-diastereoselectivity of the product oxetane in the photocycloaddition of spiro [4.2]-heptadiene **126** to benzaldehyde was reduced substantially to an *exo/endo*-ratio of 3.5:1 by the spiro-cyclopropane ring. The opposite exclusive endo-diastereoselectivity was found in the reaction with methyl ester of phenyl pyruvate **127**. Possibly *endo*-orientation of the large carbomethoxy group reduces the population of the *exo*-conformer and only the *endo*-conformer undergoes the ISC process and forms the carbon–carbon bond in the latter case [71].

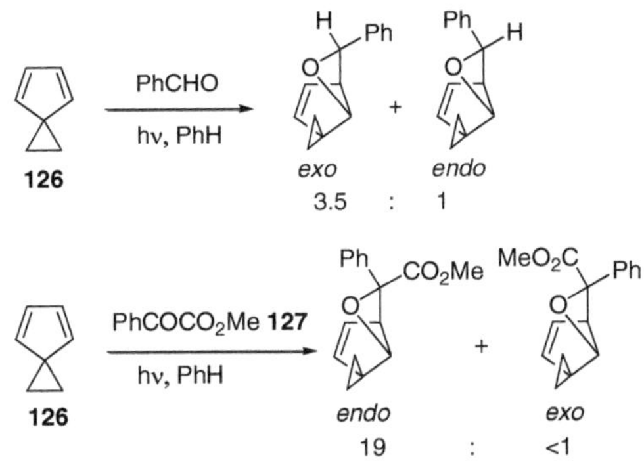

The reaction of non-symmetrical allene with aliphatic aldehyde gives little regioselectivity. For example, 3-methyl-1,2-butadiene (dimethyl allene) **128** with propanal **121** gives **129** and **130** in the ratio of 2:1 [72].

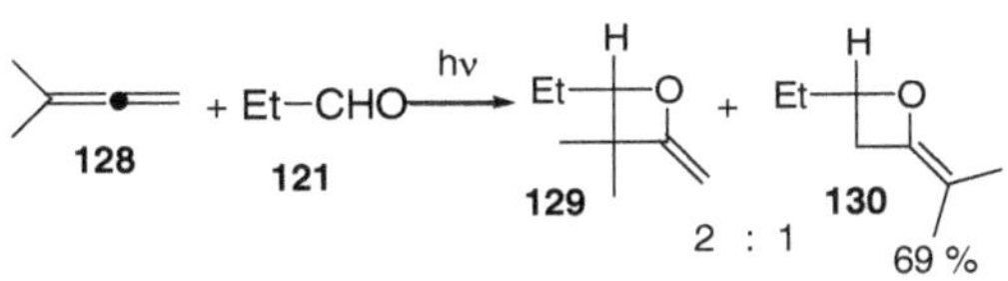

Possibly both steric and electronic factors play the key roles in the stability of intermediate diradical.

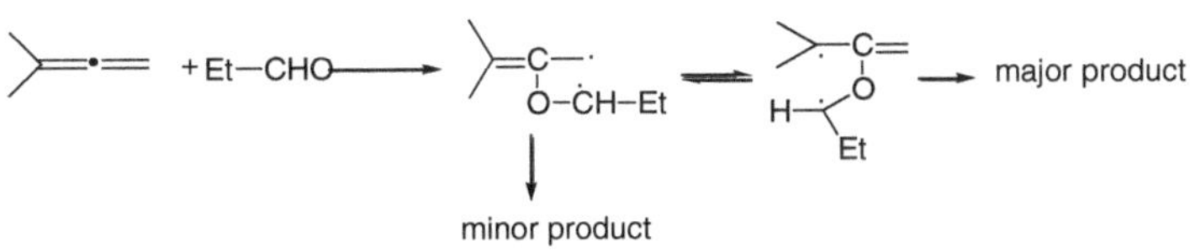

The reaction of 1-ethoxy-allene **131** with 3-methylbutanal **132** gives regioselective product **133** as a 2:1 mixture of diastereoisomers [72].

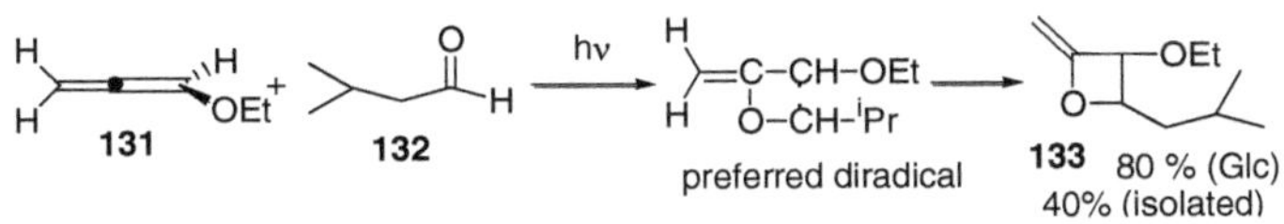

Some other typical examples [71, 73–78] of the Paterno–Büchi reaction are:

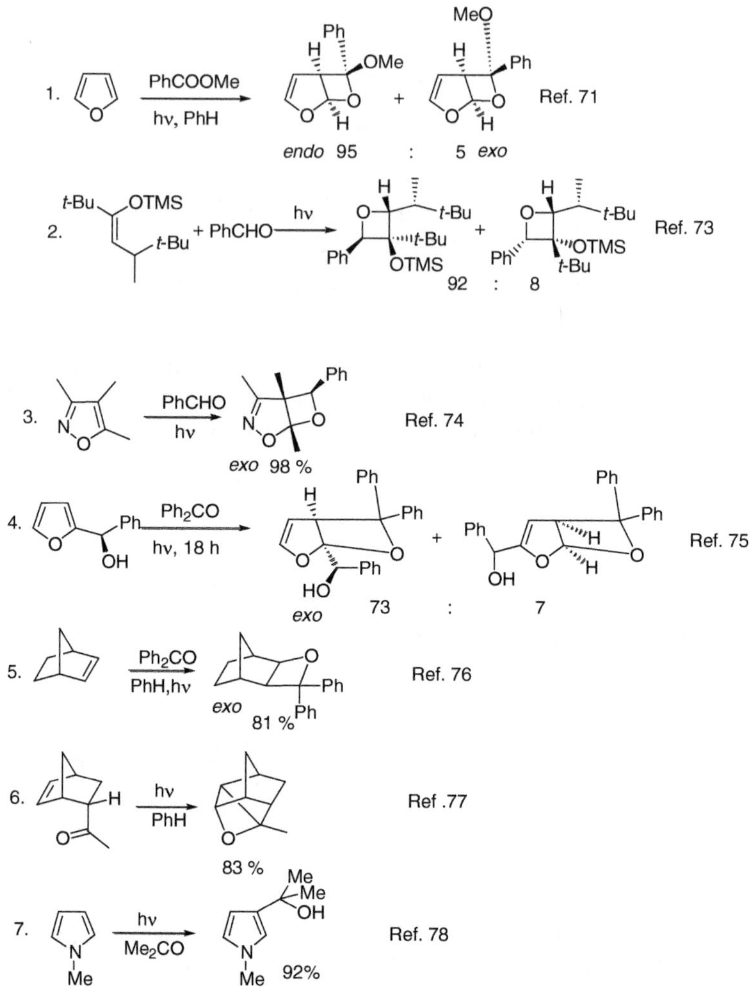

8.5.1 Limitations

The Paterno–Büchi reaction of carbonyl compounds to alkenes fails when the energy difference between the triplet and ground states of carbonyl compound is greater than that between the corresponding states of the alkene. In such case, the excited triplet state of carbonyl compound transfers its excess energy to the alkene and returns to its ground state. The generated triplet excited state of the alkene undergoes dimerization. For example, the irradiation of acetophenone (E_T = 74 kcal/mol) in presence of norbornene **134** gives mainly norbornene dimers **135** and **136** because the energy difference between the triplet and ground states of acetophenone is greater than that of the corresponding states of norbornene. Acetophenone serves here as photosensitizer, whereas the reaction of norbornene with benzophenone (E_T = 69 kcal/mol) gives cycloaddition product **137** [79].

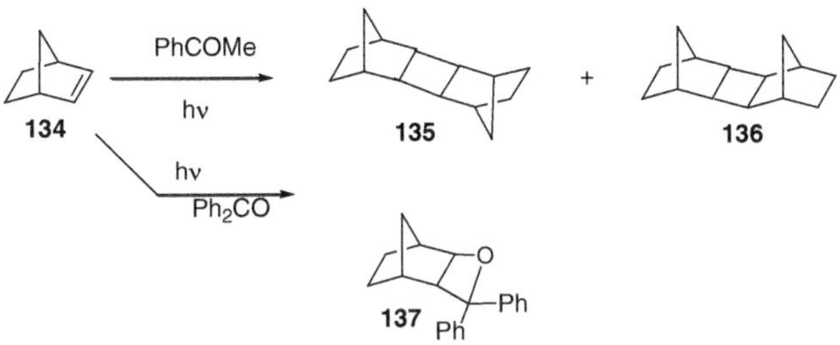

Photo-induced addition reaction of aromatic carbonyl compounds with alkynes gives unstable oxetanes. For example, the reaction of benzaldehyde with 2-butyne **138** gives α,β-unsaturated carbonyl compound **139** by cleavage of intermediate oxetane [80].

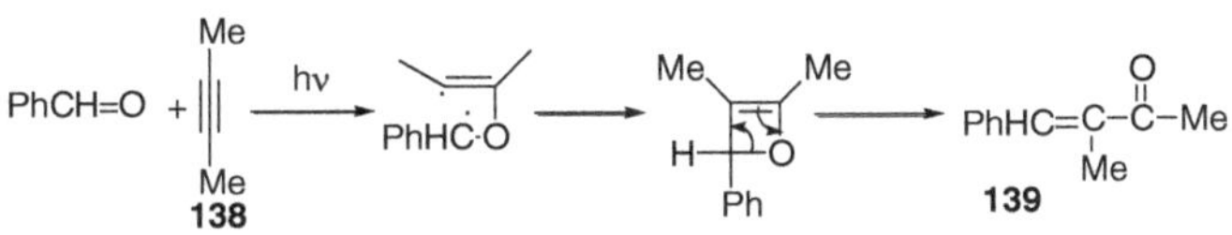

Benzaldehyde reacts with 1-hexyne **140** to give **141–143** without cycloaddition product [81].

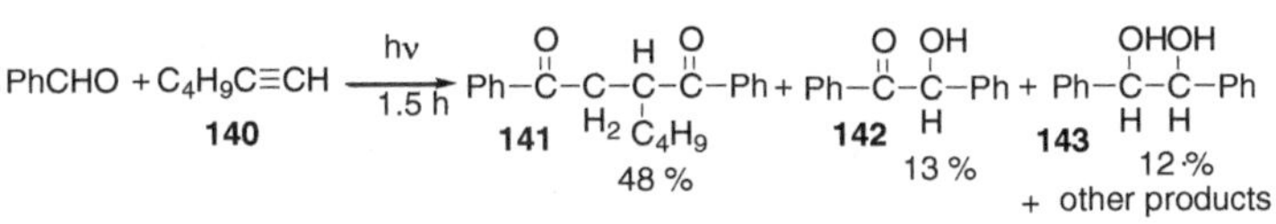

Possibly this reaction of benzaldehyde takes place by radical pair formation and hydrogen abstraction mechanism to give major product **141**.

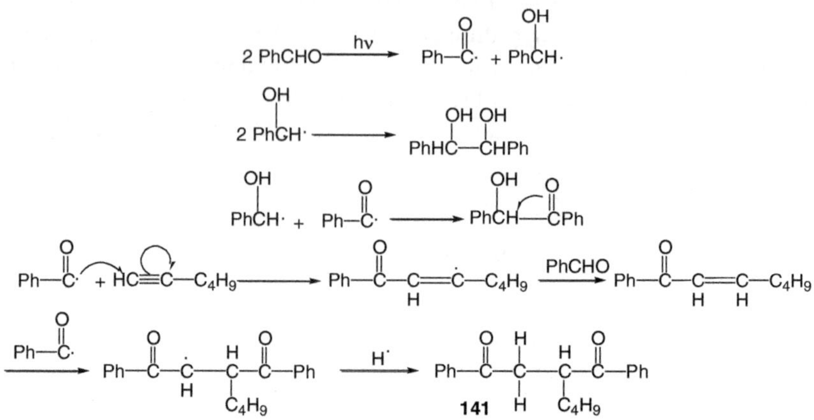

The photosensitized reaction of benzaldehyde with *p*-benzoquinone **144** in supercritical CO_2 does not give oxetane. It gives 2-benzoyl-1,4-hydroquinone **145** in a radical–radical coupling, followed by enolization through triplet excited state of the quinone [82]. This method is effective as environmentally benign method for synthesis of 2-acyl-1,4-hydroquinone without using benzene or other hazardous solvent.

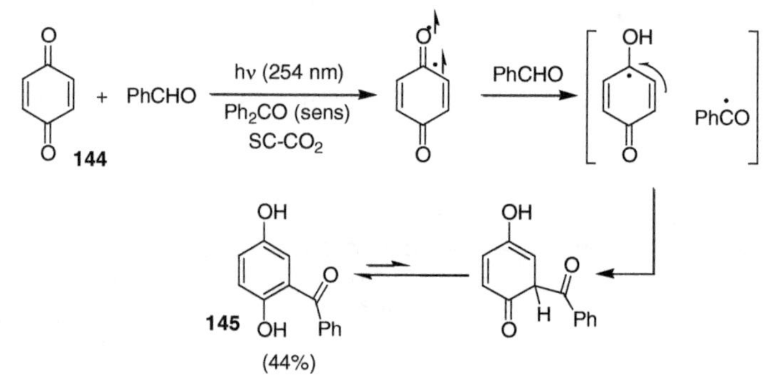

8.6 Problems

8.6.1. Predict the structure including stereochemistry of the expected product(s) for the following reactions. For the reactions, where more than one product is formed, indicate the major and minor products with justification.

(a) 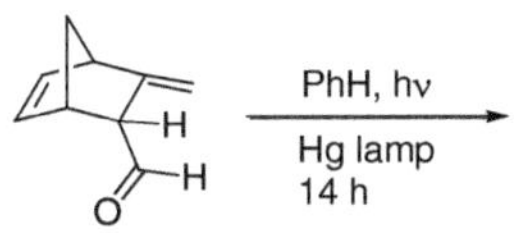

(b) 

(c) 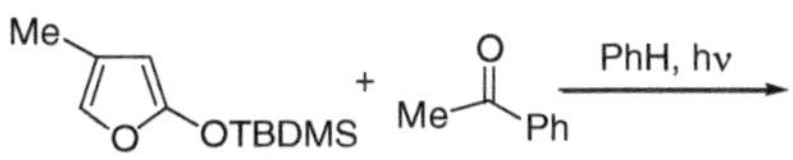

(d) 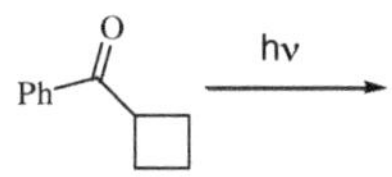

(e) 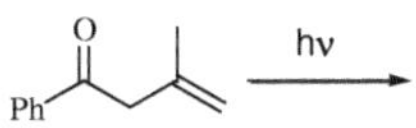

(f) 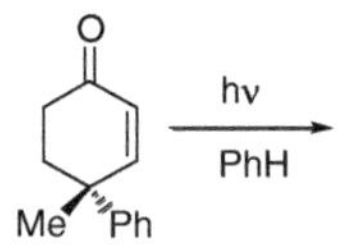

(g) 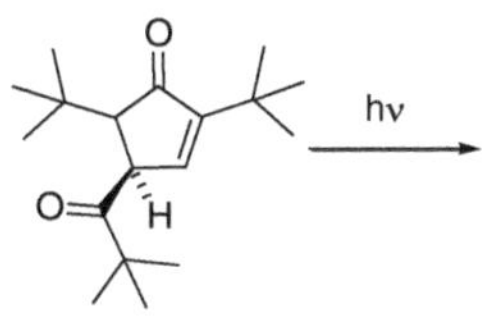

(h) 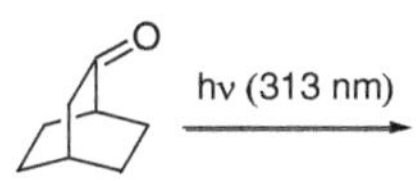

(i) 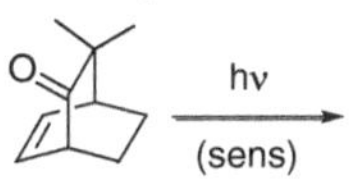

(j) 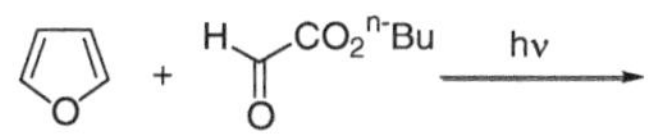

(k) 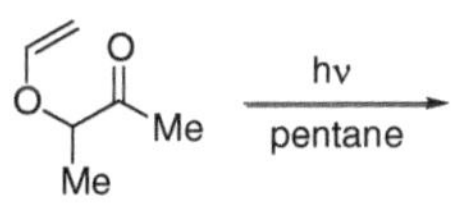

(l) 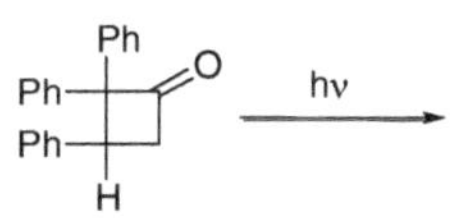

(m) 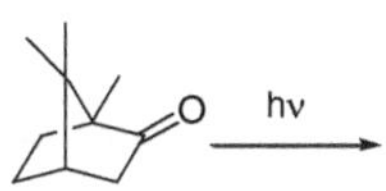

(n) 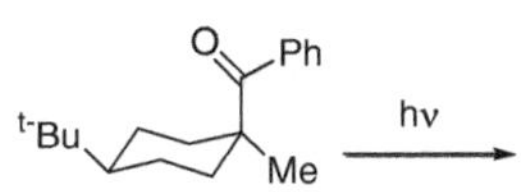

8.6.2. Provide a mechanistic rationalization for each of the following reactions:

(a) 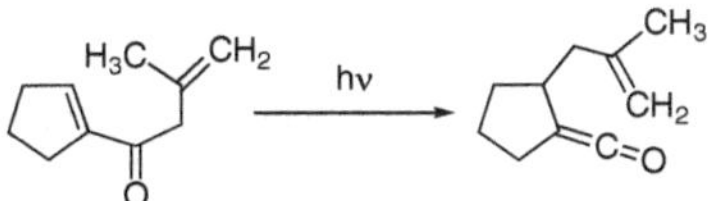

(b) 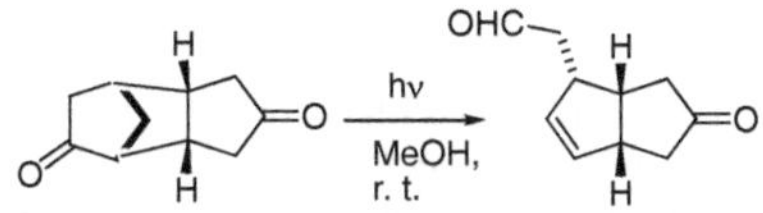

(c) 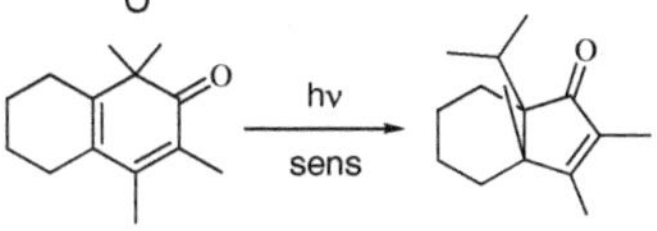

(d) 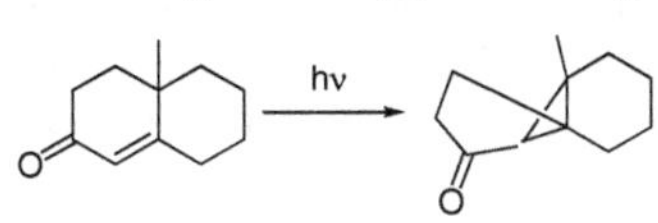

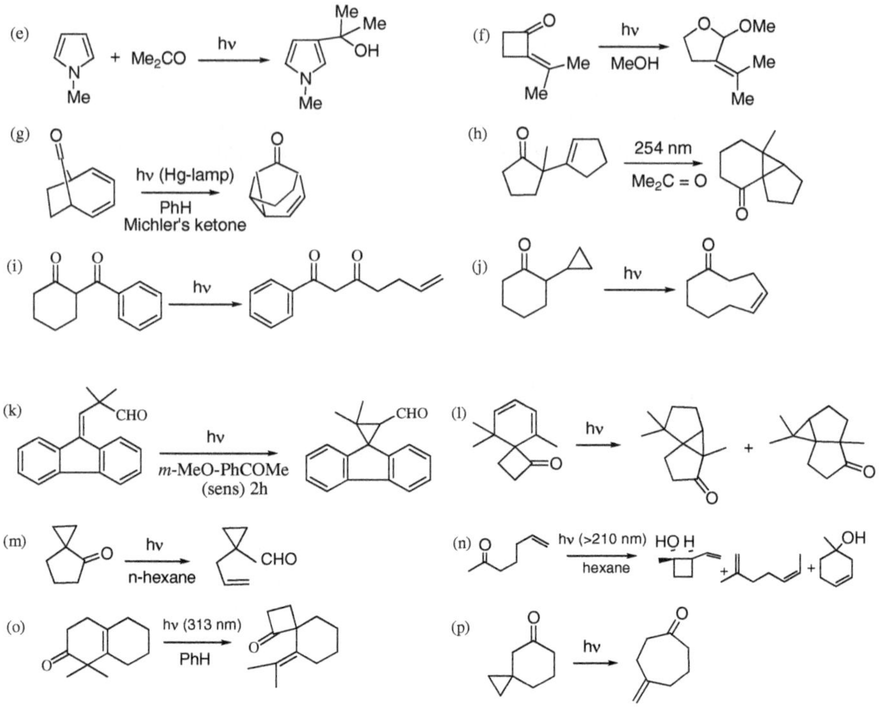

8.7 Further Reading

1. Zimmerman HE (1982) Topic in photochemistry, Top Curr Chem 100:45
2. Zimmerman HE, Armesto D (1996) Synthetic aspects of the di-π-methane rearrangement, Chem Rev 96:3065
3. Hixon SS, Mariano PS, Zimmerman HE (1973) The di-π-methane and oxa di-π-methane rearrangements, Chem Rev 73:531

References

1. Brand JCD, Williamson DG (1963) Adv Phys Org Chem 1: 365; Freeman DE, Klemperer W (1966) J Chem Phys 45: 52
2. Bell JA, Linschitz H (1963) J Am Chem Soc 85: 528; Turro NJ (1965) Molecular photochemistry. WA Benjamin, New York pp 143–144
3. Sammes PG (1976) Tetrahedron 32:405

4. Wagner PJ, Subrahmayam D, Park BS (1991) J Am Chem Soc 113: 709; Norman ROC, Coxon JM (1993) Principles of organic synthesis. ELBS, Chapman & Hall, Oxford, p 500
5. Norrish Bamford (1937) Nature 140:195
6. Quinkert G, Opitz K, Wiersdorff WW, Weinlich J (1963) Tetraheron Lett 1863
7. Yang NC, Feit ED (1968) J Am Chem Soc 90:504
8. Barton DHR, Charpiot B, Ingold KU, Johnston LJ, Motherwell WB, Scaiano JC, Stanforth S (1985) J Am Chem Soc 107:3607
9. Badcock CC, Perona MJ, Pritchand GO, Rickborn B (1969) J Am Chem Soc 91:543
10. Schenck GO, Schaller F (1965) Chem Ber 98: 2056; Agosta WC, Schreiber WL (1971) J Am Chem Soc 93: 3947
11. Yates P, Loutfy RO (1975) Acc Chem Res 8:209
12. Wagner PJ (1980) In: Mayo Pde (ed) Rearrangements in ground and excited states, vol. 3. Academic Press, New York; Wagner PJ, Klan P (2004) In: Horspod W, Lenci F (eds) CRC Handbook of photochemistry and photobiology. CRC Press, Boca Raton, Florida
13. Wagner PJ (1971) Acc Chem Res 4: 168; Wessig P, Muhling O (2001) Angew Chem Int Ed Engl 40: 1064
14. Yang NC, Elliott SP, Kim B (1969) J Am Chem Soc 91: 7551; Yang NC, Trap DM (1966) Tetrahedron Lett 3671
15. Wagner PJ, Kelso PA (1969) Tetrahedron Lett 4151; Coldwell RA, Fink P (1969) Tetrahedron Lett 2987
16. Matsui RT, Komatsu A, Moroe T (1967) Bull Chem Soc Jpn 40:2204
17. Mori D, Matsui K, Nozaki H (1970) Tetrahedron Lett 1175
18. Wagner PJ, Zand A, Park BS (1996) J Am Chem Soc 118:12856
19. Zepp RG, Wagner PJ (1970) J Am Chem Soc 92: 7466; Turro NJ, Lee TJ (1970) ibid 92: 7467
20. Cormier RA, Schreiber WL, Agosta WC (1973) J Am Chem Soc 95: 4873; Cormier RA, Agosta WC (1974) J Am Chem Soc 96: 618
21. Smith AB, Agosta WC (1972) J Org Chem 37:1259
22. Weigl W, Wagner PJ (1996) J Am Chem Soc 118:12858
23. Barnard M, Yang NC (1958) Proc Chem Soc 302; Haswegawa AT (1997) Chem Abst 127: 121403
24. Hasegawa T, Nishimura M, Yoshioka M (1990) J Phys Org Chem 3:230
25. Hasegawa T, Nishimura M, Kodama Y, Yoshioka M (1990) Bull Chem Soc Jpn 63:935
26. Hornback JM, Poundstone ML, Vadlamani B, Graham SM, Gabay J, Patton ST (1988) J Org Chem 53:5597
27. Ricard R, Sauvage P, Wan CSK, Weedon AC, Wong DF (1986) J Org Chem 51:62
28. Sato H, Furutachi N, Nakanishi K (1972) J Am Chem Soc 94: 2150; Paquette LA, Eigember RF, Cox O (1968) J Am Chem Soc 90: 5153
29. Engel PS, Schnexnayder MA (1975) J Am Chem Soc 97:145
30. Wolff S, Schreiber WL, Smith AB III, Agosta WC (1972) J Am Chem Soc 94:7797
31. Schuster DI, Yang JM, Woning J, Rhodes TA, Jensen AW (1995) Can J Chem 73:2004
32. Andrew D, Hastings DJ, Oldroyed DL., Rudolph A, Weedon AC, Wong DF, Zhang B (1992) Pure Appl Chem 64: 1327; Andrew D, Hastings DJ, Weedon AC (1995) J Am Chem Soc 117: 5647
33. Eaton PE (1962) J Am Chem Soc 84:2344
34. Bauslaugh PG (1970) Synthesis 287
35. Schuster DI (1980) In: Pde Mayo (ed) Rearrangements in ground and excited states, vol 3. Academic Press, New York
36. Schuster DI, Rao JM (1981) J Org Chem 46: 1515; Schuster DI, Brown RH, Resnick BM (1978) J Am Chem Soc 100: 4504

37. Bellus M, Kearns DR, Schaffner K (1969) Helv Chim Acta 52:971
38. Chapman Ol, Sieja JB, Welstead WJ, Jr (1966) J Am Chem Soc 88: 161
39. Zimmerman HE, Kneese R, Nasielski J, Swenton JS (1966) J Am Chem Soc 88: 4895; Zimmerman HE (1974) Tetrahedron 30:1617
40. Zimmerman HE, Rieke RD, Scheffer JR (1967) J Am Chem Soc 89:2033
41. Dauben WG, Kellog MS, Seeman JI, Spitzer WA (1970) J Am Chem Soc 92: 1786; Zimmerman HE, Armesto D (1996) Chem Rev 96: 3065
42. Williams JR, Ziffer H (1967) Chem Commun 469
43. Armesto D, Ortiz MJ, Romano S (1995) Tetrahedron Lett 36:965
44. Ipaktschi J (1970) Tetrahedron Lett 3179; (1972) Chem Ber 105:1840
45. Reimann B, Sadler DE, Schaffner K (1986) J Am Chem Soc 108: 5527; Su MD (1996) J Org Chem 61: 3080
46. Givens RS, Oettle WP, Coffin RL, Carison RG (1971) J Am Chem Soc 93: 3957; Givens RS, Oettle WF (1971) J Am Chem Soc 93: 3963
47. Van der Veen RH, Kruk C, Cerfontain H (1983) J Am Chem Soc 105: 4367; Zimmerman HF, Kamath A (1988) J Am Chem Soc 110: 900
48. Su MD (1996) J Org Chem 61: 3080; Oren J, Fuchs B (1986) J Am Chem Soc 108: 4881
49. Viskin R, Oven J, Fuchs B (1985) Tetrahedron Lett 26:2365
50. Knott PA (1970) Tetrahedron Lett 1829; Knott PA, Mellor JM (1972) J Chem Soc, Perkin Trans 1, 1030
51. Bagiolini E, Schaffner K, Jeger O (1969) Chem Commun 1103
52. Ipaktschi J (1969) Tetrahedron Lett 2153; (1972) Chem Ber 105: 1840
53. Armesto D, Ortiz MJ, Romano S, Agarrabeitia AR, Gallego MG, Ramos A (1996) J Org Chem 61:1459
54. Paquette LA, Eizember RF, Cox O (1968) J Am Chem Soc 90:5153
55. Demuth M, Raghavon PR, Carter C, Nakano K, Schaffner K (1980) Helv Chim Acta 63:2434
56. Zimmerman HE, Schuster DI (1961) J Am Chem Soc 83:4486
57. Zimmerman HE, Crumrine DS (1968) J Am Chem Soc 90:5612
58. Zimmermann HE, Solomon RD (1986) J Am Chem Soc 108:6276
59. Dauben WG, Koch K, Smith SL, Chapman OL (1963) J Am Chem Soc 85:2616
60. Griffiths J, Hart H (1968) J Am Chem Soc 90:5298
61. Paterno E, Chieffi G (1909) Gazz Chim Ital 39: 341; Buchi G, Inman CG, Lipinsky ES (1954) J Am Chem Soc 76: 4327
62. Arnold DR, Hinman RL, Glick AH (1964) Tetrahedron Lett 1425
63. Arnold DR (1968) Adv Photochem 6:301
64. Griesbech AG, Stadtmuller S (1990) Chem Ber 123:357
65. Griesbeck AG, Mauder H, Stadtmuller S (1994) Acc Chem Res 27: 70; D'Auria M, Emanuele L, Racioppi R (2006) Int J Photoenergy 3: 244
66. Griesbeck AG, Mauder H, Peters K, Peters EM (1991) Chem Ber 124:407
67. Park SK, Lee SJ, Baek K, Yu CM (1998) Bull Korean Chem Soc 19:35
68. Kossanyi J, Jost P, Furth B, Daccord G, Chaquin P (1980) J Chem Res Synop 368
69. Griesbeck AG, Bondock S (2001) J Am Chem Soc 123:6191
70. Shima K, Kubota T, Sakurai H (1976) Bull Chem Soc Jpn 49:2567
71. Griesbeck AG, Buhr S, Fiege M. Schmickler H, Lex J (1998) J Org Chem 63: 3847
72. Howell AR, Fan R, Truong A (1996) Tetrahedron Lett 37:8651
73. Bach T, Jodicke K, Kather K, Hecht K (1995) Angew Chem Int Ed Engl 34: 2271; Bach T, Jodicke K, Kather K, Frohlich R (1997) J Am Chem Soc 119: 2437
74. Griesbeck AG, Franke M, Neudorfl J, Kotaka H (2011) Beilstein J Org Chem 7:127
75. D'Auria M, Racioppi R, Romaniello G (2000) Eur J Org Chem 3265
76. Arnold DR, Glick AH, Abraitys VY (1971) Org Photochem Synth 1:51

77. Sauers RR, Schinski W, Sickles B (1971) Org Photochem Synth 1:76
78. Jones G, Gilow HM, Low J (1979) J Org Chem 44:2949
79. Arnold DR, Treeker DJ, Whipple EB (1965) J Am Chem Soc 87:2596
80. Friedrich LE, Bower JD (1973) J Am Chem Soc 95:669
81. Bradshaw JS, Knudsen RD, Parish WW (1975) J Org Chem 40:529
82. Pacut R, Grimm ML, Kraus GA, Tanko JM (2001) Tetrahedron Lett 42:1415

Chapter 9
Photochemistry of Aromatic Compounds

9.1 Introduction

Aromatic compounds on absorption of light undergo photoexcitation to produce excited singlet and triplet states. Each of these excited states may undergo different chemical reactions in proceeding back to the ground state. The triplet excited state of relatively long lifetime, frequently undergoes photochemical reactions such as isomerization, cycloaddition, and di-π-methane rearrangement. These reactions are useful for synthesis of various types of strained compounds for industrial applications.

9.2 Photoisomerization Reactions of Aromatic Compounds

Liquid benzene on photoirradiation gives a very small amount of benzvalene (tricyclo[3.1.0.0^{2,6}]-hex-3-ene) through formation of a diradical as an intermediate [1].

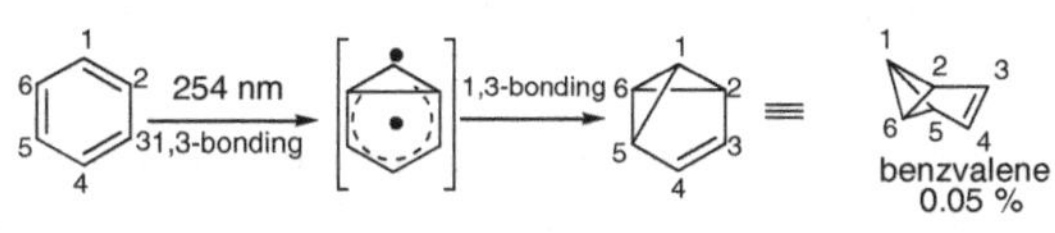

The presence of the diradical intermediate was supported by the fact that irradiation of benzene in acidic hydroxylic solvents gives addition products with solvent [2].

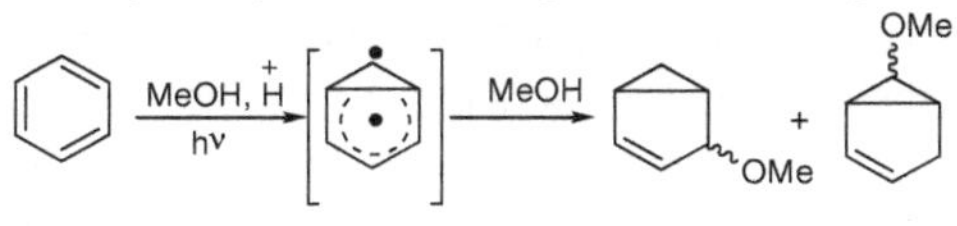

B. Dinda, *Essentials of Pericyclic and Photochemical Reactions*,
Lecture Notes in Chemistry 93, DOI 10.1007/978-3-319-45934-9_9

1,3,5-Tri-*t*-butylbenzene undergoes photoisomerization to give various products of valence isomers [3]. Possibly bulky *tert*-butyl group in the aromatic ring induces steric interaction to facilitate this isomerization. It is difficult to predict the exact mechanism of formation of all these photoproducts.

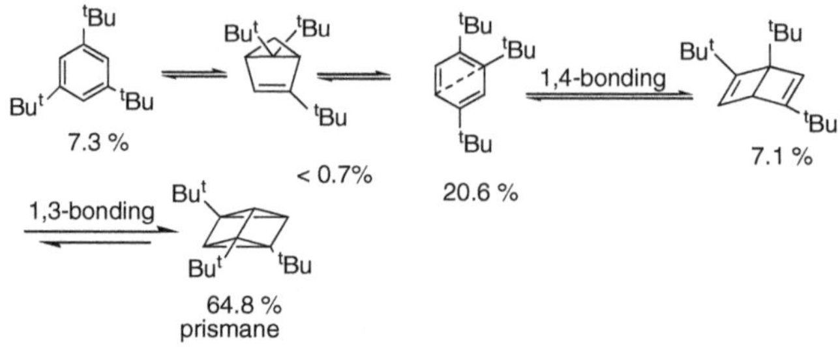

Similarly 1,2,4-tri-*t*-butylbenzene gives a Dewar benzene.

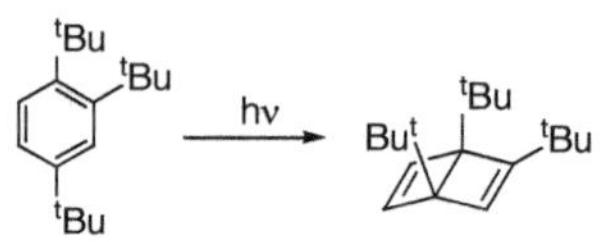

9,10-Dimethoxy-octamethylanthracene on photoirradiation isomerizes to 9,10-Dewar isomer [4].

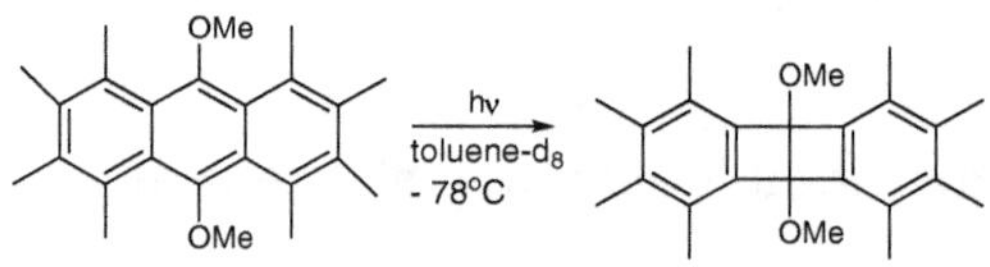

Ref. 4

9.3 Photocycloaddition Reactions of Aromatic Compounds with Unsaturated Compounds

Photoirradiation of benzene and its derivatives with alkenes give *ortho*-, *meta*-, *para*-cycloaddition products. In most cases, either *meta*- or *ortho*-adducts are obtained as major products [5]. Bryce-Smith and Gilbert suggested a prefulvene type diradical intermediate mechanism (Path A) for the *meta*-adduct [6], whereas Morrison and Srinivasan groups [7, 8] proposed the exciplex mechanism (Path B) for the formation of these adducts (Scheme 9.1). The exciplex intermediates undergo photo-induced electron transfer processes between donor (D) and acceptor (A) to produce radical ion pairs as intermediates, stabilized by coulombic interactions to give adducts [9].

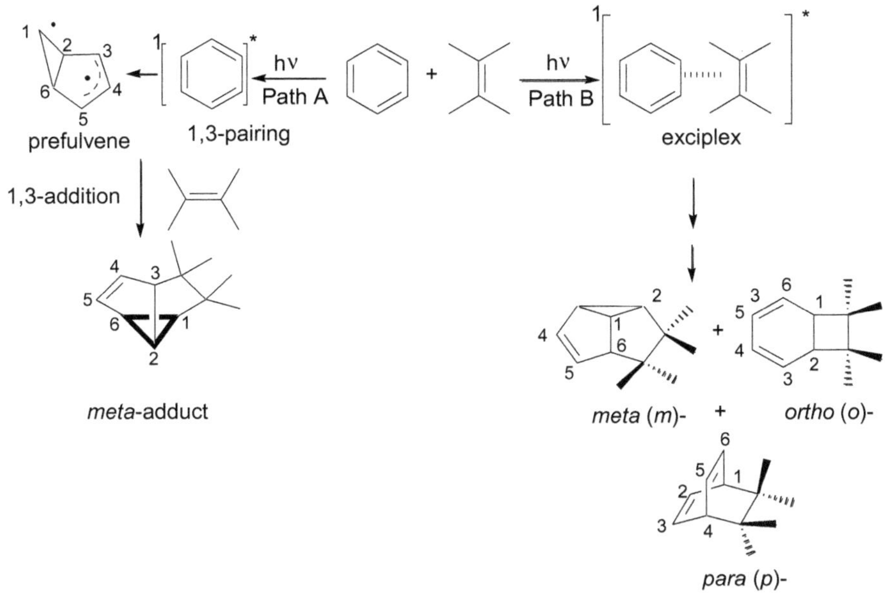

$$A \xrightarrow{h\nu} A^* \xrightarrow{D} [AD]^* \xleftarrow{A} D^* \xleftarrow{h\nu} D$$

$$[AD]^* = \{ [A^*D] \longleftrightarrow [AD^*] \longleftrightarrow A^{\bullet -}\ D^{\bullet +} \} \longrightarrow A\text{-}D$$

exciplex

Scheme 9.1 Mechanism for formation of photochemical adducts from the reaction of aromatic compounds with alkenes

The relative efficiencies of *ortho*- and *meta*-cycloadditions of ethenes to arenes depend on the stability of the exiplex and the polarity of the solvent. In general, higher stability of the exciplex favors *meta*-addition and ground-state ion pair formation between electron donor–electron acceptor favors *ortho*-addition. In *meta*-additions of benzene, *endo*-selectivity is preferred because of greater stability of *endo*-sandwich exciplex, whereas in *ortho*-additions, electron donor alkenes give *endo*-products and electron acceptor alkenes give *exo*-products preferentially. In substituted benzenes, *ortho*-additions are preferred with polar arene substituents than for alkyl substituents. The lifetime of exciplex is longer in solvent of medium polarity such as in diethyl ether and dimethoxyethane, but decreases in polar solvents due to formation of ion pair.

$$A \xrightarrow{h\nu} A^* \xrightarrow{D} [AD]^* \xleftarrow{A} D^* \xleftarrow{h\nu} D$$

$$\underset{\text{exciplex}}{[AD]^*} = \{ [A^*D] \longleftrightarrow [AD^*] \longleftrightarrow \overset{\bullet -}{A}\ \overset{\bullet +}{D} \} \longrightarrow A\text{-}D$$

The charge transfer process between donor and acceptor molecules influences the mode of the addition reaction. This charge transfer process is closely related to the free energy change ΔG of the radical ion pair formation and can be calculated from the Rehm–Weller equation [9] using the oxidation potential of the donor (D), the reduction potential of the acceptor (A), and the excitation energy of the excited species and coulombic interaction energy by the radical ion pair at the encounter distance for their interaction.

$$\Delta G = E^{Ox}_{1/2}(D) - E^{Red}_{1/2}(A) - \Delta E_{excit} + \Delta E_{Coul}$$

The value of ΔE_{Coul} depends on the dielectric constant of the solvent used.

The photoreaction of benzene with alkenes depends on the electron donor and electron acceptor ability of the alkene. Both poor electron donor and electron acceptor olefins react with benzene to give preferentially *meta*-cycloadducts, where ΔG values are greater than 1.4–1.6 eV. For example, propene, isobutene, cyclobutene, cyclopentene, and cyclohexene give *meta*-cycloadducts as major products [10]. All other olefins having strong electron donor and electron acceptor abilities mainly give *ortho*-cycloadducts. In *ortho*-cycloadduct, strong electron donor alkenes such as 2,3-dihydrofuran and 3,4-dihydropyran give *endo*-isomers as major products, whereas strong electron acceptor alkenes such as acrylonitrile give *exo*-isomers as major products. Alkenes such as 1,3-dioxoles and 1,3-dioxol-2-one (vinylene carbonate) give *meta*-adducts as major or sole products with benzene, where ΔG values are very low. Only exception is that the tetramethyl 1,3-dioxole gives *ortho*-adduct only with benzene because of destabilization of the intermediate exiplex by the steric crowding and formation of ion pair complex. 2,2-Dimethyl-1,3-dioxole gives *ortho*-adduct as major adduct with *ortho*:*meta* ratio of 3.5:1. One interesting feature of the *meta*-adducts of benzene is that cycloalkenes give predominantly the *endo*-isomers due to greater stability of *endo*-sandwich exciplexes by an electronic interaction, whereas 1,3-dioxoles give predominantly *exo*-isomers because of repulsive effects of the oxygen atoms of electron-rich *cis*-

enediol ethers with the partly negative charged arene in the *endo*-configuration of the zwitterionic species of exciplex. Vinylene carbonate, where oxygen atoms in the five-membered ring are electron deficient due to carbonyl group, enjoys an electronic attraction with the negatively charged allylic moiety of the arene to stabilize the *endo*-configuration and gives preferentially *endo*-isomer [11]. As for examples, 1,1-dimethoxyethene, 2,4-dihydrofuran, 3,4-dihydro-2*H*-pyran, tetramethyl-1,3-dioxole give only *ortho*-cycloadducts [12]. The ΔG values of these olefins are less than 1.4 eV.

Electron donor-substituted benzenes such as anisole and toluene on reaction with neutral and weak electron acceptor alkenes usually give *meta*-cycloadducts as major products along with minor *ortho*-cycloadducts [13]. If the alkene is a strong electron acceptor such as acrylonitrile or acrylate, maleic anhydride, and maleimide, the *ortho*-cycloadducts are the major products [13]. In such cases, ΔG values are negative. Photocycloaddition of 1,3-dioxoles and vinylene carbonate to anisole gives *meta*-adduct preferentially on similar grounds of selectivities as described for benzene with electron donor substituent at C-1 carbon atom of arene [14]. The solvent is an important factor in the photochemical reactions of anisole with electron-withdrawing alkenes. In aprotic solvent, it gives cycloadduct as major product, whereas in protic solvent gives substituted product as major product. For example, photoirradiation of anisole with acrylonitrile in acetonitrile solvent gives 1,2-*ortho*-cycloadduct as major product (73 %). When this reaction is carried out in methanol, *para*-substituted product was found as major product (49 %) along with a small amount of *ortho*-substituted product (10 %) [15]. Such contrasting results in methanol can be explained from the formation of ion pair complex in protic solvent and its stabilization as cation radical, which favors the substitution in *para*- and *ortho*-positions [16]. The relative ratio of the cycloadduct is markedly affected by the polarity of the solvent. It is evident in the reactions of acrylonitrile with anisole in presence of cyclohexane and acetonitrile as solvent. In cyclohexane, the ratio of 1,2- and 3,4-cycloadducts is 2:1, whereas in acetonitrile it is 20:1. Possibly, polar solvent stabilizes the ground-state ion pair or excited state complex for 1,2-*ortho*-orientation of arene and alkene [17].

Electron acceptor-substituted benzenes such as benzonitrile and α,α,α-trifluorotoluene generally give *ortho*-cycloadducts with olefins [18]. Some exceptions are also found. For example, photo addition of 1,3-dioxol-2-one to benzonitrile gives only *meta*-adduct of *endo*-configuration having electron acceptor substituent at C-2 and C-4 carbon atoms. Possibly the electron acceptor ability of benzonitrile is higher at *ortho*- and *para*-positions [19]. The following examples are illustrative:

1. 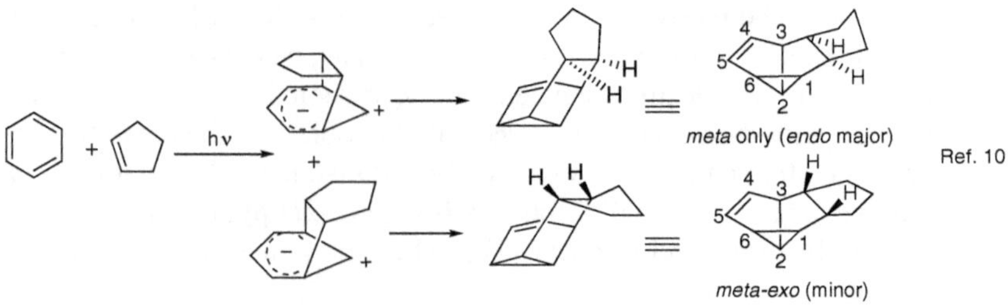
hν
meta only (endo major)
meta-exo (minor)
Ref. 10

2. 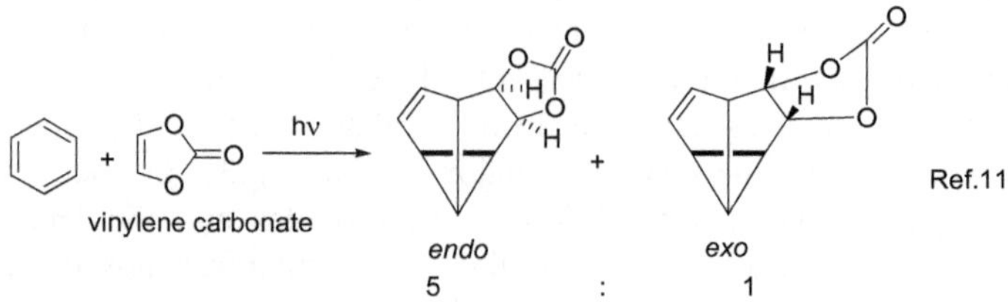
hν
vinylene carbonate
endo
exo
5 : 1
Ref.11

3. 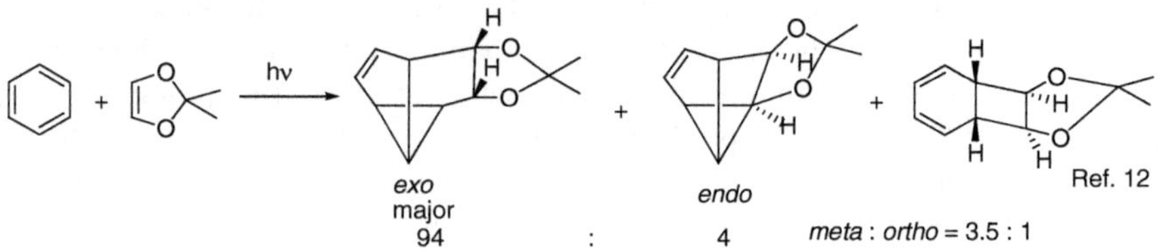
hν
exo
major
94 : 4
endo
meta : ortho = 3.5 : 1
Ref. 12

4. 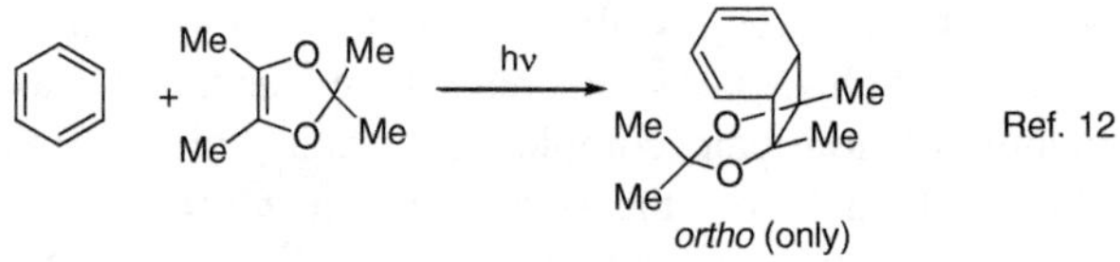
hν
ortho (only)
Ref. 12

5. 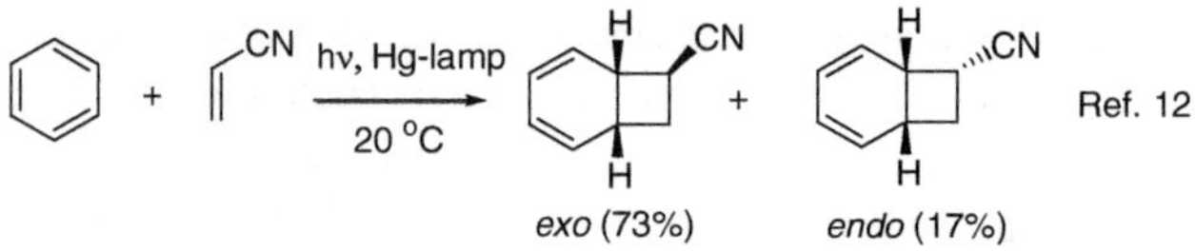
hν, Hg-lamp
20 °C
exo (73%)
endo (17%)
Ref. 12

6. 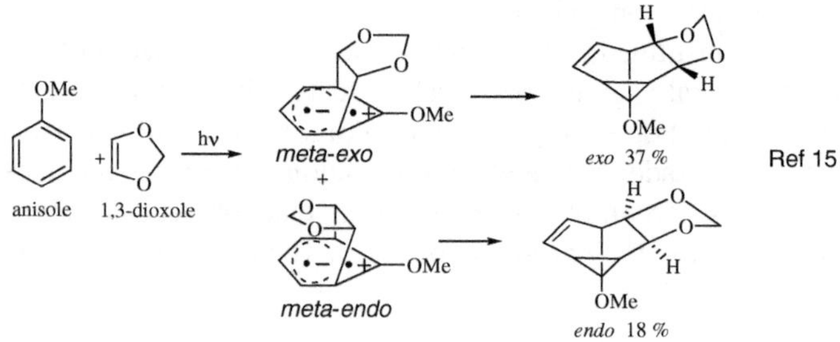
anisole
1,3-dioxole
hν
meta-exo
meta-endo
exo 37 %
endo 18 %
Ref 15

7. 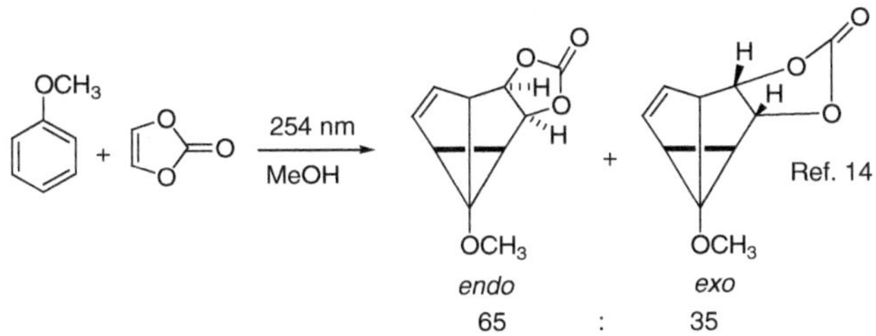
OCH3
254 nm
MeOH
endo
exo
65 : 35
Ref. 14

8. 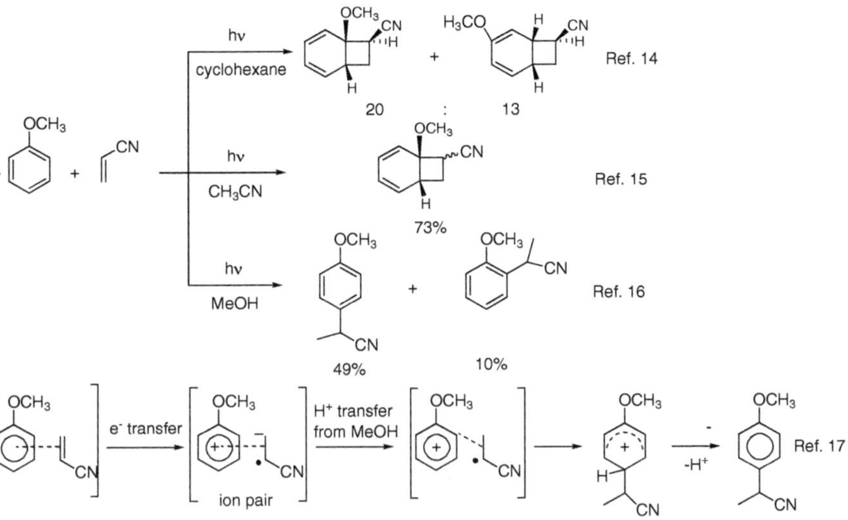
hν
cyclohexane
20 : 13
Ref. 14
hν
CH3CN
73%
Ref. 15
hν
MeOH
49%
10%
Ref. 16
e⁻ transfer
ion pair
H⁺ transfer
from MeOH
-H⁺
Ref. 17

9. 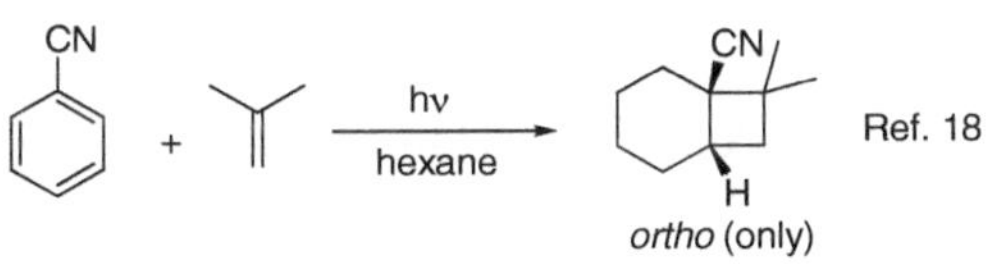
hν
hexane
ortho (only)
Ref. 18

10. 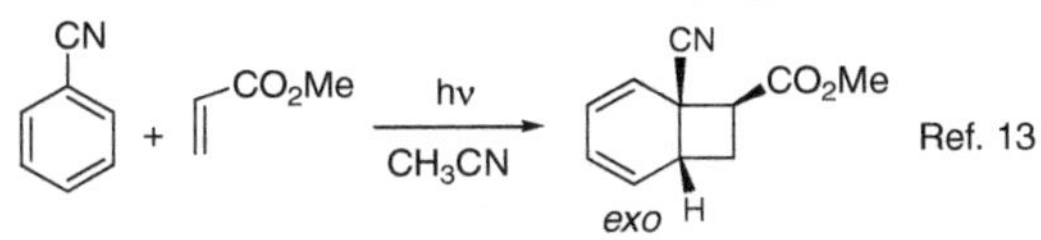
hν
CH3CN
exo
Ref. 13

11. 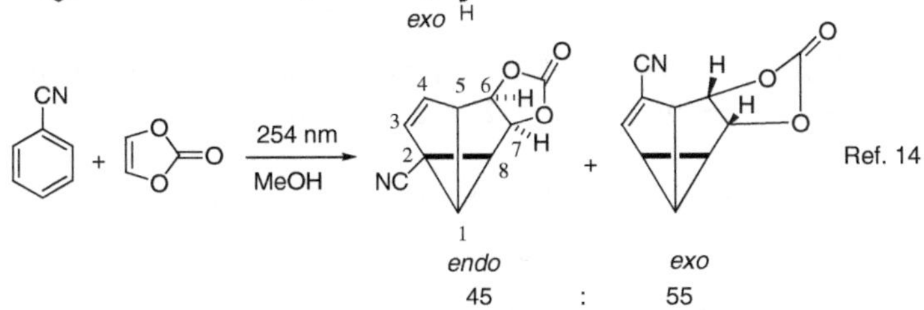
254 nm
MeOH
endo
exo
45 : 55
Ref. 14

Benzene and other aromatic hydrocarbons undergo photoaddition reactions with furan, dienes, and allenes to give *para*-like adducts as major products by allowed ($\pi^4s + \pi^4s$) cycloaddition process. For examples, benzene reacts with furan in molar ratio 1:1 gives *para*-like adduct as major product [19] and with allene gives *para*- and *meta*-adducts in a ratio of 2:1 [20].

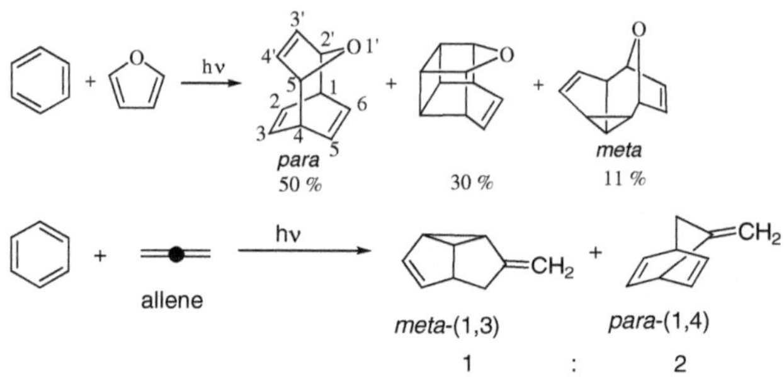

Ref. 20

Naphthalene on direct photoirradiation with 1,3-diene gives [4 + 4]-cycloaddition product, which undergoes [3,3]-sigmatropic shift to give stable product [21].

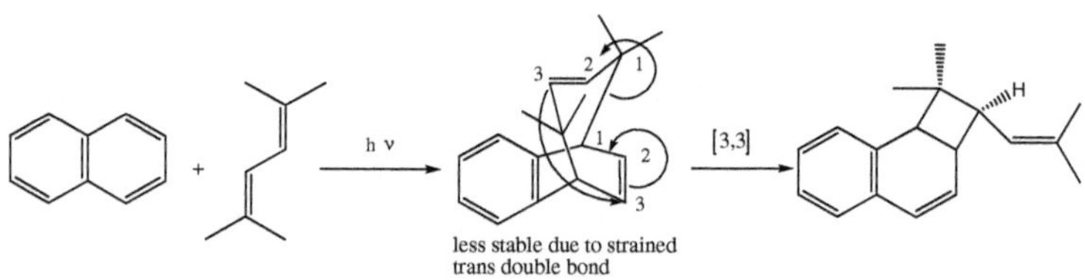

1-Cyanonaphthalene undergoes photoaddition with furan to give [4 +4]-cycloadduct [22].

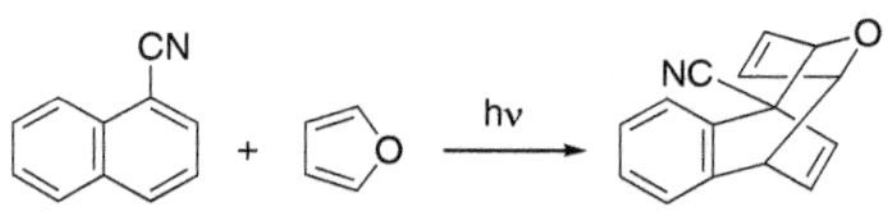

Ref. 22

The presence of an excess amount of furan in the reaction with benzene resulted in *ortho*-adduct as minor product by [2 + 2]-cycloaddition process [19].

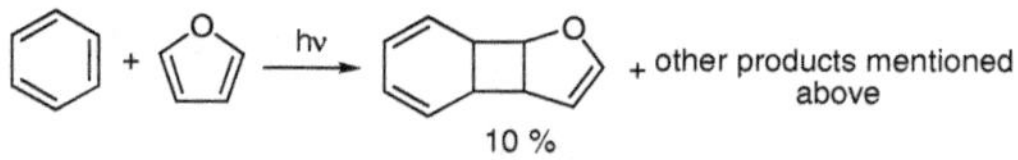

[2 + 2]-Photoadducts of aromatic hydrocarbons, benzene and naphthalene, and N-methylindole with acetylenic compounds often undergo electrocyclic ring opening to give stable products. The following examples are illustrative:

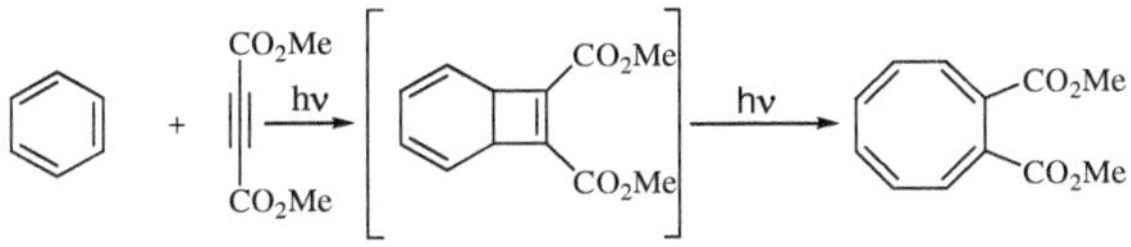

Ref 23

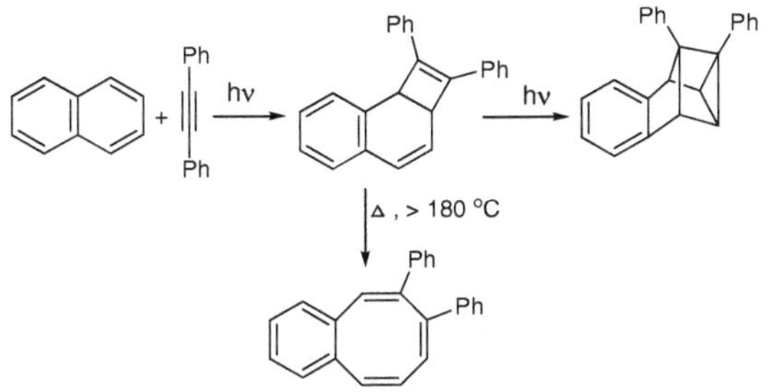

Ref 24

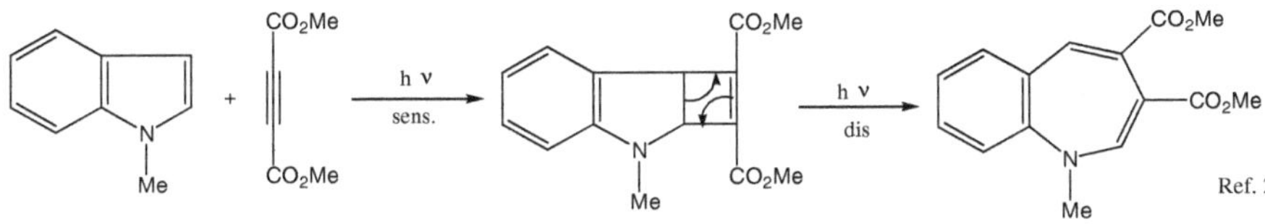

benzazepine (major)

Ref. 25

Intramolecular photoaddition of alkene part with aromatic ring takes place in non-conjugated aryl olefins when two π systems in a molecule are in close proximity and are separated by four sigma bonds. For example, *cis*-6-phenyl-2-hexene **1** in solution undergoes intramolecular 1,3-cycloaddition to give two *meta*-adducts **2** and **3** by the formation of exciplex [26].

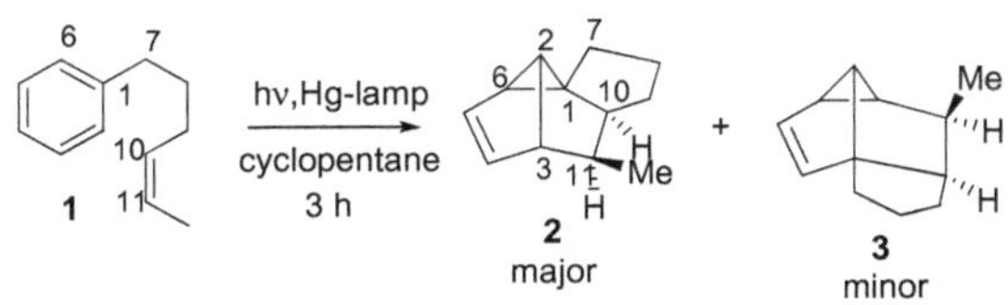

Similarly, 3-methoxy-4-butenyloxyacetylbenzene **4** gives *ortho*-cycloadduct **5** [27].

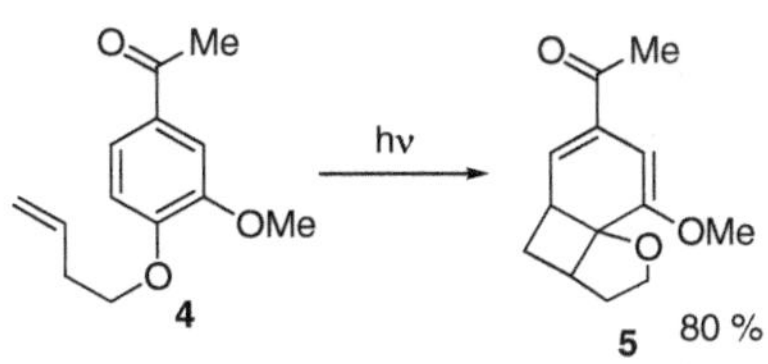

Ref 27

Indene derivative **6** undergoes intramolecular [2 + 2]-cycloaddition and rearrangement reactions through triplet excited state [28].

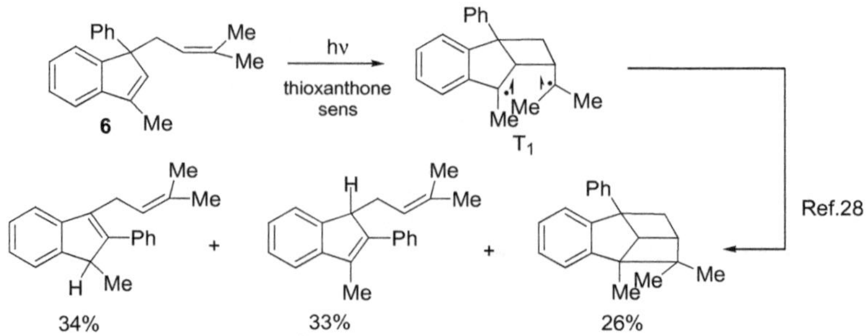

Photo-induced intramolecular [2 + 2]-cycloaddition reaction also occurs in **7** bearing a non-conjugated aromatic ring and acetylenic π system [26].

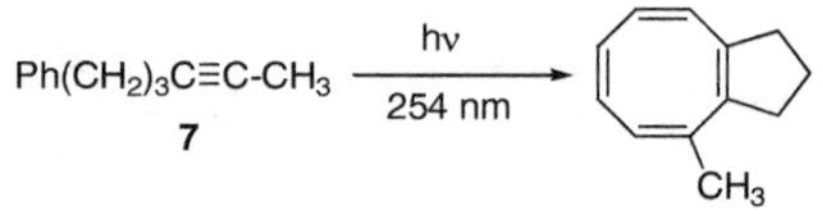

Aromatic compounds of aromatic rings in close proximity undergo photo-induced [2 + 2]-cycloadditions, e.g., **8** gives **9**.

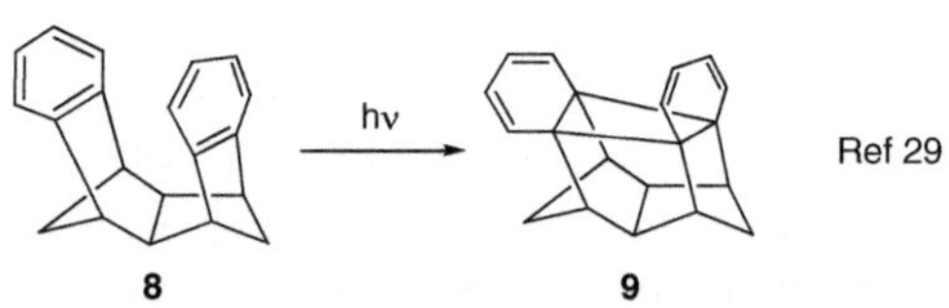

Regioselective [2 + 2]-photoaddition of 4-methyl-1-cyanonaphthalene **10** with TME gives [2 + 2]-cycloadduct **11** selectively at 1,2-position [30].

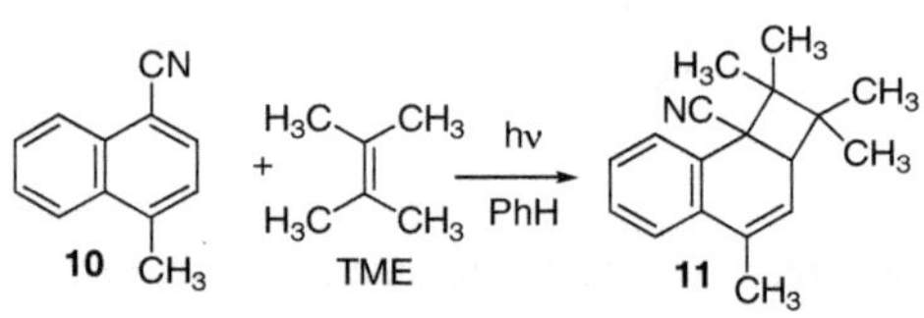

Both regio- and stereo-selectivities of [2 + 2]-photocycloaddition of ethene **12** to 1-cyano-4-hydroxymethylnaphthalene **13** is observed to get **14** by the hydrogen bonding between the addends [31].

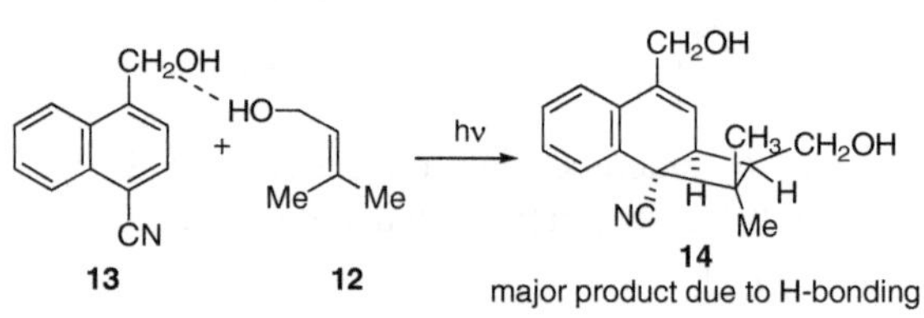

Phenanthrene **15** undergoes [2 + 2]-photoaddition reaction with maleic anhydride at 9,10-positions to give **16** [32].

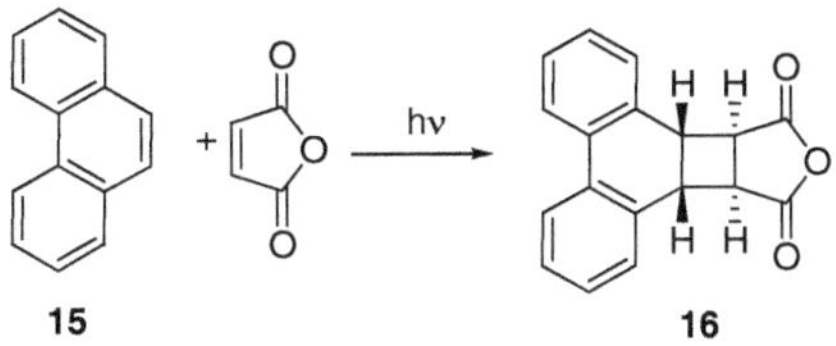

9.3.1 Photo-Diels–Alder Cycloaddition Reactions of Aromatic Compounds

Indole **17** undergoes photo-induced radical cation Diels–Alder reaction with cyclohexa-1,3-diene **18** in the presence of triphenylpyrylium tetrafluoroborate to give diastereoselective product **19** [33].

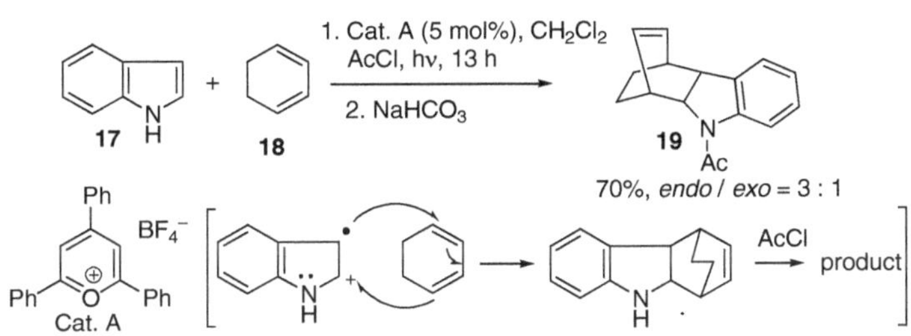

Naphthalene undergoes [4 + 2]-cycloaddition reaction with reactive dienophile, N-methylthiazolinedione (MTAD) **20** to give cycloadduct **21** via both singlet and triplet excited MTAD and also by direct irradiation of ground-state charge transfer (CT) complex at 458 nm [34]. The cycloadduct **21** on irradiation with triplet sensitizer xanthone at 300 nm gives di-π-methane rearrangement product **22**, which may be utilized as a potential source of diazasemibullvalene **23**.

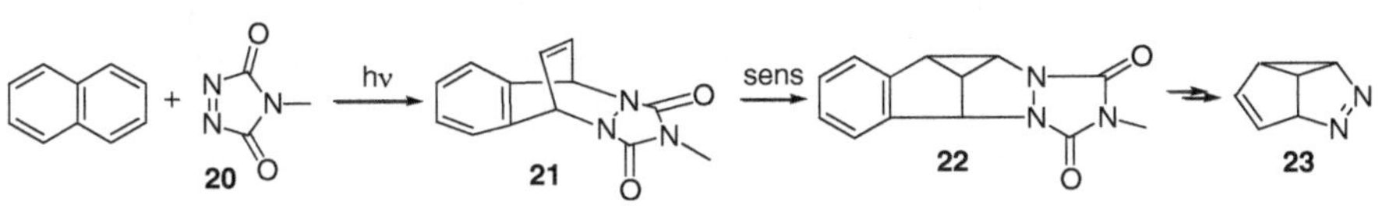

1-Acetyl naphthalene **24** on photoirradiation undergoes [4 + 2]-cycloaddition reaction with chiral electron acceptor alkene, (*S*)-(2-methoxymethyl-1-pyrrolidinyl)-acrylonitrile and its (*R*)-enantiomer **25** to form chiral (+)- and

(−)-cycloadducts **26** as sole products, which on hydrolysis afford 1,4-diketones **27** in ~97 % enantiomeric excess. The absolute configuration of **26** at C-1, C-4 and C-9 carbons has not yet determined, but the methoxymethylpyrrolidinyl moiety had probably *syn*-geometry to the benzenoid ring [35].

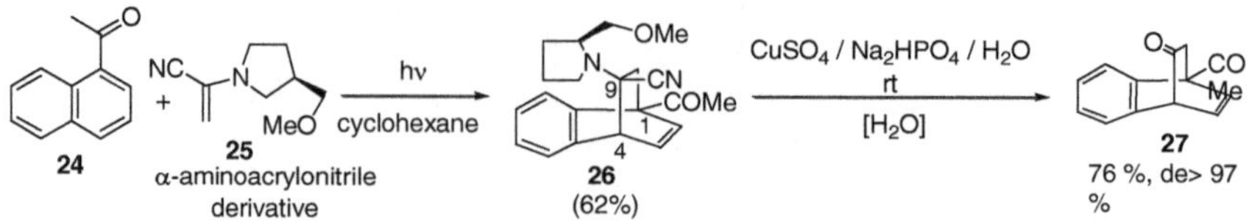

1,1-Diphenylethene derivative **28** (as electron donor) undergoes photo-[4 + 2]-cycloaddition reaction with 1,4-dicyanobenzene (as electron acceptor) in the presence of phenanthrene sensitizer to give isoquinoline derivative **29** in a PET process. The reaction takes place in a highly polar exciplex and/or a contact radical ion pair generated in a PET process, followed by air oxidation [36].

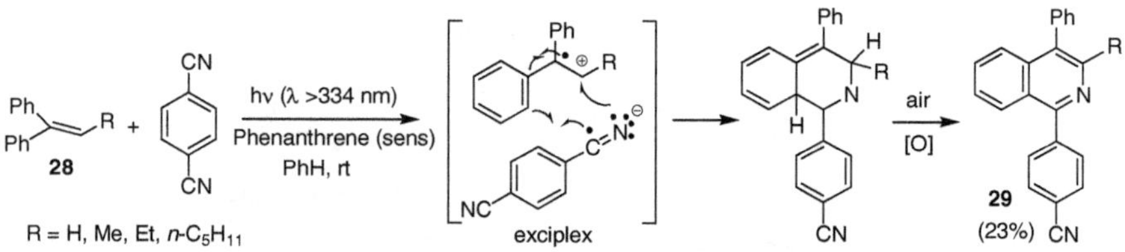

9.4 Photo-Induced Hydrogen Abstraction and Addition Reactions of Aromatic Compounds

Photoexcited aromatic hydrocarbons undergo hydrogen abstraction and 1,4-addition reactions with cycloalkanes and hydroxylic compounds. These reactions are believed to take place through an excited singlet state of arene to form an exciplex or ion pair as an intermediate, which on back electron transfer dissociates into triplet diradical and undergoes proton abstraction from solvent or amines, followed by addition of an alkyl or aryl unit to give the products [37]. The following examples of this reaction are illustrative:

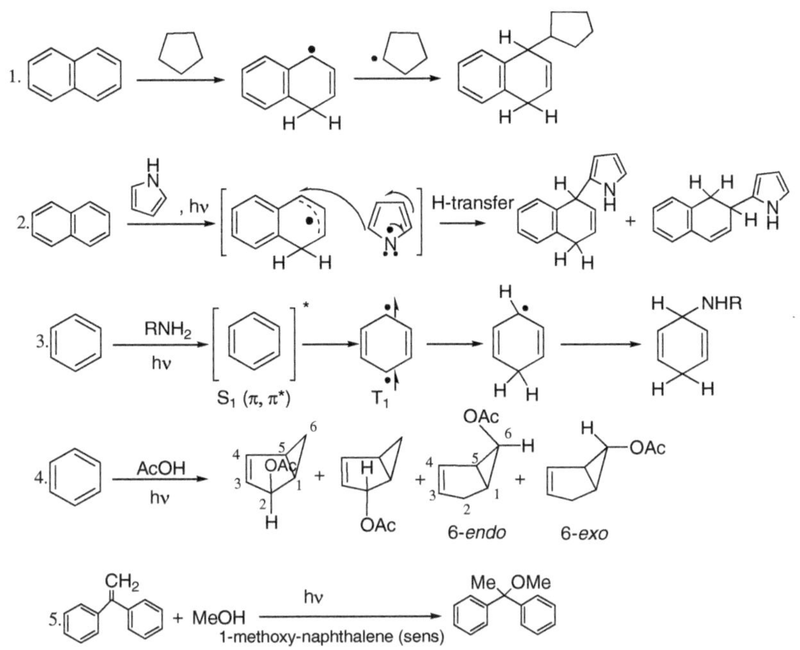

In entry 4, the reaction proceeds through a prefulvene diradical intermediate.

9.5 Photocyclization Reactions of Aromatic Compounds

cis-3-Styrylpyridine **30** undergoes photocyclization to give azaphenanthrene **31** [38].

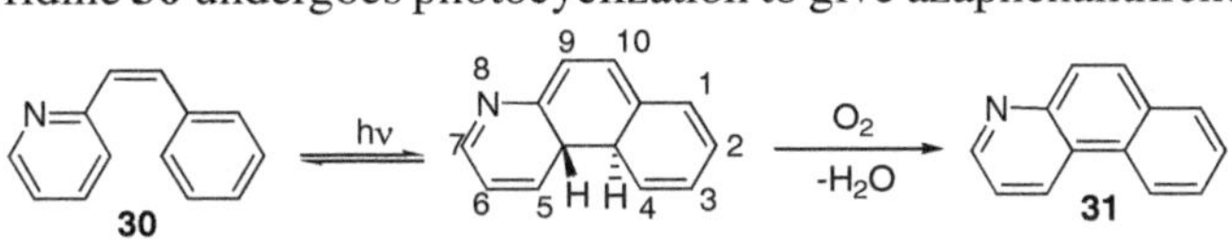

Irradiation of diphenylethynyl cycloalkene **32** in isopropanol gives cyclic product, diphenylbenzo cycloalkene **33** up to 21 % dependent on the alkene ring size [39].

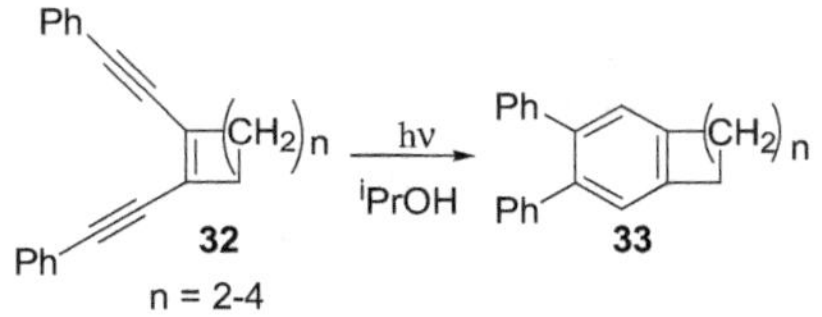

9.6 Photorearrangement Reactions of Aromatic Compounds

Irradiation of 3-methoxyphenol **34** in presence of $AlBr_3$ gives lumiketone type product, 4-methoxybicyclo[3.1.0]-hex-3-en-2-one **35**. This reaction provides a useful method for synthesis of bicyclo[3.1.0]-hex-3-en-2-ones [40].

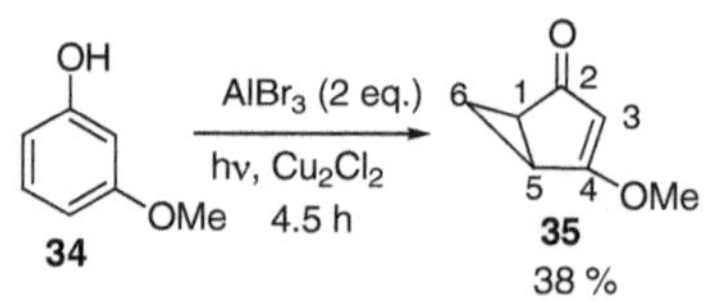

The reaction is believed to proceed through a *para*-protonated Lewis complex. The presence of excess $AlBr_3$ prevents the cyclo-reversion of the product.

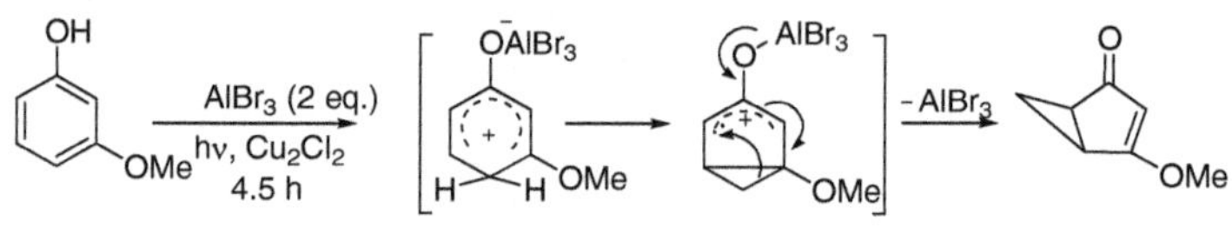

The light-induced rearrangement of phenyl esters **36** into hydroxy aryl ketones **37** and **38** is called the photo- *Fries rearrangement* [41]. Because of low yields, this procedure has little commercial importance. The reaction is usually carried out in an aprotic solvent and involves a radical mechanism by the homolytic cleavage of the *O*-acyl bond from its singlet excited state [41].

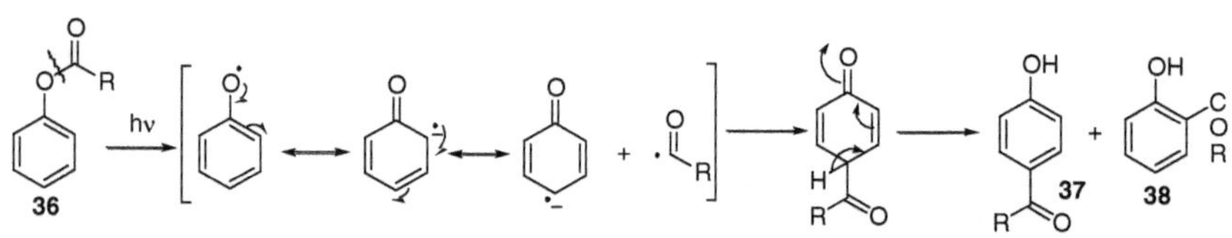

The rearrangement is found to be of intramolecular nature. The radical pair remains in the solvent cage and their recombination in the cage affords the acyl migration products, while hydrogen abstraction by the phenoxy radical from the solvent leads to the formation of phenol as by-product. When the reaction of phenyl acetate was carried out with deuterated phenol in methanol, the major products were *ortho*- and *para*-hydroxy acetophenones and phenol. Only trace amounts of crossover products were obtained [42]. The presence of methoxy substituent at *meta*- and *para*-positions increases the yield of *ortho*-Fries product [43]. For example, **39** gives **40**.

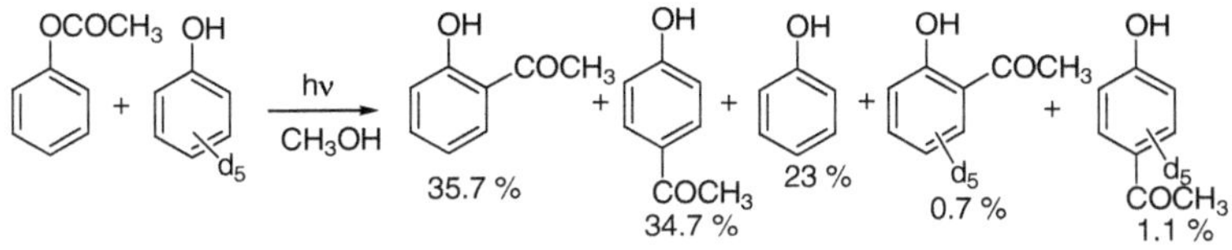

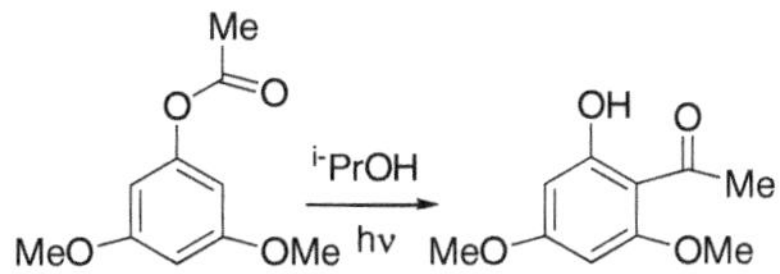

39 **40** **63.6%**

This reaction occurs in nature when a plastic bottle made of polycarbonate (polyethyleneterephthalate) exposed to sunlight (wavelength 310 nm) leads to leaching of phthalate from the plastic [44].

The reaction rate of aromatic esters in photo-Fries rearrangements is greatly influenced by the steric and electronic factors of the substituents. For example, the reaction rates of 4-*t*-butylphenyl-4-substituted benzoates **41** are influenced by the substituent in 4-position. Usually EWG accelerates the rate of the reaction and ERG retards the rate [45].

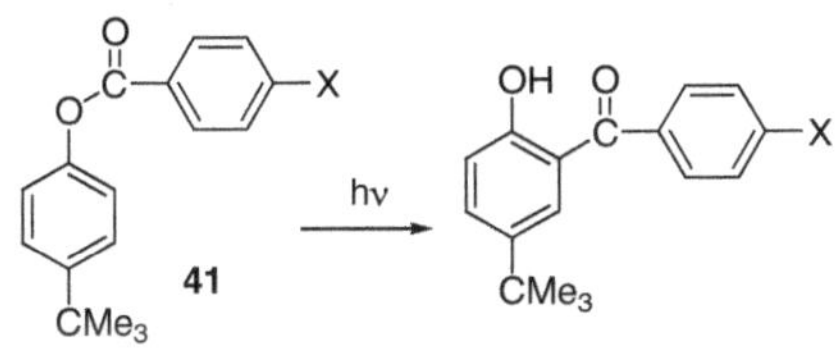

41

X	Reaction rate ϕI_0 mole.cm^{-2}sec^{-1} x 10^8
CMe$_3$	46.6 ± 2.2
CN	135.6 ± 6.1

It indicates that both the radicals remain associated as pair at all stages of the reaction because of attractive forces between them.

3,5-Dimethoxybenzyl derivatives **42** (X = OAc, O(PO)(OEt)$_2$) undergo photorearrangement in alcohol solution to give 1,3-dimethoxy-5-methylenecyclohexa-1,3-dienes **43** in low yields (~16 %) [46].

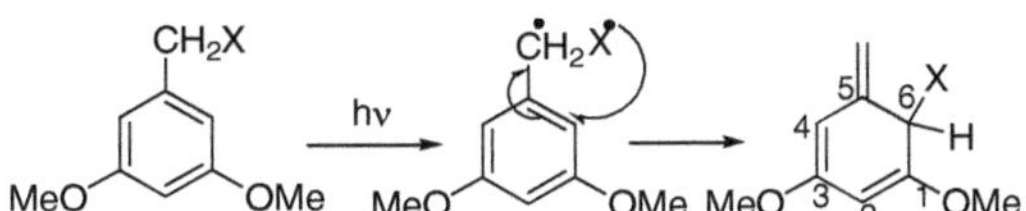

42 **43**

Dibenzocyclopentanemethylene derivative **44** undergoes photosensitized-*oxa-di-π-methane rearrangement* to give **45** in 96 % yield [47].

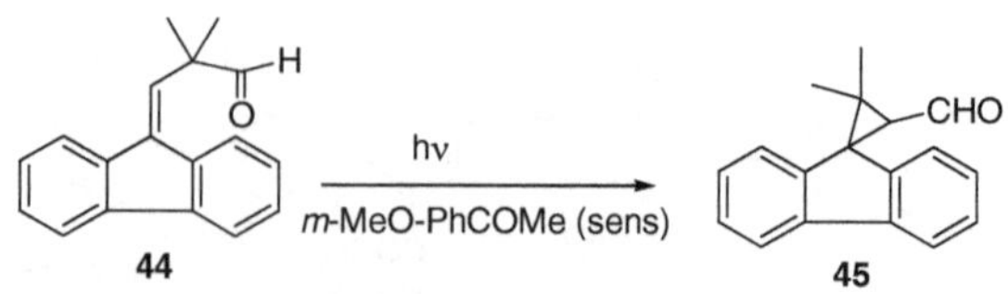

Aromatic β,γ-unsaturated oximes, oxime esters, acyl hydrazones and semicarbazones **46** undergo photochemical rearrangements into their cyclopropane derivatives **47**. These rearrangements are known as the *aza-di-π-methane (ADPM) rearrangements* [48].

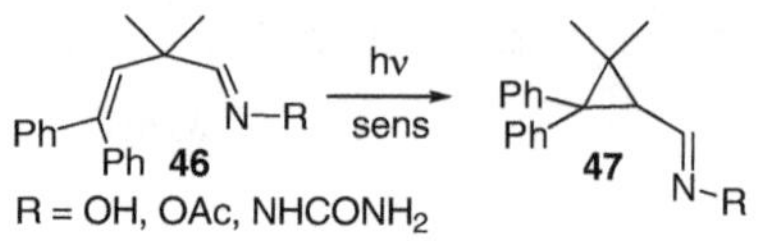

For example, oxime **48** gives **49**.

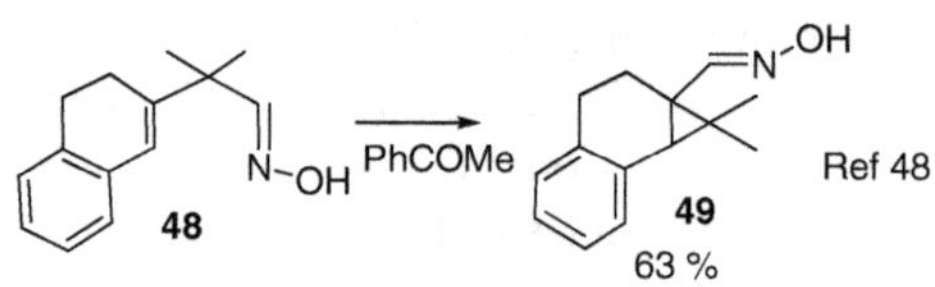

9.7 Photooxidation Reactions of Aromatic Compounds

Aromatic hydrocarbons **50** and **51** react with oxygen under the influence of light to give cyclic peroxides. Oxygen in its singlet state serves as a dienophile.

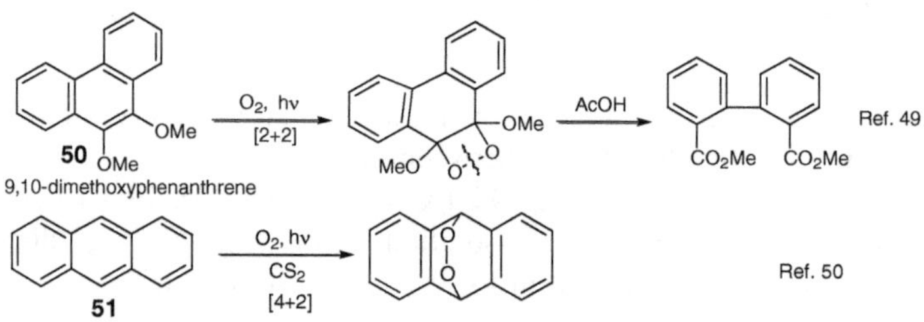

9.8 Photodimerization Reactions of Aromatic Compounds

Anthracene and 9-substituted anthracene undergo photodimerization via a singlet excimer, which collapses to dimer in a symmetry allowed $[\pi^4_s + \pi^4_s]$-cycloaddition process. Substituted anthracenes with different substituents at C-9 position give dimers having head-to-head geometry of the substituents [51, 52].

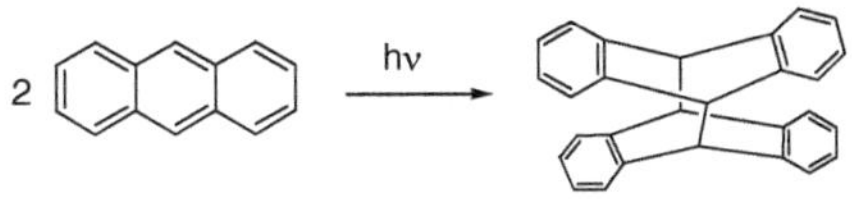

Ref. 51

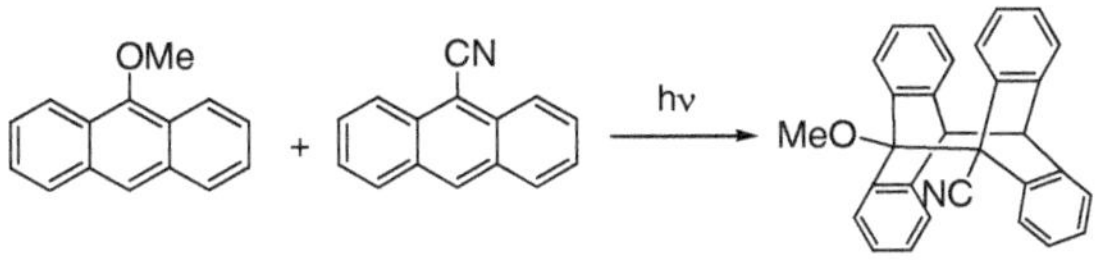

Ref 51, 52

The kinetic study indicated that the reaction takes place in the following steps:

$$A + h\nu \longrightarrow {}^{1}A^{*}$$

$${}^{1}A^{*} + A \longrightarrow {}^{1}(A-A)^{*} \quad \text{(eximer)}$$

$${}^{1}A^{*} \longrightarrow A + h\nu_{f} \quad \text{(fluorescence)}$$

$${}^{1}(A-A)^{*} \longrightarrow 2A + h\nu_{e} \quad \text{(emission)}$$

$${}^{1}(A-A)^{*} \longrightarrow A-A$$

A = anthracene

9-Substituted anthracene **52** gives dimer **53** of head-to-tail regiochemistry [51].

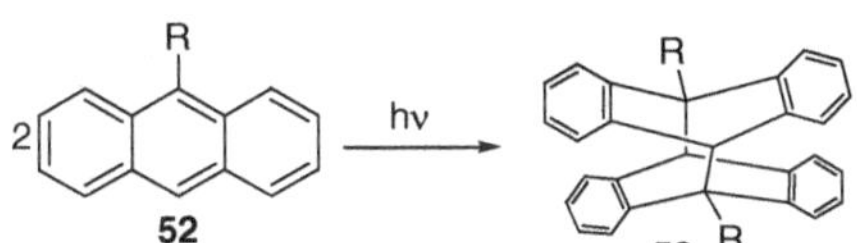

52 R = Me, Et, OMe, CN, Cl, Br; **53**

Similarly, 2-methoxynaphthalene **54** undergoes photodimerization to give two isomeric products **55** and **56** [53, 54].

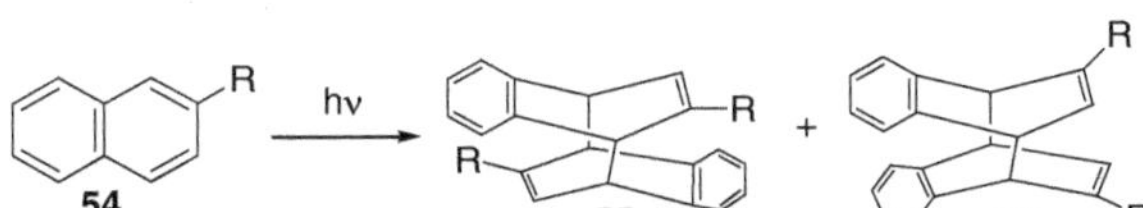

54 R = OMe; **55**; **56**

Methyl naphthalene 2-carboxylate and 2-cyanonaphthalene **57** undergo photodimerization followed by [2 + 2]-cycloaddition to give cage compounds **58** [55].

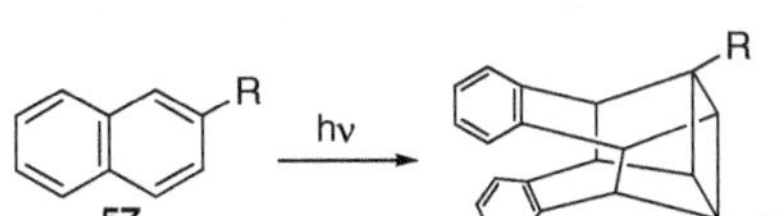

57 R = CO_2Me, CN; **58**

meta-Methoxystyrene **59** undergoes photodimerization in the presence of acceptor sensitizer to give tetralin derivatives **60** and **61** via the cyclization of dimeric cation radical [56].

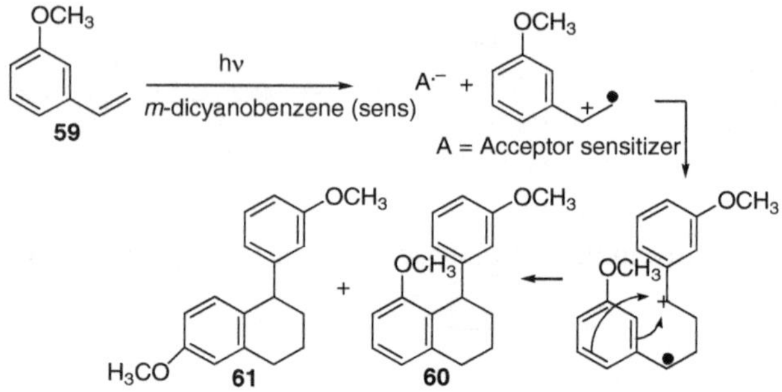

9.9 Photosubstitution Reactions of Aromatic Compounds

Photo-induced aromatic substitution reactions occur through an electron transfer process, which creates an aromatic radical anion or aromatic radical cation as intermediate. This intermediate couples with the electrophile or nucleophile radical to give the product. This mechanism is called $S_{RN}1$ (where the abbreviations stand for substitution, radical, nucleophilic, and first order). Photoirradiation of aromatic compounds in the presence of nucleophiles gives nucleophilic-substituted products different from those of thermal reaction. For example, 3,4-dimethoxynitrobenzene on UV irradiation in presence of hydroxide ion gives 3-hydroxy-substituted product, while on heating gives 4-hydroxy-substituted product [57].

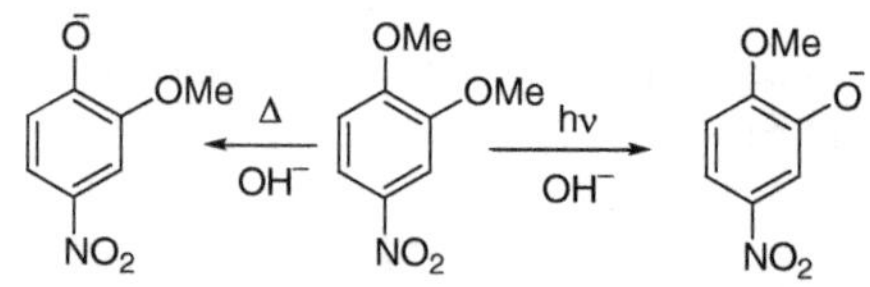

Usually electron-withdrawing group such as nitro group makes the *ortho-* and *meta*-positions positive relative to the *para*-position in the excited state, whereas the *ortho-* and *para*-positions are positive relative to the *meta*-position in the ground state. With electron-releasing substituent, the reverse is observed, i.e., the *ortho-* and *meta*-positions are negatively charged relative to *para*-position in the excited state. For example, 4-methoxynitrobenzene gives 4-methoxyphenol [57].

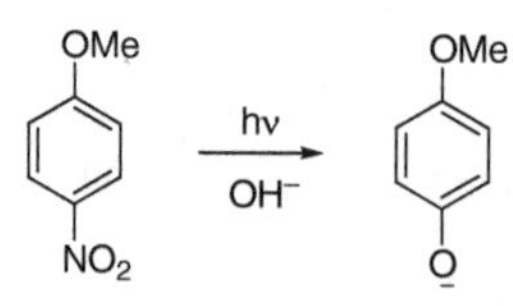

Photoirradiation of phenylacetate dianions **62** with aryl bromides and iodides in liquid ammonia gives isomeric arylated phenyl acetic acids **63** and **64** [58].

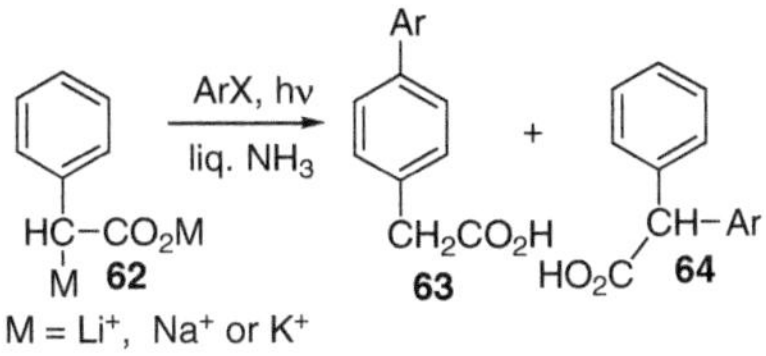

Irradiation of azulene **65** in presence of aryl iodide gives 1-arylazulene **66** [59].

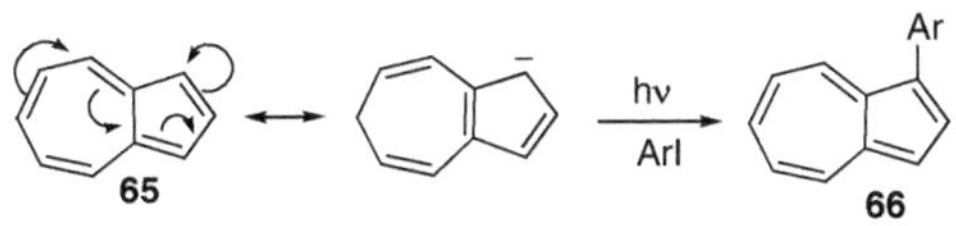

Photochemical nucleophilic-substitution reactions of cyanobenzene with allenes take place in a radical coupling process at the less heavily substituted radical site by donor–acceptor property. For example, 1,2,4,5-tetracyanobenzene **67** reacts with 1,1-dimethylallene **68** in the presence of diphenyl to give **69** as major product [60].

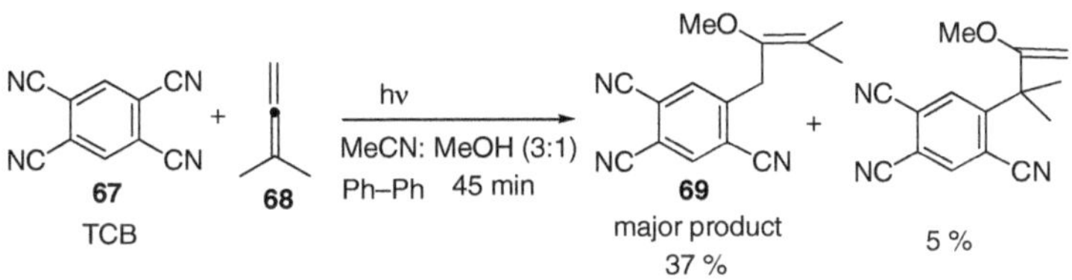

Possibly diphenyl acts as a co-donor to drive the reaction in the forward direction. The major product of the reaction is formed in a stepwise process as follows:

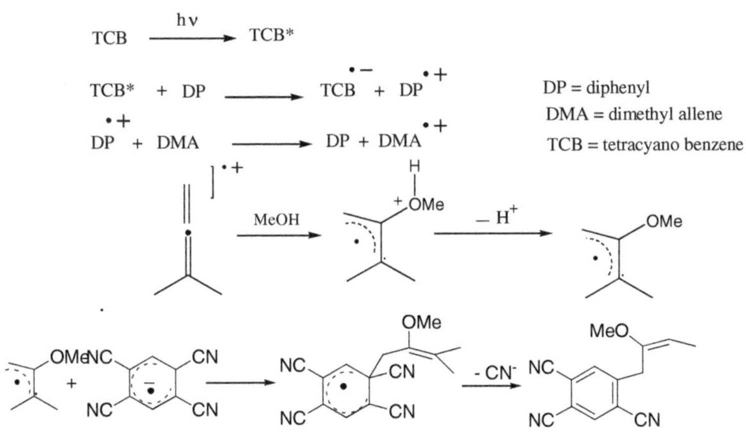

Similarly, 1,4-dicyanobenzene **70** on reaction with tetramethylallene **71** gives substituted product **72** [60].

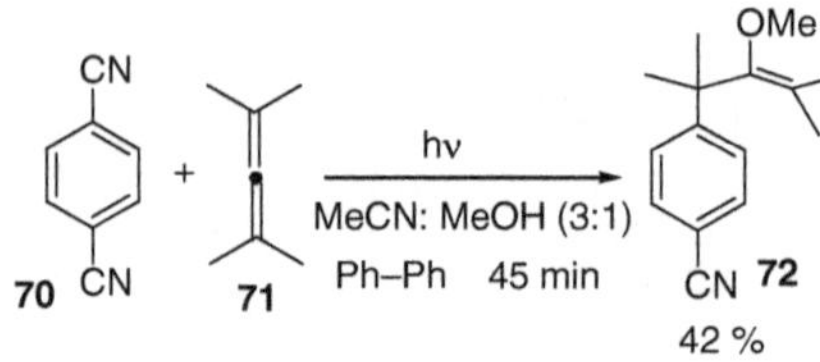

Aromatic amine **73** undergoes photo-induced alkylation with pentafluoro iodobenzene to give **74** and **75** through a radical coupling mechanism in a photo-induced electron transfer process [61].

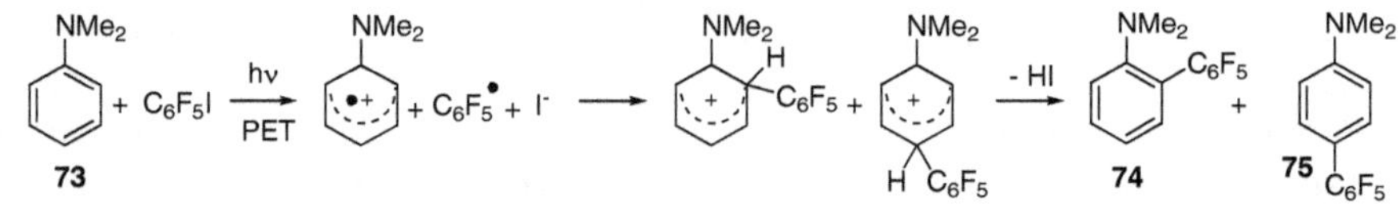

9.10 Problems

9.10.1. Predict the structure (s) including stereochemistry of the product (s) of the following reactions with plausible mechanism of formation:

(a) 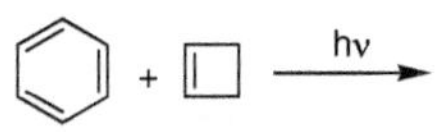

(b) 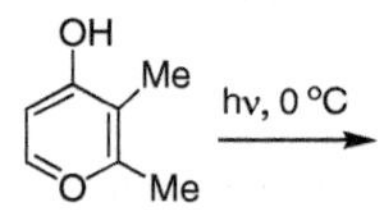

(c) 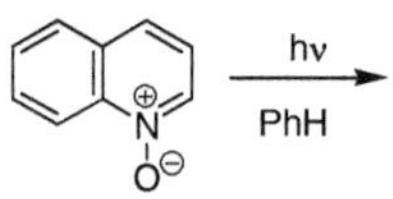

(d) 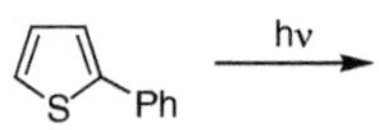

(e) 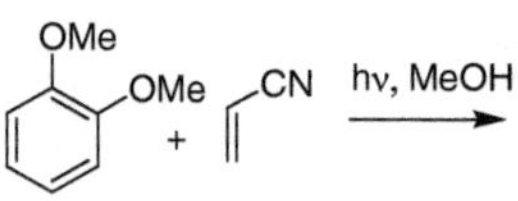

(f) 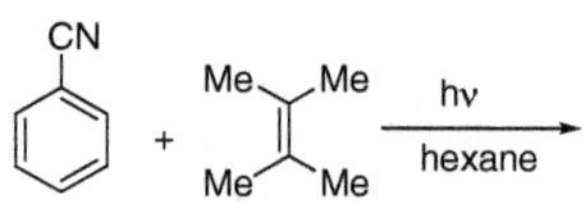

(g) 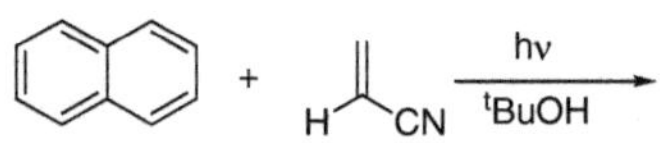

(h) 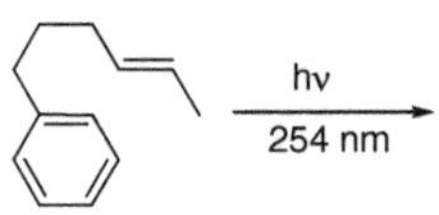

(i)

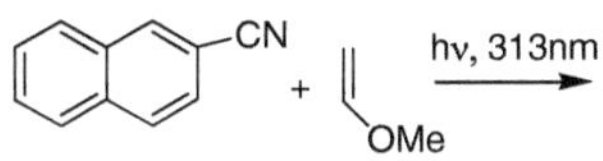

(j)

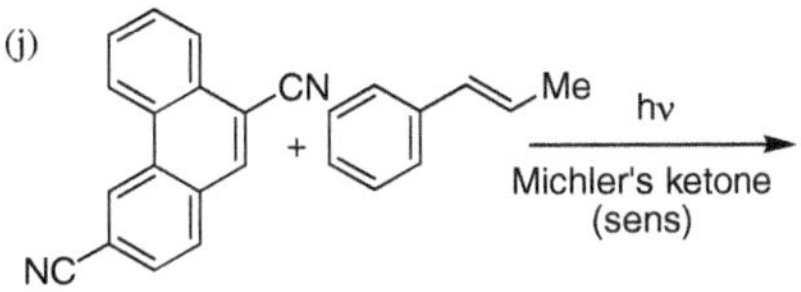

(k)

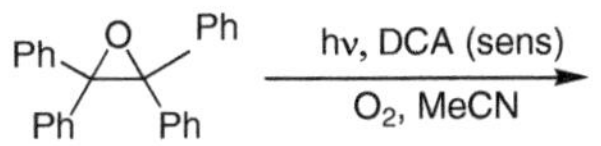

DCA = 9,10-dicyanoanthracene

(l)

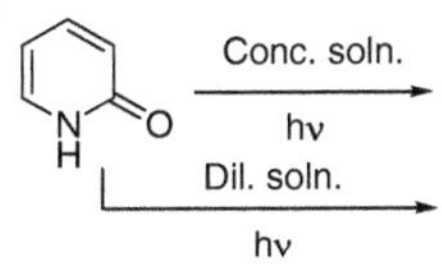

(m)

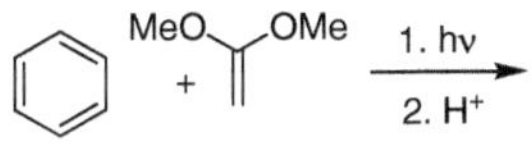

9.10.2. Suggest the plausible mechanism for each of the following reactions:

(a)

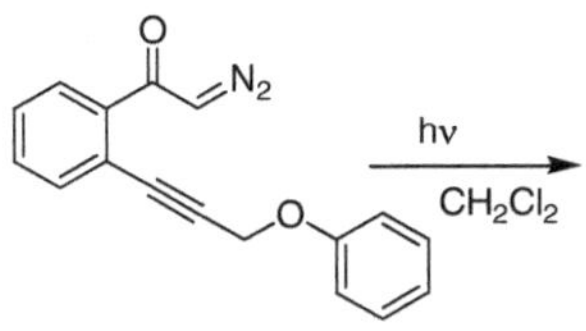

(b)

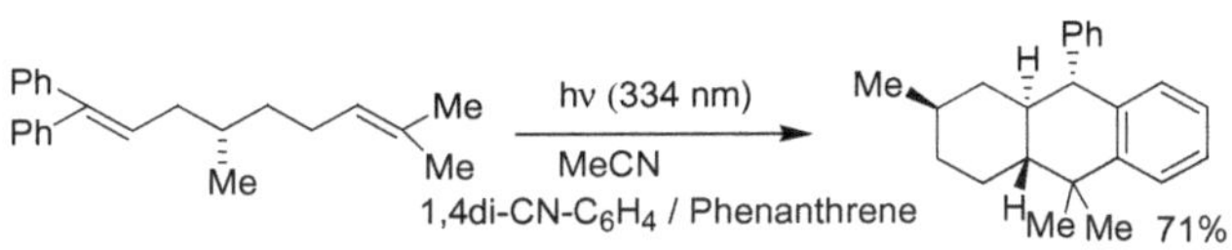

9.11 Further Reading

1. Turro NJ (1978) Modern molecular photochemistry. Benjamin-Cummings, Mento Park, CA, USA
2. Gilbert A, Baggott J (1991) Essentials of molecular photochemistry. CRC Press, Boca Raton, Florida, USA
3. McCullough JJ (1987) Photoadditions of aromatic compounds. Chem Rev 87:811
4. D'Auria M, Racioppi R (2013) Oxetane synthesis through the Paterno–Buchi reaction. Molecules 18:1138

References

1. Wilzbach KE, Ritscher JS, Kaplan L (1967) J Am Chem Soc 89:1031
2. Kaplan L, Rausch DJ, Wilzbach KE (1972) J Am Chem Soc 94:8638
3. Wilzbach KE, Kaplan L (1965) J Am Chem Soc 87:4004
4. Malkin J (1992) Photophysical and photochemical properties of aromatic compounds. CRC Press, Boca Raton, Ann Arbor, p 141
5. Wilzbach KE. Kaplan L (1966) J Am Chem Soc 88:2066; Cornelisse J, Merritt VY. Srinivasan R (1973) J Am Chem Soc 95:6197; Gilbert A, Yianni P (1981) Tetrahedron 37:3275; Bryce- Smith D, Gilbert A, Orger BH (1966) J Chem Soc Chem Commun 512; Wagner-Jauregg T, (1980) Synthesis 165:769; Cornelisse J (1993) Chem Rev 93:615
6. Bryce- Smith D, Gilbert A, Orger BH (1974) J Chem Soc Chem Commun 334
7. Ferree W, Grutzner GB, Morrison H (1971) J Am Chem Soc 93:5502
8. Cornelisse J, Merritt VY, Srinivasan R (1973) J Am Chem Soc 95: 6197; Cornelisse J, Srinivasan R (1973) Chem Phys Lett 20:278
9. Weller A (1982) Z Phys Chem NF 133:93
10. Gilbert A (1980) Pure Appl Chem 52:2669; Bryce-Smith D, Foulger B, Forrester J, Gilbert A, Orger BH, Tyrrell HM (1980) J Chem Soc, Perkin Trans 1, 55; Cornelisse J, Srinivasan R (1973) Chem Phys Lett 20:278
11. Heine HG, Hartmann W (1975) Angew Chem 87: 708; Mattay J, Runsink J, Leisman H, Scharf HD (1982) Tetrahedron Lett 23:4919
12. Gilbert A, Taylor GN, Samsudin MWB (1980) J Chem Soc Perkin Trans 1, 869; Mattay J (1985) Tetrahedron 41:2405; Scharf HD, Mattay J (1977) Tetrahedron Lett 401
13. Gilbert A, Yianni P (1982) Tetrahedron Lett 4611; Idem (1981) Tetrahedron 37:3275
14. Osselton EM, Eyken CP, Jans AWH, Cornelisse J (1985) Tetrahedron Lett 26:1577
15. Mattay J, Leismann H, Scharf HD (1979) Chem Ber 112:577
16. Ohashi M, Tanaka Y, Yamada S (1976) J Che Soc Chem Commun 800
17. Ohashi M, Tanaka Y, Yamada S (1977) Tetrahedron Lett 3629
18. Cantrell TS (1977) J Org Chem 42:4238
19. Cantrell TS (1981) J Org Chem 46:2674
20. Bryce-Smith D, Foulgar BE, Gilbert A (1972) J Chem Soc Chem Commun 664
21. Yang NC, Libman J, Savitzky MF (1972) J Am Chem Soc 94:9226
22. Pac C, Sugioka T, Sakurai H (1972) Chem Lett 39; Kan K, Kai Y, Yasuoka N, Kasai N (1979) Bull Chem Soc Jpn 52:1634
23. Bryce-Smith D, Lodge JE (1963) J Chem Soc 695; Grovenstein E, Rao DV (1961) Tetrahedron Lett 148
24. Sasse WHF (1969) Aust J Chem 22:1257; Kowala C, Poppleton BJ, Sasse WHF (1976) Aust J Chem 29:1851; Sasse WHF, Collin PJ, Sugowdz G (1965) Tetrahedron Lett 3373
25. Davis PD, Neckers DC (1980) J Org Chem 45:456; Davis PD, Neckers DC, Blount JR (1980) J Org Chem 45:462
26. Morrison H (1979) Acc Chem Res 12:383; Pallmer M, Morrison H (1980). J Org Chem 45:798
27. Ohashi M, Tanaka Y, Yamada S (1977) Tetrahedron Lett 18:3629; Ferree W Jr, Grutzner JB, Morrison H (1971) J Am Chem Soc 93:5502
28. Padwa A, Goldstein S, Pulwer M (1982) J Org Chem 47:3893
29. Sedelmeir G, Fessner WD, Grund C, Spurr PR, Fritz H, Prinzbach H (1986) Tetrahedron Lett 27:1277
30. McCullough JJ, MacInnis WK, Lock CJL, Faggiani R (1982) J Am Chem Soc 104:4644
31. Yokoyama A, Mizuno K (2000) Org Lett 2:3457
32. Creed D, Caldwell RA, Ulrich MM (1978) J Am Chem Soc 100:5831
33. Gieseler A, Steckhan E, Wiest O, Knoch F (1991) J Org Chem 56:1405
34. Kjell DP, Sheridan RS (1985) J Photochem 28:205
35. Dopp D, Pies M (1987) J Chem Soc Chem Commun 1734

36. Ishii H, Imai Y, Hirano T, Maki S, Niwa H, Ohashi M (2000) Tetrahedron Lett 41:6467
37. McCullough JJ, Wu WS, Huang CW (1972) J. Chem Soc Perkin Trans 2:371
38. Lewis FD, Kalgutkar RS, Yang JS (2001) J Am Chem Soc 123:3878
39. Jones GB, Wright JM, Plourde G, Purohit AD, Wyatt JK, Hynd G, Fouad F (2000) J Am Chem Soc 122:9872
40. Kakiuchi K, Ue M, Yamaguchi B, Nishimoto A, Tobe Y (1991) Bull Chem Soc Jpn 64:3468
41. Bellus D (1971) Advances in photochemistry. Vol 8, Wiley, Chichester, p 109–159; Stenberg (1967) Org Photochem 1:127
42. Yoon YJ, Ko SH, Ko MK, Chae WK (2000) Bull Korean Chem Soc 21:901
43. Hagman HJ (1969) Tetrahedron 25:6015
44. Searle N (2003) Environmental effects on polymeric materials. In: Andrade A (ed) Plastic and environment. Wiley, pp 313–358
45. Coppinger GM, Bell ER (1966) J Phys Chem 70:3479
46. de Costa DP, Howell N, Pincock AL, Pincock JA, Rifai S (2000) J Org Chem 65:4698; Bogdanova A, Popik VV (2001) Org Lett 3:1885
47. Armesto D, Ortiz MJ, Romano S, Agarrabeita AR, Gallego MG, Ramos A (1996) J Org Chem 61:1459
48. Armesto D, Ramos A, Mayoral EP (1994) Tetrahedron Lett 35:3785; Armesto D, Ortiz MJ, Ramos A, Horspool WM, Mayoral EP (1994) J Org Chem 59:8115
49. Rio G, Berthelot J (1972) Bull Soc Chim Fr 822
50. Bowen EJ (1953) Discuss Faraday Soc 14:143; Foote CS (1968) Acc Chem Res 1: 104
51. Cowan DO, Drisko RL (1976) Elements of organic photochemistry. Plenum, New York
52. Bouas- Laurent H, Castellan A, Desvergne JP (1980) Pure Appl Chem 52:2633
53. Kowala C, Sugowdz G, Sasse WHF, Wunderlich JA (1972) Tetrahedron Lett 4721
54. Seliger BK, Sterns M (1969) J Chem Soc Chem Commun 978
55. Teitei T, Wells D, Spurling TH, Sasse WHF (1978) Aust J Chem 31:85; Teitei T, Wells D, Sasse WHF (1976) Aust J Chem 29:1783
56. Yamamoto M, Yoshikawa H, Gotoh T, Nishijima Y (1983) Bull Chem Soc Jpn 56:2531
57. Norman ROC, Coxon JM (1993) Principles of organic synthesis. Chapman & Hall, ELBS ed., Oxford, p 513
58. Nwokogu GC, Wong JW, Greenwood TD, Wolfe F (2000) Org Lett 2:2643
59. Ho TI, Ku CK, Liu RSH (2001) Tetrahedron Lett 42:715
60. Mangion D, Arnold DR, Cameron TS, Robertson KN (2001) J Chem Soc Perkin Trans 2:48
61. Chen QY, Li ZT (1993) J Chem Soc Perkin Trans 1:1705

Chapter 10
Photofragmentation Reactions

10.1 Introduction

The light-induced reactions of organic compounds for introduction of functional groups at unactivated carbons through free radical attack are known as the photofragmentation reactions. The fragmented free radicals are generated by the cleavage of heteroatomic bonds such as C–H, C–N, C–O, C–S, O–N, etc. The most common reactions of this class are the *Barton reaction, the hypohalite reaction* and the *Hofmann-Loffler-Freytag reaction*. A variety of reagents and conditions have been employed in these reactions to improve the yields of the reactions. Each of these procedures follows a generalized mechanism (Scheme 10.1). The first step of each procedure is the homolytic cleavage of an heteroatomic bond, X–Y, where X is a heteroatom, O or N, and Y is H, halogen or NO, to form either an oxygen or a nitrogen radical. The generated heteroatom radical undergoes hydrogen abstraction normally from the δ-carbon in a quasi-chair- like six-atom transition state to generate a new carbon radical, which in absence of any competing external radical, adds the radical Y to form a C–Y bond. The presence of a good competing radical source, such as iodine, diverts the reaction to produce an iodohydrin, which can cyclize to form a tetrahydrofuran derivative in most cases. The most common means of generating the requisite oxygen radical (**1**: X=O) is by thermolysis or photolysis of a nitrite ester (Barton reaction), a hypochlorite, or a hypoiodite. Sometimes, lead tetraacetate-iodine or mercuric oxide-iodine variation of the hypoiodite reaction is carried out to improve the yield of the product. The requisite nitrogen radicals (**1**: X=N) are generated by heating or photoirradiation of appropriate N-haloamines with sulfuric acid or other acid to afford pyrrolidines or piperidines (Hofmann-Loffler-Freytag reaction). Sometimes, very weakly basic or neutral medium is employed to increase the stability of nitrogen radicals. Extensive

B. Dinda, *Essentials of Pericyclic and Photochemical Reactions*,
Lecture Notes in Chemistry 93, DOI 10.1007/978-3-319-45934-9_10

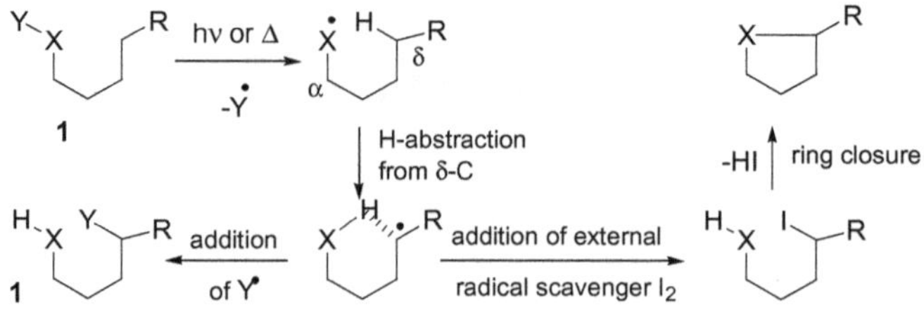

X = O; Y = NO : The Barton reaction
= O; = Cl : Pyrolysis of hypochlorite
= O; = I : Pyrolysis of hypoiodite
= O; = H : $Pb(OAc)_4$-I_2/HgO-I_2 variation of hypoiodite reaction
= H; = Cl : The Hoffmann-Loffler-Freytag reaction

Scheme 10.1 Generalized pathway for photofragmentation reaction

study on these reactions indicated that these free radical functionalizations work best with rigid substrates that hold the reacting atoms in fixed geometries. Many of these intramolecular free radical based functionalizations have been reviewed [1, 2].

10.2 The *Barton Reaction*

The photolysis reactions of nitrite esters into δ-nitrosoalcohols are known as the *Barton reactions*. The reaction involves the cleavage of O–N bond of nitrite ester, followed by an intramolecular abstraction of δ-H by the alkoxy radical and formation of carbon-nitrogen monoxide bond by radical coupling. The abstraction of hydrogen is very fast and takes place through a six-membered transition state. In many cases, nitroso alcohols are either tautomerize into oximes or dimerize. Labeling studies using nitrogen-15 revealed that the nitrosyl radical is transferred intermolecularly but not in a cage process [3]. The oxime derivatives are further hydrolyzed or oxidized.

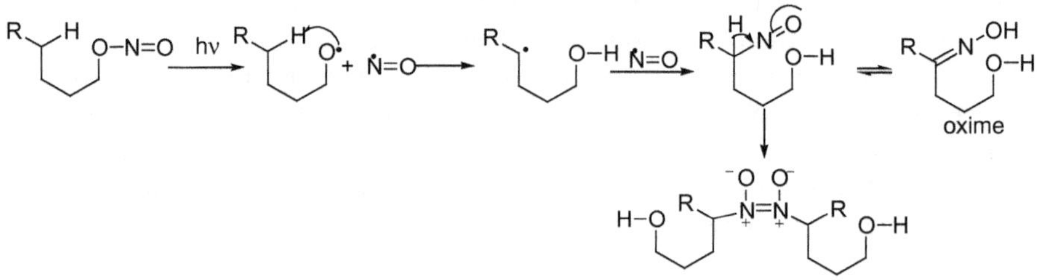

This reaction has been extensively utilized in the functionalizations of steroids and other bioactive organic compounds. For example, 11-hydroxycorticosterone acetate **2** was converted into aldosterone-21-acetate **3** [4].

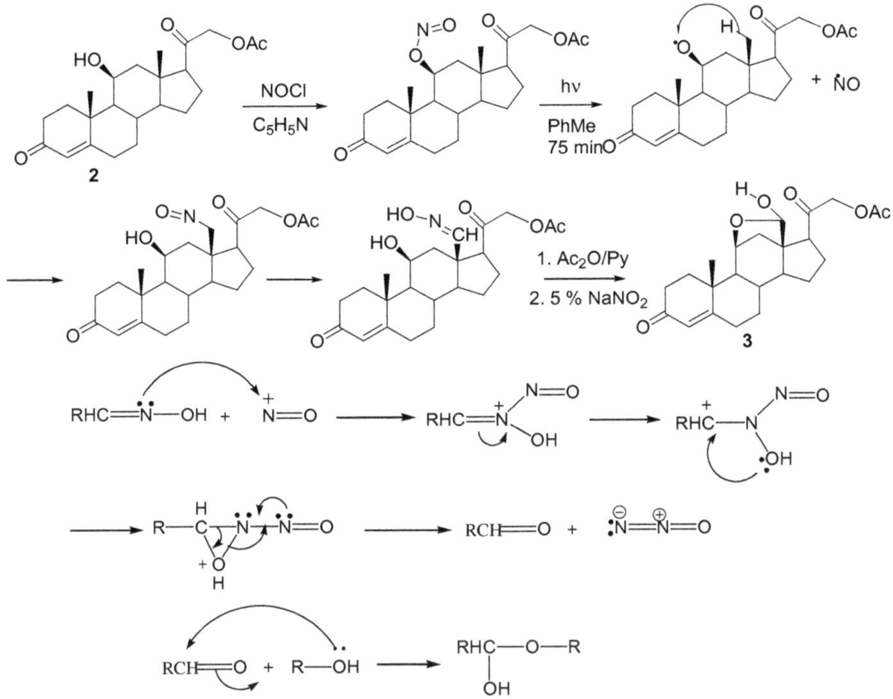

In steroid **4**, δ-hydrogen abstraction takes place from C(19)-methyl instead of C (18)-methyl to give oxime **5** [5].

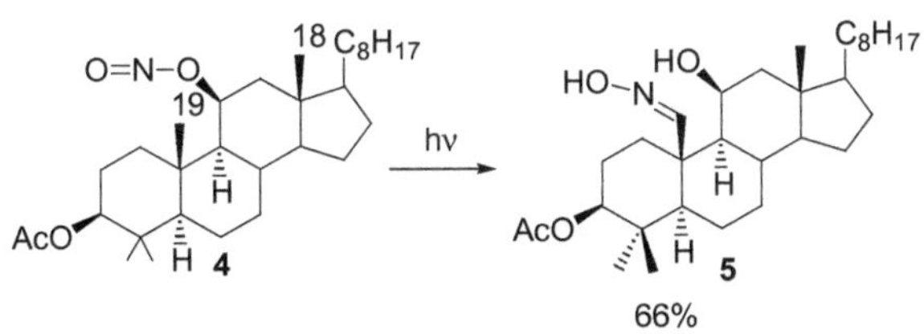

Other important applications of Barton reaction are in the synthesis of alkaloids and terpenoids. For example, the photolysis of nitrite **6** was used in a crucial step in the synthesis of alkaloid perhydrohistrionicotoxin **7** [6]. The oxime **8**, formed in about 20 % yield, gives the spirolactam ring of the alkaloid **7** on Beckmann rearrangement.

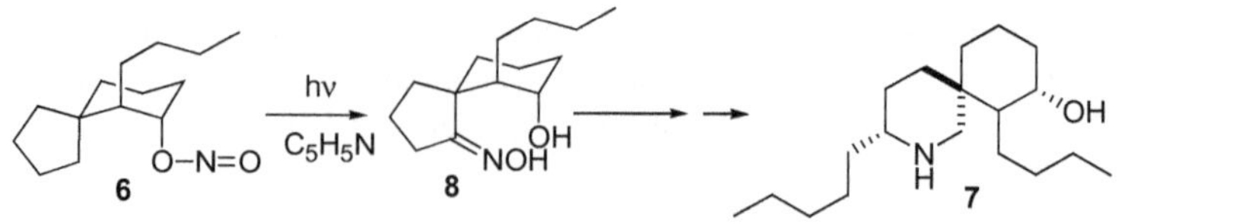

The Barton reaction (BR) is also used in key steps in the synthesis of tetracyclic triterpenoid, azadiradione **9** [7] and 3(α-hydroxyethyl)-quinuclidine **10** for synthesis of cinchona alkaloids [8].

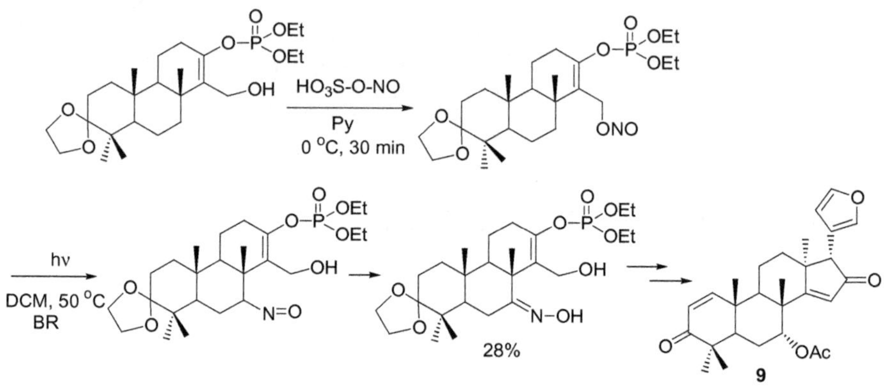

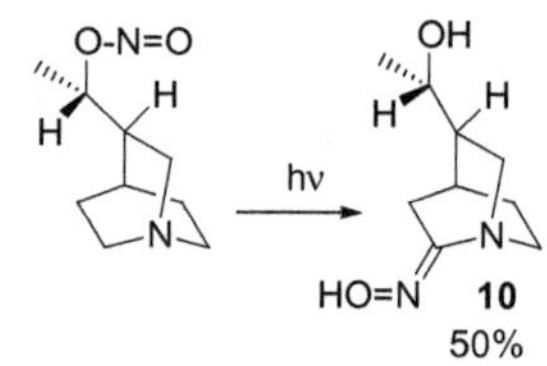

10.3 The *Hypohalite Reactions*

The photolysis of a tertiary hypohalite (readily prepared from the corresponding alcohol) generates alkoxy radical, which abstracts hydrogen from δ-carbon to produce an alkyl radical. The resulting alkyl radical abstracts halogen atom from a second molecule of the hypohalite to give δ-haloalcohol, which is cyclized in presence of a base to form a tetrahydrofuran derivative [9]. These reactions are very similar to the *Barton reaction* and proceed through a six-membered cyclic transition state. For example, the hypochlorite **11** of 2-methyl-2-hexanol gives tetrahydrofuran derivative **12**.

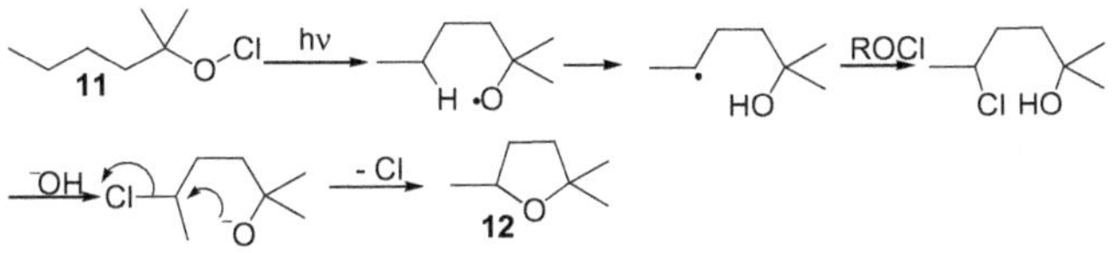

This method has been used in the synthesis of steroids for introduction of functionality at angular methyl groups. For example, steroid **13** gives **14** [10].

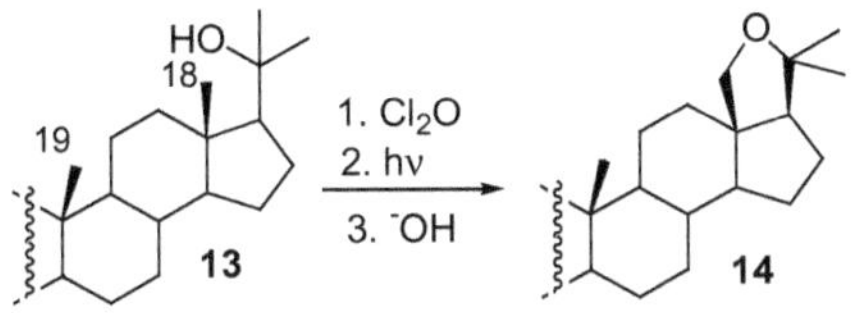

In the hypochlorite reactions, intramolecular hydrogen abstraction competes with β-cleavage and other reactions. For example, tertiary hypochlorite **15** on photolysis gives ketones and alkyl chlorides from β-cleavage and radical coupling reactions [10].

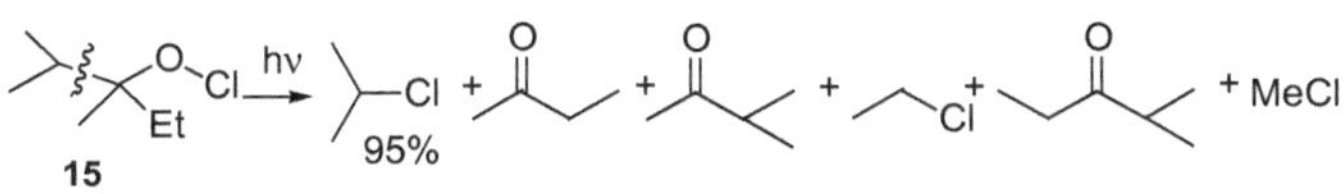

Thus, the hypochlorite reactions can produce several products by multiple functionalizations. Moreover, the stabilities of the hypochlorites are much less compared to hypoiodites. Hence, the photolysis of hypoiodites is more convenient method for generation of oxyradicals. The generated oxyradicals on hydrogen abstraction from δ-C give the corresponding alcohols. The hypoiodites are unstable at reaction temperatures and are prepared in situ by irradiation of the solutions of the corresponding alcohols in presence of N-iodosucinimide (NIS), acyl hypoiodites or diacetoxyiodobenzene (DIB), lead tetraacetate, mercuric acetate or mercuric oxide and iodine. The resulting alcohols form either five-membered ring oxides or acetals [11]. For example, the steroid acetate **16** gives cyclic ether **17** in 75 % yield [12].

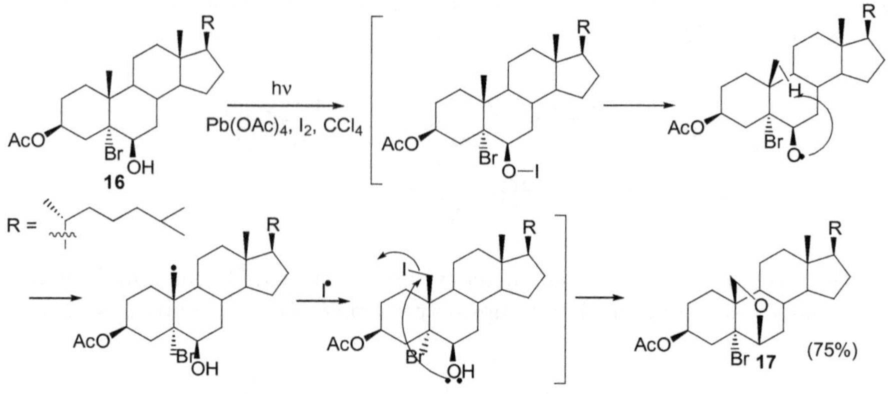

The hypohalite method of lead tetraacetate and iodine has been applied in the preparation of a key intermediate **18** for the synthesis of labdane diterpene, manool **18a** [13].

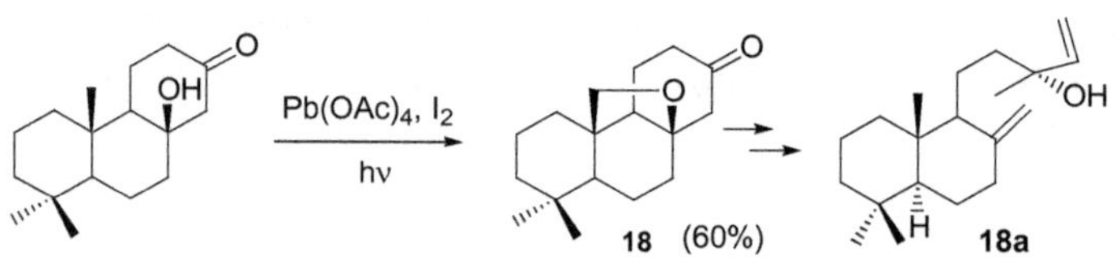

Sesquiterpene, cedrol **19** and its derivatives are functionalized at remote carbons to produce **20** and **21** using DIB, diphenylselenium acetate and iodine or HgO and bromine. The use of HgO/Br_2 and selenium acetate improves the yield of the reaction [14, 15].

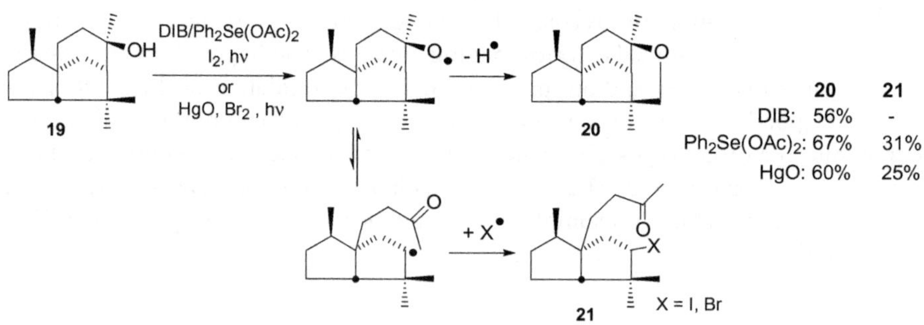

The hypoiodite method was also applied in a key step in the synthesis of antibiotic talaromycin **22** from **22a** [16].

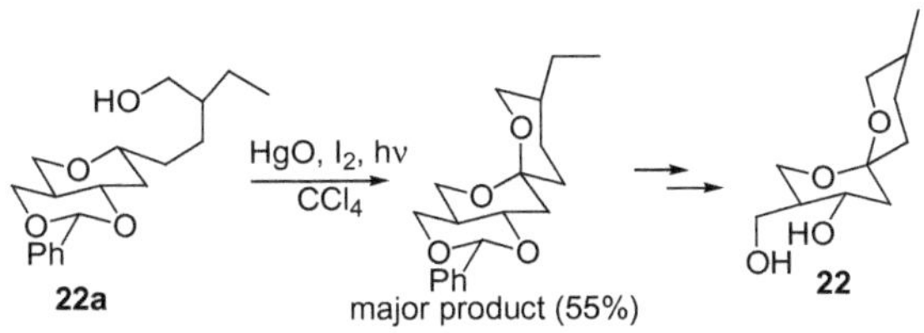

10.4 The *Hofmann-Löffler-Freytag Reaction*

The thermal or photolytic decompositions of N-halogenated amines in acid (H_2SO_4 or CF_3CO_2H) solutions followed by basification of the reaction mixture to produce pyrrolidines or in some cases piperidine derivatives are known as the *Hofmann-Löffler-Freytag(HLF) reactions*. For example, N-chlorobutylamine **23** gives N-butylpyrrolidine **24** [17].

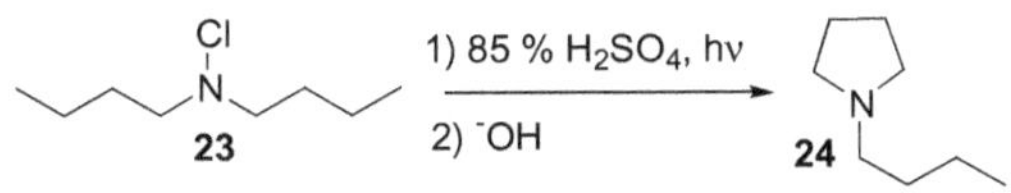

The reaction proceeds by homolytic cleavage of protonated halogenated amine to a nitrogen cation-radical, which undergoes an intramolecular abstraction of hydrogen atom from the δ-carbon via a six-membered cyclic transition state to form an alkyl radical. The resulting alkyl radical abstracts a halogen atom from another molecule of halogenated amine to form a protonated δ-halo amine **23a**. The protonated δ-halo amine on basification undergoes intramolecular cyclization to form pyrrolidine derivative by elimination of hydrohalic acid [18].

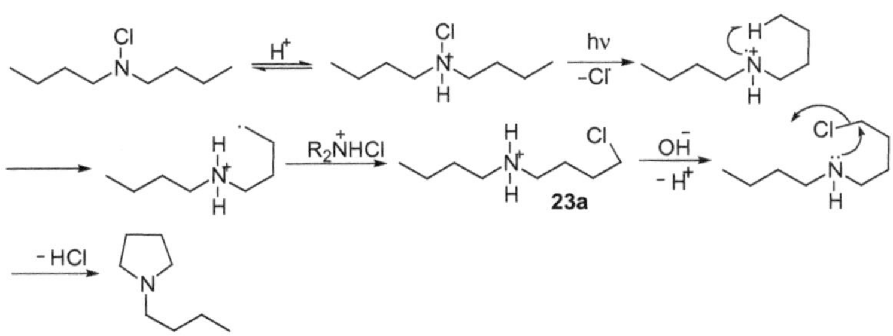

When an N-halogenated amine contains two δ carbons, the abstraction of hydrogen atom preferably takes place from secondary and tertiary δ-carbons. For example, N-chloro N-butyl N-pentyl amine **25** gives only 1-*n*-butyl-2-methylpyrrolidine **26**. 1-Amylpyrrolidine **27** is not detected.

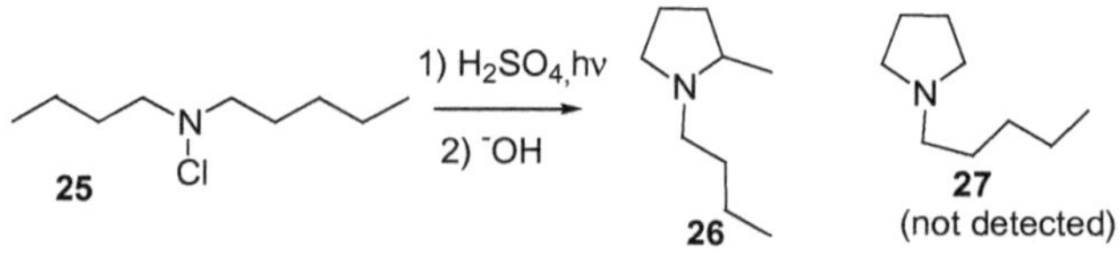

This reaction was utilized by Hofmann for synthesis of ö-coneceine **28** from N-bromoconiine **29** [19].

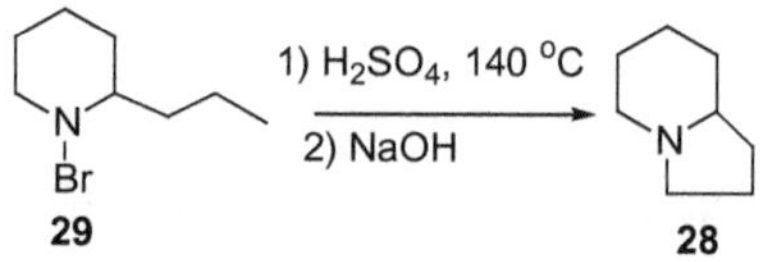

Loffler reported the synthesis of alkaloid nicotine **30** using this reaction [20].

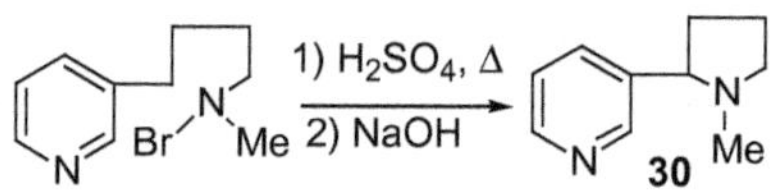

This reaction has been used extensively for introduction of functionality in different classes of nitrogen heterocycles. For example, this reaction has been used in the synthesis of alkaloid, N-methylgranatinine **31** from N-chloro-N-methylcyclooctylamine **32** [21] and in a key step in the synthesis of steroidal alkaloid, dihydroconessine **33** [22].

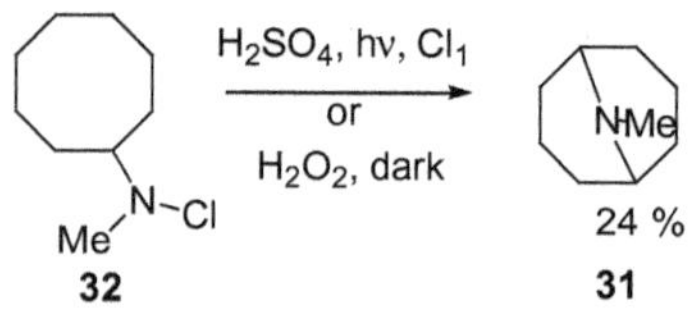

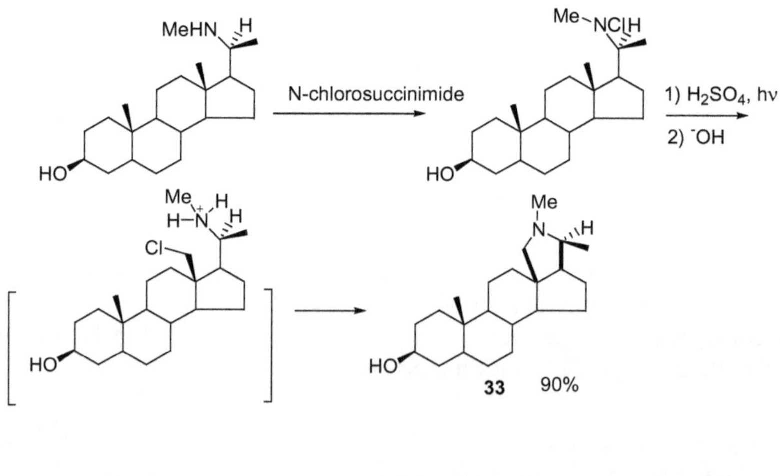

The HLF reactions have been applied in the synthesis of different classes of organic compounds using usual acidic medium. The following examples [23–28] are illustrative:

1. 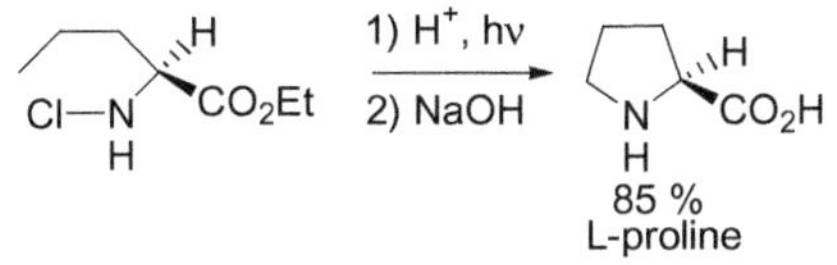

Ref. 23

2. 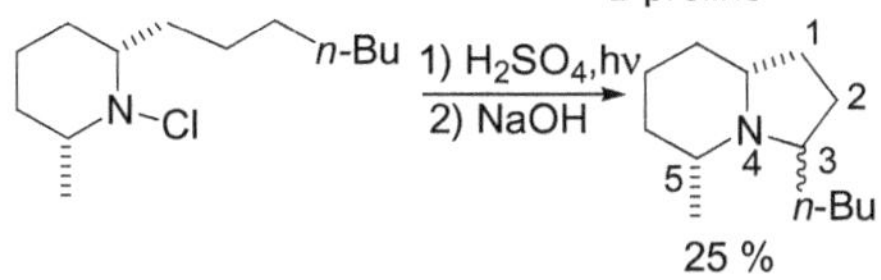

25 %
3-butyl-5-methyloctahydroindolizine
Ref. 24

3. 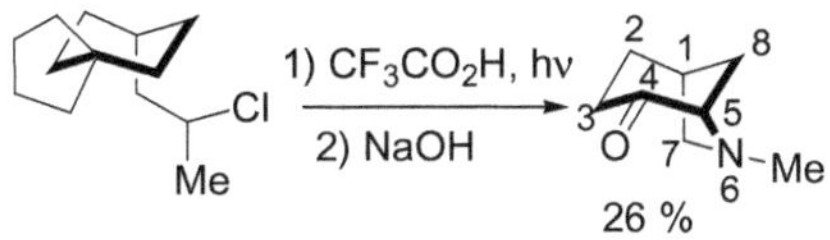

6-methyl-6-azabicyclo[3.2.1]-octan-4-one
Ref. 25

4. 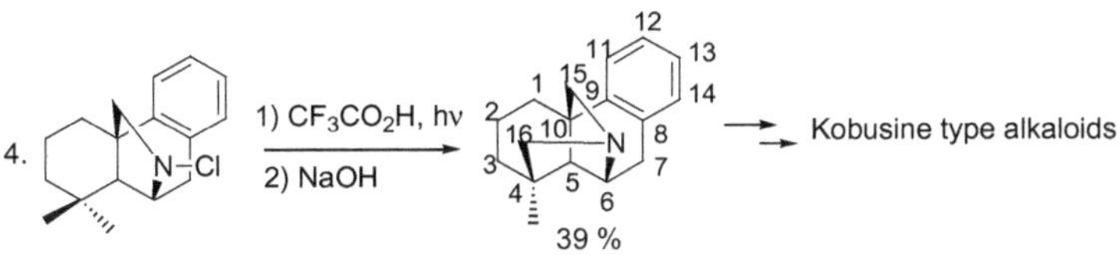

(±) 6,15,16-iminopodocarpane-8,11,13-triene
Ref. 26

5. 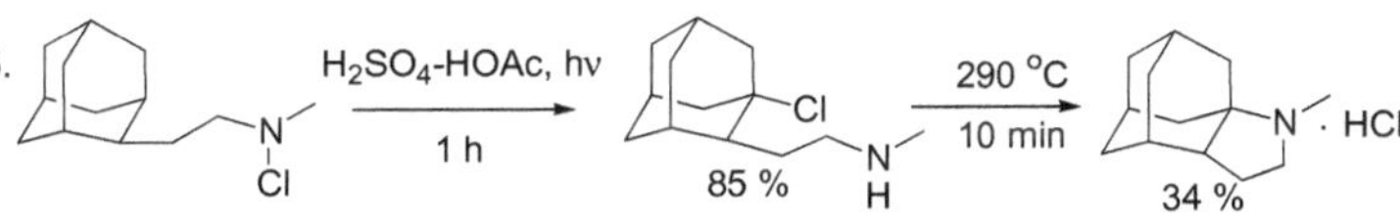

Ref. 27
1-methyladamantano-[1,2b]-pyrrolidine

6. 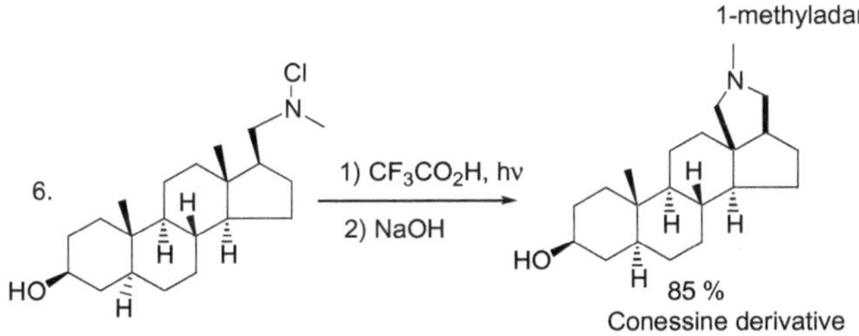

Ref. 28

Kimura and Ban modified the condition of HLF reaction in weakly basic medium to improve the yield of the reaction [29]. The authors suggested that weakly basic medium increased the stability of the intermediate alkyl radical. They reported the synthesis of dihydrodeoxyepiallocernuine **34** using this reaction condition [29].

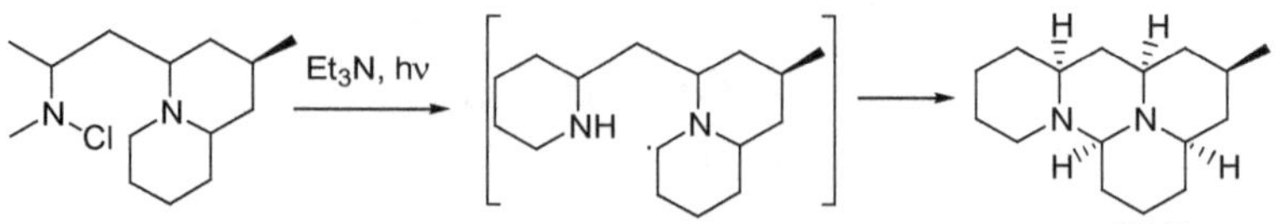

30 %
34

Baldwin and Doll modified the HLF reaction condition in neutral medium to prepare a key tricyclic intermediate **35** for the synthesis of alkaloid gelsemicine **36** and found better yield of the product [30].

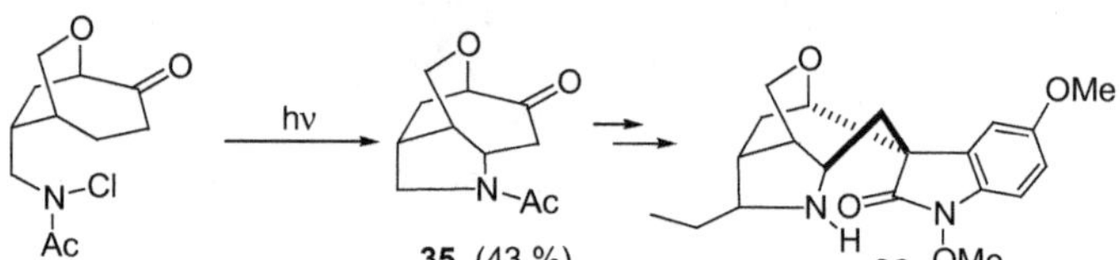

35 (43 %) **36**
2-aza-7-oxatricyclo[4.3.2.0^{4,8}]-undecane

Suarez et al. modified the HLF reaction condition in neutral medium for the substrates, N-nitroamides, N-cyanamides, N-phosphoramidates and N-carbamates of steroids **37** for synthesis of five membered nitrogen ring compounds **38** using iodine and DIB or LTA as oxidizing agent, and a tungsten lamp for irradiation with visible light [31]. All these substrates react with hypervalent (trivalent) iodine reagents to generate nitrogen-radicals via homolytic cleavage of iodoamide intermediate. The nitrogen radicals undergo hydrogen atom abstraction from δ-carbons, followed by addition of iodine radical to δ-C and ring formation via elimination of HI (Scheme 10.2). N-Phosphoramidate substrate gives an excellent yield and other substrates give fair to good yields. The following examples are illustrative [31–33].

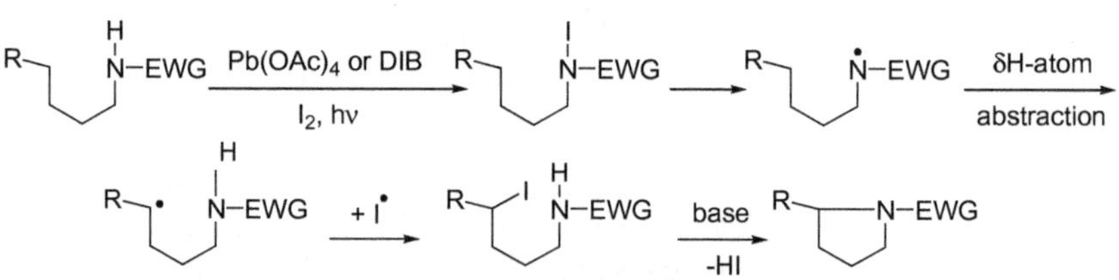

EWG = NO_2, CN, $P(O)(OR)_2$, CBz, Boc

Scheme 10.2 The generalized pathway of HLF reactions in neutral medium

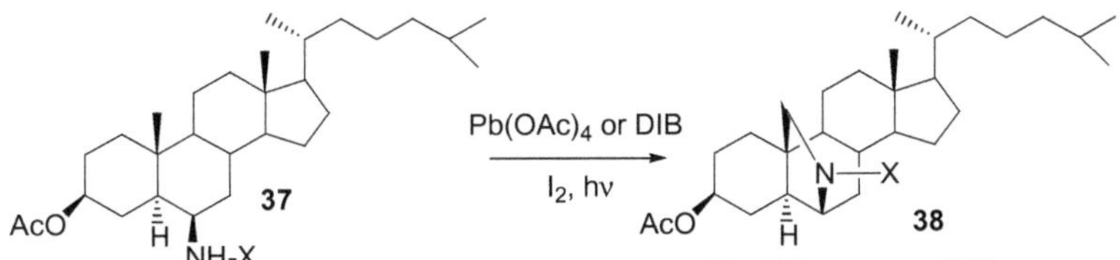

X = NO_2 63%
= $P(O)(OEt)_2$ 99%
= CN 64%

Using this modified reaction condition, Suarez et al. synthesized oxoindolizidine **39** in 82 % yield [34].

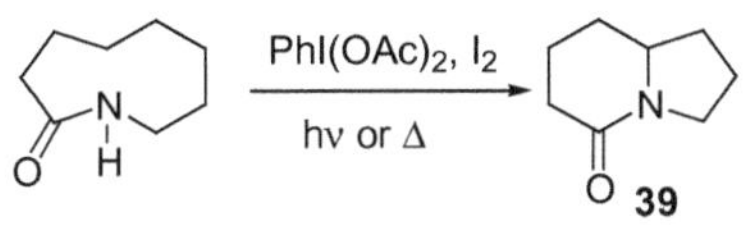

39
1-aza-bicyclo-[4.3.0]nonan-2-one

Yokoyama et al. applied the HLF reaction in neutral medium to sulfonamides **40** and found excellent yield of **41** [35].

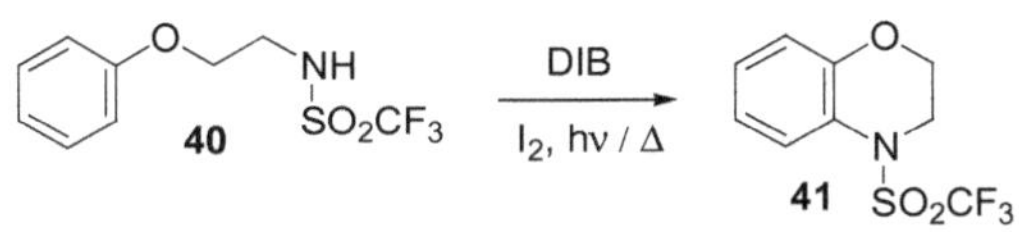

72%

10.5 Problems

10.5.1. Predict the product(s) with structure and stereochemistry of the following reactions:

(a)

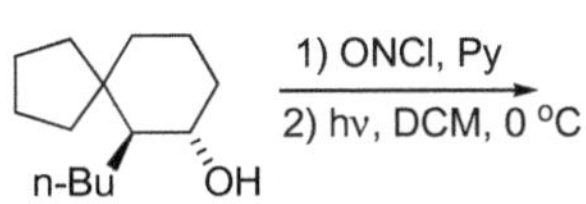

(b)

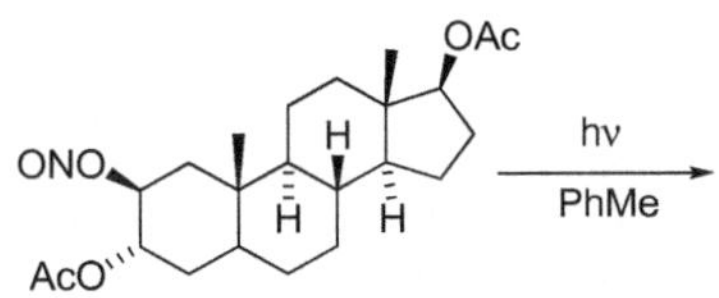

(c)

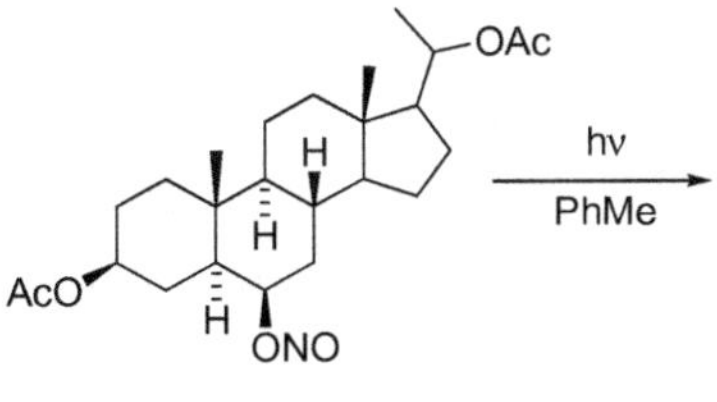

(d)

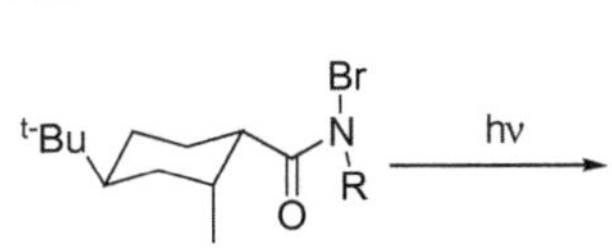

(e)

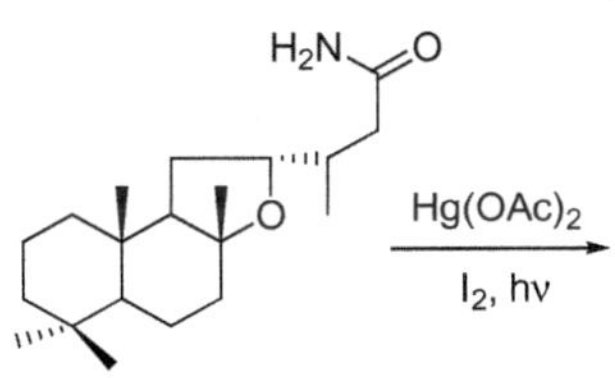

10.5.2. Suggest the mechanism of the following reactions:

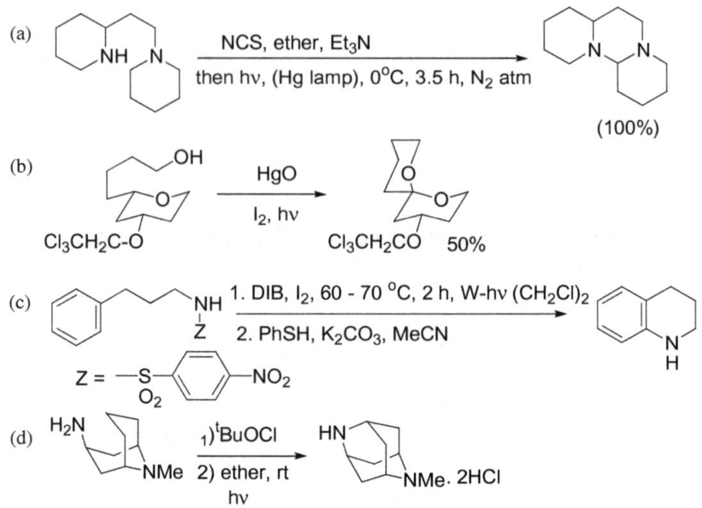

10.6 Further Reading

1. Kopecky J (1992) Organic photochemistry. Wiley VCH, New York.
2. Majitich G, Wheless K (1995) Remote intramolecular free radical functionalizations: An update. Tetrahedron 51:7095–7129.

References

1. Heusler K, Kalvoda J (1964) Angew Chem 3:525
2. Majetich G, Wheless K (1995) Tetrahedron 51:7095
3. Barton DHR, Hesse RH, Pechet MM, Smith LC (1979) J Chem Soc Perkin Trans 1:1159
4. Barton DHR, Beaton JM (1960) J Am Chem Soc 82:2641
5. Barton DHR, Budhiraja RP, Mc Ghie JF (1969) J Chem Soc C 336
6. Corey EJ, Arnett JF, Widiger GN (1975) J Am Chem Soc 97:430
7. Corey EJ, Hahl RW (1989) Tetrahedron Lett 30:3023
8. Stotter PL, Hill KA, Friedman MD (1987) Heterocycles 25:259
9. Walling C, Padwa A (1963) J Am Chem Soc 85:1597; Akhter A, Barton DHR (1964) J Am Chem Soc 86:1528
10. Norman ROC, Coxon JM (1993) Principles of organic synthesis, 3rd edn. ELBS with Chapman & Hall, Oxford, p 555

11. Akhtar A, Barton DHR (1964) J Am Chem Soc 86:1528; Heusler K, Kalvoda J (1964) Angew Chem Int Ed 3:525; Mihailovic ML, Gojkovic S, Konstantinovic S (1973) Tetrahedron 29:3675
12. Hadd HE (1978) Steroids 31:453
13. Ceccherelli P, Curini M, Marcotuillio MC, Mylari BL, Wenkert E (1986) J Org Chem 51:1505
14. Brun P, Pally M, Waegell B (1970) Tetrahedron Lett 331; Brun P, Waegell B (1976) Tetrahedron 32: 1137
15. Dorta RL, Francisco CG, Freire R, Suarez E (1988) Tetrahedron Lett 29:5429
16. Kay IT, Bartholomew D (1984) Tetrahedron Lett 25:2035
17. Wolff ME (1963) Chem Rev 63:55; Wawzonck S, Culbertson TP (1959) J Am Chem Soc 81:3367
18. Corey EJ, Hertler WR (1960) J Am Chem Soc 82:1657
19. Hofmann AW (1885) Ber 18:109; Lellmann E (1890) Ber 23:2141
20. Loffler K, Freytag C (1909) Ber 42:3427; Loffler K, Kober S (1909) Ber 42:3431; Loffler K (1910) Ber 43:2035
21. Wawzonek S, Thelan PJ (1950) J Am Chem Soc 72:2118
22. Van de Woude G, van Hove L (1973) Bull Soc Chim Belg 82: 49; ibid (1975) Bull Soc Chim Belg 84:911; van de Woude G, Biesemans M, van Hoe L (1980) Bull Soc Chim Belg 89:993
23. Titouani SL, Lavergne JP, Viallefont P, Jacquier R (1980) Tetrahedron 36:2961
24. Sonnet PE, Oliver JE (1975) J Heterocycl Chem 12:289
25. Esposito G, Furstoss R, Waegell B (1971) Tetrahedron Lett 12:899
26. Shibanuma Y, Okamoto T (1985) Chem Pharm Bull 33:3187
27. Narayanan Vl, Setescak L (1971) J Org Chem 3: 4127
28. Hora J, Sorm F (1968) Collect Czech Chem Commun 33:2059; Van De Woude G, Van Hore L (1973) Bull Soc Chim Belg 82:49
29. Kimura M, Ban Y (1976) Synthesis 201; Ban Y, Kimura M, Oishi T (1976) Chem Pharm Bull 24:1490
30. Baldwin SW, Doll RJ (1979) Tetrahedron Lett 3275
31. Betancor C, Concepcion JI, Hernandez R, Salazar JA, Suarez E (1983) J Org Chem 48:4430
32. Carrau R, Hernandez R, Suarez E, Betancor C (1987) J Chem Soc Perkin Trans 1:937
33. De Armas P, Francisco CG, Hernandez R, Salazar JA, Suarez E (1988) J Chem Soc Perkin Trans 1:3255
34. Dorta RL, Francisco CG, Suarez E (1989) Chem Commun 1168
35. Togo H, Hoshina Y, Muraki T, Nakayama H, Yokoyama M (1998) J Org Chem 63:5193

Chapter 11
Photochemistry in Nature and Applied Photochemistry

11.1 Introduction

Several photochemical processes take place in nature for the evolution of life and their existence on earth. Such evolutionary processes of life have been brought about by photochemical reactions of supramolecular devices. Among these photochemical reactions, photosynthesis of plants, vision of animals, and depletion of stratospheric ozone layer are significant. Based on these photo-induced reactions, several devices have been developed to meet the needs of our daily life. For example, zeolites have been used in the field of sunscreens and purification of drinking water, while porphyrin, metalloporphyrin, and fullerene have been used in the building block of triads for the harvesting of solar energy as electrical energy. Artificial photosynthesis systems have been developed for the most attracting fuel-producing reaction in photolytic cleavage of water.

11.2 Depletion of Stratospheric Ozone Layer from Photochemical Degradation

The stratospheric region of the atmosphere is located above 15–35 km from the surface of the earth. It contains a deep layer of ozone that acts as a filter of harmful UV radiation of sunlight to reach the earth's surface and thus protects us from hazardous effect of UV radiation. The massive loss of ozone in the stratosphere occurs daily by atmospheric pollutants, UV-induced photolysis of ozone in the presence of man-made chlorofluorocarbons (CFCs), hydrochlorofluorocarbons (HCFCs), halons (brominated hydrocarbons), CCl_4, and methylchloroform (CH_3CCl_3). These halocarbons generate halogen radicals which have active roles for photolysis of ozone. These are also derived from gaseous chlorine and hydrochloric

B. Dinda, *Essentials of Pericyclic and Photochemical Reactions*,
Lecture Notes in Chemistry 93, DOI 10.1007/978-3-319-45934-9_11

acid, which are ejected from volcanoes. The massive loss of ozone has created several holes in ozone layer, known as ozone holes.

$$CFCl_3 \xrightarrow{h\nu} \dot{C}FCl_2 + \dot{C}l$$
$$CH_3Br \xrightarrow{h\nu} \dot{C}H_3 + \dot{B}r$$
$$\dot{X} + O_3 \longrightarrow O_2 + X\dot{O}$$
$$(\dot{X} = \dot{C}l, \dot{B}r)$$
$$CCl_4 \xrightarrow{h\nu} \dot{C}Cl_3 + \dot{C}l$$
$$Cl_2 \xrightarrow{h\nu} 2\dot{C}l$$
$$X\dot{O} + O_3 \longrightarrow \dot{X} + 2O_2$$

It is a chain reaction and thus one chlorine radical can break down more than 100,000 molecules of ozone. Bromine radical is more destructive than chlorine radical because of its extended chain cycle. These CFCs are widely used as coolants in refrigerators, cold cleaning solvents, aerosol spray cans, and foaming products. CCl_4 and halons are widely used in fire extinguishers and air conditionings. The lifetime of CFCs and halons are more than a century. The emission of CFC-113a (1,1,1-trichloro-2,2,2-trifluoroethane) in the atmosphere has jumped to 45 % in 2010–2012 [1]. The ozone layer protects the UV-B (280–320 nm) radiation from sunlight to reach the earth's surface. Due to these ozone holes, the UV radiation from sun reaches the earth's surface and causes malignant melanoma, corneal damage, cataract, DNA mutation, and enormous growth of harmful cyanobacteria. To reduce the levels of different halocarbons in the atmosphere, an International Treaty, known as Montreal Protocol was signed in 1987 by different countries for the use of less hazardous chemicals as alternative to halocarbons.

11.3 Photochemical Smog in Polluted Zones of Troposphere

The photochemical smog (polluted smoke) in the industrial areas is the product of photochemical reactions of primary air pollutants such as nitrogen oxides (NO_2,NO) and hydrocarbons in the presence of bright sunlight. These pollutants are generated from emissions of vehicles and industrial plants. These primary pollutants undergo complex reactions to give secondary pollutants such as ozone, aldehydes, peroxy-acyl nitrate (PAN), peroxybenzoyl nitrate (PBzN), and particulate matter which are the major constituents of smog. The following are the major reactions:

$$NO_2 \xrightarrow{h\nu} NO + O$$

$$O + O_2 \longrightarrow O_3$$

$$O_2 + h\nu \longrightarrow 2\dot{O}$$

$$O + h\nu \longrightarrow \dot{O} + e$$

$$O_3 + h\nu \longrightarrow \dot{O} + O_2$$

$$\dot{O} + H_2O \longrightarrow 2\,\dot{O}H$$

$$\dot{O}H + \underset{\text{hydrocarbon}}{RCH_3} \longrightarrow R\dot{C}H_2 + H_2O$$

$$R\dot{C}H_2 + O_2 \longrightarrow RCH_2\dot{O}_2$$

$$RCH_2\dot{O}_2 + NO \longrightarrow RCH_2\dot{O} + NO_2$$

$$RCH_2\dot{O} + O_2 \longrightarrow RCHO + H\dot{O}_2$$

$$H\dot{O}_2 + NO \longrightarrow \dot{O}H + NO_2$$

$$RCHO + \dot{O}H \longrightarrow R\dot{C}{=}O + H_2O$$

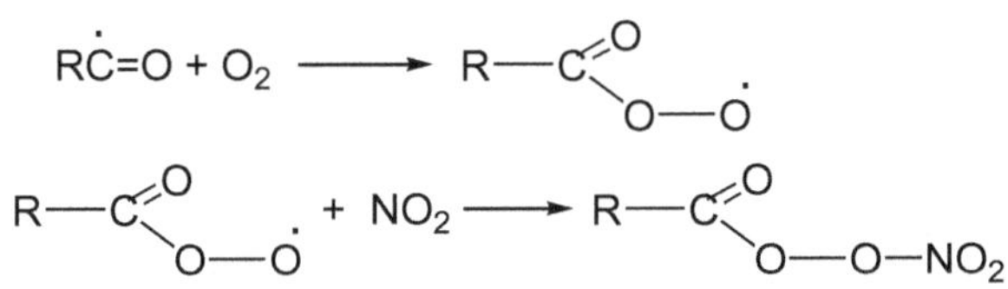

PAN, R = CH_3
PBZN, R = C_6H_5

The formation of photochemical smog in the industrial belt reduces the visibility and causes health hazards such as headache, eye irritation, cough, bronchial, other respiratory problems, and pulmonary edema (accumulation of fluids in lungs). These pollutants also create irritations in lungs.

11.4 Photochemistry of Vision: Geometrical Isomerisation of Retinal

Our vision involves a photochemical process in the visible light (400–800 nm). The retina of the eye is lined with millions of photoreceptor cells, called rods and cones. The rod cells are sensitive to dim light, whereas the cone cells are sensitive to bright light and color vision. Hence, rod cells are black and white receptors while cone cells are color receptors. In color vision, there are three types of cone cells corresponding to red ($\sim$622 nm), green ($\sim$535 nm), and blue ($\sim$455 nm) light receptors. In both rod and cone cells, small guest molecules, 11-*cis*-retinal **1** is held within the internal cavity of the much larger protein host molecule (opsin) as a result of noncovalent

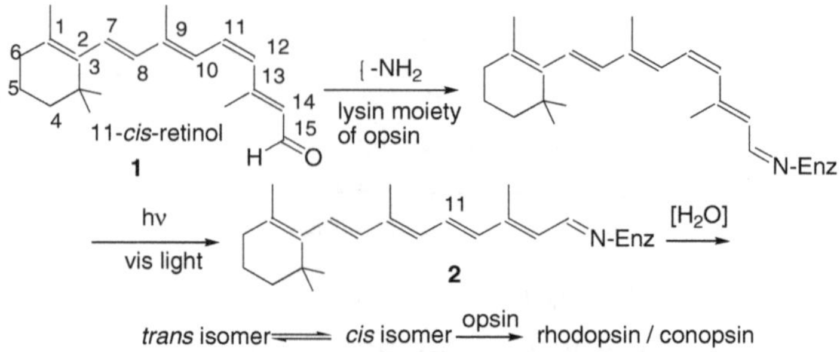

Fig. 11.1 Photochemical reaction in the vision process

bonding and thus the light-absorbing part of rod cells is called rhodopsin and of cone is conopsin. After absorbing a photon, the 11-*cis*-retinal undergoes photoisomerization into its geometrical 11-*trans*-retinal **2** having all *trans*-double bonds (Fig. 11.1). This change in molecular geometry of the retinal does not fit well in the protein opsin, and so a series of geometrical changes occur in the protein and the attached plasma membrane resulting to set a reverse potential difference across the nerve cell membrane. This potential difference is passed along the adjoining nerve cell as an electrical impulse. The nerve cell then carries the impulse to the brain, where the visual information is interpreted. Thus, eye functions as a transducer as does CCD (charge coupled device) camera by the photochemical reactions between 11-*cis*-retinal and opsin. The time taken for this whole process is of the order of a few picoseconds. After the process, the *trans*-isomer is hydrolyzed and converted to *cis*-isomer and attached well to opsin for the recyclic process.

11.5 Phototherapy of Neonatal Jaundice

Neonatal jaundice occurs in a newborn baby due to deposition of yellow pigment bilirubin in brain cells and skin because of abnormal liver function. It occurs due to rapid breakdown of hemoglobin to bilirubin in red blood cells compared to breakdown of bilirubin in liver. Untreated baby suffers from the damage of central nervous system. For treatment of this disease, the affected baby is subjected to phototherapy with visible bluish-green light.

In bilirubin molecule **3**, two isomerizable double bonds at C(4) and C(15) normally exist as *cis,cis*-isomer. On exposure to visible blue-green light, isomerization of one or both double bonds takes place to produce *trans,cis*-(4*E*,15*Z*) (major) **4** and *trans,trans*-(4*E*,15*E*) (minor) bilirubins (Fig. 11.2) [2]. These compounds can form hydrogen bonds with water molecules and become highly soluble in water and are excreted in urine, relieving the baby from this toxic effect. The isomerization reaction is very fast and occurs in femtosecond range.

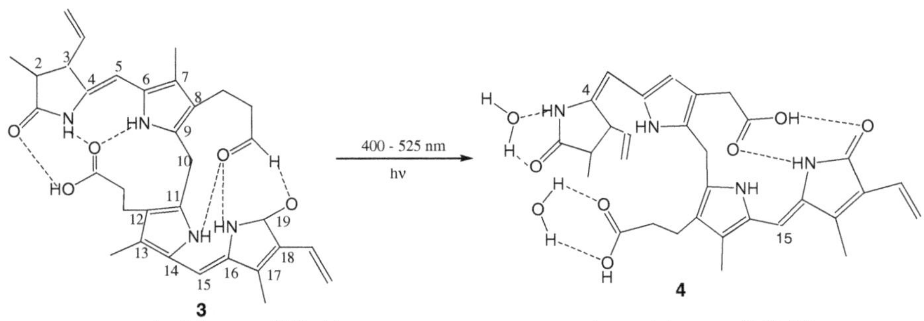

3 *cis,cis*-isomer of bilirubin

4 *trans,cis*-isomer of bilirubin

Fig. 11.2 *Cis–trans*-isomerisation of bilirubin

11.6 Photosynthesis of Plants and Bacteria

Photosynthesis is a photochemical process by which green plants, sea weeds, algae and certain bacteria adsorb solar energy and utilize it to convert the atmospheric carbon dioxide to carbohydrates in the presence of water. The overall reaction for all cases except for photosynthetic bacteria may be written as:

$$nCO_2 + nH_2O + h\nu \rightarrow (CH_2O)_n + nO_2,\ \Delta G = 500\,kJ\,mol^{-1}(CO_2)$$

Photosynthetic process occurs on a large scale, fixing more than 2×10^{11} tons of carbon from atmospheric CO_2 as carbohydrates per annum.

Photosynthetic process occurs in two stages, namely light reactions and dark reactions.

Light reactions: Solar light energy is converted into short-term chemical energy, producing oxygen from water as a by-product. Certain reducing agents, such as NADPH is formed by hydrogen atoms of water with simultaneous phosphorylation of ADP to ATP. NADPH and ATP are considered as chemical energy produced in the light reaction of photosynthesis [3].

$$2H_2O + 2NADP^+ + 2ADP + 2Pi + h\nu \rightarrow O_2 + 2NADPH + 2H^+ + 2ATP$$

Dark reactions: The short-term chemical energy from light reactions is utilized for reduction of carbon dioxide into glucose in the absence of light [4].

$$6CO_2 + 18ATP + 12NADPH + 12H^+ \rightarrow C_6H_{12}O_6 + 18ADP + 18Pi + 12NADP^+ + 6H_2O$$

Dark reactions do not imply that these reactions take place in the dark. It implies that these reactions occur without light but take place simultaneously with light reactions in daytime.

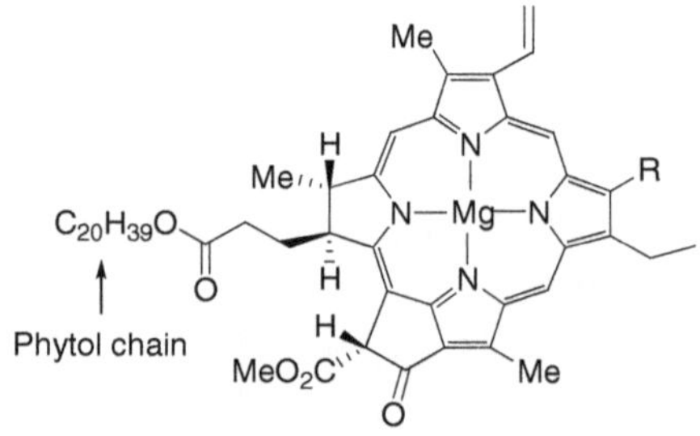

Chlorophyll a: R = Me
Chlorophyll b: R = CHO

Fig. 11.3 Structures of chlorophyll a and chlorophyll b

We will concern about the light-induced photochemical reactions of the supramolecules, chlorophylls. The light reaction takes place in the molecular device located mainly in the leaves of the plants. The leaves of plants are green because they contain the light-absorbing pigments, called chlorophylls, which absorb the blue and red regions of visible white light, leaving the intermediate green light to be reflected to our eyes. A chlorophyll molecule (Fig. 11.3) consists of two major parts, one rigid, planar and conjugated porphyrin ring co-ordinated with Mg^{2+} ion, and a long hydrophobic phytyl chain, which keeps the chlorophyll molecule embedded in the photosynthetic membrane of leaves. Accessory pigments called carotenoids (such as β-carotene; Fig. 11.4) and blue or red phycobilins (such as phycoerythrobilin; Fig. 11.4) are also found in plants. These accessory pigments are essential for photoprotective mechanism employed by the plants to dissipate excess photon energy absorbed by chlorophyll as heat, thus preventing the formation of highly reactive oxygen species and protecting the biological system of leaves from photochemical damage. Light reactions of photosynthetic process occur in the following sequence:

Light harvesting Several hundred chlorophyll pigment molecules act together as the photosynthetic unit, which is made up of two basic sections, light-harvesting antenna and the reaction center. The light-harvesting antenna allows the absorption of light of a broad range of wavelengths and its rapid transfer to the reaction center. The energy transfer occurs by means of the Coulombic long-range mechanism. Excitation of the reaction center is over within a few femtoseconds. The harvested light is transferred as electrons through an electron transport chain of Z-shape, called the Z-scheme (Fig. 11.5).

Reaction centers Light-dependent reactions take place at two reaction sites, photosystem-I (PS-I) and photosystem-II (PS-II) and these are connected in Z-scheme. Both PS-I and PS-II are associated with chlorophylls a and b in different proportions. PS-I is characterized by its absorption maxima at longer wavelength (700 nm) and is designated as P700, P meaning pigment and 700 being the wavelength in nanometre at which the dimer absorbs most strongly. Similarly, PS-II is designated as P680 (maximum absorption at 680 nm). The chemical reactions occur within a few picoseconds after absorption of light.

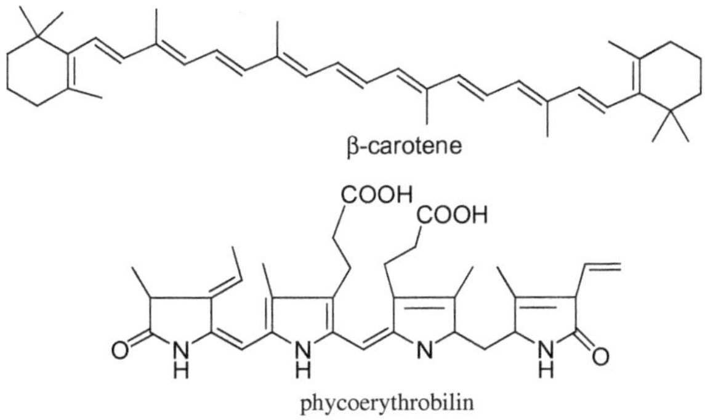

Fig. 11.4 Structures of β-carotene and phycoerythrobilin

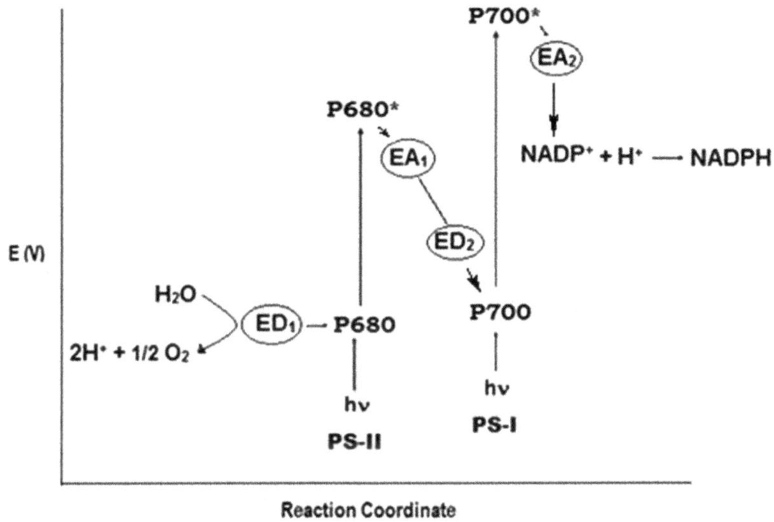

Fig. 11.5 Photochemical electron transport chain in a Z-scheme during light-dependent reactions of photosynthesis. *EA* and *ED* refer to the electron acceptor and electron donor of the two photosystems. Adapted with permission from (Wardle B, 2009 Principles and Applications of Photochemistry, Wiley, p. 226). Copyright (2009) John Wiley & Sons

P700 chlorophyll unit absorbs light and is excited to P700*. P700* loses an electron to the electron acceptor EA_2 (P-430). EA_2 transfers its electron to $NADP^+$ through a number of electron carriers including ferredoxin. After transferring its electron, P700* becomes $P700^+$ (in the oxidized state) and it requires one electron to reach its ground state.

$$\mathbf{P700} \xrightarrow{h\nu} \mathbf{P700^*} \xrightarrow[\mathbf{EA_2} \;\rightarrow\; \mathbf{EA_2^-}]{} \mathbf{P700^+}$$

Then the PS-II comes into operation. P680 absorbs light and is excited to P680*. The excited P680* loses an electron to an unidentified electron acceptor EA_1. EA_1 transfers its electron to ED_2 (plastocyanin). ED_2 transfers its electron directly to $P700^+$ and $P700^+$ returns to its ground-state P700 and continues the process of light absorption. After losing an electron, P680* becomes $P680^+$, which is strongly oxidizing and its redox potential is very close to that of $O_2/2H_2O$ system (E^0, +0.816 V). So, water molecule in the presence of Mn(II)-protein absorbs light and splits into O_2 and H^+ with the liberation of one electron. The liberated electron is received by $P680^+$ and returns to its ground-state P680 for continuation of photosynthetic process.

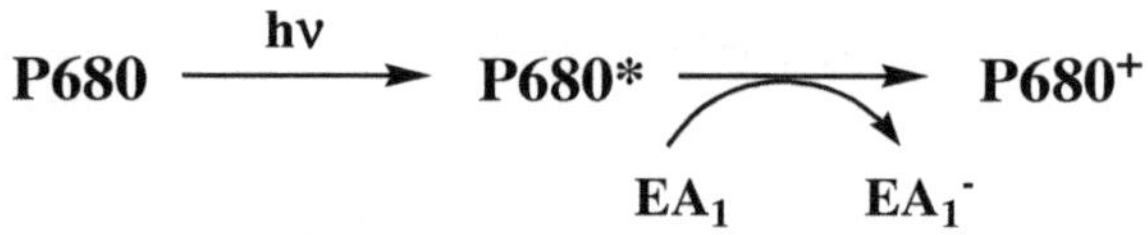

$$H_2O + h\nu \rightarrow 2H^+ + {}^1/_2O_2 + 2e^-$$

$NADP^+$ after receiving two electrons, one from PS-I and another from PS-II, is converted to electron-rich NADPH in the presence of flavoprotein {Fd(II)} called ferredoxin-NADP oxidoreductase.

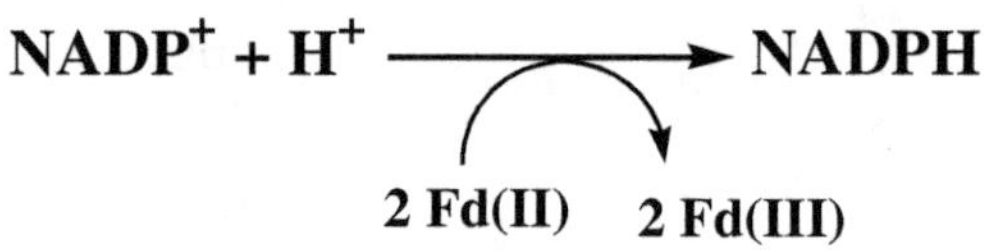

Thus, NADPH is formed along with O_2, H^+ and ATP in the light phase of photosynthesis. The products, NADPH and ATP are utilized in the dark phase of photosynthesis for fixation of CO_2 as glucose.

$$nCO_2 + nH_2O + h\nu \rightarrow (CH_2O)_n + nO_2$$

For the assimilation of one molecule of CO_2 by green plants, two molecules of NADPH are needed. To utilize H_2O, both PS-I and PS-II are to be activated four times each to produce the four electrons required to reduce 2 $NADP^+$. Therefore, a total of eight quanta of light will be required for conversion of one mole of CO_2 into carbohydrate.

$$2NADP^+ + 2H_2O + h\nu \rightarrow 2NADPH + 2H^+ + O_2$$

Archaebacteria use a simpler method using a pigment, archaea rhodopsin similar to vision process. The pigment changes its configuration in response to sunlight acting as a proton pump producing a proton gradient for conversion of light energy into chemical energy. This process does not involve CO_2 fixation and does not release O_2 [5].

11.6.1 *Artificial Photosynthesis*

The multi-step electron transfer process in natural photosynthesis has been utilized in the construction of various triads using porphyrin, metalloporphyrin, fullerene, and imide as basic components for harvesting solar energy as electrical energy and for photoreduction of water to get clean fuel hydrogen [6, 7]. Recently, tetrads, pentads and hexads have been constructed using porphyrin, fullerene, and a chromophoric unit as basic components for fast energy transfer process.

11.7 Photo-Induced DNA-Damage and Its Repair

Pyrimidine base, thymine **5** present in DNA, on exposure to UV light undergoes dimerization to give **6** and **7** by [2+2]-cycloaddition reaction [8].

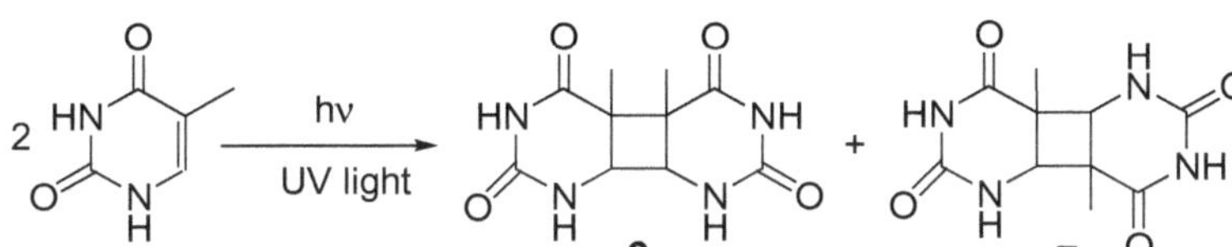

5 **6** **7**

Due to this dimerization, this dimeric thymine alters the structure of DNA and consequently inhibits replication of DNA. Such mutation of DNA results in cell death in some instances. Repairing of this mutation may be done by exposing the DNA at longer wavelength (>300 nm) of light [9].

11.8 Conservation of Solar Energy as Electrical Energy: Photovoltaic Solar Cells

The conversion of solar energy into direct current electricity can be achieved by the use of photovoltaic solar cells. Photovoltaic solar cells are electrical cells based on semiconductors that produce electricity from sunlight and deliver electricity to an external load. It is now the third important renewable energy source after hydro and wind power resources. More than 100 countries use solar cells for the harvest of solar energy. According to the estimate of International Energy Agency, about 177 GW (giga-watt) of electricity has been produced in 2014 from solar photovoltaic installations and Germany is the world's largest producer of electricity from solar resource, contributing about 7 % of its annual domestic electricity consumption [10]. Commercial solar cells were initially developed using silicon-based p–n junctions as semiconductors. Silicon-based p–n junctions were made by n-type doping with arsenic or phosphorous atom (five valence electrons) in a silicon crystal and p-type doping with gallium or indium (three valence electrons) in a silicon crystal. The n-type doping provides an extra loosely bound electron that is more easily excited into the conduction band (CB) than in case of pure silicon and results in electrical conductivity of CB electrons; whereas p-type doping creates a hole in the silicon crystal from which thermally excited electrons from the valence band (VB) can move, leaving behind mobile holes. When these n-doped and p-doped silicons are joined together into a single semiconductor crystal, the electrons in CB and VB have an average potential energy, known as Fermi level. Excess electrons move from the n-type side to the p-type side, resulting in a buildup of negative charge along the p-type side and a buildup of positive charge along the n-type side at the interface. When light is absorbed by the p–n junction, it acts as a photovoltaic cell resulting in the promotion of electrons from the VB to CB, forming an electron hole pair. When these p- and n-type silicon sites are connected to an external load, the electron–hole pair tends to separate, resulting in the production of current from the flow of electrons from the n-region to p-region. Thus, the n-type and p-type silicon sites become the negative and positive pole, respectively, of the solar cell (Fig. 11.6).

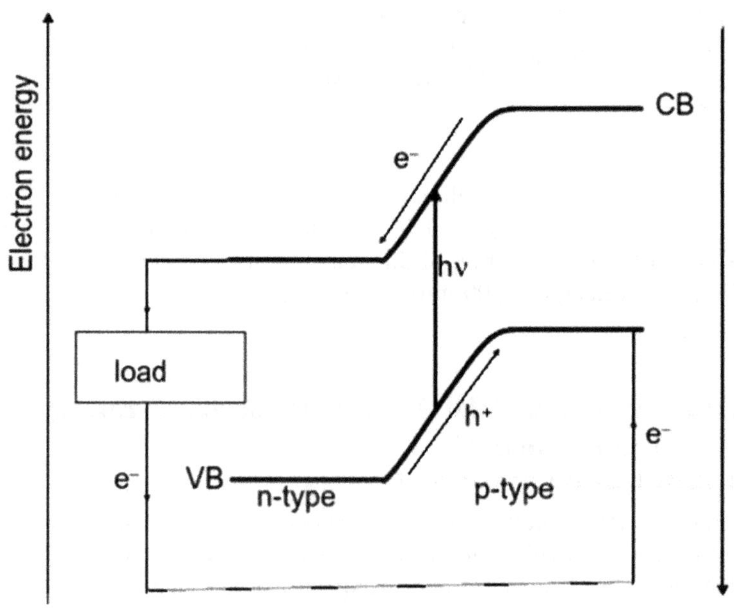

Fig. 11.6 The working mechanism of a silicon p–n junction solar cell. Adapted with permission from (Wardle B, 2009 Principles and Applications of Photochemistry, Wiley, p. 217). Copyright (2009) John Wiley & Sons

The original silicon-based p–n junction solar cells have been replaced by recent organic solar cells to reduce the cost of production.

The dye-sensitized solar cells (DSSCs) have received more attention from the industry because of high efficiency, low cost, environment friendliness, low incident light angle dependence, flexibility, etc. In these dye-sensitized solar cells, nanometer-sized TiO_2 (or Nb_2O_5) particles are allowed to absorb large amounts of colored organic dye based on Ru(II) having a broad absorption range of visible light. For example, ruthenium-polypyridine dye, Ru(4,4′-dicarboxy-2,2′-bipyridine)$_2$ (NCS)$_2$ **8** is used. The nanoparticles of TiO_2 are deposited on a glass support covered with a transparent layer of Sn-doped indium oxide (ITO) having electron conduction property. To regenerate the sensitizing dye from its oxidized form, a liquid electrolyte solution of iodide/triiodide mixture is used as a mediator in between the electrodes. Usually ITO and Pt are used as working and counter-electrode, respectively. As in a conventional alkaline battery, ITO acts as anode and Pt as cathode in the redox shuttle electrolyte, I_3/I. The schematic structure of a nanocrystalline DSSC is shown in Fig. 11.7. Photoexcitation of the dye results the flow of electrons into CB of working electrode TiO_2 and from CB to external circuit via working electrode ITO to counter-electrode platinum, which reduces triiodide to iodide. The generated iodide reduces Ru(III) to Ru(II) for continuation of light absorption process. The following reactions take place at the electrodes:

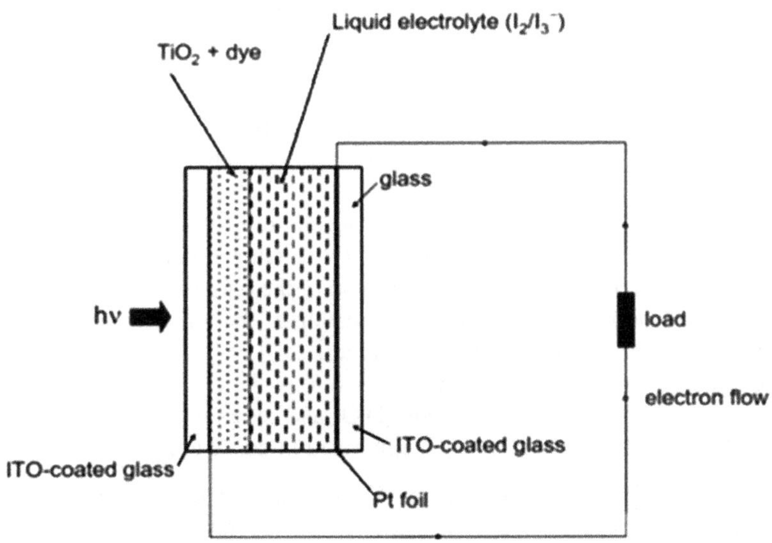

Fig. 11.7 Schematic diagram of a dye-sensitized solar cell where semiconductor TiO_2 nanoparticles are coated with Ru(II)-based dye. Adapted with permission from (Wardle B, 2009 Principles and Applications of Photochemistry, Wiley, p. 202). Copyright (2009) John Wiley & Sons

At ITO electrode:

$$\mathrm{Ru(II)} + \mathrm{hv} \rightarrow \mathrm{Ru(II)}*$$

$$\mathrm{Ru(II)}* \rightarrow \mathrm{Ru(III)} + \mathrm{e}^- (\text{injected to CB of } \mathrm{TiO_2})$$

At Pt electrode:

$$\mathrm{I_3^-} + 2\mathrm{e}^- \rightarrow 3\mathrm{I}^-$$

$$3\mathrm{I}^- + \mathrm{Ru(III)} \rightarrow \mathrm{I_3^-} + \mathrm{Ru(II)}$$

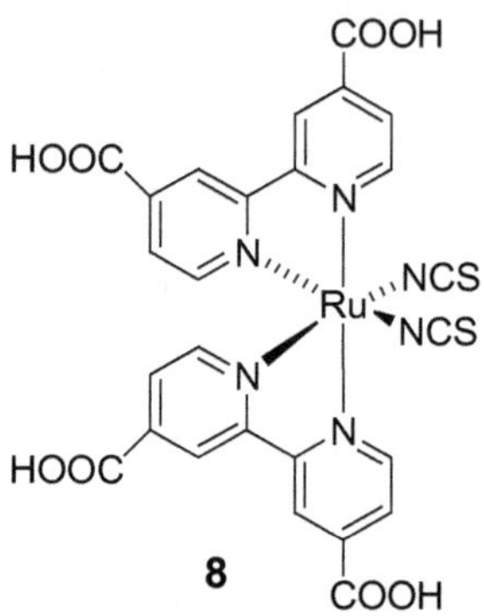

8

The overall efficiency of the DSSC depends on the following factors:

a. The HOMO-LUMO gap of the photosensitizer (dye): the smaller the size of this gap, the larger will be the photocurrent, due to the ability of the dye to absorb light of longer wavelength regions.
b. The LUMO-HOMO energy levels of the photosensitizer: the LUMO energy must be higher than that of the CB of TiO_2 to allow efficient electron injection into CB. Similarly, the HOMO energy level of the photosensitizer must be lower than that of the redox couple of the mediator so that efficient electron transfer can occur from mediator to photosensitizer.
c. The maximum voltage of the DSSC is given by the energy gap between Fermi level of the semiconductor electrode (ITO) and redox potential of the mediator. It is about 0.7 V (V_{oc}) under solar illumination conditions.

The DSSC differs substantially from silicon p–n junction solar cells by the fact that no holes are formed in the VB of the semiconductor.

The modern type of dye-sensitized solar cell known as the Gratzel cell was invented by O' Regan and Gratzel [11]. Its overall quantum efficiency for green light is about 90 % and power conversion efficiency is about 11 % [12].

The major disadvantages of the dye-sensitized solar cells are the stability problem of liquid electrolyte, high cost of ruthenium dye and platinum catalyst and volatile solvent of liquid electrolyte. At low temperatures, the electrolyte can freeze. The solvents used for the preparation of electrolyte are hazardous to human health and environment. Replacing the liquid electrolyte by a solidified melted salt has shown some promise but suffers from higher degradation during continued operations and is not flexible.

An alternative inexpensive organic polymer-based photovoltaic solar cell has been invented. In this device, p-type and n-type semiconductors are sequentially stacked on top of each other. In such devices, absorption of a photon by a π-conjugated polymer results in the formation of an excited state, where coulombically bound electron–hole pair (exciton) is created. This exciton diffuses to a region of interface of n-type semiconductor where exciton dissociation takes place and transport of charge to the respective electrodes occurs. For example, the photo-induced electron transfer from a donor layer (p-type) to acceptor layer (n-type) takes place in a polymer/fullerene-based organic bilayer solar cell, MDMO-PPV: PCBM, with power conversion efficiency of 2.5 % (Fig. 11.8) [13].

PCBM [(6,6)-phenyl-C_{61}-butyric acid methyl ester] has been widely used as electron acceptor in polymer/fullerene solar cells due to its greater solubility than C_{60}. The polymer MDMO-PPV [poly-{2-methoxy-5-(3,7-dimethyloctyloxy)}-*para*-phenylenevinylene] has been used as electron donor polymer for better absorption of solar light. This heterojunction solar cell has 80 wt% of PCBM, which is supposed to be the main light absorber. The change of solvent from toluene to chlorobenzene increases the efficiency by a factor of 3 [13].

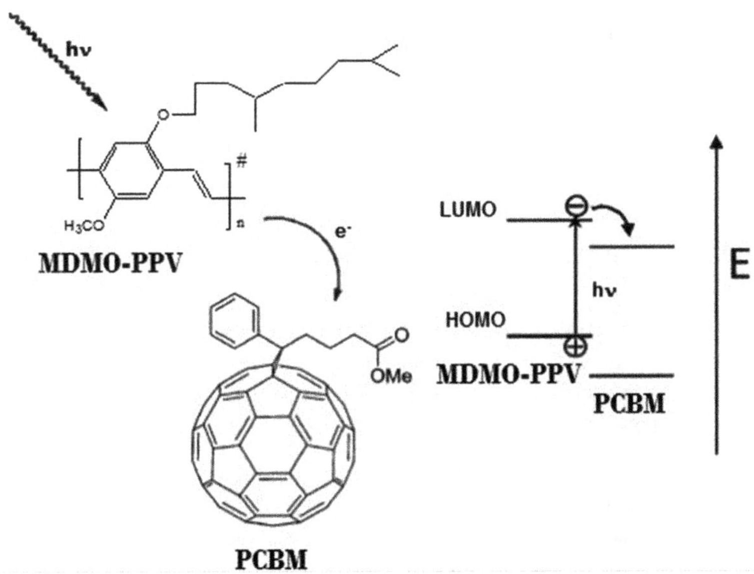

Fig. 11.8 Photo-induced electron transfer from excited MDMO-doped PPV to PCBM

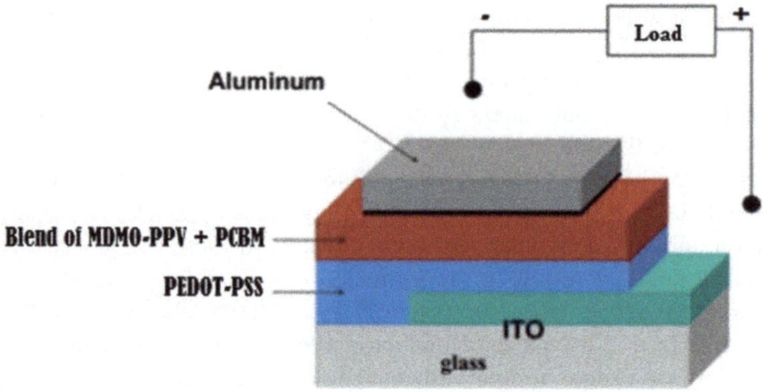

Fig. 11.9 Schematic device structure for polymer/fullerene bulk heterojunction solar cells. Adapted with permission from (Gunes et al. 2007 Chem Rev 107:1324). Copyright (2007) American Chemical Society

The general structure of the bilayer solar cells is similar to the light-emitting diodes (LEDs). The devices are fabricated in sandwich geometry (Fig. 11.9). The active layer is sandwiched between two contacts: an indium-tin-oxide electrode (ITO) (cathode) coated with a hole transport layer/glass or plastic foil. The blend polymer/PCBM solution was doctor-bladed on top of the hole conductor PEDOT: PSS [poly(3,4-ethylenedioxythiophene) doped with polystyrene sulfonic acid]. It improves the surface quality of ITO electrode. On the top of this polymer, about 100-nm aluminum layer is placed as an electrode (anode).

The efficiency of bilayer solar cells is limited by its exciton (charge) generation in a 10–20 nm scale around the donor–acceptor interface. It leads to the loss of absorbed photons further away from the interface and results in low quantum efficiencies [14].

11.9 Photo-Induced Supramolecular Devices

Photo-induced supramolecular devices may be designed in the field of information processing and construction of chemical-based computers in future. The operations of these devices are monitored by their luminescence. These devices are mainly of three types:

1. **Devices based on photo-induced energy transfer**. Molecular wires can be designed to send an electrical signal by connecting the receiver components over a long distance. For example, carbon nanotubes can be soldered in an electron microscope to get connectivity over a long distance. Photosensitive dendrimers may be used as antenna systems for harvesting solar energy.

2. **Devices based on photochemical or photo-physical process**. Molecular-level two or more inputs may be designed to get a single output by performing basic logic operations in information processing. For example, the molecular-level two-level AND logic gate can be designed using two covalently linked receptors that are able to quench the luminescence of a fluorophore. When one of the receptors acts as a host to a suitable group, it becomes inactive to quench the luminescence of the fluorophore, but the luminescence may be quenched by the other free receptor. The output signal as luminescence of the fluorophore can only be observed when both the receptors are bound to suitable chemical groups.
3. **Devices based on photo-induced nuclear movements**. Molecular machines may be designed based on mechanical movements in supramolecular structures on absorption of light. Such movements occur when the charge-transfer interactions between electron donor and electron acceptor groups are weakened; for example, a six-component bistable [2]-rotaxane that has been designed to work as a light-driven molecular machine based on photo-induced electron transfer reactions (Fig. 11.10) [15]. The rotaxane is composed of an electron-rich macrocycle, [bis-*p*-phenylene [34]-crown-10, BPP34C10] (M) and a dumbbell-shaped component. The dumbbell-shaped component contains a $[Ru(bpy)_3]^{2+}$-based light-harvesting unit (P) with a rigid spacer (S) (*p*-terphenyl

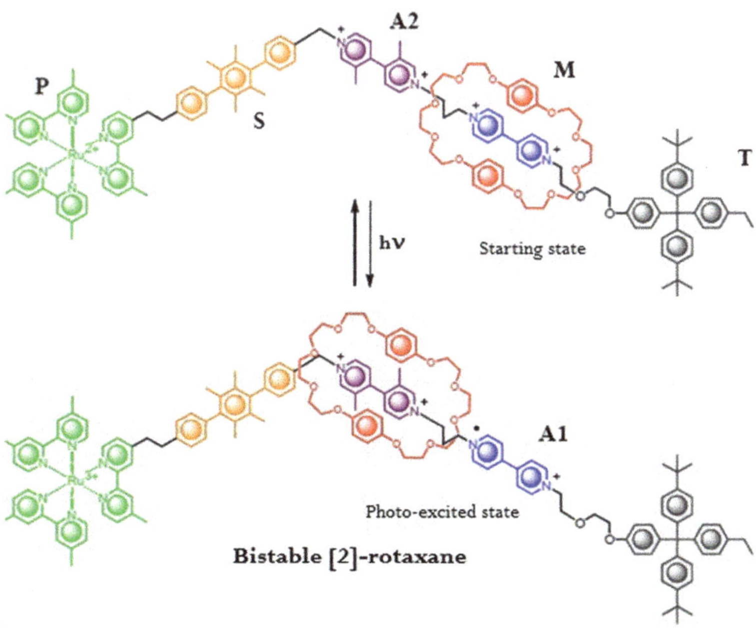

Fig. 11.10 Molecular structures of the components for a light-driven molecular scale machine. Adapted with permission from (Bolzani et al. 2006 Aust J Chem 59:193). Copyright (2006) CSIRO Publishing

type ring) which separates the P unit from the mechanical switching moiety and serves as a stopper. The mechanical switching moiety is composed of a 4,4′-bipyridinium component (A_1) as a strong primary π electron-accepting unit, a 3,3′-dimethyl-4,4′-bipyridinium component (A_2) as a weak secondary π electron-accepting unit and a tetraarylmethane group as the second stopper (T). The A_1 and A_2 components act as stations for the macrocyclic unit M. M encircles the A_1 station in the starting state and can be moved automatically to A_2 station on absorption of light by P unit. The intramolecular mechanism for this ring movement can be explained as follows. The light absorption of the photoactive unit P promotes it into the excited state. The excited state transfers an electron to the station A_1 and deactivates the station. Due to deactivation, the ring moves from station A_1 to station A_2. The reduced station A_1^- then transfers an electron back to the oxidized P^+ unit and becomes activated as electron acceptor. Due to this electronic reset of A_1 station, the ring moves back to this station [15]. These molecules can be driven at a frequency of about 1 kHz and are stable for about 10^3 Hz in solution at an ambient temperature. The ring motion generates a mechanical power of 3×10^{-17} W/molecule [16].

Several artificial molecular machines have been designed to work as functional elements in molecular electronics, NEMS (nano-electro-mechanical systems), nanophotonics as nanoscale optical integrated circuits for electronic display and light-harvesting process and in nanomedicine for drug delivery process [17]. In our body several protein molecules act as bio-motors and machines for performing various functions of the cells such as powering of skeletal muscles, synthesis of ATP, and templating of DNA/RNA.

11.10 Further Reading

1. Blackenship RE (2002) Molecular mechanisms of photosynthesis, 1st edn. Blackwell Science, Oxford
2. Pessarakli M (2005) Handbook of photosynthesis, 2nd edn. CRC Press, Boca Raton
3. Hara K, Arakawa H (2003) Dye-sensitized solar cells. In: Luque A, Hegedus S (eds) Handbook of photovoltaic science and engineering. Wiley, pp 663
4. Foyer CH (1984) Photosynthesis. Wiley, New York

References

1. McGrath M (2014) Mysterious new man made gases pose threat to ozone layer. BBC News, March 10
2. Mreihil K, McDonagh AF, Nakstad B, Hansen TWR(2010) Pediatric Res 67: 656; Lightner DA, Woolridge TA, McDonagh AF(1979) Proc Natl Acad Sci 76:29

3. Photosynthesis (2007) McGraw Hill encyclopedia of science and technology. McGraw Hill, New York; Duysens JNM, Amesz J, Kemp BM (1937) Nature 139:881; Arnon DI, Allen MB, Whatley FR (1954) Nature 174:394
4. Raven PH, Evert RF, Eichhorn SE (2005) Biology of plants, 7th edn. WH Freeman and company, New York, pp 124–127
5. Ingrouille MJ, Eddie B (2006) Plants: diversity and evolution. Cambridge University Press, Cambridge
6. Styring S (2011) Faraday Discuss 155:357
7. Kalyanasundaram K, Gratzel M (2010) Curr Opin Biotechnol 21 :298
8. Setlow RB (1966) Science 153:379
9. Essen LO, Klar T (2006) Cell Mol Life Sci 63: 1266
10. Snapshot of Global PV 1992–2014. International Energy Agency, Photovoltaic power systems programme, March 30, 2015
11. O' Regan B, Gratzel M (1991) Nature 353:737
12. Gao F, Wang Y, Zhang J, Shi D, Wang M, Humphry-Baker R, Wang P, Zakeeruddin Sm, Gratzel M (2008) Chem Commun 2635
13. Shaheen S, Brabec CJ, Sariciftci NS, Padinger F, Fromherz T, Hummelen JC (2001) Appl Phys Lett 78:841
14. Winder C, Sariciftci NS (2004) J Mater Chem 14:1077
15. Balzani V, Credi A, Venturi M (2003) Pure Appl Chem 75:541
16. Balzani V, Clemente-Leon M, Credi A, Semeraro M, Venturi M, Tseng HR, Weger S, Saha S, Stoddart JF (2006) Aust J Chem 59:193
17. Zheng YB, Hao Q, Yang YW, Kiraly B, Chiang IK (2010) J Nanophotonics 4:042501

Appendix

Answers to Problems

Chapter-2

2.6.1.

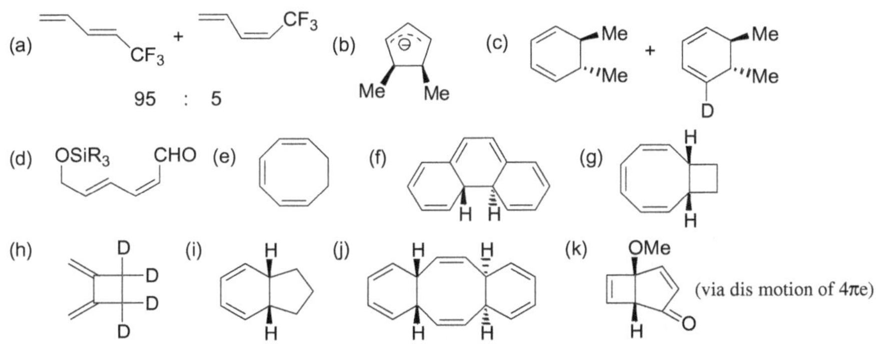

2.6.2. (a) Carbanion formation followed by cyclization and proton abstraction.
(b) Electrocyclic ring opening of cyclobutene, then nucleophilic addition of diazo acetic ester, electrocyclic ring closure of oxy anion (8 π e) gives the major product and keto-enol tautomerism and cyclization gives the minor product(major: minor, 86:14, 84 %).
(c) Electrocyclic ring opening and ring closure of cyclopropyl cation by dis-motions.
(d) Electrocyclic ring closure by dis-motion, followed by H abstraction and addition of methoxyl radical.

B. Dinda, *Essentials of Pericyclic and Photochemical Reactions*,
Lecture Notes in Chemistry 93, DOI 10.1007/978-3-319-45934-9

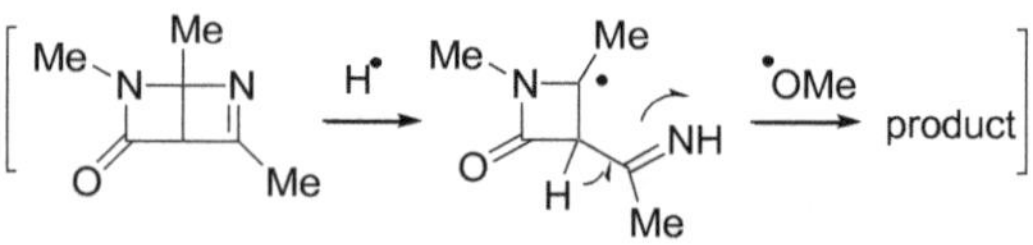

(e) Conrotatory ring opening and closure in the same direction
(f) Inward disrotatory ring opening of cyclopropane ring with loss of fluoride ion in a concerted process and addition of formate ion to the resultant cyclohexenyl cation.
(g) Aziridine ring opening by con-motion and cyclization of the resultant dipolar ion by con-motion.
(h) One electrocyclic ring opening of cyclobutene to cyclohexadiene by allowed dis-motion due to the formation of Ag+-π complex and then opening of cyclohexadiene.
(i) Electrocyclic ring opening of cyclobutene, formation of cyclobutene with exo-methylene group, opening of cyclobutene and finally cyclization of 6πe.
(j) Electrocyclic ring opening of six-membered ring to an unstable product, which on double bond isomerization gives the product.

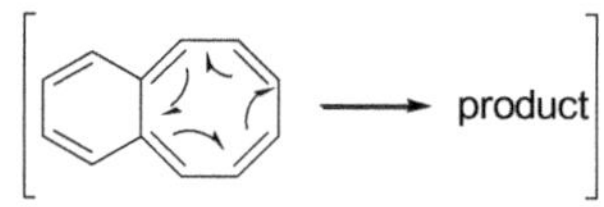

2.6.3. (a) Both A and B undergo ring opening by allowed con-motion, whereas C undergoes ring opening by forbidden dis-motion.
(b) 2-Vinylcyclopropanols in the presence of organolithium generate cyclopropyl anions, which on conrotatory ring opening produce anions of long lifetime. These anions undergo ring closure by dis-motion as 6e process and abstract proton from solvent to give the products.
(c) Cyclobutene ring undergoes inward conrotatory ring opening to produce a conjugated triene. The triene undergoes cyclization by dis-motion to give the product.
(d) The cyclopropene ring opening by inward dis-motion in D accelerates the removal of Cl atom as leaving group by E_2-like process by participation of p-orbitals to produce cyclohexenyl cation much faster. The resultant cyclohexenyl ion forms acetate with solvent. Such inward ring opening in E does not favor the removal of Cl atom because generated p-orbital is *syn* to Cl atom.

Chapter-3

3.7.1. (a) Electrocyclic ring opening of cyclobutene followed by D–A cycloaddition.
(b) D–A cyclization followed by *retro* D–A cleavage.
(c) D–A cyclization and then loss of CO_2 and MeOH from the adduct.
(d) Electrocyclic ring opening of oxirane to produce a carbonyl ylide, which undergoes 1,3-dipolar cycloaddition.
(e) D–A cycloaddition followed by *retro* D–A cleavage.
(f) D–A cycloaddition of 1,3-diphenyl-2-azallyl anion.

(g) [2+2]-cycloaddition.

(h) 1,3-Dipolar cycloaddition gives *cis*-isomer as major product through *E-endo*-TS.

(i) Electrocyclization of cycloheptatriene to a norcaradiene and then its D–A cycloaddition.

(j) Electrocyclization of benzocyclobutene, then D–A cyclization and loss of H_2.

(k) Loss of MeOH gives a ketene, which undergoes [2+2]-cycloaddition and *retro*-ene reaction.

(l) D–A cycloaddition and then *retro* D–A reaction and electrocyclic ring opening of *trans*-3,4-diacetoxycyclobutene.

(m) Electrocyclization of 6e process by dis-motion and then D–A cycloaddition.

(n) [6+4]-Cycloaddition reaction using *exo*-methylene group of the fulvene as 4π electron system.

3.7.2.

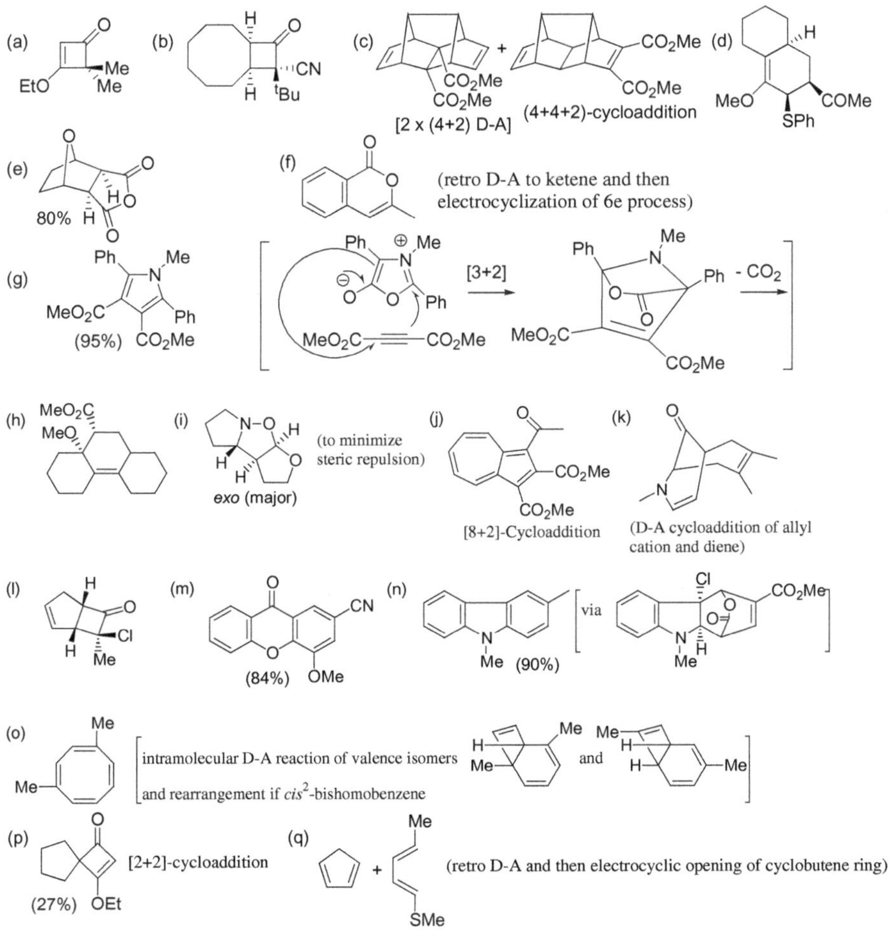

(r) 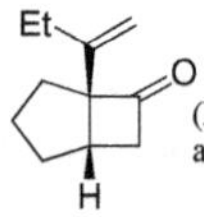 (Electrocyclic ring opening of cyclobutene to ketene and the intramolecular [2+2]-cycloaddition)

(s) 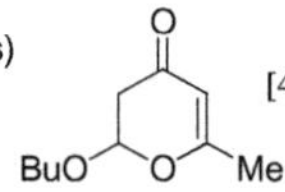 [4+2]-cycloaddition

(t) 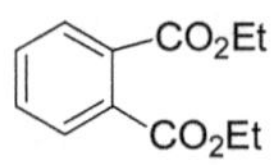

(u) 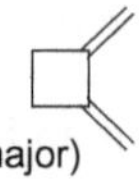

(major)

(v) 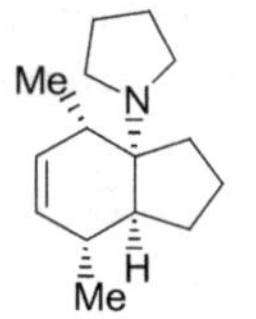

(w) 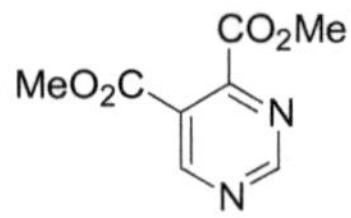

3.7.3.

(a) 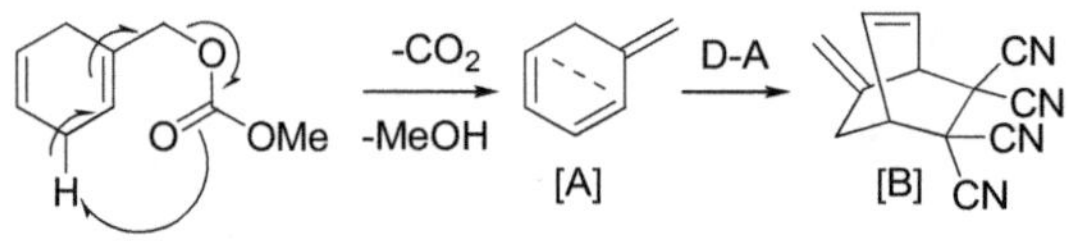

(b) 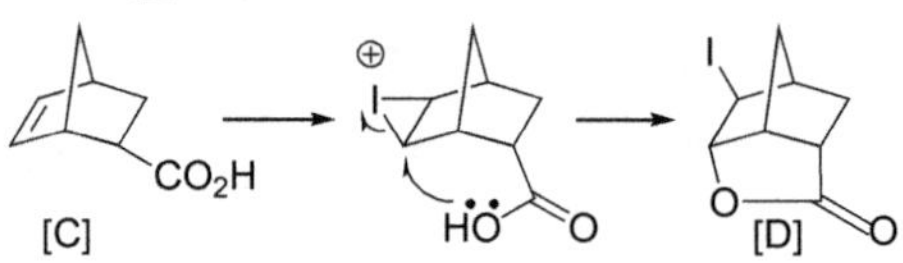

(c) 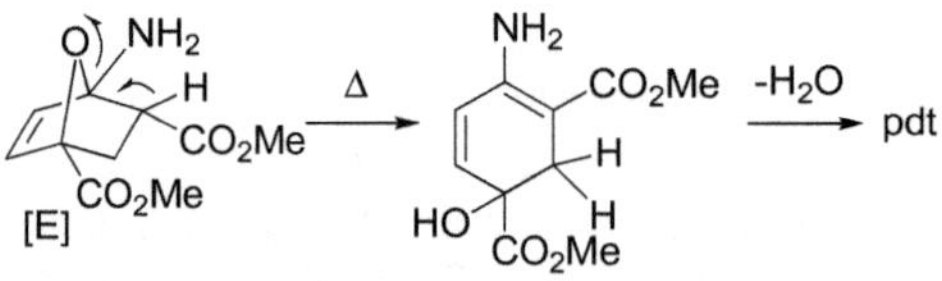

(d) 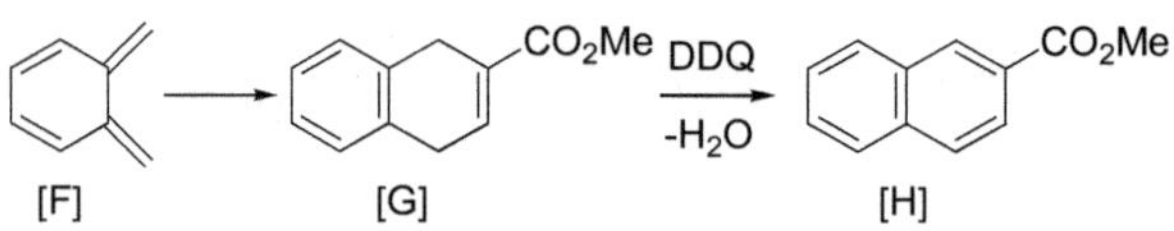

(e) 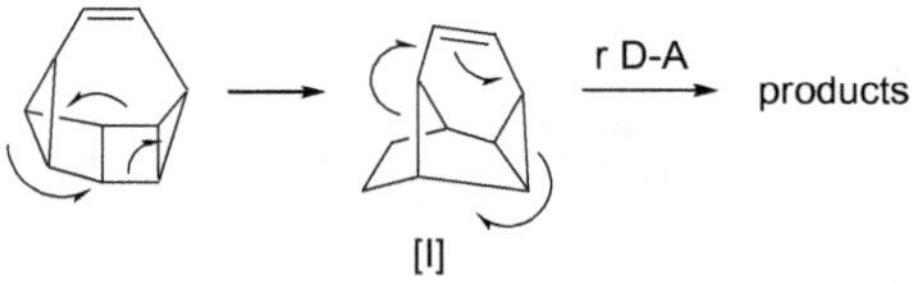

3.7.4.

(a) 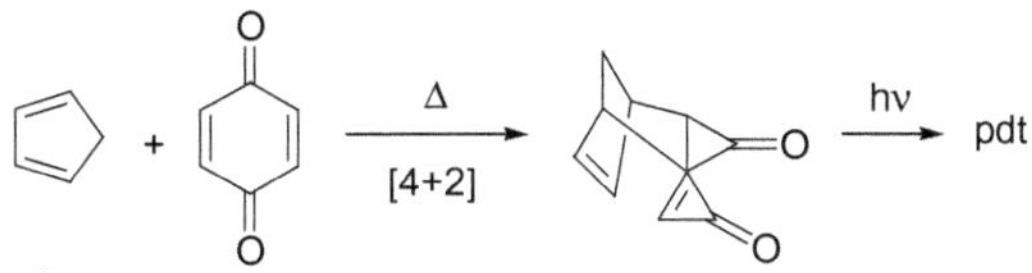

(b) 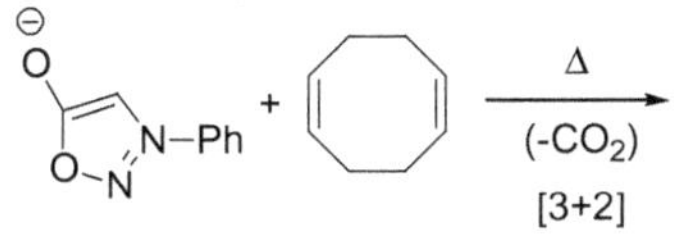

(c) 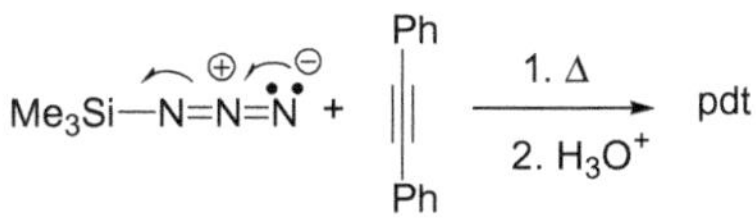

(d) 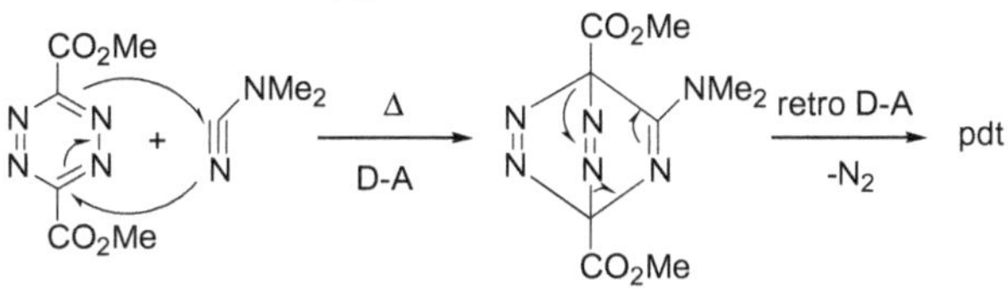

(e) 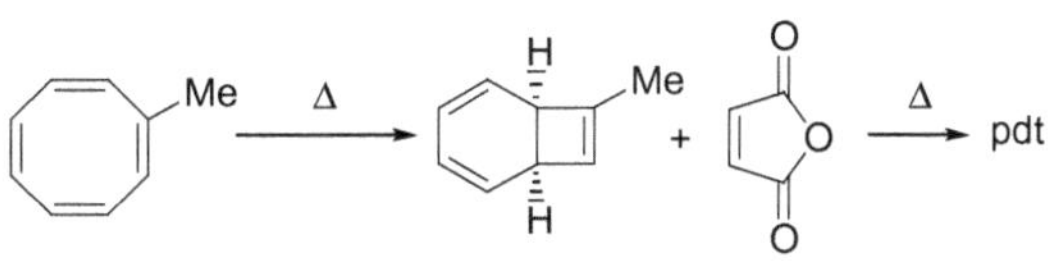

(f) 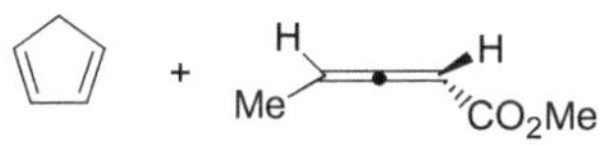

(g) 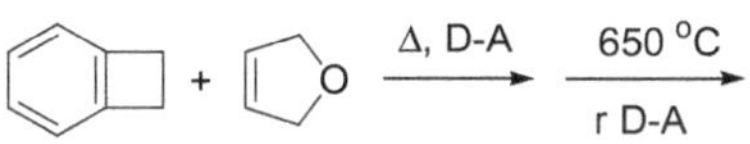

Chapter-4

4.10.1. (a) [1,5]-CH_3 shift and then [1,5]-H shift, orbital symmetry allowed process.
(b) [3,3]-Sigmatropic shift (Cope rearrangement), allowed.
(c) [1,5]-R shift, then [1,5]-H shift, allowed.
(d) The *Claisen rearrangement* and then two times of Cope rearrangements, allowed.
(e) [1,3]-Alkyl shift with inversion of configuration at chiral carbon, allowed.
4.10.2. (a) Electrocyclic ring opening of cyclobutene and then [1,5]-H shift.
(b) The *oxy anionic Cope rearrangement.*
(c) The *Cope rearrangement.*
(d) [2,3]-Sigmatropic shift.
(e) The *Cope rearrangement.*
(f) The *Cope rearrangement.*
(g) Electrocyclization of 10 π electrons and then [1,5]-sigmatropic H shift.

(h) The *Cope rearrangement* of chair form to *cis*-1,2-di-*trans*-1′-propenylcyclobutane, then Cope rearrangement of boat form to *cis*-1,2-dimethyl.
(i) [1,5]-Sigmatropic alkyl shift, then [1,5]-sigmatropic H shift.
(j) The *Cope rearrangement*, then Claisen rearrangement.
(k) The *Claisen rearrangement.*
(l) The *Cope rearrangement.*
(m) [1,5]-H shift, followed by electrocyclization of 6πe system.

4.10.3.

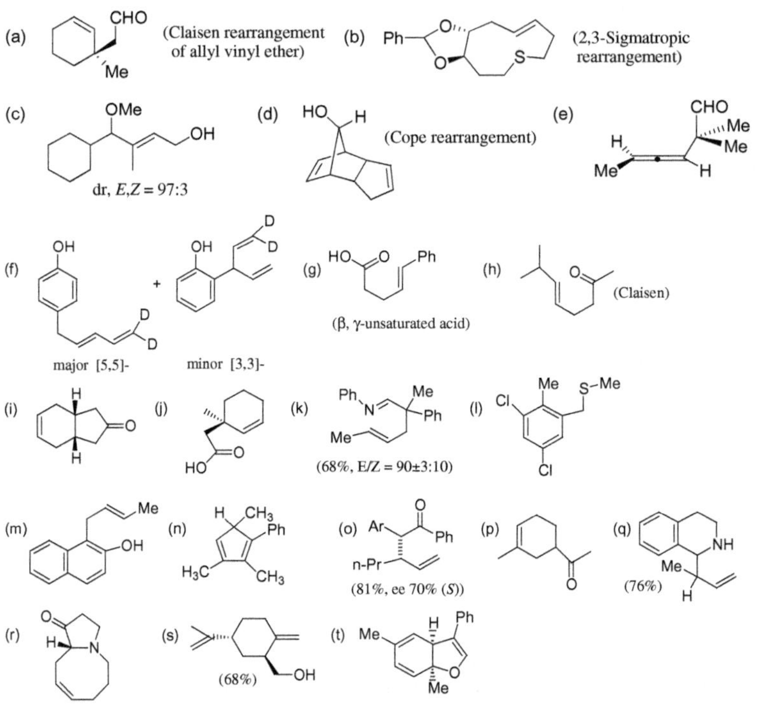

4.10.4. (a) [1,5]-H shift and then *retro* D–A reaction.
(b) Electrocyclic ring opening of 6 π e system, followed by [1,7]-H shift and electrocyclization of 6 π e system.
(c) [1,5]-H shift and then [4+2]-cycloaddition.
(d) Electrocyclic ring opening of 8 π e system, then electrocyclic ring closure of 6 π e system to produce *trans*-bicyclic compound, which on [1,5]-H shift gives the product.
(e) Aldol condensation to produce an allyl vinyl ether, which on Claisen and Cope rearrangements gives the product.
(f) Electrocyclization of 6 π e system (dis) and then [1,5]-H shift.
(g) [1,5]-Alkyl shift, then [1,5]-H shift.

(h) Electrocyclic ring opening of 6 π e system, then [1,7]-H shift and electrocyclization of 6 π e system.
(i) Two times [1,5]-H shifts, then Claisen rearrangement.
(j) The *Cope rearrangement* using cyclobutene, then electrocyclic ring opening of cyclobutene.
(k) [4+2]-Cycloaddition followed by two successive [1,5]-H shift.

Chapter-5

5.7.1 (a) Major and minor products are derived from favored TS and disfavoured TS.

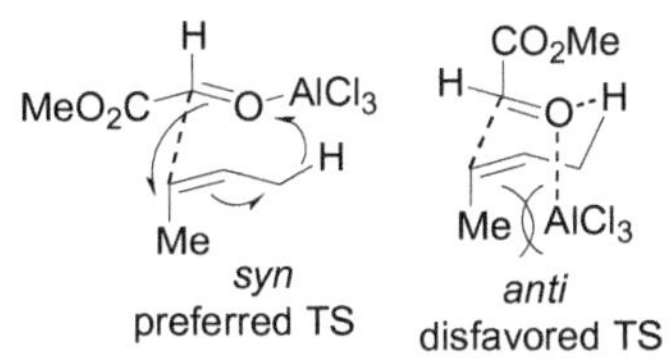

(b) Ene reaction of O_2.
(c) Intramolecular ene of H delivery, then *retro*-ene of D-delivery with cyclization.
(d) Intramolecular ene reaction *via* enol form of ketone.

5.7.2

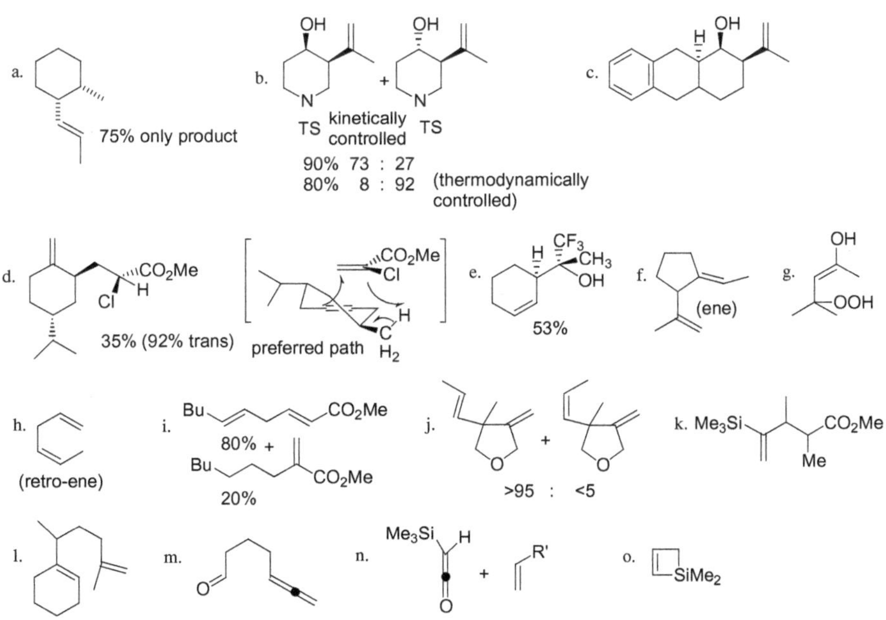

Chapter-7

7.6.1 (a) The reaction proceeds preferentially through a triplet excited state in violation of free rotor hypothesis. The major product is derived from the bond formation between C(4) and C(6), followed by cleavage of C(5)–C(6) bond to give a stable diradical, whereas minor product is derived by cleavage of C(4) and C(5) bond.

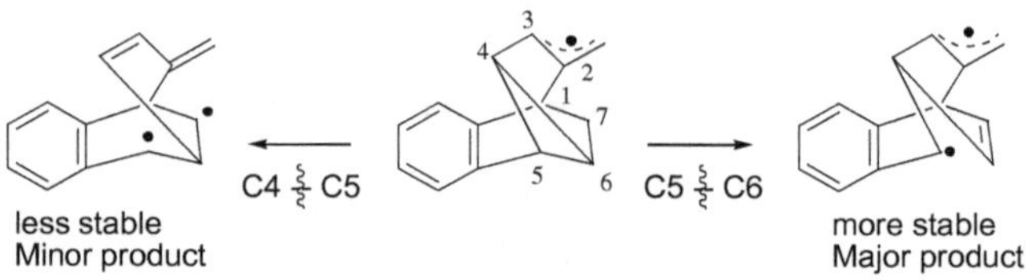

(b) Reaction takes place by *cis–trans*-isomerization about the terminal C(1)–C(2) bond followed by [4+2]-*supra-antara* cycloaddition.

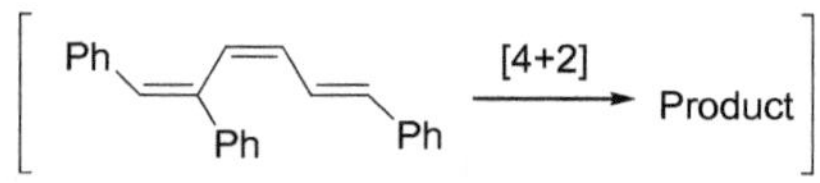

(c) Photochemical allowed [1,3]-H shift.

(d) Rearrangement occurs through *cis–trans*-isomerization, followed by rapid intramolecular photo D–A reaction.

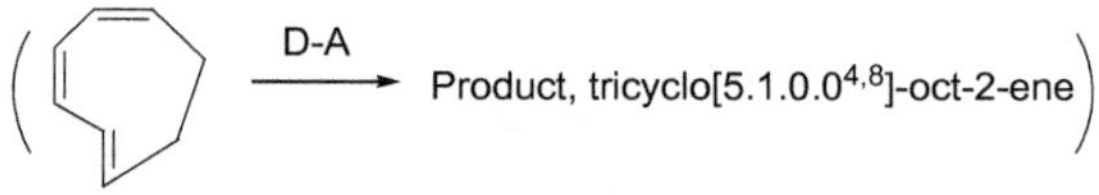

(e) Electrocyclic ring closure to highly strained 1,2-cyclobutadiene, which on opening of ring gives skeletal rearrangement product.

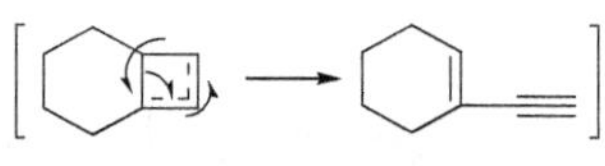

(f) The photorearrangement occurs through *cis–trans*-isomerization and intramolecular photo D–A reaction (*supra-antara* path).

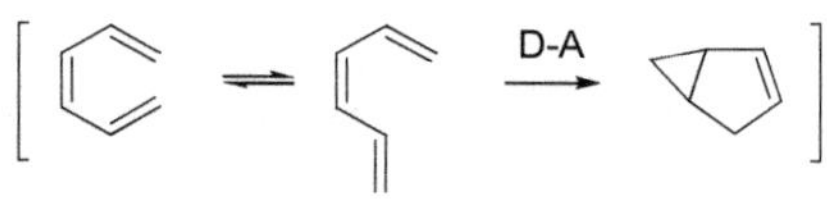

(g) Both the bicyclic compounds (bicyclo[3.2.0]-hept-2-enes) were formed as major products in a concerted process by orbital symmetry allowed 1,3-alkyl shift in both singlet and triplet excited states. The minor product (m-hydroxymethyllimonene) was formed by cleavage of cyclopropane ring followed by γ-H abstraction.

7.6.2.

(a) 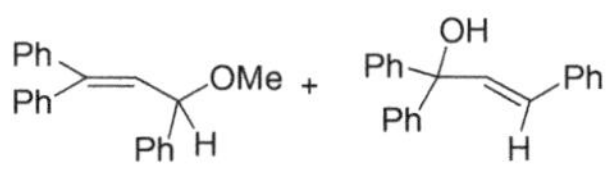

5 : 1

(reaction proceeds through the singlet excited state to form exciplex with solvent, then hydrogen abstraction and addition of methoxy radical)

(b) 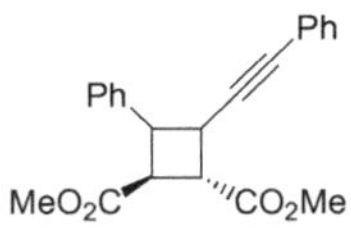

([2+2]-cycloaddition)

(c) 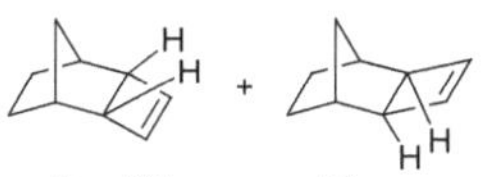

endo 72 : 28 *exo*

(reaction takes place through singlet excited state)

(d) 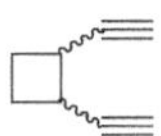

(*cis* and *trans*)

([2+2]-cycloaddition through triplet excited state)

(e) 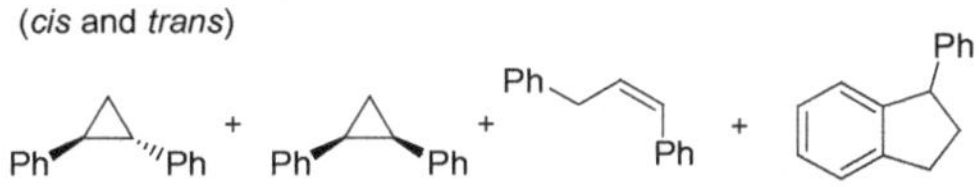

1-phenylindane

(di-π-methane rearrangement is the favoured path through singlet excited state to produce *trans*-1,2-diphenylcyclopropane as major product)

(f) 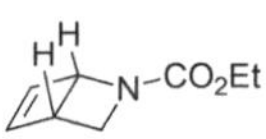

50%

(g) 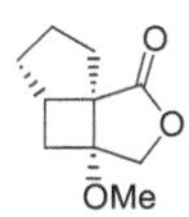

Chapter-8

8.6.1.

(a) 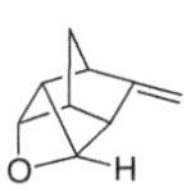

(Paterno-Buchi (PB) reaction)

(b) 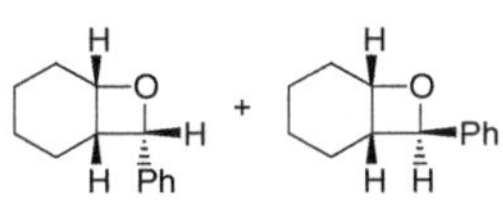

80 : 20

(c) 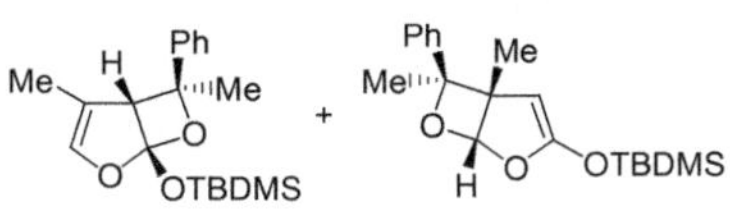

58 (exo/endo, 82:18) 42 (exo/endo, 95:5)

(d) 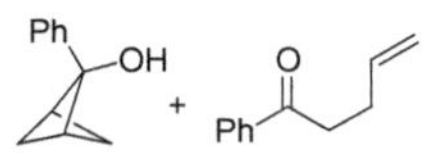

60% 40%

(Norrish type-II-cleavage, p-orbital of cyclobutyl radical is almost perpendicular for maximum ovelapping to give cyclic product)

(e) 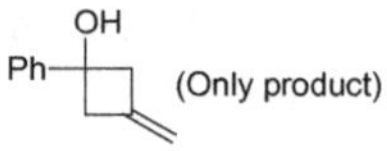

(f) 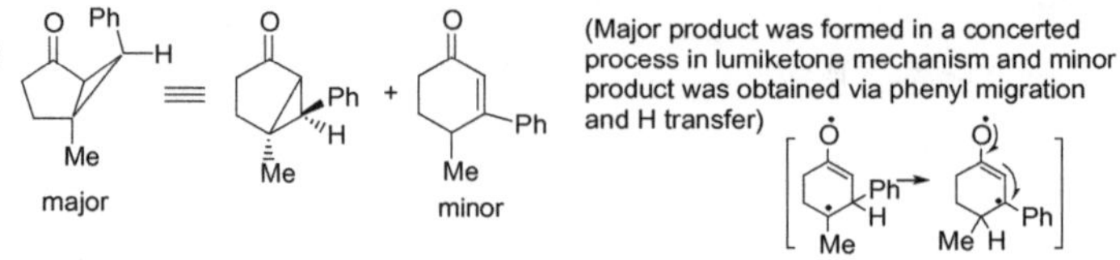

(g) 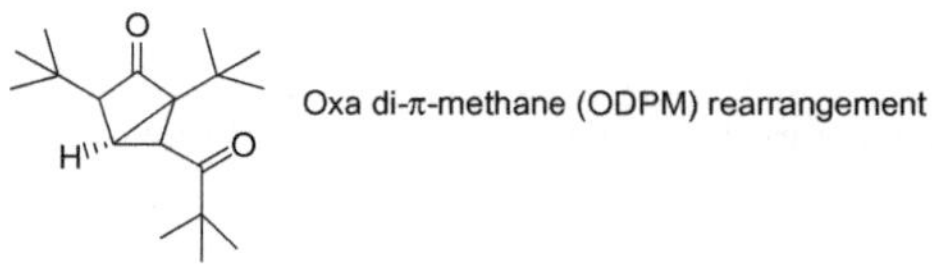

(h) 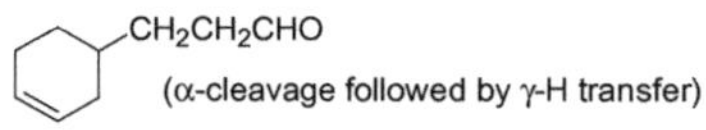

(i) 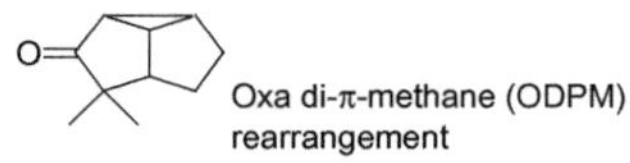

(j) 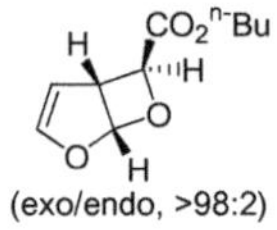

(k) 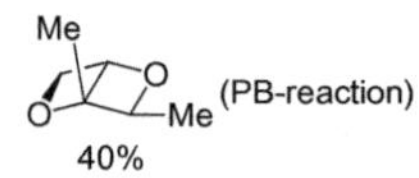

(l) 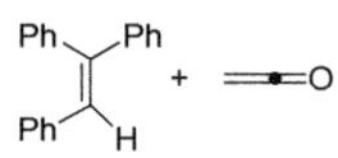

(m) 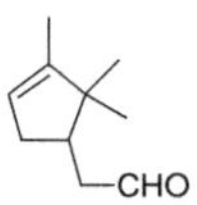

(n) 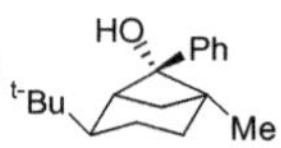

8.6.2. (a) Alpha cleavage and addition to allylic radical to cycloalkene radical.
(b) Alpha cleavage at the bridge side, then γ-H transfer from the bridge γ-C and formation of π bond.
(c) Di-π-methane rearrangement through triplet excited state of the reactant.
(d) Lumiketone-like rearrangement, which involves the shift of C(4)-C(5) bond to C(3) followed by formation of bond between C(2) and C(4).
(e) The reaction proceeds via regioselective less stable oxetane. The oxetane on ring opening gives the product.

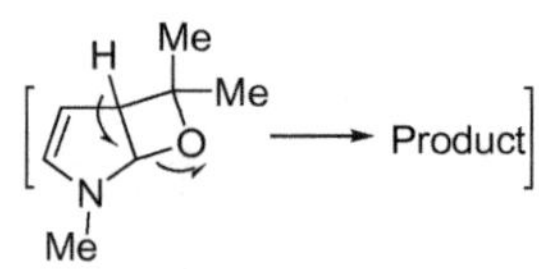

(f) The reaction involves an α-cleavage of excited singlet state of cyclobutanone to an acyl alkyl diradical, which undergoes electronic rearrangement to afford an oxacarbene. The oxacarbene intermediate was trapped in a ring expanded acetal.

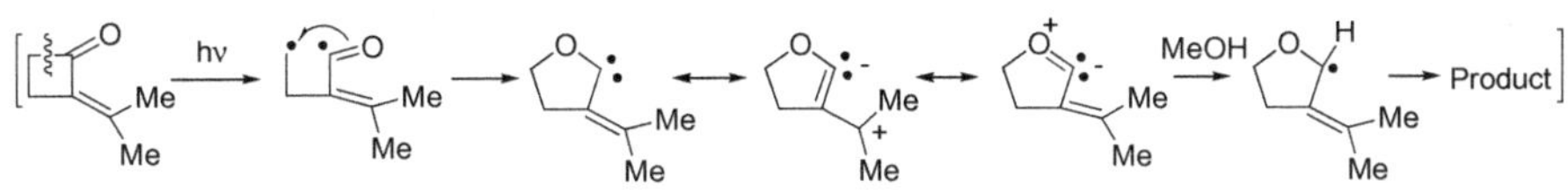

(g) 1,2-Acyl shift or oxa-di-π-methane rearrangement of triplet state of reactant gives the only product, dihydrobarbaralone.
(h) 1,3-Acyl migration followed by cyclopropane ring formation.
(i) Norrish type-II cleavage due to δ-H abstraction by benzoyl carbonyl and β-cleavage of the triplet diradical.
(j) Alpha cleavage followed by generation of homoallyl carbinyl diradical and cyclization.

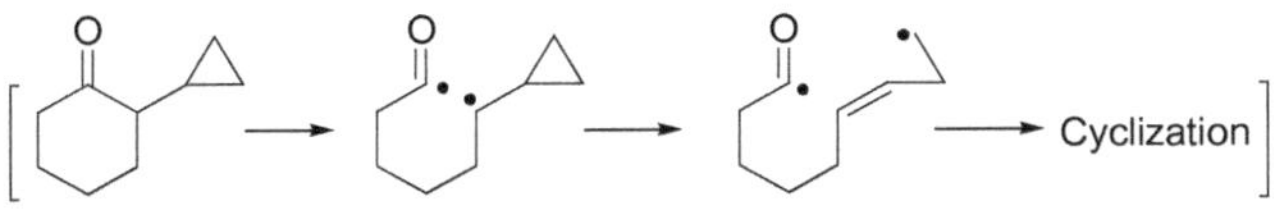

(k) ODPM rearrangement
(l) Normal ODPM rearrangement involving the first double bond (C(1)-C(2)) of the diene system gives major product via triplet excited state whereas vinylogous ring closure using extended double bond gives minor product. The major product is obtained from larger spin–orbit coupling interaction due to less atomic motion.

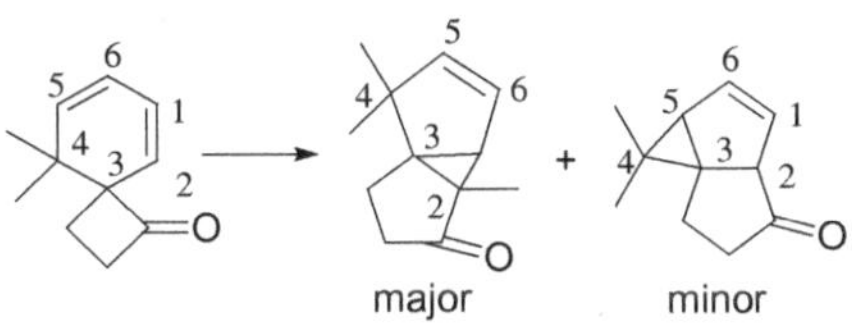

(m) Alpha cleavage followed by H abstraction from the β carbon.
(n) γ-H Abstraction by photoexcited singlet state of alkanone gives Norrish type-II cleavage product.

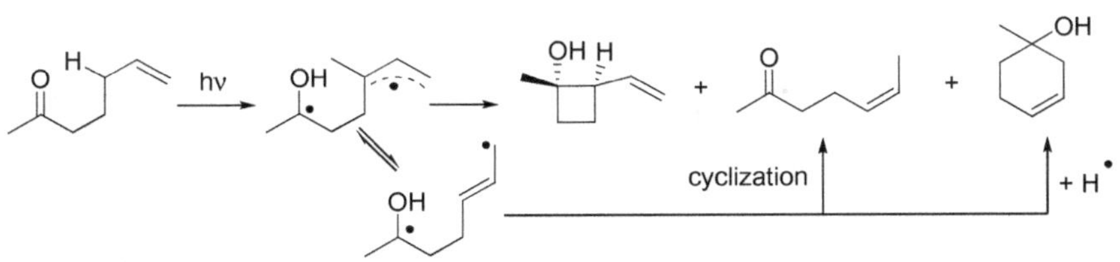

(o) In direct irradiation, the reaction proceeds through singlet excited state to give cyclobutane derivative by concerted 1,3-acyl shift, whereas in sensitized irradiation, the reaction proceeds through triplet excited state to give cyclopropane

derivative by ODPM rearrangement. In direct irradiation, ISC is prevented by steric interaction of the α-methyl groups with carbonyl oxygen nonbonding electrons.
(p) Alpha cleavage to produce diradical, which induces cleavage of the cyclopropane ring by conjugative interaction to generate another diradical. The latter diradical undergoes cyclization.

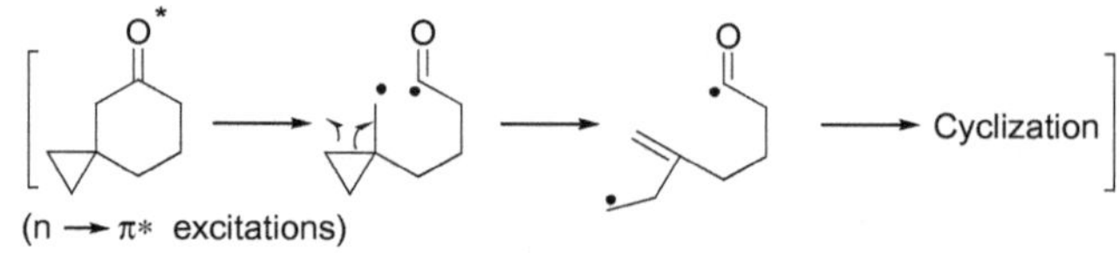

Chapter-9

9.10.1.

(a) 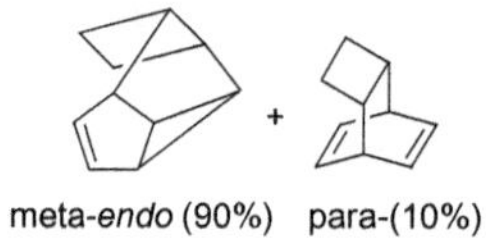

(b) 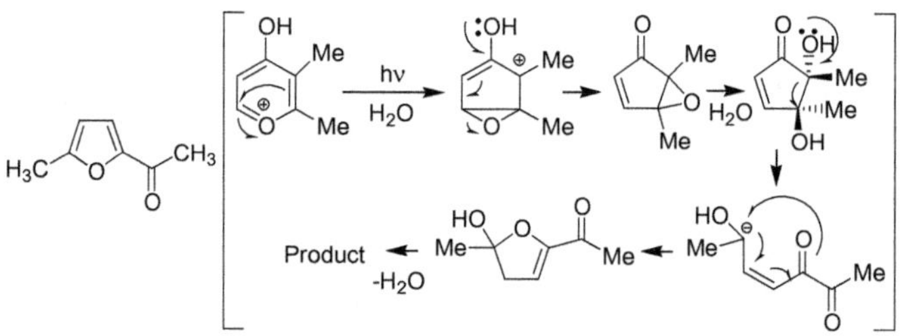

(c) 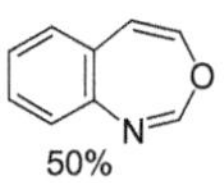

(d) 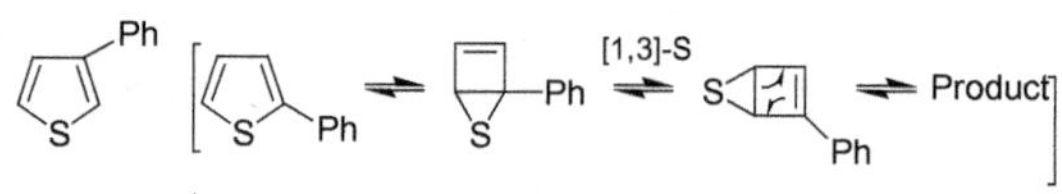

(e) 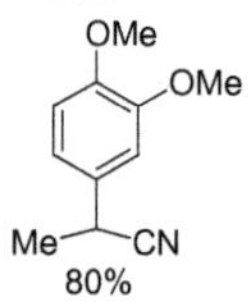

(f) 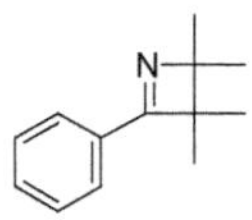

(g) 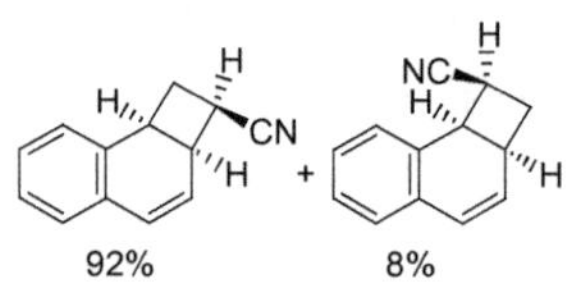

(h) 

(i) 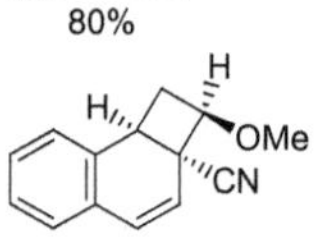

(j) 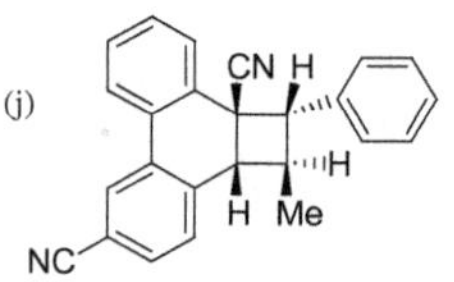

(k) 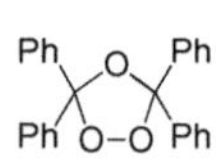

(l) 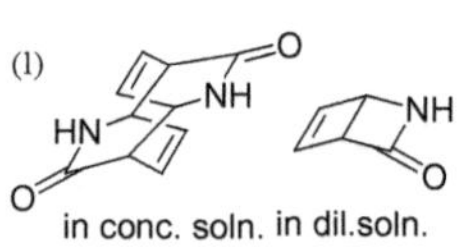

(m) 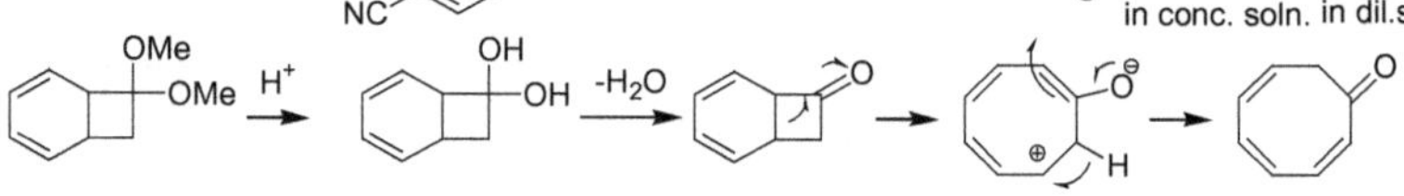

9.10.2. (a) The reaction takes place in a Wolff rearrangement via ketene. The ketene on cyclization gives a diradical, which attacks neighboring Ph gr to give product.

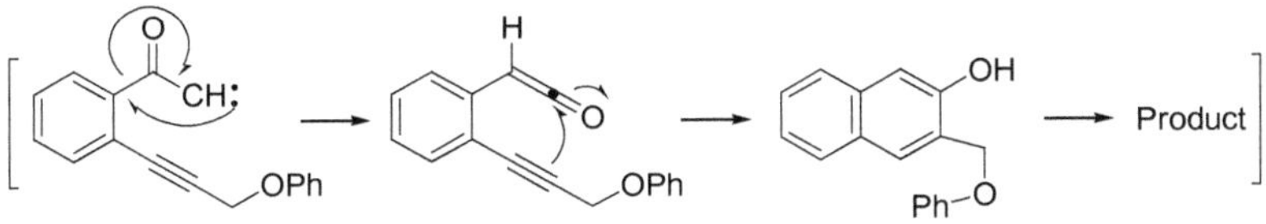

(b) Intramolecular [4+2]-tandem cyclization through a chair-like TS of a radical cation in a PET process, where alkene serves as electron donor and DCNB as electron acceptor.

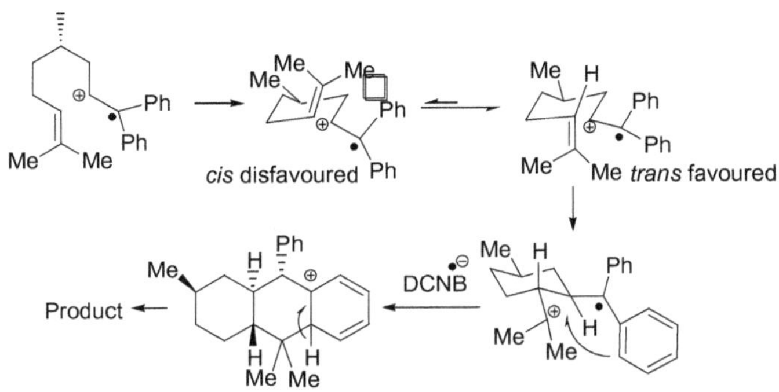

Chapter-10

10.5.1

a.

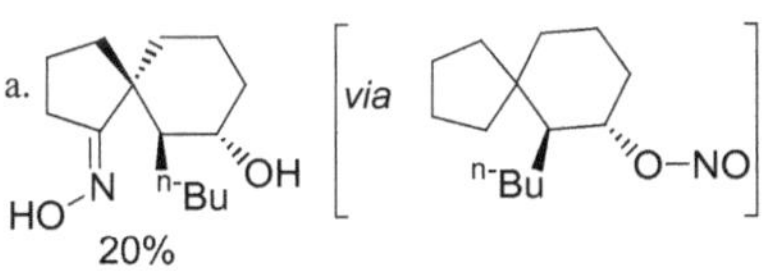

b.

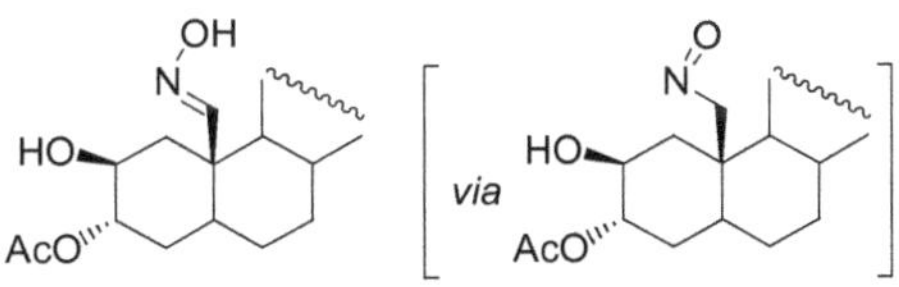

c.

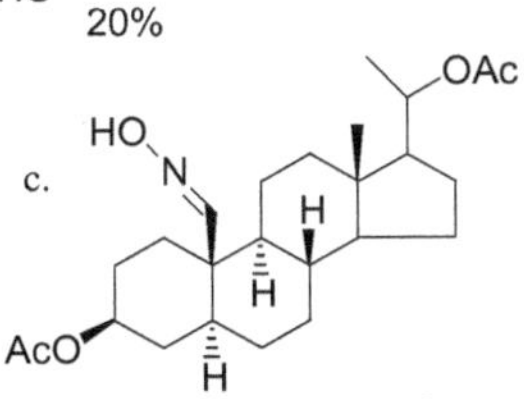

d.

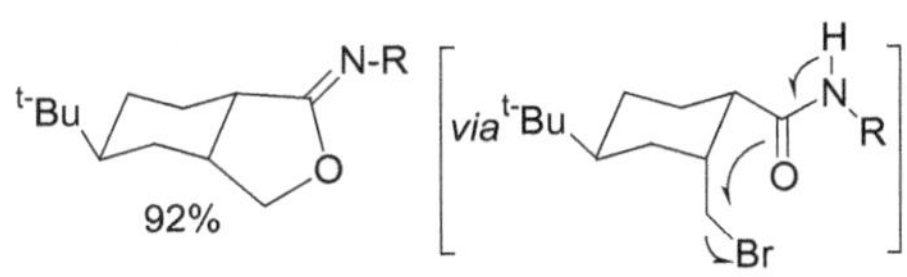

e.

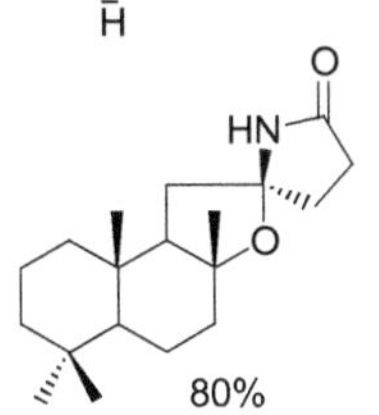

10.5.2 (a)

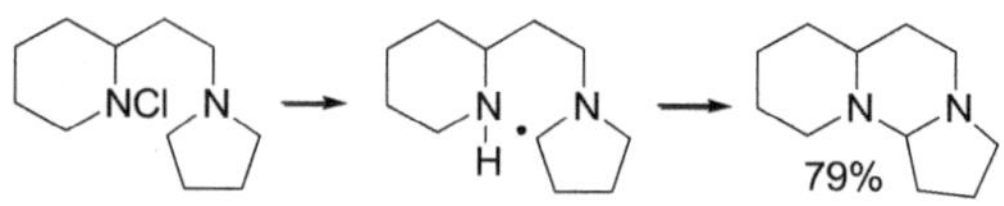

(b) Intramolecular amidation at C-19 methyl *via* phosphonamidyl radical.

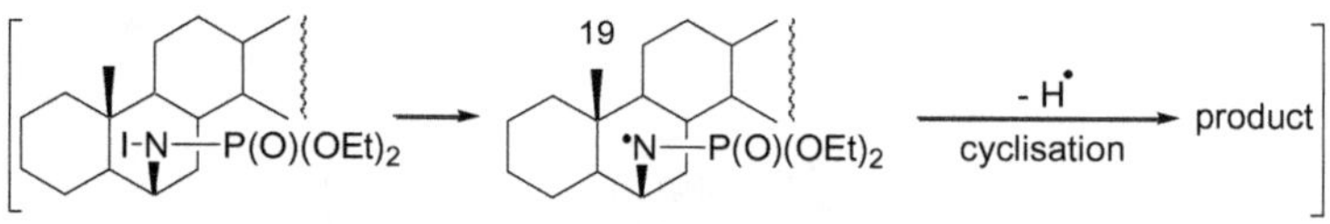

(c) HLF reaction, where intramolecular amidation on aromatic ring *via* sulfonamidyl radical gives the product.

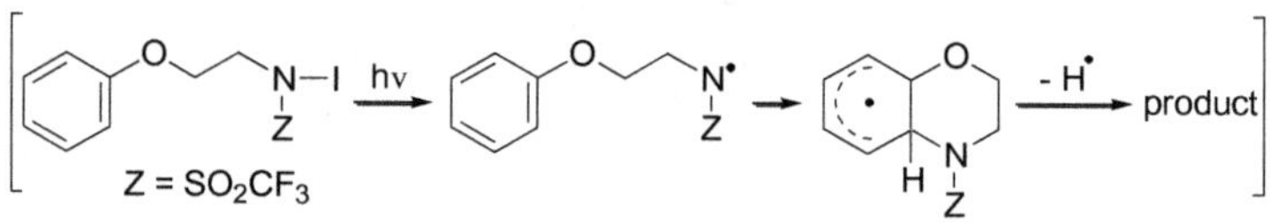

(d) HLF reaction, where N-chloroamine undergoes cleavage of Cl bond and then ε-H abstraction and cyclization.

Index

B. Dinda, *Essentials of Pericyclic and Photochemical Reactions*,
Lecture Notes in Chemistry 93, DOI 10.1007/978-3-319-45934-9

S

T

V

W

X

Y

Z

CPSIA information can be obtained
at www.ICGtesting.com
Printed in the USA
LVHW081002230220
647903LV00001B/135

9 783319 459332